BIOFILMS: RECENT ADVANCES IN THEIR STUDY AND CONTROL

BIOFILMS: RECENT ADVANCES IN THEIR STUDY AND CONTROL

Edited by

L.V. Evans

The Clore Laboratory for Life Sciences
University of Buckingham
United Kingdom

harwood academic publishers
Australia • Canada • France • Germany • India • Japan • Luxembourg
Malaysia • The Netherlands • Russia • Singapore • Switzerland

Printed in Singapore.

Amsteldijk 166
1st Floor
1079 LH Amsterdam
The Netherlands

British Library Cataloguing in Publication Data

Biofilms : recent advances in their study and control
1. Biofilms
I. Evans, L. V. (Leonard Vernon), 1937–
571.6′29

ISBN: 90-5823-093-7

Contents

Preface

In editing the journal *Biofouling* I am greatly struck by the amount of fundamental new knowledge that is emerging about biofilms, largely as a result of the sophisticated technologies that have become available for their detailed study. This book brings together information representing recent advances which have been made in our state of knowledge, from a wide range of different areas of biofilm research in the context of the techniques which are being used to gain this knowledge, and couples this with how the new information may be used to devise improved methods of control.

This book would not have been possible without the willing co-operation of many colleagues, and I would like to thank most warmly all those who have contributed chapters, and also those who helped review the contributions. In the words of Karl Popper, "All science is provisional", and in reporting the current state of the art in this field, it is my hope that this will stimulate ideas for further interdisciplinary research initiatives, and that these will lead to better ways of manipulating and controlling these ubiquitous and successful biological communities.

Contributors

Allison, David G.
School of Pharmacy and
Pharmaceutical Sciences
University of Manchester
Oxford Road
Manchester, M13 9PL
UK

Angles, Mark L.
Australian Water Technologies
51 Hermitage Road
West Ryde, NSW 2114
Australia

Bayston, Roger
Biomaterials-related Infection Group
Division of Microbiology
The University of Nottingham
Nottingham, NG5 1PB
UK

Beech, Iwona B.
School of Pharmacy and Biomedical
Sciences
University of Portsmouth
St. Michael's Building
White Swan Road
Portsmouth, PO1 2DT
UK

Blackall, Linda L.
Advanced Wastewater Management
Centre
Department of Microbiology and
Parasitology
The University of Queensland
St. Lucia, QLD 4072
Australia

Brecx, M.
Department of Periodontology
Université Libre de Bruxelles
Lennik Road 808
CP 622, B-1070 Brussels
Belgium

Buret, Andre G.
Department of Microbiology and
Infectious Diseases
Faculty of Medicine
University of Calgary, Alberta
Canada, T2N 4NI

Busscher, Henk J.
Department of Biomedical Engineering
University of Groningen
Antonius Deusinglaan 1
9713 AV Groningen
The Netherlands

Callow, Maureen E.
School of Biosciences
University of Birmingham
Edgbaston
Birmingham, B15 2TT
UK

Camper, Anne K.
Center for Biofilm Engineering
Montana State University
366 EPS Building
Bozeman, MT 59717
USA

Ceri, Howard
Department of Microbiology and
Infectious Diseases
Faculty of Medicine
University of Calgary, Alberta
Canada, T2N 4NI

Donlan, Rodney M.
Center for Disease Control and Prevention
Building 1, Clifton Road
Atlanta, GA 30333
USA

Flemming, Hans-Curt
Department of Aquatic Microbiology
University of Duisberg
Geibelstraße 41
D-47057 Duisberg
Germany

Geesey, Gill G.
Department of Microbiology and Center for Biofilm Engineering
Montana State University
366 EPS Building
Bozeman, MT 59717
USA

Gibson, Hazel
Campden and Chorleywood Food Research Association
Chipping Campden
Gloucestershire, GL55 6LD
UK

Gilbert, Peter
School of Pharmacy and Pharmaceutical Sciences
University of Manchester
Oxford Road
Manchester, M13 9PL
UK

Goodman, Amanda E.
School of Biological Sciences
The Flinders University of South Australia
GPO Box 2100
Adelaide, SA 5001
Australia

Griebe, Thomas
Institute for Physical and Theoretical Chemistry
University of Duisberg
Lotharstraße 1
D-47057 Duisberg
Germany

Gubner, Rolf J.
School of Pharmacy and Biomedical Sciences
University of Portsmouth
St. Michael's Building
White Swan Road
Portsmouth, PO1 2DT
UK

Hamilton, W. Allan
Department of Molecular and Cell Biology
Institute of Medical Sciences
University of Aberdeen
Aberdeen, AB25 2ZD
UK

Holah, John
Campden and Chorleywood Food Research Association
Chipping Campden
Gloucestershire, GL55 6LD
UK

Holmström, Carola
School of Microbiology and Immunology
University of New South Wales
Sydney, NSW 2052
Australia

Kjelleberg, Staffan
School of Microbiology and Immunology
University of New South Wales
Sydney, NSW 2052
Australia

Lewandowski, Zbigniew
Center for Biofilm Engineering
Montana State University
310 EPS Building
Bozeman, MT 59717
USA

McFeters, Gordon A.
Center for Biofilm Engineering
Montana State University
310 EPS Building
Bozeman, MT 59717
USA

Maira-Litran, Tomas
School of Pharmacy and Pharmaceutical Sciences
University of Manchester
Oxford Road
Manchester, M13 9PL
UK

Mayer, Christian
Institute for Physical and Theoretical Chemistry
University of Duisberg
Lotharstraße 1
D-47057 Duisberg
Germany

Morck, Douglas W.
Department of Microbiology and Infectious Diseases
Faculty of Medicine
University of Calgary, Alberta
Canada, T2N 4NI

Nielsen, Per Halkjær
Environmental Engineering Laboratory
Aalborg University
Sohngaardsholmsvej 57
DK-9000, Aalborg
Denmark

Olson, Merle E.
Department of Microbiology and Infectious Diseases
Faculty of Medicine
University of Calgary, Alberta
Canada, T2N 4NI

Quirynen, M.
Department of Periodontology
Catholic University of Leuven
Capucijnenvoer 7
B-3000 Leuven
Belgium

Read, Ronald R.
Department of Microbiology and Infectious Diseases
Faculty of Medicine
University of Calgary, Alberta
Canada, T2N 4NI

Suci, Peter A.
Department of Microbiology and Center for Biofilm Engineering
Montana State University
310 EPS Building
Bozeman, MT 59717
USA

Tapper, Rudi C.
School of Pharmacy and Biomedical Sciences
University of Portsmouth
St. Michael's Building
White Swan Road
Portsmouth PO1 2DT
UK

van der Mei, H.C.
Department of Biomedical Engineering
University of Groningen
Antonius Deusinglaan 1
9713 AV Groningen
The Netherlands

van Hoogmoed, C.G.
Department of Biomedical Engineering
University of Groningen
Antonius Deusinglaan 1
9713 AV Groningen
The Netherlands

van Steenberghe, D.
Department of Periodontology
Catholic University of Leuven
Capucijnenvoer 7
B-3000 Leuven
Belgium

White, David C.
Center for Environmental Biotechnology
University of Knoxville
10515 Research Drive
Knoxville, TN 37932
USA

Wimpenny, Julian
Cardiff School of Biosciences
Cardiff University
Cathays Park
Cardiff, CF1 3TL
UK

Wingender, Jost
Department of Aquatic Microbiology
University of Duisberg
Geibelstraße 41
D-47057 Duisberg
Germany

Wright, J. Barry
Westaim Biomedical Corp.
10102 — 114th Street
Fort Saskatchewan, AB
Canada, T8L 3W4

Yu, F. Philip
Microbiology Department
Nalco Chemical Company
One Nalco Center
Naperville, IL 60563-1198
USA

Zimmerman, Richard C.
Moss Landing Marine Laboratories
Hopkins Marine Station
Pacific Grove, CA 93950
USA

Zinn, Manfred
Division of Engineering and Applied Sciences
Harvard University
40 Oxford Street
Cambridge, MA 02138
USA

1 Structure and Function of Biofilms

Zbigniew Lewandowski

Biofilms consist of microcolonies separated by interstitial voids and are heterogeneous in many respects *e.g.* structurally, chemically, and physiologically. A new model of biofilm structure, declaring microcolonies as building blocks of biofilms, is used to interpret experimental results and to verify hypotheses about the relations between biofilm structure and function. Microscale chemical profiles, intrabiofilm hydrodynamics, and intrabiofilm mass transport mechanisms are all affected by biofilm heterogeneity.

KEY WORDS: biofilm structure, chemical gradients, intrabiofilm mass transport, intrabiofilm hydrodynamics

INTRODUCTION

During the last decade it became obvious that the researchers who studied biofilms at the microscale accumulated a collection of experimental results that were impossible to interpret using the traditional conceptual model of biofilms where microorganisms are uniformly distributed in a continuous matrix of extracellular polymers. Some of these results are discussed later in this chapter. As a solution, a new conceptual model of heterogeneous biofilms has been suggested. There are several versions of that model now, all conveying the same message, *viz.* that biofilms consist of microcolonies separated by interstitial voids. It soon became evident that the basic declaration of the new model, that the building blocks of biofilms are microcolonies, may have implications going much further than initially expected. Reported evidences of spontaneous microbial coaggregation and cell-cell recognition (Kolenbrander and London, 1992; 1993), and cell-cell communication in biofilms (Davies *et al.*, 1998) were quickly associated with the new biofilm model and hypotheses were suggested regarding the possible role of microcolony structure and internal cell organization in biofilm activity and survival. The journal *New Scientist* expanded on these hypotheses in an article published in August 1996, and emphatically compared biofilms to cities built by microorganisms. General expectation among biofilm researchers is that these elaborate microorganism-formed structures have meaning. However, notions endowing biofilms with abilities to intentionally control these structures, and their environment, should be approached with caution, as there is little experimental evidence to support them. Nevertheless, the unusual propensity of microorganisms to form complex structures on surfaces has been noted by many researchers (Keevil and Walker, 1992; Costerton *et al.*, 1994, Massol-Deya *et al.*, 1994; Wolfaardt *et al.*, 1994; Bishop and Rittmann, 1995), although the reasons for this remain unclear. In natural and engineered systems a spectrum of structurally heterogeneous biofilms is observed ranging from dense, amorphous biofilms, which are less structurally heterogeneous, to biofilms demonstrating robust, well developed structures.

This chapter discusses the relations between the structure and function in biofilms. The popular term biofilm structure means, more often implicitly than explicitly, spatial distribution of biomass density in biofilms, or, sometimes, the complementary distribution of biofilm porosity. Biofilms with well-developed microcolonies, separated by wide interstitial voids are considered "structurally heterogeneous", and most of their biomass is concentrated within microcolonies. The terms structurally heterogeneous and morphologically heterogeneous are used interchangeably and should be considered synonyms. Usually a heterogeneous biofilm means a structurally heterogeneous biofilm while other heterogeneities, when referred to, carry appropriate adjectives, *e.g.* physiological, chemical, ecological. Defining biofilm function is more challenging because it means different things to different people, depending on their interest in biofilms. To construct such a definition biofilms may be looked at as agents of certain activities and the term function associated with the main human activity the biofilm impacts; biofilm function is then realized through the agency of the biofilm. For example, the main function of biofilms forming dental plaque is tooth decay, while the main function of biofilms contributing to microbially influenced corrosion is accelerated anodic dissolution of metals, and the main function of biofilms responsible for bioremediation of toxic compounds is converting toxic compounds to a more benign form. Biofilm function may also refer to their ability to exhibit specific physiological reactions, *e.g.* nitrification, denitrification, or sulfate reduction (function: removal of ammonia, nitrate, sulfate). Clearly, biofilm function is an obscure term covering a broad spectrum of meanings and should be defined each time it is used. For the purpose of this chapter biofilm function is identified with substrate conversion. From such a definition biofilm function can be evaluated as the substrate conversion rate, overall or local, and quantitatively related to biofilm structure.

Many concepts and all the experimental results presented in this chapter originated in the Biofilm Structure and Function group of the Center for Biofilm Engineering, Bozeman, Montana. Experimental results have been selected to illustrate the opinions of that research group which, consequently, may be biased toward promoting their definition of biofilm function and toward exposing relations between biofilm structure and biofilm activity, the main area of study. Important aspects of biofilm research, related to other definitions of biofilm function, as well as relationships between biofilm structure, biofilm ecology, and genetics have been omitted.

BIOFILM STRUCTURE AND THE CONCEPT OF HETEROGENEOUS BIOFILMS

The model of heterogeneous biofilms was needed to interpret experimental results difficult to explain using the model of homogeneous biofilms. Examples of such results follow.

Example 1

Drury (1992) introduced small (1 μm diameter), fluorescent latex particles during biofilm growth and studied their fate. The expectation was that these particles, after settling on the biofilm surface, would be pushed off by the growing and exfoliating bacteria. According to one of the assumptions accepted for biofilm modeling, the displacement would be perpendicular to the substratum (Wanner and Gujer, 1986). In a sense the beads should have imitated bacteria having a growth rate equal to zero and should have been pushed out by faster growing microrganisms. After the experiment was terminated, the biofilm was sectioned and the beads were recovered. Many of them were found at the bottom of the biofilm, near the substratum. This contradicted one of the assumptions of the model. If the biofilms were a continuous gelatinous layer, as the model stipulated, how did the beads get to the bottom of the biofilm? Clearly, the model of homogeneous biofilms could not give a satisfactory explanation. On the other hand, in heterogeneous biofilms such an effect is expected; the beads penetrate and are trapped in interstitial channels that frequently reach the bottom of the biofilm. This technique is now used to study flow rate within biofilm (see Figure 1a); fluorescent beads are injected into individual voids/channels in a biofilm and their movement followed with confocal scanning laser microscopy (CSLM) (DeBeer *et al.*, 1994a; Stoodley *et al.*, 1994).

Example 2

An argument often used when discussing microbially influenced corrosion (MIC) was that if biofilms are forming continuous layers on metal surfaces, they should decrease the corrosion rate, not increase it, because they deplete oxygen, the principal cathodic reactant, near the surface. The experimental observations indicating the opposite are difficult to interpret. In heterogeneous biofilms, on the other hand, oxygen freely penetrates to the substratum through the voids and, therefore, causes formation of differential aeration cells and accelerates corrosion. Differential aeration cells cannot explain the whole complexity of MIC but they are a known phenomenon, and do not raise immediate objections. Besides, their role has been demonstrated experimentally (Roe *et al.*, 1996). Oxygen consumption within microcolonies also promotes growth of SRB whose sulfidogenic activity influences the chemistry and electrochemistry near metal surfaces and initiates a second, longer phase of active corrosion (Lee, W. *et al.*, 1995; Hamilton, 2000).

Example 3

The use of microelectrodes to measure chemical profiles in biofilms was expected to help verify models of biofilm activity. Microelectrodes with a tip diameter of the order of a few microns are driven across biofilms, measuring concentration profiles with high spatial resolution. These measurements provide valuable insight into the chemistry of the inner space of a biofilm. However, their use to verify mathematical models of biofilm activity has been impeded by the fact that profiles measured at different locations were different to an extent that could not be justified by experimental error only. As long as single profiles were analyzed, which was the case in most early

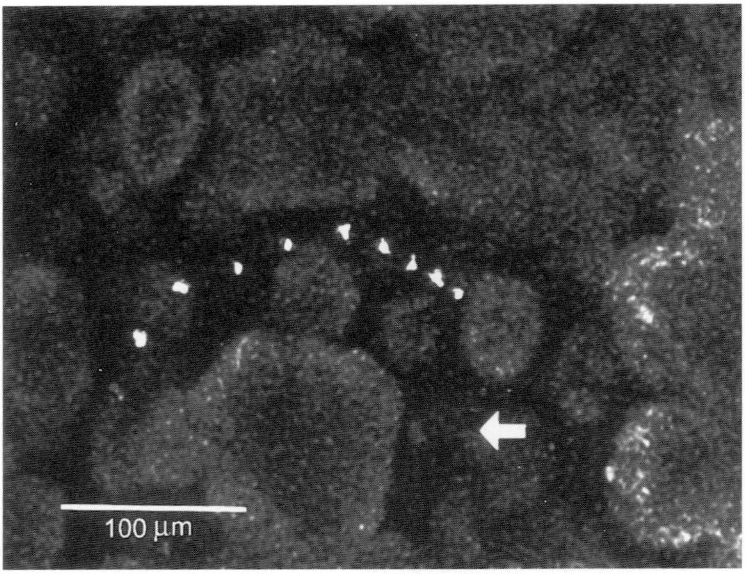

Figure 1a Confocal image of biofilm structure. A small fluorescent latex bead was injected into the biofilm and has moved along the network of channels. The arrow indicates the direction of water flow. (Reproduced from Stoodley *et al.*, 1994, with permission.)

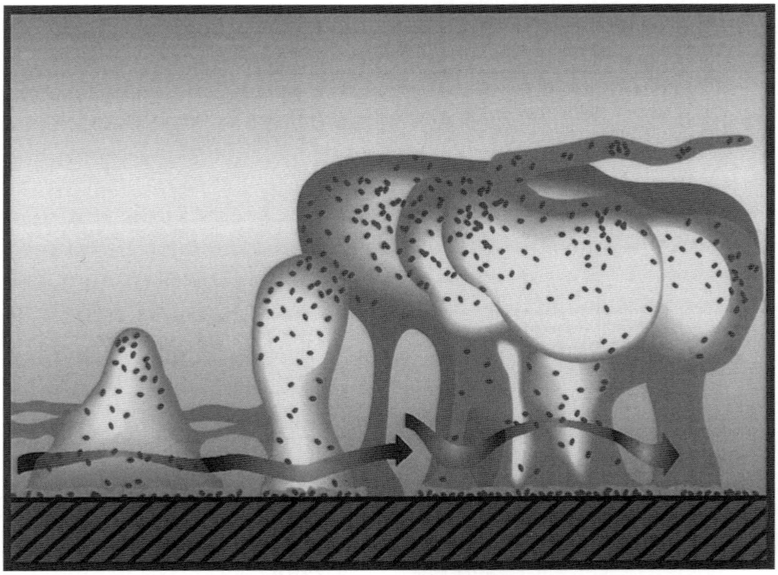

Figure 1b Diagrammatic representation of the structure of a hypothetical bacterial biofilm drawn from CSLM examination of a large number of mixed-species biofilms. The discrete microcolonies of microorganisms are surrounded by a network of interstitial voids filled with water. The arrows indicate convective flow within the water channels.

publications, everything worked as expected. Mathematical models of biofilm activity, however, describe biofilm activity over a certain surface area, not only at a single point. Attempts to find an average substrate consumption rate over a certain area often generated unexpected variability in that parameter. Since there was no independent technique to verify the microelectrode measurement, the suspicion was that the microelectrodes were not accurate. Therefore, the first confocal images of the inner space of biofilms were most welcomed by those who had invested time and effort in developing microelectrode technology; the biofilms on those images appeared to be quite heterogeneous (Lawrence *et al.*, 1991, see also Figure 1a). This could, possibly, explain why chemical profiles were so different at different locations in biofilms. Indeed, biofilm heterogeneity strongly influenced the shape of chemical profiles measured by microelectrodes, as was demonstrated by combining microelectrode measurements with confocal microscopy (DeBeer *et al.*, 1994b).

Example 4

Flow velocity distribution in biofilms was first studied using nuclear magnetic resonance imaging (NMRI) (Lewandowski *et al.*, 1992, 1993a; 1993b). It was expected that the biofilms covering walls of conduits would behave as if they were merely decreasing the dimensions of the conduits. Flow velocity was expected to decrease on approaching the biofilm surface and to finally reach zero at the biofilm surface. However, careful measurements of flow velocity distribution in biofilms demonstrated that the flow velocity was reaching zero at the conduit surface, instead of the biofilm surface as expected (Lewandowski *et al.*, 1993b). Again, such an effect is expected in heterogeneous biofilms where water can move in the interstitial voids. Flow velocity is now routinely measured in individual pores of heterogeneous biofilms (DeBeer *et al.*, 1994a; Stoodley *et al.*, 1994; Xia *et al.*, 1998).

The foregoing examples are samples of experimental results which lead to the change in the conceptual model of biofilm structure. According to the new model, biofilms are made of microcolonies separated by interstitial voids. Microcolonies are compact aggregates of extracellular polymers with densely packed microorganisms. Voids between these aggregates are filled with water and, perhaps, with strands of extracellular polymers connecting individual microcolonies. The shape of the microcolonies appears to be different in different biofilms, a fact that became an issue among different research groups. However, despite some differences in opinion about the shape of microcolonies, the extent of heterogeneity, and the somewhat different terminology used by different research groups the general conclusion is that biofilms are structurally heterogeneous. Figure 1a shows an image of a heterogeneous biofilm with microcolonies separated by interstitial voids, while Figure 1b shows an idealized model of biofilm structure derived from images such as the one in Figure 1a, *i.e.* mushroom shaped microcolonies separated by interstitial voids.

According to Figure 1b, heterogeneous biofilms are composed of 1) densely compact sublayers, 2) roundly-shaped microcolonies, 3) streamers, which are long strands of extracellular polymers extending the microcolonies, and 4) interstitial voids. The sublayer is not continuous and, at places, exposes the substratum. Above the sublayer are dense, roundly-shaped microcolonies, filled with extracellular polymers, densely packed with microorganisms, and finished with elongated streamers extending downstream. The microcolonies are separated by interstitial voids forming a network of interconnected channels, giving biofilms their characteristic, porous structure. Water can freely move within the network of these channels.

It is now known that as biofilms get older the semicontinuous sublayer tends to get denser, thicker, and accumulates various particles from the system while the upper layer consisting of characteristic roundly-shaped microcolonies and streamers remains the same. New advances in quantifying biofilm structure (see later) show that, as time progresses, the porosity of the bottom layers is much lower than the porosity of the upper layers. It is not clear whether that decrease in porosity is the result of biofilm growth and expansion near the bottom or the effect of accumulated debris and pieces of biofilm sloughed upstream. Free flowing particles can be trapped in biofilm pores, in the same way as fluorescent beads were trapped in the experiment of Drury (1992).

Structural heterogeneity presents different challenges for microbial ecologists and for those who mathematically model biofilm activity. For microbial ecologists it is important to define how, and why, the microorganisms are organized into such structures. For those who model biofilm activity it is important to determine the relations between structural heterogeneity, mass transport rates and mechanisms, and microbial activity. At present, mathematical description of heterogeneous biofilms is inadequate, which impedes progress in understanding of the significance of biofilm heterogeneity. For modeling purposes, it does not seem useful, nor even possible, to describe the spatial distribution of activity in each microcolony and in each interstitial void. Some simplifying assumptions will have to be made. The important issue is that these simplifying assumption are made as a result of careful experimentation, not merely for computational convenience. Ultimately, this is mathematical modeling that will quantify the relations between biofilm processes at the microscale and biofilm system performance at the macroscale.

SUBSTRATE CONCENTRATION PROFILES IN HETEROGENEOUS BIOFILMS

Dissolved substrates are transported from the bulk solution to the biofilm along concentration gradients, and are used in relevant biochemical reactions. The waste products of these reactions are transported back to the bulk solution *via* the same path. Consequently, substrate utilization in biofilms may be limited by the biofilm activity, (rate limiting reactions) or by the intensity of substrate delivery to the cells (mass transport limited to the surface of the biofilm or through the biofilm matrix). Most biofilms are mass transport limited because the dissolved substrates are delivered at a slower rate than they can be consumed. The intricate interplay between factors influencing substrate concentration profiles is best interpreted using one-dimensional profiles in homogeneous biofilm, for example oxygen concentration profiles. Mechanisms determining substrate concentration profiles in heterogeneous biofilm are the same; they act in three dimensions because of the microcolony structure. Assuming that the biofilm matrix is continuous, homogeneous, flat, and that there is no substrate consumption above the biofilm surface, three major factors influence oxygen concentration profiles, *viz.* hydrodynamics, mass transport, and biofilm activity. Transport of oxygen within the system is caused by convection and by molecular diffusion; the ratio of those two processes varies between locations and depends on the local flow rate. Spatial distribution of oxygen across the system, from the bulk liquid to the bottom, is presented as oxygen concentration profiles. At steady state, the concentration of oxygen at each point within the system is constant and determined by the rate of oxygen delivery to that point from the bulk solution equal to the rate of oxygen removal from that point, along concentration gradient. There are two boundary layers above the biofilm surface, *viz.* the hydrodynamic boundary layer and the mass transfer (diffusion) boundary layer; their presence is a direct consequence of water viscosity and substrate diffusivity. The kinematic viscosity of water solutions exceeds oxygen diffusivity (they both have the same dimension and can be numerically compared) by a few orders of magnitude and, therefore, the hydrodynamic boundary layer is thicker than the mass transfer boundary layer. Consequently, the latter is imbedded in the hydrodynamic boundary layer and remains proximate to the biofilm surface. Oxygen must travel across these two boundary layers to reach the biofilm. In the main stream, away from the hydrodynamic boundary layer, flow velocity is high, convective mass transport rate is high, and oxygen is uniformly distributed throughout the liquid. Near the biofilm surface, within the hydrodynamic boundary layer, and even more so within the mass transfer boundary layer, flow velocity slows down and the rate of convection decreases. As a result, the oxygen transport rate decreases in that zone, which is reflected by the curvature of the oxygen profile just above the biofilm surface. Below the biofilm surface a new factor, consumption of oxygen, changes the curvature of oxygen profiles from concave up, above the biofilm surface, to concave down below the biofilm surface. Mathematically, these two parts of the oxygen profile, above and below the biofilm surface, are described by different equations bound together by the requirement that oxygen flux across the common boundary, the biofilm surface, must be the same. Below the biofilm surface mass transport is said to be entirely due to molecular diffusion. Using these requirements, the shape of oxygen profiles can be analyzed numerically and relevant procedures have been developed to calculate the kinetic parameters of biofilm reactions from oxygen profiles (Lewandowski, 1991; 1994).

One-dimensional conceptual models of biofilms, such as the one described, assume that the dissolved substrates are transported perpendicularly to the substratum and that mass transport rates along the substratum are negligible. This is not true in heterogeneous biofilms. In these each microcolony acts as an entity, and the shape of boundary layers and local concentration profiles are influenced by the activity and spatial distribution of individual microcolonies, and by local hydrodynamics. That can make the substrate concentration profiles convoluted, as shown in Figure 2, using oxygen concentration profiles as an example.

HYDRODYNAMICS IN HETEROGENEOUS BIOFILMS

Hydrodynamics influences biofilms at all stages of their development. Biofilm accumulation is a net effect of cell attachment, detachment, and growth, and hydrodynamics influence all these processes. Bouwer (1987) points out that increased surface irregularity due to biofilm formation can influence particle transport rate and biofilm

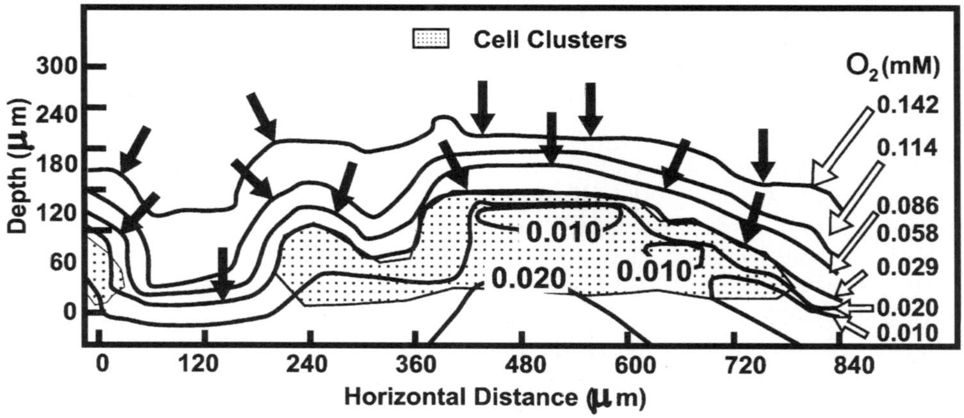

Figure 2 Oxygen concentration around a microcolony. Continuous lines = isobars; arrows = the direction of oxygen fluxes, always perpendicular to the active surface. Note that the microcolony is anoxic in the middle while oxygen is still detectable at the bottom, which demonstrates that oxygen near the bottom was transported there *via* channels and voids, not merely by diffusion through the microcolony. (Reproduced from DeBeer *et al.*, 1994b, with permission.)

attachment rate by 1) increasing convective mass transport near the surface, 2) providing shelter from shear forces, and 3) increasing surface area for attachment. In mature biofilms, the rate of substrate metabolism is controlled by the mass transport rate, which is influenced by hydrodynamics. Hydrodynamics is known to influence biofilm erosion and, possibly, sloughing.

Mathematical models of biofilm activity make assumptions about adjacent flow velocity distribution. It is important to verify these assumptions experimentally, using NMRI and flat plate biofilm reactors – rectangular conduits. The flow velocities employed were on the order of a few centimeters s^{-1}, because NMRI could not be used for higher velocities. Figure 3 shows the distribution of flow velocity in two identical reactors, one with a biofilm.

It was reasonable to expect that if biofilm accumulation increased surface roughness it should also increase the entry length to the reactor. (Entry length is the distance water must flow from the reactor entrance to the place where flow is fully developed, which, in the case of laminar flow, means parabolic distribution of flow velocity.) Entry length in the reactor with biofilm was expected to be longer than that in the sterile reactor for the same flow velocity. The flow velocity profiles were measured in both reactors at the same distance. Figure 3 shows that the flow velocity profile in the reactor with the biofilm is already parabolic while in the reactor without biofilm there still is a jet in the middle of the conduit. This result shows that the flow is already stable in the biofilm reactor while it remains unstable in the absence of the biofilm; this is opposite to expectations. This effect is believed to be caused by the viscoelastic nature of biofilms. By definition, biofilms are microorganisms imbedded in viscoelastic extracellular polymers. The expectation that biofilm increases surface roughness was based on the concept of rigid roughness elements. However, this concept does not

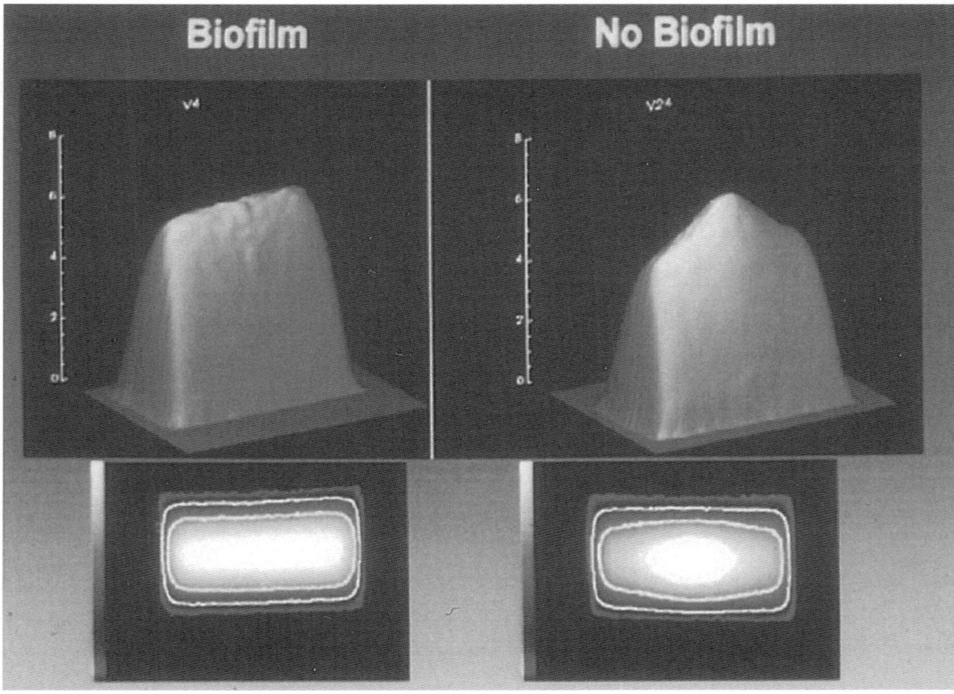

Figure 3 NMRI image of flow velocity distribution in rectangular conduits with and without biofilm. (Reproduced from Lewandowski and Altobelli, 1994, with permission.)

seem to apply to biofilms and the viscoelastic polymers do not behave as rigid roughness elements. They do not dissipate kinetic energy because the roughness elements protrude through the boundary layer. Instead, being elastic, they have the ability to move in the stream of water. Such surfaces, actively interacting with boundary layers, are called compliant surfaces and their deformations have the ability to stabilize flow in conduits (Lee T. *et al.*, 1995). Therefore, biofilms may delay the onset of turbulence in the main conduit.

When the image in Figure 3 was digitized, another unexpected result was noticed. The thickness of the biofilm in the reactor was on the order of 200 to 250 μm and it was expected that the flow velocity would reach zero at that distance from the wall. Instead, it reached zero at the reactor surface, some 200 μm below the surface of the biofilm (Lewandowski *et al.*, 1993b). Consequently, it was concluded that water was moving within the biofilm. This conclusion was later corroborated by injecting fluorescent beads to a biofilm and tracking them with CSLM.

Most biofilms, particularly those in industrial installations, are exposed to turbulent flow, while most biofilms in laboratories are grown in laminar flow. To fully appreciate the effects of hydrodynamics on biofilm growth, several flat plate reactors were operated at high flow velocities of a few m s^{-1} (Lewandowski and Stoodley, 1995). As biofilm accumulated the pressure drop across the rector was monitored for several days.

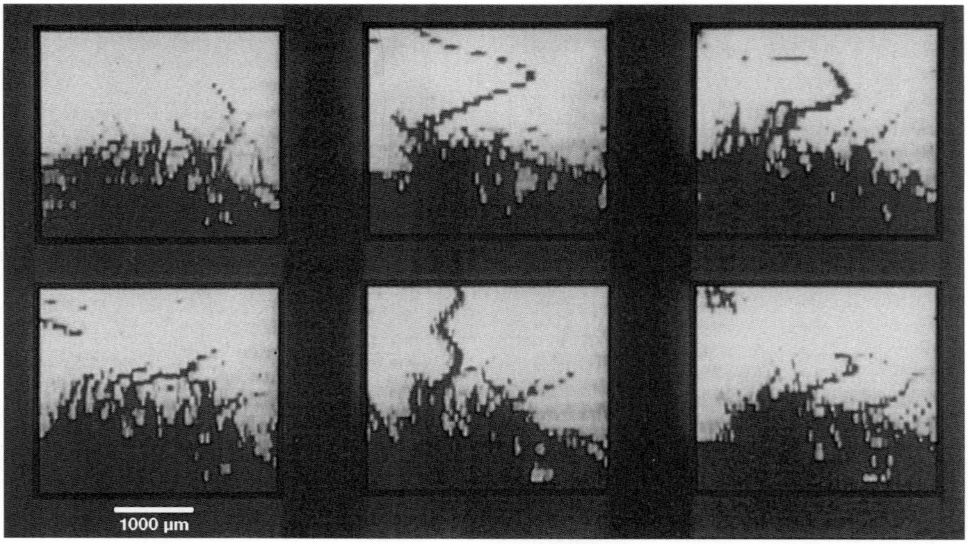

Figure 4 Surface of a biofilm with well developed streamers. The time sequence of images gives a planar view looking at the biofilm attached to the side wall of a reactor. (Reproduced from Lewandowski and Stoodley, 1995, with permission.)

As expected, the pressure drop increased with time. However, some reactors operated at lower flow velocities showed pressure drops higher than reactors operated at higher flow velocities (Lewandowski and Stoodley, 1995). This effect was, again, a consequence of the elastic and viscoelastic properties of biofilms. Microcolonies are made of bacterial cells imbedded in gelatinous extracellular polymers that can change shape under stress. At high flow velocities the hydrodynamic boundary layer separates from the microcolonies causing pressure drag which pulls the microcolonies downstream. The microcolonies, being made of viscoelastic materials, slowly flow under the strain, forming elongated shapes called streamers (Figure 4). Such streamers are often seen when biofilms grow at high flow velocities. The streamers move rapidly and dissipate the kinetic energy of the flowing water, which is reflected by the pressure drop. The movement of the streamers is transmitted to the underlying microcolonies, which oscillate rhythmically (Stoodley *et al.*, 1998). It would be expected that as the flow velocity increases, the streamers, eventually, lose mechanical stability, and separate from the parent microcolonies. Because of that, it would also be expected that streamers formed at high flow velocities should be shorter than streamers formed at lower velocities. Further, if the pressure drop is proportional to the streamers' length, that, hypothetically, explains why biofilm reactors operated at low flow velocities can sometimes show higher pressure drop than reactors operated at high flow velocity.

In conclusion, it is important, when discussing hydrodynamics in biofilms, to remember that 1) biofilms are made of viscoelastic polymers and 2) that hydrodynamics can actively change the biofilm structure. Based on present understanding, at low flow velocities biofilms would be expected to behave as compliant surfaces, to hydraulically smooth surfaces, and to stabilize the flow. However, when these biofilms are exposed to high flow velocities they oscillate faster and, eventually, the frequency of their oscillation can not follow the frequency of the eddies. At that point biofilm oscillation is "out of phase" and the biofilm not only fails to damp the flow instabilities but actively contributes to them by randomly oscillating at a different frequency. At that point the pressure drop rapidly increases, exceeding that expected for rigid roughness elements be a few times. Such an effect was monitored before it was properly explained (Picologlou *et al.*, 1980).

MASS TRANSPORT IN HETEROGENEOUS BIOFILMS

Most mathematical models of biofilm activity assume that the transport of dissolved substrates in the bulk liquid is controlled by convection and within the biofilm by molecular diffusion (Rittmann and McCarty, 1980). This assumption, reflecting the fact that flow velocity decreases near surfaces, causing convection to become negligible just above the biofilm, was also used to interpret chemical gradients measured in biofilms by microelectrodes (Lewandowski *et al.*, 1991). However, this assumption is not consistent with experimental evidence or with the model of heterogeneous biofilms. Considering relations between intra-biofilm mass transport and microbial activity, two important factors distinguish heterogeneous biofilms from homogeneous biofilms, *viz.* 1) heterogeneous biofilms have a much larger active surface area than the surface they cover, and 2) water can move within heterogeneous biofilms delivering nutrients to the deeper layers. Quantification of the dynamics of mass transport in such biofilms has been attempted.

To study the mechanisms of mass transport in biofilms microelectrodes have been developed based on the limiting current technique, to measure local mass transport coefficients (Yang and Lewandowski, 1995), and local effective diffusivity (Beyenal *et al.*, 1998). The term local refers to the mass transport rate in the vicinity of the microelectrode tip. To evaluate local effective diffusivity, cathodically polarized microelectrodes with a tip diameter of 10 μ are used to measure local consumption of ferricyanide introduced to the biofilm reactor. The solution of ferricyanide, $Fe(CN)^{-3}{}_6$, dissolved in a suitable electrolyte, replaces the nutrient solution in the reactor, and is allowed to equilibrate with the biomass until there is no detectable reduction of the ferricyanide by the biomass. Then, cathodically polarized microelectrodes are driven across the biofilm and the ferricyanide is reduced at their tips to ferrocyanide, $Fe(CN)^{-4}{}_6$, which generates current. The current is proportional to the reaction rate (ferricyanide reduction), which is proportional to the mass transport rate in the vicinity of the microelectrode tip. From such measurements local mass transport coefficient is calculated. Local limiting current density can be also directly related to the local effective diffusivity by calibrating the microelectrodes in layers of agar of known densities and effective diffusivities for the ferricyanide (Beyenal *et al.*, 1998). Using microeletrodes is simpler and less time consuming than using microinjections of fluorescent dyes, a technique used previously (DeBeer *et al.*, 1997) to evaluate local mass transport rates in biofilms. The result of such measurements is a map representing spatial distribution of effective diffusivity of ferricyanide to the tip of the microelectrode. It would be expected that molecules of similar size to ferricyanide would have similar diffusivities in biofilms. Having independent measurements of mass transport coefficients in biofilms helps understanding chemical gradients in

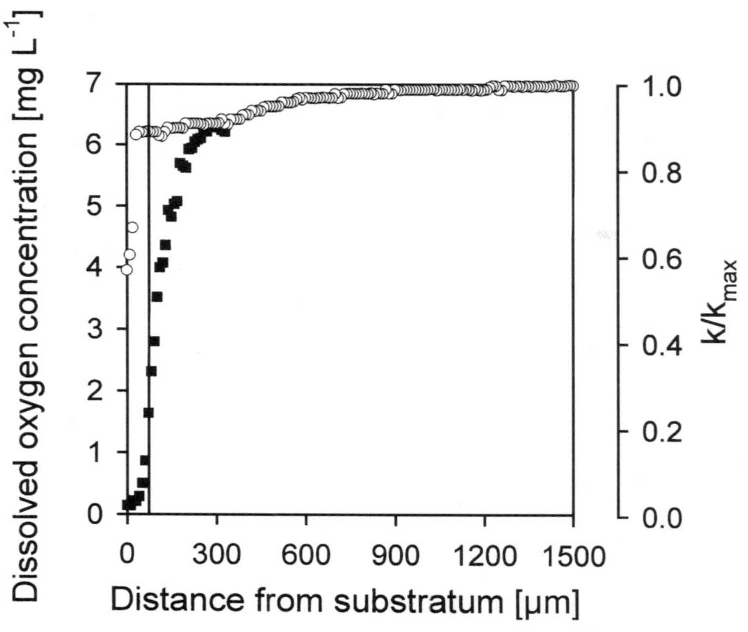

Figure 5 Profiles of dissolved oxygen (■) and local mass transfer coefficient (○) through a thin biofilm cluster. The vertical line marks the observed thickness of the biofilm. At distances less than 30 μm the wall effect caused the local mass transport coefficient to decrease. The biofilm thickness was 70 μm in this location. k/k_{max} was only slightly affected by the presence of the biofilm until a distance less than 30 μm from the substratum. (Reproduced from Rasmussen and Lewandowski, 1998, with permission.)

biofilms. Also, these measurement have helped in understanding the fact that the unusual resistance of biofilms to antimicrobial agents is not necessarily due to the high diffusion resistance of biofilm; signs of unusually low diffusivity that could be responsible for such an effect have not been seen. Parameters determining intra-biofilm mass transport can be also measured at a microscale level using fluorescence recovery after photobleaching (Bryers and Drummond, 1996), and confocal microscopy (Lawrence *et al.*, 1994).

A comparison of profiles of local mass transport coefficient and profiles of oxygen in a biofilm (Figure 5) shows that, for thin biofilms, the decrease in oxygen concentration near the biofilm surface is not correlated with the decrease in mass transport coefficient. The oxygen concentration decreases rapidly near the biofilm surface, due to oxygen consumption by the biofilm, while the mass transport coefficient remains relatively constant, due to convective mass transport within the mass transport boundary layer and water movement within the biofilm. For thicker biofilms these profiles are more complicated but, generally, mass transport coefficient profiles are poorly correlated with oxygen profiles near the biofilm surface (Rasmussen and Lewandowski, 1998).

Microelectrodes, similar to those used to measure local mass transport rates, have been calibrated to measure local flow velocity in interstitial voids of heterogeneous

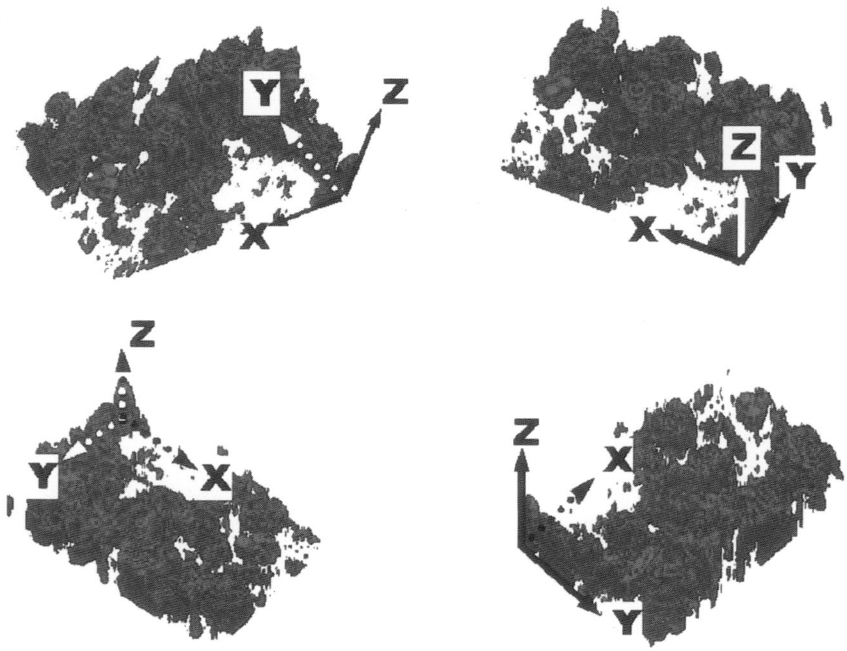

Figure 6 A digitally reconstructed 3-D image of a biofilm. To examine biofilm structure the image can be rotated around horizontal axes X and Y, and the vertical axis Z.

biofilms. This is a particularly useful and promising development since the procedures previously used to study flow velocity distribution in biofilms were either expensive (NMRI) or tedious (fluorescent particle velocimetry with confocal microscopy). Using such microelectrodes sensitive to water movement it is possible to see differences in flow velocity in biofilm channels differently oriented with respect to the flow direction in the main conduit (Xia *et al.*, 1998).

RECONSTRUCTING BIOFILMS AND QUANTIFYING HETEROGENEITY

Transport of substrates, hydrodynamics, and microbial activity are all interrelated with the structure of biofilms (Van Loosdrecht *et al.*, 1995). To study these relations, biofilms are reconstructed in three dimensions from their confocal images and chemical gradients and local diffusivity profiles superimposed on these constructs. A large number of confocal images is collected at different distances from the substratum and collated with an aid of a computer (Figure 6).

Such collated images can be rotated and sliced digitally to provide insight into the inner space of the biofilm and to understand chemical profiles measured by microelectrodes. Figure 7 shows how a convoluted effective diffusivity profile can be explained when it is superimposed on the biofilm structure. Without information about the biofilm structure such a profile could have been rejected as an artifact.

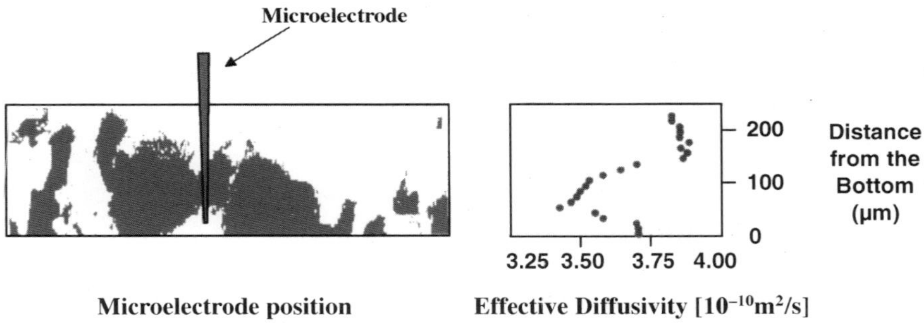

Figure 7 An effective diffusivity profile (right) is analyzed by comparing it side by side with an image of the internal structure of the biofilm (left). The internal structure can be visualized at any location by digitally slicing images such as the one in Figure 6. The results demonstrate that when density increases (within the microcolony), effective diffusivity decreases.

To study the relationships between local biofilm activity and biofilm heterogeneity it is necessary to quantify biofilm heterogeneity, and to use heterogeneity as a variable that can be correlated with other system variables such as mass transport rates and flow velocity. Attempts to quantify biofilm heterogeneity have already been made with respect to biofilm density, porosity, pore structures, tortuosity (Zang and Bishop, 1994a; 1994b) and fractal structure (Hermanowicz *et al.*, 1995). Also, attempts have been made to introduce some of these parameters to mathematical modeling of biofilm activity (Wimpenny and Colasanti, 1997; Picioreanu *et al.*, 1998).

Biofilm heterogeneity is best visualized by confocal microscopy combined with image analysis and 3-D reconstruction (Figure 6). To extract quantitative parameters from confocal images a computer program has been designed for the Unix computer operating system (Yang X *et al.*, 2000). Using these parameters, biofilm heterogeneity can be placed on a numerical scale and correlated, for example, with mass transport rates. Several parameters characterizing the shape of microcolonies and interstitial voids are calculated; four are particularly promising, *viz.* areal porosity, diffusional distance, fractal dimension, and textural entropy. Areal porosity is the ratio of the void area to the total area of an image and ranges in value from 0 to 1. The lower the value, the higher the surface coverage. Diffusion distance is the minimum distance from each cluster pixel to its nearest void pixel. The average diffusion distance and the maximum diffusion distance are calculated for a given image. Fractal dimension reflects the degree of raggedness of biofilm cell clusters and for a two dimensional image it can range from 1 to 2, where the higher the number, the more ragged the edge of the cluster. Textural entropy measures biofilm homogeneity; it has a minimum value of 0 and no theoretical maximum value. High values represent more heterogeneous structures. It is conjectured that there exists a finite number of parameters that uniquely describe the structure of a biofilm and contain enough information either to reflect variations in the growth dynamics or to predict the functional characteristics of the biofilm. To find these parameters reliance is placed on the fact that biofilms achieve steady state conditions, where the physical structure is dynamic at the molecular level, but static at the scale corresponding to the microscopic field of view. If a parameter appears to approach steady state, it is behaving in the expected manner, and it is accepted as describing biofilm heterogeneity. Initial experiments indicate that porosity, fractal dimension, difusional distance, and fractal dimension are such parameters and appear to reach steady state for some biofilms (Lewandowski *et al.*, 1998).

CONCLUDING REMARKS

Microscale biofilm research has demonstrated that biofilms are structurally heterogeneous. However, except for the simple case of differential aeration cells in microbially influenced corrosion, it is still not clear how these structures influence the overall biofilm performance at the macroscale. The most exciting hypotheses presently tested include the active role of biofilms in controlling their own structural heterogeneity; greater structural heterogeneity encourages delivery of substrates to deeper layers of the biofilm. Therefore, if substrate concentration decreases in the bulk water, biofilms may actively increase their heterogeneity, conceivably, using cell-cell communication. Whether this is true or not is another matter. The state of the art in microscale biofilm research is such that hypotheses of this kind can be put forward and there are technical means to verify them experimentally.

ACKNOWLEDGEMENTS

This work was partially funded by Cooperative Agreement #EEC-8907039 between the National Science Foundation Engineering Research Centers Program and Montana State University, and by the Industrial Associates of the Center for Biofilm Engineering.

REFERENCES

Beyenal H., Tanyolac A., Lewandowski Z. (1998). Measurement of local effective diffusivity in heterogeneous biofilms. *Water Sci Technol*, **38**, 171–178.

Bishop P.L., Rittmann B.E. (1995). Modelling heterogeneity in biofilms: report of the discussion session. *Water Sci Technol*, **32**, 263–265.

Bouwer E.J. (1987). Theoretical investigation of particle deposition in biofilm systems. *Water Res*, **21**, 1489–1498.

Bryers D.J., Drummond F. (1996). Local mass transfer coefficients in bacterial biofilms using fluorescence recovery after photobleaching (FRAP). In: Wijffels R.H., Buitelaar R.M., Bucke C. and Tramper J. (eds) *Progress in Biotechnology 11, Immobilized Cell: Basics and Applications*. Elsevier, New York, pp. 196–204.

Costerton J.W., Lewandowski Z., DeBeer D., Caldwell D., Korber D., James G. (1994). Minireview: biofilms, the customized microniche. *J. Bacteriol*, **176**, 2137–2142.

Davies D.G., Parsek M.R., Pearson J.P., Iglewski B.H., Costerton J.W., Greenberg E.P. (1998). The involvement of cell-to-cell signals in the development of a bacterial biofilm. *Science*, **280**, 295–298.

DeBeer D., Stoodley P., Lewandowski Z. (1994a). Liquid flow in heterogeneous biofilms. *Biotechnol Bioeng*, 44, 636–641.

De Beer D., Stoodley P., Lewandowski Z. (1997). Measurement of local diffusion coefficients in biofilms by microinjection and confocal microscopy. *Biotechnol Bioeng*, **53**, 151–158.
DeBeer D., Stoodley P., Roe F., Lewandowski Z. (1994b). Effects of biofilm structures on oxygen distribution and mass transport. *Biotechnol Bioeng*, **43**, 1131–1138.
Drury W.J. (1992). Interactions of 1 micron latex microbeads with biofilm. PhD Thesis, Montana State University, Bozeman, USA.
Hamilton W.A. (2000). Microbially influenced corrosion in the context of metal microbe interactions. In: Evans L.V. (ed) *Biofilms: Recent Advances in their Study and Control*. Harwood Academic Publishers, pp. 419–434.
Hermanowicz S.W., Schindler U., Wilderer P. (1995). Fractal structure of biofilms: new tools for investigation of morphology. *Water Sci Tech*, **32**, 99–105.
Keevil C.W., Walker J.T. (1992). A Normarski DIC microscopy and image analysis of biofilm. *Binary*, 4, 93–95.
Kolenbrander P.E., London J. (1992). Ecological significance of coaggregation among oral bacteria. *Adv Microb Ecol*, **12**, 183–217.
Kolenbrander P.E., London J. (1993). Adhere today, here tomorrow: oral bacterial adherence. *J Bacteriol*, **175**, 3247–3252.
Lawrence J.R., Wolfaardt G.M., Korber D.R. (1994) Determination of diffusion coefficients in biofilms by confocal laser microscopy. *Appl Environ Microbiol*, **60**, 1166–1173.
Lawrence J.R., Korber D.R., Hoyle B.D., Costerton J.W., Caldwell D.E. (1991). Optical sectioning of microbial biofilm. *J Bacteriol*, **173**, 6558–6567.
Lee T., Fisher M., Schwarz W.H. (1995). Investigation on the effects of a compliant surface on boundary-layer stability. *J Fluid Mech*, **288**, 37–58.
Lee W., Lewandowski Z., Nielsen P.H., Hamilton W.A. (1995). Role of sulfate-reducing bacteria in corrosion of mild steel: a review. *Biofouling*, **8**, 165–194.
Lewandowski Z. (1994). Dissolved oxygen gradients near microbially colonized surfaces. In: Geesey G., Lewandowski Z., Flemming H-C. (eds) *Biofouling and Biocorrosion in Industrial Water Systems*. CRC Press Incorporated, Lewis Publishers, Boca Raton, pp. 175–189.
Lewandowski Z., Altobelli S. (1994). Water flow in a narrow conduit covered with biofilm. Int Assoc Water Quality Res Seminar *Biological Degradation of Organic Chemical Pollutants in Biofilm Systems*. May 1994, Kollekolle, Copenhagen, Denmark.
Lewandowski Z., Stoodley P. (1995). Flow induced vibrations, drag force, and pressure drop in conduits covered with biofilm. *Water Sci Tech*, **32**, 19–26.
Lewandowski Z., Walser G., Characklis W.G. (1991). Reaction kinetics in biofilms. *Biotechnol Bioeng*, **38**, 877–882.
Lewandowski Z., Altobelli S.A., Fukushima E. (1993a). NMR and microelectrode studies of hydrodynamics and kinetics in biofilms. *Biotechnol Prog*, **9**, 40–45.
Lewandowski Z., Altobelli S.A., Majors P.D., Fukushima E. (1992) NMR imaging of hydrodynamics near microbially colonized surfaces. *Water Sci Tech*, **26**, 577–584.
Lewandowski Z., Stoodley P., Altobelli S., Fukushimam E. (1993b). Hydrodynamics and kinetics in biofilm systems – recent advances and new problems. *Proc 2nd IAWQ Int Spec Conf Biofilm Reactors*. September/October 1993, Paris, France, pp. 313–319.
Lewandowski Z., Webb D., Hamilton M., Harkin G. (1998). Quantifying biofilm structure. Int Conf *Microbial Ecology of Biofilms: Concepts, Tools and Applications*. Lake Bluff, Illinois, October 1998. *Water Sci Technol* (In press).
Massol-Deya A.A., Whallon J., Hickey R.F., Tiedje J.M. (1994). Channel structures in aerobic biofilms of fixed-film reactors treating contaminated groundwater. *Appl Environ Microbiol*, **61**, 769–777.
Picioreanu C., VanLoosdrecht M.C.M., Heijnen J.J. (1998). Mathematical modeling of biofilm structure with a hybrid differential-cellular automaton approach. *Biotechnol Bioeng*, **58**, 101–116.

Picologlou B.F., Zelver N., Characklis W.G. (1980). Biofilm growth and hydraulic performance. *J Hydraul Div Am Soc Civ Eng*, **106**, No HY5 733–746.

Rasmussen K., Lewandowski Z. (1998). Microelectrode measurements of local mass transport rates in heterogeneous biofilms. *Biotechnol Bioeng*, **59**, 302–309.

Rittmann B.E., McCarty P.L. (1980). Model of steady-state biofilm kinetics. *Biotechnol Bioeng*, **22**, 2343–2357.

Roe F., Lewandowski Z., Funk T. (1996). Simulating microbially influenced corrosion by depositing extracellular biopolymers on mild steel surfaces. *Corrosion*, **52**, 744–752.

Stoodley P., DeBeer D., Lewandowski Z. (1994). Liquid flow in biofilm systems. *Appl Environ Microbiol*, **60**, 2711–2716.

Stoodley P., Lewandowski Z., Boyle J.D., Lappin-Scott H.M. (1998). Oscillation characteristics of biofilm streamers in turbulent flowing water as related to drag and pressure drop. *Biotechnol Bioeng*, **57**, 536–544.

Van Loosdrecht M.C.M., Eikelboom D., Gjaltema A., Mulder A., Tijhuis L., Heijnen J.J. (1995). Biofilm Structures. *Water Sci Tech*, **32**, 35–43.

Wanner O., Gujer W. (1986). A multi-species biofilm model. *Biotechnol Bioeng*, **28**, 314–328.

Wimpenny J.W., Colasanti R. (1997). A unifying hypothesis for the structure of microbial biofilms based on cellular automaton models. *FEMS Microbiol Ecol*, 22, 1–16.

Wolfaardt G.M., Lawrence J.R., Robarts R.D., Caldwell D.E. (1994). A multicellular organization in a degradative biofilm community. *Appl Environ Microbiol*, **60**, 434–446.

Xia F., Beyenal H., Lewandowski Z. (1998). An electrochemical technique to measure local flow velocity in biofilms. *Water Res*, **32**, 3631–3636.

Yang S., Lewandowski Z. (1995). Measurement of local mass transfer coefficient in biofilms. *Biotechnol Bioeng*, **48**, 737–744.

Yang X., Beyenal H., Harkin G., Lewandowski Z. (2000). Quantifying biofilm structure using image analysis. *J Microbiol Methods*, **39**, 109–119.

Zhang T.C., Bishop P.L. (1994a). Evaluation of tortuosity factors and effective diffusivities in biofilms. *Water Res*, **28**, 2279–2287.

Zhang T.C., Bishop P.L. (1994b). Density, porosity and pore structure of biofilms. *Water Res*, **28**, 2267–2277.

2 Physico-chemical Properties of Biofilms

Hans-Curt Flemming, Jost Wingender, Thomas Griebe and Christian Mayer

Microorganisms in aggregates such as flocs, film and sludge do not only display biochemical and biological, but also physical and physico-chemical properties. Among these are mechanical stability, binding of water, diffusion, sorption, mass transport and optical properties, and friction resistance. These properties are chiefly caused by the extracellular polymeric substances (EPS) which fill the space between the cells and account for a considerable proportion of the organic carbon content of biofilms. The EPS consist not only of polysaccharides but also of considerable amounts of protein; nucleic acids and lipids are also found in the EPS. Above all, the EPS form the morphology and internal structure of biofilms, including surface pores and channels. The EPS provide a matrix which allows the cells to maintain their position for a much longer period of time compared to the planktonic mode. This facilitates the formation of synergistic microconsortia of different species which can perform orchestrated degradation processes. Mechanical stability of biofilms includes aspects such as sloughing of the biomass in biofilm reactors, resulting in possibly adverse effects to the process. On the other hand, when biofilms have to be removed as biofouling layers, it is the cohesive and adhesive forces which have to be overcome. Three types of weak interactions have to be considered, *viz.* hydrogen bonds, electrostatic interactions and van der Waals interactions. As EPS contain many groups capable of different forms of these interactions, the binding force between macromolecules is multiplied by the number of interacting groups, which increases the overall binding force by several orders of magnitude. Interactions between extracellular enzymes and polysaccharides are known which stabilize the enzymes and possibly enhance their activity.

INTRODUCTION

Most microorganisms live and grow in aggregates such as biofilms, flocs ("planktonic biofilms"), and sludge. This form of microbial life is described by the somewhat inexact but generally accepted term "biofilm". The feature which is common to all these phenomena is that the microorganisms are embedded in a matrix of extracellular polymeric substances which are responsible for the morphology, structure, coherence and physico-chemical properties of these aggregates. Biofilms are ubiquitously distributed in natural soil and aquatic environments, on tissues of plants, animals and man as well as in technical systems such as filters and other porous materials, reservoirs, pipelines, ship hulls, heat exchangers, and separation membranes (Costerton *et al.*, 1987; Flemming and Schaule, 1996). Biofilms develop adherent to a solid surface (substratum) at solid-water interfaces, but can also be found at water-air and at solid-air interfaces. They are composed of accumulations of microorganisms (prokaryotic and eukaryotic unicellular organisms), extracellular polymeric substances (EPS), multivalent cations, biogenic and inorganic particles as well as colloidal and dissolved compounds. EPS are mainly responsible for the structural and functional integrity of biofilms and are considered as the key components that determine the physico-chemical properties of biofilm. EPS form a three-dimensional, gel-like, highly hydrated and locally charged biofilm matrix, in which the microorganisms are embedded and more or less immobilized. In technical systems, not only the biological but also the physico-chemical properties of biofilms are of practical importance. For example, if a biofilm has to be removed from a surface, the forces which keep the matrix of the EPS together (cohesion) and attached to the surface (adhesion), have to be overcome. In other words, the mechanical stability of the microbial aggregate plays a key role in cleaning processes, regardless of the viability of the biomass. Compared to the biological and biochemical properties of biofilms, the consideration of the physico-chemical properties have been much less in the focus of biofilm research. However, they include such important aspects (examples in parentheses) as frictional resistance (*e.g.* on ships' hulls and in water pipes), mechanical stability (in cleaning and for prediction of biofilm sloughing), mass transfer resistance and diffusion (in biofilm reactors), heat transfer resistance (on heat exchangers), hydraulic resistance (on separation membranes), sorption properties, mechanisms and capacity (as a sink and source of pollutants), binding of water (in sludge dewatering and change of surface properties), optical properties (in colouring of surfaces), and tolerance against biocides (in surface disinfection).

Since biofilms are formed by microorganisms, the regulation of the production of EPS and other microbial products is determined biologically. However, once biomass is formed, regardless of whether it is as a film, floc or sludge, the aggregates are physical bodies and display properties which will influence technical and natural processes.

THE CRUCIAL ROLE OF EPS

The EPS represent the construction material of biofilms. This group of biopolymers consists mainly of polysaccharides and proteins, but other macromolecules such as DNA, lipids and humic substances have also been found in wastewater biofilms and activated sludge (Neu, 1992; Urbain *et al.*, 1993; Jahn and Nielsen, 1996; Nielsen *et al.*, 1997). Most bacteria are able to produce EPS, whether they grow in suspension or in biofilms. Cell surface polymers and EPS are of major importance for the development and structural integrity of flocs and biofilms. They mediate interactions between microorganisms and form the matrix in which the microorganisms are immobilized and kept in a three-dimensional arrangement. In general, the proportion of EPS in blofilms can vary between 50 and 90% of the total organic matter (Christensen and Characklis, 1990; Neu, 1992; Nielsen *et al.*, 1997). It must be pointed out that polysaccharides are not necessarily the main EPS component. However, not much is known about synergistic gelling of polysaccharides, proteins and humic subtances. In many cases of environmental biofilm samples, proteins prevail, and humic substances are also integrated in the EPS matrix, being considered by some authors as belonging

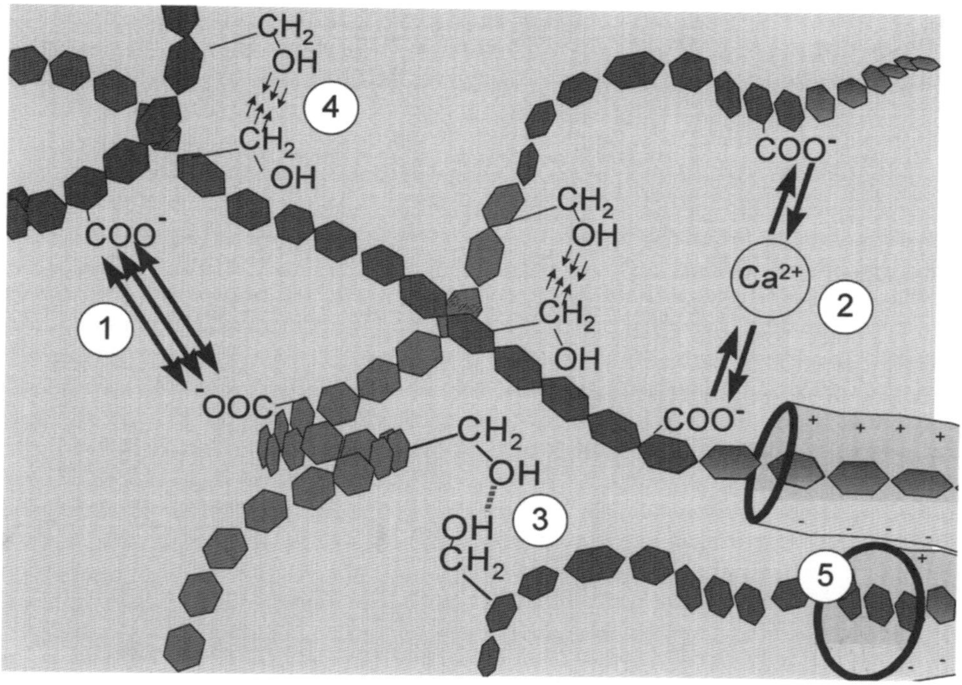

Figure 1 Proposed model for dominating intermolecular interactions which contribute to mechanical stability in a biofilm. Five different phenomena are considered. 1 = repulsive electrostatic interactions between ionic residues; 2 = attractive electrostatic forces, typically in the presence of divalent cations; 3 = hydrogen bonds; 4 = other electrostatic interactions, *e.g.* between dipoles; 5 = London (dispersion) interactions.

to the EPS (see Wingender *et al.*, 1999). In activated sludges and sewer biofilms 85–90% and 70–98% respectively, of the total organic carbon was found to be extracellular, indicating that cell biomass constitutes only a minor fraction of the organic matter in microbial aggregates (Jahn and Nielsen, 1998; Frølund *et al.*, 1996). Although mostly a minor component, lipids can make up a significant proportion of the EPS in some cases. This has been shown in the case of strongly acidophilic organisms, colonizing and leaching pyrite (Gehrke *et al.*, 1998).

Binding Forces in EPS

As already mentioned, EPS are involved in the formation of microbial aggregates and provide the maintenance of their stability. However, these processes are not mediated by the formation of covalent C—C bonds between the EPS molecules, but by weak physico-chemical intermolecular interactions. Three types of noncovalent interactions have to be considered as cohesive forces between the components within the EPS matrix of microbial aggregates, *viz.* London (dispersion) forces, electrostatic interactions and hydrogen bonds (Flemming *et al.*, 1998). In Figure 1, these interactions are schematically depicted.

Dispersion forces act intra- and intermolecularly (*e.g.* within and between proteins) and are not dependent on functional groups. Dispersion forces provide a major contribution to the interaction forces within the hydrophobic regions of molecules or between molecules known as the "hydrophobic interactions". The binding energy is about 2.5 kJ mol^{-1}.

Electrostatic interactions occur between charged molecules (ions) and between permanent or induced dipoles. Repulsion is expected between functional groups such as carboxyl groups of proteins and polysaccharides; however, divalent cations such as Ca^{2+} can act as bridges, contributing significantly to the overall binding force. Positively charged groups from amino sugars in polysaccharides or from amino acids in proteins can also interact with negatively charged groups, providing cohesion forces. Electrostatic interactions seem to be of major importance for the stability of the EPS matrix. The binding energy of nonionic electrostatic interactions range between 12 and 29 kJ mol^{-1}. The binding force is strongly dependent on the distance between the partners of the bond and the water concentration.

Hydrogen bonds form mainly between hydrogen atoms of hydroxyl groups and the more electronegative oxygen or nitrogen atoms that are abundant in polysaccharides and proteins. Hydrogen bonds are active within (*e.g.* in the maintenance of the secondary and tertiary structure of proteins) and between macromolecules, but are also involved in binding of water to EPS. The binding energy ranges between 10 and 30 kJ mol^{-1} and reaches only a short distance.

However, the main partner in polysaccharide hydrogen bond binding sites is water. Therefore, only a small proportion of the possible hydrogen bonds will establish between polysaccharides. But bearing in mind the macromolecular character of the molecule and the large number of potential binding sites, this still can contribute significantly to the overall binding energy.

The role of entanglement in the gelling process of biofilms has not yet been elucidated. Generally, the influence of entanglement on the stability of gels is strongly temperature dependent. Thus the proportion of entanglement in the overall binding energy would decrease with rising temperature. Although experience shows that biofilms cannot be "melted", perhaps the increase in molecular mobility caused by increasing temperature contributes to the efficacy of biofilm removal by application of steam as sometimes practised in ultrapure water systems.

The individual binding force of any of these interactions is relatively small compared to a covalent C—C bond (about 250 kJ mol^{-1}). However, if an EPS molecule possesses 106 functional groups and only 10% of these are involved in bonding, the total binding energy adds up to values in the range of several covalent C—C bonds. All three types of binding forces contribute to the overall stability of floc and biofilm matrices, probably to various extents.

Cleaning formulations mainly address dispersion interactions and electrostatic interactions, respectively, by application of surface active substances, or of acid, base or complexing agents (Mayer *et al.*, 1999). Hydrogen bonds, providing a considerable part of the overall binding energy, are not addressed by any components of commercial cleaners. EPS are the key components in a number of models explaining the aggregation of microorganisms as well as the physico-chemical properties of the extracellular matrix in flocs and biofilms (*e.g.* Pavoni *et al.*, 1972; Harris and Mitchell, 1973; Nielsen *et al.*, 1997). In activated sludge flocs, EPS have been implicated in determining floc structure, floc charge, the flocculation process, floc settleability and dewatering properties. In the polymer-bridging model, floc formation is considered as the result of the interaction of high-molecular-weight, long-chain EPS with microbial cells and other particles as well as with other EPS molecules, so that EPS bridge the cells into a three-dimensional matrix. Flocculation is associated with the formation of EPS. Cellular aggregation was found to depend on the physiological state of the microorganisms; flocculation of cultures of mixed populations from domestic wastewater did not occur until they entered into a restricted state of growth (Pavoni *et al.*, 1972). There was a direct correlation between microbial aggregation and EPS accumulation; the ratio of EPS to microorganism mass rapidly increased during culture aggregation. The major EPS were polysaccharides, proteins and nucleic acids (RNA, DNA). Surface charge was not considered a necessary prerequisite for flocculation, since it remained constant throughout all growth phases regardless of the flocculability of the culture. Bacteria washed free of EPS formed stable dispersions, but readdition of extracted EPS again resulted in flocculation. In batch cultures with *Zoogloea* it was also shown that production of an extracellular polysaccharide was accompanied by flocculation of the bacteria (Unz and Farrah, 1976). Polymer formation was initiated in mid-logarithmic growth phase and the quantity produced appeared to be influenced by the level of carbon and nitrogen in the medium. The detection of extracellular polymeric fibrils in natural and wastewater flocs by high-resolution TEM confirmed the role of EPS as structural support to the microbial aggregates (Liss *et al.*, 1996).

In addition to EPS, divalent cations are regarded as important constituents of microbial aggregates, since they bind to negatively charged groups present on bacterial surfaces, in EPS molecules and on inorganic particles entrapped in flocs and biofilms. It has been reported that extraction of Ca^{2+} from flocs and biofilms by displacement with monovalent cations or by chelation with the more general complexing agent EDTA or the more Ca^{2+}-specific chelant EGTA resulted in the destabilisation of flocs (Bruus *et al.*, 1992; Higgins and Novak, 1997) and biofilms (Turakhia *et al.*, 1983). The practical implications are that weakening of activated sludge structure by removal of Ca^{2+} leads to an increase in the number of small particles with a subsequent decrease in filterability and dewaterability. These observations suggest that divalent cations may be important for the maintenance of floc and biofilm structure by acting as bridging agents within the three-dimensional EPS matrix. Bruus *et al.* (1992) also integrated the role of divalent cations into their sludge floc model. The floc structure was proposed to be a three-dimensional EPS matrix kept together by divalent cations with varying selectivity to the matrix ($Cu^{2+} > Ca^{2+} > Mg^{2+}$). It was argued that approximately half of the Ca^{2+} pool was associated with EPS, forming a matrix that resembled gels of carboxylate-containing alginates. Fe^{3+} ions may also be of importance in floc stabilisation. Specific removal of Fe^{3+} from activated sludge flocs caused a weakening of floc strength resulting in release of particles to bulk water, dissolution of EPS and partial floc disintegration (Nielsen and Keiding, 1998). On the basis of investigations on laboratory-scale activated sludge reactors, Higgins and Novak (1997) emphasized the role of structural proteins in conjunction with divalent cations in flocculation. Increasing the concentrations of Ca^{2+}, or Mg^{2+} resulted in an increase in bound protein, whereas there was little effect on bound polysaccharides. Addition of high concentrations of Na^+ led to a decrease in bound protein. It was believed that the monovalent sodium ions displaced divalent cations from within the flocs. This displacement would reduce binding of protein within the floc and result in solubilization of protein. Further support for the involvement of extracellular protein in the aggregation of bacteria into flocs came from the observation that treatment of activated sludge flocs with a proteolytic enzyme (pronase), resulted in deflocculation, with a shift to smaller particles in the 5–40 μm range and a release of polysaccharide. Gel electrophoretic analysis of extracted EPS from municipal, industrial and laboratory activated sludge revealed the presence of a single protein with a molecular mass of approximately 15,000 Daltons. Analysis of amino acid composition and sequence indicated that this protein displayed similarities to lectins; binding site inhibition studies demonstrated the lectin-like activity of the 15,000-Dalton protein (Higgins and Novak, 1997). On the basis of these results a model of flocculation was proposed, *viz.* lectin-like proteins bind polysaccharides that are cross-linked to adjacent proteins. Divalent cations bridge negatively charged functional groups on the EPS molecules. The cross-linking of EPS and cation bridges leads to the stabilization of the biopolymer network mediating the immobilization of microbial cells. Urbain *et al.* (1993) concluded from their studies on 16 activated sludge samples from different origins that internal hydrophobic bondings were involved in flocculation mechanisms and their balance with hydrophilic interactions determined the settling properties of the sludge. Hydrophobic areas in between the cells were considered as essential adhesives within the floc structure. Cell surface hydrophobicity was shown to be important for adhesion of bacteria to activated sludge flocs (Olofsson *et al.*, 1998). Cells with high cell surface hydrophobicity attached in higher numbers to the flocs than bacteria with a more hydrophilic surface. The hydrophobic cells attached not only to the surface of the flocs, but also penetrated the flocs through channels and pores, whereas hydrophilic cells did not behave in this way. It was assumed that adhesion of hydrophobic bacteria within flocs would increase the potential of the flocs to clear free-living cells from the water phase (Olofsson *et al.*, 1998).

EPS and Biofilm Morphology

EPS are considered essential matrix polymers responsible for the integrity of the three-dimensional structure of biofilms. In addition, EPS may be involved in the interaction between microbial cells and the substratum, leading to irreversible adhesion and surface colonization. The chemical composition of EPS largely determines the physical properties of biofilms (Christensen and Characklis, 1990) and, moreover, the morphology. Biofilms occur in natural and technical systems in a wide morphological variety, ranging from smooth slimy layers to thick filamentous deposits. Figures 2 and 3 show SEM micrographs of biofilm samples from two irreversibly biofouled reverse osmosis (RO) membranes originating from RO plants treating surface water from different rivers. Although the preparation process requires dehydration and the resulting SEM picture, showing the desiccated sample, must be considered in that respect an artifact, both samples were prepared in the same way. They show a different morphology. In Figure 2, the organisms are embedded in the EPS matrix while in Figure 3 only traces of EPS in the form of irregular fibres can be seen after desiccation, although the original sample was similarly slimy.

Although both biofilms have been exposed to numerous cleaning and disinfection cycles they could not be removed or killed but continued to grow until the membrane

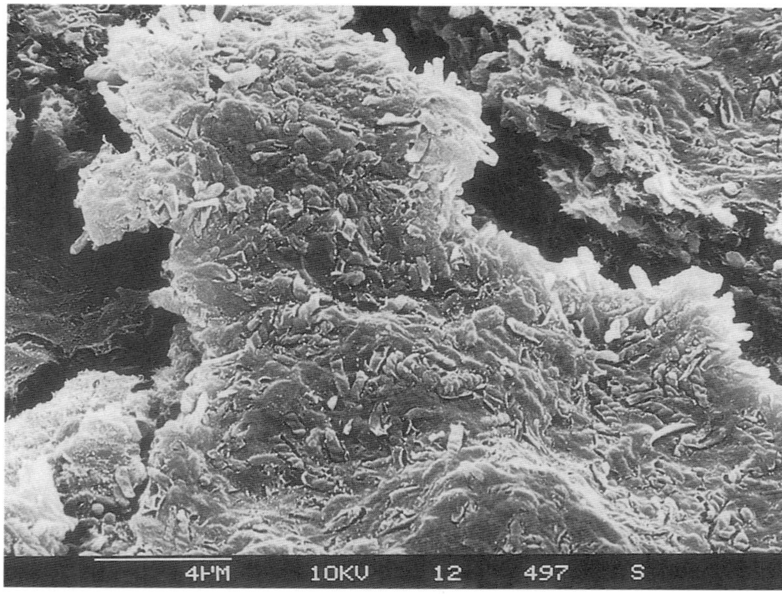

Figure 2 SEM photograph of a biofilm on an irreversibly biofouled RO membrane from a river Main RO water treating plant. (Reproduced from Flemming and Schaule, 1989, with permission.)

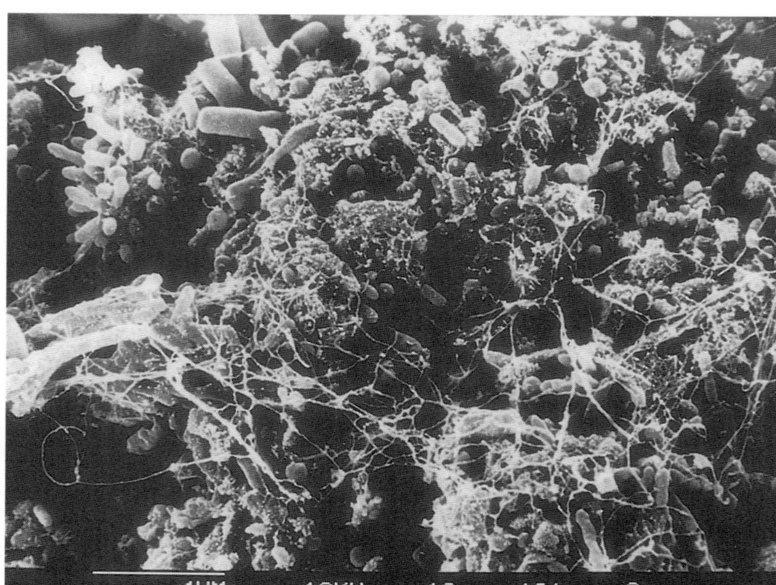

Figure 3 SEM photograph of a biofilm on an irreversibly biofouled RO membrane from a river Seine RO water treating plant. (Courtesy of G. Schaule.)

was irreversibly fouled. This increased tolerance to biocides and cleaners, compared to planktonic organisms, is well known (LeChevallier *et al.*, 1988; Morton *et al.*, 1998). Whether it is due to a physico-chemical effect is not yet clear; diffusion hindrance has been discussed (Costerton *et al.*, 1987), but the diffusion coefficients of small molecules in biofilms are about the same as in water (De Beer *et al.*, 1994; Beyenal *et al.*, 1998). Morton *et al.* (1998) speculate that some biofilm organisms grow slowly and thus, are less susceptible to biocides which usually attack fast-growing planktonic cells. Another explanation is that biocides such as chlorine react with the EPS matrix and, thus, are consumed before they reach the cells.

Much information has been gathered on the chemical and physical properties of extracellular polysaccharides, since they are abundant in bacterial EPS. Specific polysaccharides (*e.g.* xanthan) are only produced by individual bacterial strains, whereas nonspecific polysaccharides (*e.g.* levan, dextran or alginate) are found in a variety of bacterial strains or species (Christensen and Characklis, 1990). Noncarbohydrate substituents like acetyl, pyruvyl and succinyl groups can greatly alter the physical properties of extracellular polysaccharides and the way in which the polymers interact with one another, with other polysaccharides and proteins, and with inorganic cations (Sutherland, 1984). The network of microbial polysaccharides displays a relatively high water-binding capacity and is mainly responsible for acquisition and retention of water, with the generation of a highly hydrated environment within flocs and biofilms (Chamberlain, 1997).

The Role of Extracellular Proteins

As mentioned above, secreted polysaccharides are believed to have mainly structural functions in forming and stabilizing the floc and biofilm matrix. The role of proteins, however, is mostly considered in terms of their enzymatic activity. Only a few authors speculate that extracellular proteins may also have structural functions (*e.g.* Dignac *et al.*, 1998), although it must be assumed that not all the extracellular proteins can be enzymes. This aspect, however, is still poorly investigated and currently under research (Wingender, Griebe and Flemming, personal observations).

Part of the extracellular proteins have been identified as enzymes. Enzyme activities in flocs and biofilms include aminopeptidases, glycosidases, esterases, lipases, phosphatases and oxidoreductases (Lemmer *et al.*, 1994; Frølund *et al.*, 1995; Griebe *et al.*, 1997). Most of these enzymes are an integrated part of the EPS matrix (Frølund *et al.*, 1995). They are believed to function in the extracellular degradation of macromolecules to low molecular weight products which can be transported into the cells and are available for microbial metabolism. The degradation of particulate matter is performed by colonization of the material and the secretion of extracellular enzymes. However, many details of this process are still obscure. The EPS matrix prevents the loss of the enzymes. Moreover, specific interactions between extracellular enzymes and other EPS components have been observed. Wingender (1990) has shown that the lipase of *Pseudomonas aeruginosa* interacts functionally with the alginate formed by the same strain. It was demonstrated that the lipase of this strain displayed a higher activity in the presence of bacterial alginate. In addition, the enzyme was less sensitive to temperature if associated with the alginate. It was shown that this effect was specific to alginate because other polysaccharides such as dextran did not interact with the enzyme. These observations suggest that the structure of the EPS matrix might not be purely random but is involved in the regulation of the activity of extracellular enzymes. Thus, the cell maintains a certain level of control over enzymes which otherwise are out of their reach. This can be considered as a strategy to a form of organization in which the cells gain control of the space around them. In 1987, Costerton *et al.* hypothesized that biofilms represent tissue-like structures. The example of the interaction between bacterial alginate and lipase support this assumption. If this is the case, the space between the cells may become a new focus of attention, providing information about the cooperative effect of biofilm cells.

Surface-active EPS

In order to degrade hydrophobic compounds, microorganisms excrete surface-active polymers. They are of considerable economic and technical interest as they are used in tertiary oil recovery in order to mobilize surface-bound oil. Neu (1996) gives an overview of the various types of EPS, their properties and their significance. Ramsay *et al.* (1987) have shown that the production of biosurfactants can be induced by hydrophobic carbon sources, indicating the potential of microorganisms to secrete EPS when required. It is well known that hydrophobic surfaces can be colonized easily as demonstrated in nature by biofilm formation on leaves during biological degradation. Also, hydrophobic technical surfaces such as reverse osmosis membranes can be colonized, leading to biofouling (Flemming and Schaule, 1988).

SORPTION PROPERTIES OF BIOFILMS

Another feature of the physico-chemical properties of biofilms is their sorption behaviour (Flemming, 1995). Biofilms play a role both as a sink and a source for pollutants. They can absorb water, inorganic and organic solutes and particles. EPS, cell walls, cell membranes and cell cytoplasm can serve as sorption sites. These sites display different sorption preferences, capacities and properties (Flemming *et al.*, 1996). In addition, biofilms may respond physiologically to sorbed substances. For example, the uptake of toluene can lead to the formation of uronic acids in the EPS and, thus, to an increased sorption capacity for cations (Schmitt *et al.*, 1995). When decomposing, biofilms will release sorbed substances. This can be of significance if the deposition of sewage water by trickling on soil is complete. The biomass will decompose, and sorbed pollutants will be remobilized and can contaminate the ground water if not retained abiotically by other soil components.

In general, at least four different sorption sites in a biofilm can be distinguished, *viz.* 1) EPS (including capsules), 2) cell walls, 3) cell membranes, offering a lipophilic region, and 4) cell cytoplasm, as a water phase separated from the surrounding water.

Water

A common feature of the EPS is that they are highly hydrated; in biofilms, a ratio of 1–2% (w/w) EPS and 98% water is not unusual (Christensen and Characklis, 1990).

This affinity for water gives a slimy consistency to biomass and serves as protection against desiccation (Roberson and Firestone, 1992). Colanic acid was identified as the dominant water binding component in *Escherichia coli*, *Acinetobacter calcoaceticus* and *Erwinia stewartii* and prevents water evaporation from their mucoid strains (Ophir and Gutnick, 1994). It is the water bound in biofilms which has to be removed, involving considerable effort, when sewage sludge is dewatered. In countries with a cold climate, water bound by biofilms in pores and crevices of concrete can cause severe damage if the temperature falls below –10 to –15°C, when the water will eventually freeze, leading to frost cracking (Blaschke, 1987). An effective method for weakening the binding force for water could be economically very interesting.

Cations

Because EPS may contain anionic groups such as carboxyl, phosphoryl, and sulphate groups (Sutherland, 1994), they offer cation exchange potential. A survey of a wide variety of marine and freshwater bacteria by Kennedy and Sutherland (1987) has shown that bacterial EPS typically contains 20–50% of their polysaccharides as uronic acids. A wide variety of metal ions is reportedly bound to EPS (Flemming *et al.*, 1996). Theoretical predictions of metal binding capacities, based on estimated numbers of available carboxyl and hydroxyl groups, suggest a very high capacity, provided in particular by the acidic polysaccharides. Harvey (1981) found a binding capacity for lead of 0.13 mMol mg^{-1} of EPS. He calculated that if EPS represented only a very small proportion of the organic matter in sediments, they could still complex all available Pb^{2+} in the surface layer sediments of a Palo Alto salt marsh. Adsorption densities as high as 22 ng mg^{-1} have been reported for copper (Kaplan *et al.*, 1987). The stability constants for Ni^{2+}, Cu^{2+}, Pb^{2+}, Cd^{2+} and Zn^{2+} complexes with EPS range between 105 and 109 (Kaplan *et al.*, 1987; Geesey and Jang, 1990). As there exists competition between H^+ and metal ions, the stability constants strongly depend upon the pH value. In studies of freshwater lakes, microbial biofilms under near-neutral pH scavenged metals up to 12 orders of magnitude higher than biofilms under lower pH (acidic) conditions (Ferris *et al.*, 1989). EPS have been shown to accumulate up to 25% their weight as metal ions (Dugan, 1975). Alginate has been proposed in biotechnological applications for absorbing dissolved copper from aqueous media (Jang *et al.*, 1990).

Anions

In general, there is very little information available concerning the binding of anions in biofilms. However, sorption of anions must occur, as amino groups in sugars, sugar acids and proteins provide positive charges which can act as anion binding sites. A biofilm which caused irreversible fouling on a reverse osmosis membrane contained accumulated sulphate at levels tenfold higher than the water with which it was in contact (Flemming and Schaule, 1989). A biofilm of *Pseudomonas diminuta* accumulated sulphate, phosphate and nitrate from the nutrient broth (Schaule, unpublished observation).

Polar Organic Molecules

Amino acids are obviously sequestered from aqueous streams by biofilms because they are utilized as substrates (Decho, 1990). The mechanism must involve the EPS as the cells are buried in the EPS matrix which has to be crossed by a molecule before it reaches the cell where it can be metabolized. Detailed data about the affinity of the EPS for these molecules and the binding capacity cannot be found in the literature.

Apolar and Hydrophobic Molecules

Some EPS such as emulsan (Rosenberg and Kaplan, 1986) exhibit surface activity, in particular during growth on hydrophobic nutrients, *e.g.* oil and fat (Cameron *et al.*, 1988). Such surface activity will change the sorption and transport properties of a biofilm in regard to traces of dissolved hydrophobic substances, such as some pesticides. Dohse and Lion (1994) coated sand in a column with EPS of phenanthrene-degrading organisms and this caused a significant enhancement of phenanthrene transport. The sorption of apolar substances by EPS is largely unexplored although it must play an important role, not only in trapping these substances in biofilms but also in the process of adhesion of microorganisms to hydrophobic surfaces, which can be compared to sorption. The first cellular components in contact with such a surface are the hydrated, hydrophilic EPS, which provide the adhesion force for the cells. The nature of the EPS components responsible for this process is still not known. However, it is likely that the apolar regions of proteins provide suitable binding sites for apolar molecules. Using confocal scanning laser microscopy, Wolfaardt *et al.* (1994) were the first to demonstrate that herbicides are accumulated in the EPS. Späth *et al.* (1998) found that benzene, toluene and xylene (BTX) were significantly accumulated

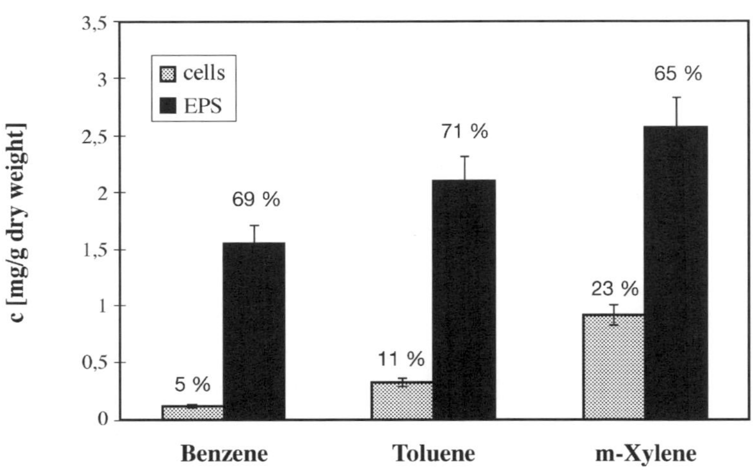

Figure 4 Distribution of benzene, toluene and m-xylene in biofilm after extraction with crown ether. Histrograms are means of triplicate determinations (± SE). The recovery of BTX was about 80%. (Reproduced from Späth *et al.*, 1998, with permission.)

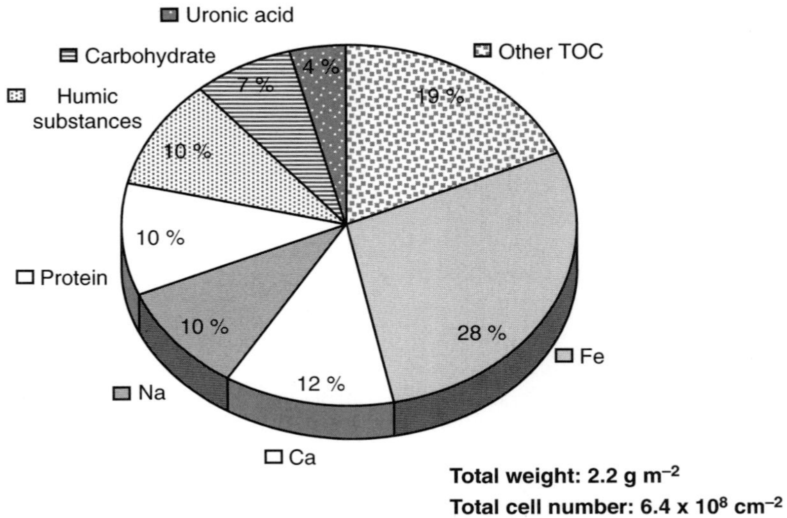

Figure 5 Typical example of the composition of a mixed biotic/abiotic deposit in a cooling water system. Note the large proportion of Fe. The deposit mass was 2.2 g cm^{-2}.

in the EPS of activated sludge, which was surprising in view of the highly hydrated status of this matrix (Figure 4). Clearly, it provides far more sorption capacity than the lipid membranes.

Particles

Due to the "sticky" surface, particles tend to be retained by biofilms. In technical water systems, corrosion products will be integrated into the biofilm matrix yielding a mixed biotic/abiotic deposit. Figure 5 shows a typical composition of such a deposit in a cooling water system.

The aspect of particle entrapment is also of interest for the transport of particulate substances in water, such as the entrainment of sand. Dade *et al.* (1990) showed that the formation of deposits is strongly influenced by the presence and the adhesive properties of biofilms.

OPTICAL PROPERTIES

The optical properties of biofilms are usually not taken into consideration. However, if pigmented biofilms develop on the walls of buildings, they may increase the heat uptake significantly, contributing to an overall increase in energy demand for air conditioning. In a case study, pigmented bacteria developed in corrosion pits on white concrete coatings in drinking water reservoirs, leading to considerable loss in the hygienic value of the coating (Herb *et al.*, 1997).

The optical properties of biofilms and entrapped particles have been exploited for monitoring biofilm development *in situ*, on-line in real time and non-destructively (Flemming *et al.*, 1998). A fiber optic device is integrated into a surface at a representative location. When reflecting material such as microorganisms and abiotic particles are deposited on the tip of the device (diameter 200 μm) an increase in the reflected light can be detected and successfully used as a signal for quantification of the deposit and the deposition rate (Tamachkiarowa and Flemming, 1996).

CONCLUSIONS

The selection of physico-chemical aspects of the EPS matrix as presented in this chapter is far from being comprehensive. However, it shows a somewhat neglected aspect of biofilms and other microbial aggregates, and reveals a considerable research requirement. Of particular interest are such problems as the measurement and engineering of frictional resistance due to biofilms. It is well known that biofilms increase frictional resistance of ships and of water in pipelines; this is exploited in pressure drop devices for biofilm monitoring (Roe *et al.*, 1994). In some instances, however, biofilms and their EPS seem to reduce friction resistance, as reported from EPS in biofilms of fish skin bacteria (Sar and Rosenberg, 1989). Further collaboration between microbiologists, physico-chemists and engineers might lead to interesting and surprising approaches to solving biofilm problems.

REFERENCES

Beyenal H., Tanyolac A., Lewandowski Z. (1998). Measurement of local effective diffusivity in heterogenous biofilms. *Water Sci Techol*, **38**, 171–178.

Blaschke R. (1987). *Bautenschutz, Bausanierung*, **1**, 28–34.

Bruus J.H., Nielsen P.H., Keiding K. (1992). On the stability of activated sludge flocs with implications to dewatering. *Water Res*, **12**, 1597–1604.

Cameron J.A., Bunch C.L., Huang S.J. (1988). Microbial degradation of synthetic polymers. In: Houghton D.R., Smith R.N., Eggins H.O.W. (eds) *Biodeterioration*, 7, Elsevier Applied Science, New York, pp. 553–561.

Chamberlain A.H.L. (1997). Matrix polymers: the key to biofilm processes. In: Wimpenny J., Handley P., Gilbert P., Lappin-Scott H., Jones M. (eds) *Community Interactions and Control*. Cardiff, Bioline, pp. 41–46.

Christensen B.E., Characklis W.G. (1990). Physical and chemical properties of biofilms. In: Characklis W.G., Marshall K.C. (eds) *Biofilms*. John Wiley & Sons Inc., New York, pp. 93–100.

Costerton J.W., Cheng K.-J., Geesey G.G., Ladd T.I., Nickel J.C., Dasgupta M., Marrie T.J. (1987). Bacterial biofilms in nature and disease. *Annu Rev Microbiol*, **41**, 435–464.

De Beer D., Srinivasan R., Stewart P.S. (1994). Direct measurement of chlorine penetration into biofilms during disinfection. *Appl Environ Microbiol*, **60**, 4339–4344.

Decho A.W. (1990). Microbial exopolymer secretions in ocean environments: their role(s) in food webs and marine processes. *Oceanogr Mar Biol Ann Rev*, **28**, 73–153.

Dade W.B., Davis J.D., Nichols P.D., Nowell A.R.M., Trexler M.B., White D.C. (1990). Effects of bacterial exopolymer adhesion on the entrainment of sand. *Geomicrobiology J*, **8**, 1–16.

Dignac M.-F., Urbain V., Rybacki D., Bruchet A., Snidaro D., Scribe P. (1998). Chemical description of extracellular polymers: implication on activated sludge floc structure. *Water Sci Technol*, **38**, 45–53.

Dohse D.M., Lion L.W. (1994). Effect of microbial polymers on the sorption and transport of phenanthrene in a low carbon sand. *Environ Sci Technol*, **28**, 541–548.

Dugan P.R. (1975). Bioflocculation and the accumulation of chemicals by floc-forming organisms. EPA-600/2-75-032, September 1975, Nat Tech Inform Service. Springfield, VA, USA.

Ferris F.G., Schultze S., Witten T.C., Fyfe W.S., Beveridge T.J. (1989). Metal interactions with microbial biofilms in acidic and neutral pH environments. *Appl Envir Microbiol*, **55**, 1249–1257.

Flemming H.-C. (1995). Sorption sites in biofilms. *Water Sci Technol*, **32**, 27–33.

Flemming H.-C., Schaule G. (1988). Biofouling on membranes – a microbiological approach. *Desalination*, **70**, 95–119.

Flemming H.-C., Schaule G. (1989). Biofouling auf Umkehrosmose- und Ultrafiltrationsmembranen. Teil II: Analyse und Entfernung des Belages. *Vom Wasser*, **73**, 287–301.

Flemming H.-C., Schaule G. (1996). Measures against biofouling. In: Heitz E., Sand W., Flemming H.-C. (eds) *Microbially Influenced Corrosion of Materials – Scientific and Technological Aspects*. Springer, Heidelberg, pp. 121–139.

Flemming H.-C., Schmitt J., Marshall K.C. (1996). Sorption properties of biofilms. In: Calmano W., Forstner U. (eds) *Environmental Behaviour of Sediments*. Lewis Publishers Chelsea, Michigan, USA, pp. 115–157.

Flemming H.-C., Tamachkiarowa A., Klahre J., Schmitt J. (1998). Monitoring systems for the detection of biofouling in technical systems. *Water Sci Technol*, **38**, 299–307.

Frølund B., Griebe T., Nielsen P.H. (1995). Enzymatic activity in the activated-sludge floc matrix. *Appl Microbiol Biotechnol*, **43**, 755–761.

Frølund B., Palmgren R., Keiding K., Nielsen P.H. (1996). Extraction of extracellular polymers from activated sludge using a cation exchange resin. *Water Res*, **30**, 1749–1758.

Geesey G., Jang L. (1990). Extracellular polymers for metal binding. In: Ehrlich H.C., Brierley C.L. (eds) *Microbial Mineral Recovery*. McGraw-Hill, New York, pp. 223–247.

Gehrke T., Telegdi J., Thierry D., Sand W. (1998). Importance of extracellular polymeric substances from *Thiobacillus ferrooxidans* for bioleaching. *Appl Environ Microbiol*, **64**, 2743–2747.

Griebe T., Schaule G., Wuertz S. (1997). Determination of microbial respiratory and redox activity in activated sludge. *J Ind Microbiol Biotechnol*, **19**, 118–122.

Harris R.H., Mitchell R. (1973). The role of polymers in microbial aggregation. *Annu Rev Microbiol*, **27**, 27–50.

Harvey R.W. (1981). Lead-bacterial interactions in an estuarine salt marsh microlayer. PhD Thesis, Stanford University, Stanford, USA, pp. 161.

Herb S., Merkl G.U., Flemming H.-C. (1997). Schäden an mineralischen Innenbeschichtungen von Trinkwasserbähltern. *Gwf Wasser Abwasser*, **138**, 137–143.

Higgins M.J., Novak J.T. (1997). Characterization of exocellular protein and its role in bioflocculation. *J Environ Eng*, **123**, 479–485.

Jahn A., Nielsen P.H. (1996). Extraction of extracellular polymeric substances (EPS) from biofilms using a cation exchange resin. *Water Sci Technol*, **32**, 157–164.

Jahn A., Nielsen P.H. (1998). Cell biomass and exopolymer composition in sewer biofilms. *Water Sci Technol*, **37**, 17–24.

Jang L.K., Geesey G.G., Lopez S.L., Eastman S.L., Wichlacz P.L. (1990). Use of a gel-forming biopolymer directly dispensed into a loop fluidized bed reactor to recover dissolved copper. *Water Res*, **24**, 889–897.

Kaplan D., Christiaen D., Shoshana A. (1987). Chelating properties of extracellular polysaccharides from *Chlorella* spp. *Appl Environ Microbiol*, **53**, 2953–2956.

Kennedy A.F.D., Sutherland I.W. (1987). Analysis of bacterial exopolysaccharides. *Biotechnol Appl Biochem*, **9**, 12–19.

LeChevallier M.W., Cawthon C.D., Lee R.G. (1988). Inactivation of biofilm bacteria. *Appl Environ Microbiol*, **54**, 2492–2499.

Lemmer H., Roth D., Schade M. (1994). Population density and enzyme activities of heterotrophic bacteria in sewer biofilms and activated sludge. *Water Res*, **28**, 1341–1346.

Liss S.N., Droppo I.G., Flannigan D.T., Leppard G.G. (1996). Floc architecture in wastewater and natural riverine systems. *Environ Sci Technol*, **30**, 680–686.

Mayer C., Moritz R., Kirschner C., Borchard W., Maibaum R., Wingender J., Flemming H.-C. (1999). The role of intermolecular interactions: studies on model systems for bacterial biofilms. *Int J Biol Macromol*, **26**, 3–16.

Morton L.H.G., Greenway D.L.A., Gaylarde C.C., Surman S.B. (1998). Consideration of some implications of the resistance of biofilms to biocides. *Int Biodeterior Biodegr*, **41**, 247–259.

Neu T. (1992). Polysaccharides in biofilms. In: Prave P., Schlingmann M., Esser K., Thauer R., Wagner F. (eds) *Jahrb Biotechnol*, **4**, 73–101.

Neu T. (1996). Significance of bacterial surface-active compounds in interaction of bacteria with interfaces. *Microb Rev*, **60**, 151–166.

Nielsen P.H., Keiding K. (1998). Disintegration of activated sludge flocs in presence of sulfide. *Water Res*, **32**, 313–320.

Nielsen P.H., Jahn A., Palmgren R. (1997). Conceptual model for production and composition of exopolymers in biofilms. *Water Sci Technol*, **36**, 11–19.

Olofsson A.-C., Zita A., Hermansson M. (1998). Floc stability and adhesion of green-fluorescent-protein-marked bacteria to flocs in activated sludge. *Microbiology*, **144**, 519–528.

Ophir T., Gutnick D.L. (1994). A role for exopolysaccharides in the protection of microorganisms from desiccation. *Appl Environ Microbiol*, **60**, 740–745.

Pavoni J.L., Tenney M.W., Echelberger W.F. (1972). Bacterial exocellular polymers and biological flocculation. *J Water Pollut Control Fed*, **44**, 414–431.

Ramsay B., McCarthy J., Guerra-Santos L., Kappeli O., Fiechter A. (1987). Biosurfactant production and diauxic growth of *Rhodococcus aurantiacus* when using *n*-alkanes as the carbon source. *Can J Microbiol*, **34**, 1209–1212.

Roberson E.B., Firestone M.K. (1992). Relationship between desiccation and exopolysaccharide production in a soil *Pseudomonas* sp. *Appl Environ Microbiol*, **58**, 1284–1291.

Roe F.L., Wentland E., Zelver N., Warwood B., Waters R., Characklis W.G. (1994). On-line side-stream monitoring of biofouling. In: Geesey G.G., Lewandowski Z., Flemming H.-C. (eds) *Biofouling and Biocorrosion in Industrial Water Systems*. Lewis Publishers, Ann Arbor, MI, USA, pp. 137–150.

Rosenberg E., Kaplan N. (1986). Surface-active properties of *Acinetobacter* exopolysaccharides. In: Inouye M. (ed) *Bacterial Outer Membranes as a Model System*. Interscience Publishers, New York, pp. 311–342.

Sar N., Rosenberg E. (1989). Fish skin bacteria: production of friction-reducing polymers. *Microb Ecol*, **17**, 27–38.

Schmitt J., Nivens D., White D.C., Flemming H.-C. (1995). Changes of biofilm properties in response to sorbed substances – an FTIR-ATR-study. *Water Sci Technol*, **32**, 149–155.

Späth R., Flemming H.-C., Wuertz S. (1998). Sorption properties of biofilms. *Water Sci Technol*, **37**, 207–210.

Sutherland I.W. (1994). Structure-function relationships in microbial exopolysaccharides. *Biotechnol Adv*, **12**, 393–448.

Tamachkiarowa A., Flemming H.-C. (1996). Glass fiber sensor for biofouling monitoring. *DECHEMA Monographs*, **133**, 31–36.

Turakhia M.H., Cooksey K.E., Characklis W.G. (1983). Influence of a calcium- specific chelant on biofilm removal. *Appl Environ Microbiol*, 46, 1236–1238.

Unz R.F., Farrah S.R. (1976). Exopolymer production and flocculation by *Zoogloea* MP6. *Appl Environ Microbiol*, 31, 623–626.

Urbain V., Block J.C., Manem J. (1993). Bioflocculation in activated sludge: an analytical approach. *Water Res*, 27, 829–838.

Wingender J. (1990). Interactions of alginate with exoenzymes. In: Gacesa P., Russell N.J. (eds) *Pseudomonas Infection and Alginates*. Chapman and Hall, London, pp. 160–180.

Wingender J., Neu T.R., Flemming H.-C. (1999). What are bacterial extracellular polymeric substances? In: Wingender J., Neu T.R., Flemming H.-C. (eds) *Microbial Extracellular Polymeric Substances*. Springer-Verlag, Berlin, pp. 1–19.

Wolfaardt G.M., Lawrence J.R., Headley J.V., Robarts R.D., Caldwell D.E. (1994). Microbial exopolymers provide a mechanism for bioaccumulation of contaminants. *Microb Ecol*, 27, 279–291.

3 Structural Determinants in Biofilm Formation

Julian Wimpenny

The ubiquity of biofilm has focused attention on the structure and function of this common mode of microbial growth. This chapter reviews the effects of resource concentration and hydrodynamic shear as well as touching on a range of other phenomena which help to determine the final structure of a biofilm. It briefly considers the importance of environmental as well as genotypic factors on biofilm formation and evolution. A construction set approach to biofilm formation is advocated as Nature's response to the spatial and temporal heterogeneity of natural ecosystems.

KEY WORDS: biofilm, structure, community, model, evolution

INTRODUCTION

There has been considerable argument as to what exactly constitutes a biofilm and a number of definitions have emerged. A clear view as to what it is not is a homogenous collection of microbes in a liquid culture. Biofilm is an amorphous group of microbial communities that fall loosely under the umbrella of microbial aggregates. The latter can be divided into those with roughly spherical co-ordinates, for example sludge floc, anaerobic digester granules, mycelial balls and marine snow. On the other hand is the family of aggregates that appear on surfaces and/or at phase interfaces. These are biofilms. They are roughly two-dimensional, ranging from one or a few micrometres (a bacterial monolayer) to millimeters or more in thickness. The structure can vary from smooth to rough with obvious frond like projections. They may be dense, opaque structures like dental plaque or translucent gelatinous associations. Coverage may range from very patchy to continuous, even and unbroken. Biofilm forms most commonly at water/solid interfaces although it can appear at an interface between two immiscible liquids like oil and water, at air-water interfaces (neuston on the surface of water bodies) and at gas-solid surfaces (lichen and other microbial associations).

It is becoming increasingly clear that most microbes from natural environments are associated with aggregates, in particular biofilms. There are a number of possible reasons for this. It has been shown from the earliest days of biofilm research that substrates tend to accumulate at surfaces, also that an inert surface provides a fixed position which, where there is fluid flow, allows groups of cells to remove nutrients from the latter. However, the most likely reason for adhesion to surfaces is the possibility of forming organised microbial communities which can interact with each other to from a structure which is optimised perhaps, to respond to the prevailing physico-chemical environment. Such structures may have emergent properties, namely that their operation as an association is greater than the sum of their individual parts.

The Prevalence of Biofilm

Biofilm forms in every environment so long as a surface, nutrients and water are (at least sometimes) available. Thus it can be found in almost all natural habitats including epilithic systems in rivers and streams, on the surfaces of plant parts including leaves (the phyllosphere) and on roots (the rhizosphere). It is associated with animals as biofilms forming on teeth and oral epithelia and on the mucosa of the digestive tract. Biofilm is closely associated with human artifacts, for example metal working systems, the hulls of marine installations including ships' bottoms and oil exploration platforms as well as at the base of oil storage tanks. In medicine it accumulates on all manner of prosthetic devices inserted in to the human body including catheters, cardiac pacemakers, replacement hips and knees. It is found in swimming pool filters, in drains, in car windscreen washers and in water conduits of all sorts. Clearly biofilm is truly ubiquitous.

There is a generally accepted life history of a 'typical' biofilm. A very clean surface is rapidly covered with organic compounds, often proteins, polysaccharides or other macromolecules. This 'conditioning film' forms in a short time (seconds or minutes) and precedes the attachment of microbes. These move randomly above the surface of the substratum and attach loosely at first by electrostatic and van der Waals forces. The second phase of attachment involves the binding of the cells more firmly by pili or fimbriae and by the secretion of adhesive polymers, The cells then start to proliferate, forming first a monolayer, then microcolonies. These produce more extracellular polymeric substances (EPS) which serve to maintain the growing biofilm on the surface. The biofilm can be colonised by secondary organisms and serves as a sink for other particulate matter in the environment. The mature biofilm increases in thickness but can then become unstable so that large sections slough off into the surrounding environment. The process can then repeat.

This chapter is concerned with all the factors that affect or control final biofilm structure. Clearly it is not always easy to see a logical pattern in biofilm formation because there are so many different types of structure that have been reported. The largest division is between the genotype of an organism and its phenotype, that is the expression of the genes as it is affected by environment.

The effect of environmental factors will be considered first, since this has been explored fairly fully. What is known about genetic control mechanisms will then be discussed, and finally the rather difficult task of reconciling the two will be attempted.

ENVIRONMENTAL FACTORS

Nutrients in the Natural Environment

Biofilm consists of living or recently living microbial biomass in addition to EPS, which often consists of one or more of a family of different polysaccharides manufactured by some or all of the biofilm organisms, although it may contain protein and nucleic acids as well. The function of these components is not fully understood but they obviously assist in attachment to surfaces, to stabilisation of the local environment and to the spatial organisation of communities which may need to collaborate to effectively use what substrates are present.

Resources range in concentration over at least six orders of magnitude (Wimpenny and Colasanti, 1997), from extremely small in some freshwater oligotrophic habitats to very high concentrations for example of sucrose at times in the oral environment.

Very Low Substrate Concentrations

There are a large family of habitats in which nutrients are present in very small amounts. These include fresh water supplies for domestic and industrial use, oligotrophic streams and lakes, distilled water containers and other devices for generating high purity water. Substrate concentrations are also very low in marine habitats, which comprise the single largest ecosystem in the world. Sea water contains a range of material from particulate to dissolved matter. Most of this resource is humic or fulvic acids which are very recalcitrant. There appears to be about 100 $\mu g\ l^{-1}$ of readily utilisable carbon present in these water bodies. It is probable that concentrations range from a few μg to a few $mg\ l^{-1}$, depending on the system under consideration.

Low and Medium Substrate Concentrations

In this category are a wide range of systems where the available nutrient concentration is $> 1\ mg\ l^{-1}$ and probably $< 1\ g\ l^{-1}$. The fact that this is a thousand fold scale confirms the huge range of resource concentrations that are available to living organisms. Such habitats include eutrophic water bodies, streams and rivers, effluent treatment plants (*e.g.* activated sludge plants, rotating drum and disc aerators, and trickling filters), household drains, septic tanks, car wash bottles as well many laboratory biofilm experimental systems.

High Substrate Concentrations

The highest substrate concentrations are almost all associated with animal surfaces. The clearest example is the oral cavity exposed to resting saliva (containing mucins and other smaller amounts of substrates including glucose) for most of the time but at times with very high concentrations of food products including glucose and sucrose, whose concentrations for brief intervals may exceed 100 $g\ l^{-1}$.

In addition decaying organic material will provide very high levels of resource as will contaminated food. Relatively little is known about the structural and functional aspects of biofilm growing on these sites.

The Effects of Nutrient Concentration on Biofilm Structure

If a value of 100 µg l^{-1} is taken as a typical low value for nutrients and 100 g l^{-1} as a high figure a range of six orders of magnitude can be seen in nutrient resources for microbial growth. However, in reality the range is even larger than this, perhaps a low value of 1 µg l^{-1} and a high value of 500 g l^{-1} (a 5×10^8 fold range), and such a range of nutrients must have a significant effect on biofilm growth. Examination of pictures of biofilm as well as of diagrammatic models of biofilm that have been proposed shows a correlation between resource concentration and structure.

At the lowest concentrations there have been several investigations by Keevil and co-workers of biofilm forming on the walls of copper pipes and on a range of surfaces in laboratory growth systems developing on unamended tap water (Keevil and Walker 1992; Keevil *et al.*, 1993; Walker *et al.*, 1995). This group examined the structure of a developing film using differential interference contrast microscopy, and clearly showed that biofilms were formed as stacks of microcolonies each well separated from its neighbours. Below the stacks adjacent to the substratum was a more or less confluent layer of film estimated to be no more than 5 µm thick. In natural water samples the film was grazed by protozoans which appeared to move between and around the stacks. This biofilm structure was called the 'heterogeneous mosaic model' by this research group.

At higher concentrations individual stacks were not seen, but microcolonies appeared as mushroom or tulip shaped structures 'tethered' to the substratum by a narrower stalk. The whole structure was porous with water channels allowing solutes to flow in and around the colonial structures. Many investigations led Costerton and co-workers to formulate such a conceptual model of a biofilm (Costerton *et al.*, 1994; 1995). Lines of evidence include, firstly, the use of the confocal laser scanning microscope (CLSM) which, because of its geometry, can scan an image with a very narrow depth of focus. If many such images are recorded at different depths the three-dimensional structure of the film can be reconstructed. The use of fluorescent labeled probes enabled the position of specific groups of organisms to be determined within the biofilm structure (Shotton, 1989; Caldwell *et al.*, 1992a; 1992b). Secondly, use of oxygen microelectrodes showed that conditions within mushroom or tulip shaped microcolonies were anaerobic, yet the electrode might enter a pore below or adjacent to part of the microcolony where oxygen tensions were measurable, suggesting that a flow of aerated water could pass through the biofilm (de Beer *et al.*, 1993; 1996; de Beer and Stoodley, 1994). Thirdly, flow in biofilm channels was monitored by nuclear magnetic resonance imaging (Lewandowski *et al.*, 1993), and fourthly, the use of fluorescent particles as markers for fluid flow, monitored using the CLSM, revealed movement at speeds that could be accurately determined within the porous structure (Stoodley *et al.*, 1994).

At the highest nutrient concentrations biofilm can appear quite uniform with few if any pores. This situation is common in biofilm forming on animal and food surfaces.

Thus transmission electron micrographs taken across the profile of three different human dental plaque samples mostly show confluent growth of microbes (Nyvad and Fejerskov, 1989).

Shear Forces in Flowing Systems

Perhaps second only to resource concentration as a structure determinant must be shear forces. This has been emphasised by Van Loosdrecht *et al.* (1995) who compared nutrient loading rate and shear as structure determinants. Their view was that structure is determined by the balance between substrate loading rate and shear. The experimental model used in this paper and more recently by Kwok *et al.* (1998) was a biofilm airlift suspension (BAS) reactor. Here, biofilm accumulates on an inert surface (basalt particles) which is kept in constant motion by gas and liquid flow. They reported that at high detachment forces the biofilm became denser and more concentrated. Also, any protuberances were removed leading to a smoother flatter biofilm. When detachment forces are reduced the biofilm becomes more heterogeneous with numerous pores and protuberances.

Blenkinsopp and Lock (1994) investigated the impact of storm flow in a microcosm model of water conduits. The increased flow rate led to a smoother, flatter biofilm.

However, shear is only one part of the detachment process. It is clear that cell related properties also influence cell removal. Thus cell hydrophobicity increases throughout the cell cycle in cultures of a pseudomonad in a steady state biofilm recirculating reactor. Increasing hydrophobicity led to easier detachment and contributed to the steady state of this system (Asconcabrera *et al.*, 1995). There are many other structural phenomena that can also affect biofilm. For example, biofilm density seems to increase with film depth. This was confirmed using a microslicing technique developed by Bishop *et al.* (1995).

Modeling Biofilm Formation

The regulation of biofilm structure by substrate concentration as well as by shear forces was first proposed by Van Loosdrecht *et al.* (1995). Resource effects have been modeled by three groups since then, each using cellular automata. Although there are many different ways of applying mathematical models to natural systems, they can be divided into two main types, *viz.* continuum models and discrete models. Continuum models involve differential and partial differential equations where one variable changes continuosly. Such models apply well to large scale systems, for example the major nutrient cycles, the dynamics of biomass growth, and even to the generation of regular patterns such as periodic rings and bands. Events taking place on the scale of microorganisms themselves are a different matter. Thus discrete events, for example the number of protons contributing to pH around a single cell, the growth and distribution of bacterial cells themselves cannot be investigated easily using continuum models. Cellular automata (CA) constitute one form of discrete modeling. They consist of an array of compartments in which objects can be located, for example bacterial cells or substrate molecules. These obey rules which act globally over the whole array but are applied individually to the contents of each compartment. The process is iterative, the computer passing over each compartment in the array in turn, as many times as is necessary. For example the Game of Life was one of the earliest CA models. Here the rules are very simple. Cells live, divide or die according to the occupation of spaces adjacent to them. CA models have so called 'emergent' properties, *i.e.* they can generate complex patterns that cannot be predicted in any other way.

A simple CA model was developed by Wimpenny and Colasanti (1997). An array was first populated with 'substrate' molecules each allowed to 'diffuse' around its origin at each iteration of the program. 'Cells' were located along the bottom edge of the array (the 'substratum'). Cells could divide if enough substrate diffused into their compartment and if there was an empty space adjacent to them. Array size, substrate diffusion rate and the yield of cells on the substrate could be varied. When substrate concentration was varied a range of structures were generated which fitted well with the observed structures discussed above. These results were confirmed first by Picioreanu *et al.* (1998) using a more realistic model, and later by Hermanowicz (1998) also using a cellular automaton. Picioreanu *et al.* (1999) developed a hybrid computer model which features numerical methods as well as CA components to reproduce solute diffusion, shear, biomass growth and cell distribution.

More recently Kreft *et al.* (1999) developed an autonomous agent model called Bacsim based on a program suite called the 'Swarm' system developed by computer scientists at the Santa Fe Institute in New Mexico, USA. This was used first to model bacterial colony growth but has recently been applied to biofilm formation (Kreft *et al.*, 1999). The Swarm system consists of agents (computer programs) which can be programmed to behave as individual bacteria, each having an appropriate (though much simplified) physiology. In this way they are a more realistic representation of 'real' bacteria than are CA based models.

This is only a brief description of the type of discrete modeling that can be applied to biofilm structure. The results, however, are unanimous in illustrating the effects of nutrient concentration as well as hydrodynamic processes on biofilm structure and any global picture of biofilm structure ought to take these processes into account.

Other Environmental Factors

Nutrient composition

Although resource concentration has been shown to be perhaps the most important determinant, nutrient composition can affect structure as well. Baillie and Douglas (1998) examined the effects of iron or glucose limitation on biofilms of *Candida albicans*. Iron limitation led to a two fold increase in the steady state cell population.

Ohashi *et al.* (1995) have shown that the prevailing C/N ratio can influence the structure of a multispecies biofilm composed of nitrifying bacteria and heterotrophs. Here, high substrate loading rates lead to detachment and a sharp fall in nitrification. If the system is set up with a low substrate loading rate at first, a stable biofilm forms which can then cope with an increase in loading rate. Substrate loading rate also affects a similar biofilm generated in a rotating biological contactor. Increase in ammonia and organic carbon led to stratification, with nitrifiers located beneath the heterotrophs (Okabe *et al.*, 1995).

Added chemical factors

It is clear that the addition of chemical agents is likely to influence biofilm structure. An obvious example is the effect of antimicrobials on biofilms. For example the antibiotic fleroxacin changes a *Pseudomonas aeruginosa* biofilm from a porous open structure, on average 42 µm thick, to a flat smooth biofilm with a depth of 29 µm (Korber *et al.*, 1994).

Grazing by eukaryotic species

A most important aspect of biofilm development which has not been investigated to any extent is the effect of predators on natural biofilm *in situ*, since it is clear that natural biofilm is commonly grazed by protozoa. Lee and Wellander (1994) investigated predation in two suspended carrier biofilm reactors carrying out nitrification. The addition of nystatin and cycloheximide to one of the fermenters led to a doubling of the nitrication rate over the controls as the predators were killed.

The gaseous environment

It is clear that biofilm can form under numerous different circumstances, for instance in the presence or absence of electron acceptors such as oxygen or nitrate. Arcangeli and Arvin (1995) investigated a toluene degrading community grown anoxically with nitrate or aerobically with oxygen as electron acceptor. Toluene oxidation was maximal in the log phase of growth where nitrate was present anoxically whilst oxidation was highest in the lag phase with oxygen. In addition there were significant structural changes. Oxic growth was irregular and filamentous whilst with nitrate growth was regular and smooth.

Too much oxygen can be disadvantageous. Villaverde *et al.* (1997) using a flat plate vapour phase bioreactor (also degrading toluene) showed that the upper layer of the film was non viable (O_2 respiration zero) but suggest that this layer protected the lower layers which were respiring.

Santegoeds *et al.* (1998) noted that oxygen was depleted in the top 200 to 400 µm of an aerobic microbial biofilm; this allowed the growth of an anaerobic sulphate-reducing community at the base of the film. Such stratification in biofilms generates domain interfaces which allow a much wider range of organisms to develop and is an important part of microbial ecology.

Osmolarity

Osmolarity affects the ability of *Pseudomonas fluorescens* WCS365 to form biofilms (O'Toole and Kolter, 1998b).

Species Related Effects

Coaggregation

Another phenomenon associated with biofilm is coaggregation, which can be defined as the specific physical association of pairs or groups of microbes. Coaggregation has been seen predominantly but not exclusively in the oral environment. Rosettes and corncob formations (Rosan *et al.* 1998) in dental plaque are specific examples. Gibbons and Nygaard (1970) noted that certain pairs of planktonic oral bacteria aggregated when mixed in a test tube whilst others did not. This observation was followed up by extensive 'mapping' of aggregating pairs by Kolenbrander and his colleagues (Kolenbrander 1988; 1989). This led to a conceptual model suggesting that by integrating all the coaggregating pairs together a network or tissue of species would theoretically be able to form. It is by no means certain that the potential to aggregate is realised in the actual plaque. However, reports by Bradshaw *et al.* (1998) illustrate the importance of coaggregation to *Fusobacterium nucleatum* in the presence of other facultative and obligate anaerobes when exposed to oxygen. There have been other examples of aggregation, for example it might also occur in herbicide degrading communities (Wolfaardt *et al.*, 1994) and between pathogenic and commensal organisims in urinary tract infections (McGroarty and Reid, 1988).

There are good arguments for assuming that, certainly in stable biofilm communities, selective inter-species binding should occur. This would certainly apply where metabolic cooperation takes place since some specific organisation of such a community would enable it to operate more efficiently as a prokaryotic 'tissue', with different organisms aligned optimally to transfer substrates and products from one to the other. This argument gains credence since it has been shown that metabolic cooperation (*e.g.* in term of mucin breakdown) takes place in dental plaque communities (Bradshaw *et al.*, 1994; Marsh and Bradshaw, 1998).

Community Structure

Schramm *et al.*(1996) investigated nitrifying biofilms using 16S rRNA probe technology and concluded that nitrification took place in the top 50 μm of the film. Ammonia oxidisers were present in dense clusters in the upper part of the film whilst the nitrite oxidisers were adjacent to these as less dense aggregates. Both species were also present in anoxic regions adjacent to the substratum. Successional changes led to a mature multilayered biofilm after 180 days in a supply line to a dental air water syringe system (Tall *et al.*, 1995). Tartakovsky and Guiot (1997) used a model to confirm the value of layered structures in a propionate degrading anaerobic granular biofilm and showed that the structure was disturbed in the presence of excess glucose. Competition for substrate and space leads to changes in species distribution in biofilm (Zhang and Bishop, 1994). One of the best defined example of community structure must be interactions in an oral biofilm system. Bradshaw *et al.*(1994) illustrated the emergent properties of a community since different members could catalyse different parts of the enzymic breakdown of mucin, a key substrate for growth of oral species in the mouth. In other words no single organism was capable of using all of the mucin molecule on its own.

GENETIC EXPRESSION

The various steps in biofilm formation are now beginning to be examined using modern molecular genetic techniques and an interesting picture is starting to emerge.

Escherichia coli can normally attach to a number of artificial surfaces forming biofilm (Pratt and Kolter, 1998). Transposon insertion mutagenesis led to mutants that were no longer able to attach to the surfaces, these were either non-motile or were unable to produce type I pili. The latter are mannose sensitive adhesins whose role in attachment was confirmed since a mannose analogue could block adhesion in the wild type strain. Motility appears to play a part in biofilm formation and it was suggested that it is needed to overcome repulsive forces near the substratum surface.

Mutants of *P. aeruginosa* have been isolated that are defective in the capacity for attachment to substrata. Two groups of such *sad* (surface attachment defective) mutants have been described; the first class was defective in flagella and motility and individuals were unable to attach to the plastic employed by the researchers. The second could not produce Type IV pili. This second group formed a monolayer on the substratum rather than aggregating into microcolonies as was the case with the wild type organisms (O'Toole and Kolter, 1998a). The type IV pilus seems to be associated with 'twitching' motility so this might be the basis for microcolony formation.

Extracellular cues, for example messenger molecules, can also lead to metabolic responses. One such cue is contact with a surface. For example Davies *et al.* (1993) found that exopolysaccharide production could be switched on when *P. aeruginosa* contacted a substratum. This organisms appears to express specific genes associated with alginate production. These include *algC, algD* and *algU::lacZ* (Davies and Geesey, 1995).

Other cues include signal molecules often associated with cell density, a process which is often described as 'quorum sensing'. Such signaling is associated *inter alia* with sporulation in *Bacillus* species, mating in *Enterococcus faecalis*, light production in *Photobacterium* spp. and virulence genes in different bacteria including pseudomonads.

Two quorum sensing systems have been investigated in *P. aeruginosa*. One (LasR-LasI) is responsible for regulating virulence and also controls the second system, Rh 1R-RH 1I, which is itself a regulator for the production of a number of secondary metabolites. Each codes for a signal molecule, Rh 1I coding for butyryl homoserine lactone and LasI for 3-oxododecanoyl-homoserine lactone. Each strain including the wild type can attach to glass slides, generating microcolonies. Only the *lasI* mutant is unable to differentiate from microcolonies to a structured thick antibiotic resistant biofilm, unless it is first exposed to the signal molecule 3-oxododecanoylhomoserine lactone. The conclusion from this experiment is that at least one quorum sensing signal system is needed for biofilm development (Davies *et al.*, 1998).

This area has recently been reviewed (Costerton *et al.*, 1999) who have suggested that the processes shown in Figure 1 take place in the development of biofilm structure.

AN INTEGRATED OPINION

The subject of biofilm structure has led to intense argument and considerable speculation. Thus it has often been considered to be a 'tissue' analogous to the tissues of

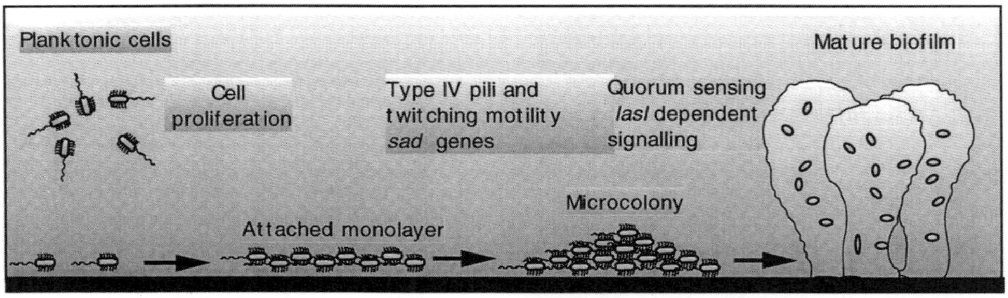

Figure 1 Scheme for the regulation of biofilm formation. (Redrawn from Costerton *et al.* 1999.)

metazoans, the assumption being that the open porous structure seen in some biofilms was generated to provide a primitive circulation system.

Pattern Formation

There are numerous examples of pattern in natural systems. The subject as far as it applies to microbial systems is reviewed by Wimpenny (1993). Pattern implies some kind of repetitive or ordered structure recognisably different from random or chaotic structures and is common in non-biological as well as biological systems. Amongst microbes bacterial colony formation can generate a range of patterns. This includes circular periodic structures as seen in swarming *Proteus* spp, or the branched radial formations made by *Bacillus subtilis* under certain circumstances. Ordered motility patterns are produced by both the eukaryotic slime moulds and the prokaryotic slime bacteria.

Pattern can be generated either by purely physico-chemical processes or by cellular regulatory processes. This poses the question, pertinent both to bacterial colony and to biofilm development, is morphogenesis genetically controlled or purely the result of non-biological processes?

It seems clear that both must occur simultaneously. This may be just one example of the sophistication of evolution. Thus if a purely physico-chemical process occurs naturally, in a direction that is deemed beneficial to a living organism or community of organisms, then it is fit to be exploited by that organism in a directed sense to make that process more efficient. Obviously there is no guiding principle here beyond the power of natural selection.

Community evolution is a topic that has received some attention in recent years. The thesis here is that a community can evolve in the same way as a species. This is the basis of Caldwell's proliferation thesis (Caldwell *et al.*, 1997). The latter defines a recursive structure which is best illustrated as follows (Gilbert and Allison, 1999). Self replicating molecules will proliferate faster when located within a cell wall and membrane structure of a prokaryotic cell; a eukaryotic cell will develop more quickly if it engulfs a prokaryote energy generator destined to become a mitochondrion; a group of eukaryotes will grow more rapidly/efficiently if they collaborate to form a metazoan; a group of species will perform better as a community than as individuals; a group of communities will proliferate faster as an ecosystem. Evolution here is seen as a question of rate rather than competition, though the difference in emphasis might seem rather small.

The application of proliferation theory seems unconvincing when applied to the biofilm community because the basic units of biofilm formation are very largely free living where a series of symbiotic associations each permanently associated to form units dedicated to solving particular environmental problems might be expected. However, they are not, because of a combination of the spatial and temporal heterogeneity, and the scale on which microbes develop. If physico-chemical conditions and the make up of the biota vary in time and space, community stability would not be an advantage, especially if variation occurred relatively quickly. For example, one source of fluctuation is diurnal variation in light and temperature as well as changes due to annual seasonal cycles. Other much more random changes can also occur. On a microscopic scale variation can be much greater. Oxgen tension can vary from saturated to zero within a soil crumb and pH gradients adjacent to clay lattice elements in the soil can shift at least 2–3 units over a few nanometers. The main determinants of community structure must be environmental spatio-temporal stability, together with the survival value of particular configurations. Thus the energy gain due to the association of a primitive eukaryote with a prokaryote carrying with it a respiratory apparatus capable of making ATP from the reduction of oxygen, would be enormous and vastly outweigh any possible negative aspects due to adverse spatio-temporal factors.

The strategy that appears to have evolved has been the 'constructor kit' approach. For example with Meccano (in the UK) a functioning crane or a car or lorry can be built from the different parts, depending on whether elevating goods or transport is the priority. The vast range of microbial species can be regarded as often autonomous agents which can collaborate in the construction of a "machine" which is best suited to gleaning advantage from the environment as it exists at any instant in time and place.

Building a biological machine is more complicated than using a construction kit for two major reasons. Firstly, the components do not necessarily and easily fall into place and they may have to compete with other 'pieces' first. In terms of proliferation theory the faster growing of two species that could appear in a community might integrate into the community leading to an increase in its net growth rate. Secondly, stochastic elements need to be incorporated. The two species suggested could both be present at different positions in the community; the slower growing of the two might colonise the community first and over the course of the life of that biofilm, occupy the particular niche perfectly effectively. What is even more relevant to a biological system is that the possible colonisation of a community by an organism detrimental to the biofilm as a whole cannot be ruled out. Notions of tightly organised structures which are completely optimised for a particular environment are too idealistic; in practice systems will often be chaotic and largely irreproducible except in the most stringently stable systems, for example the selection of stable communities like the Dalapon community described by Senior *et al.* (1976) in continuous cultures. Stability was more apparent than real even in these experiments, as out of the seven community members a secondary heterotroph mutated to become a primary converter

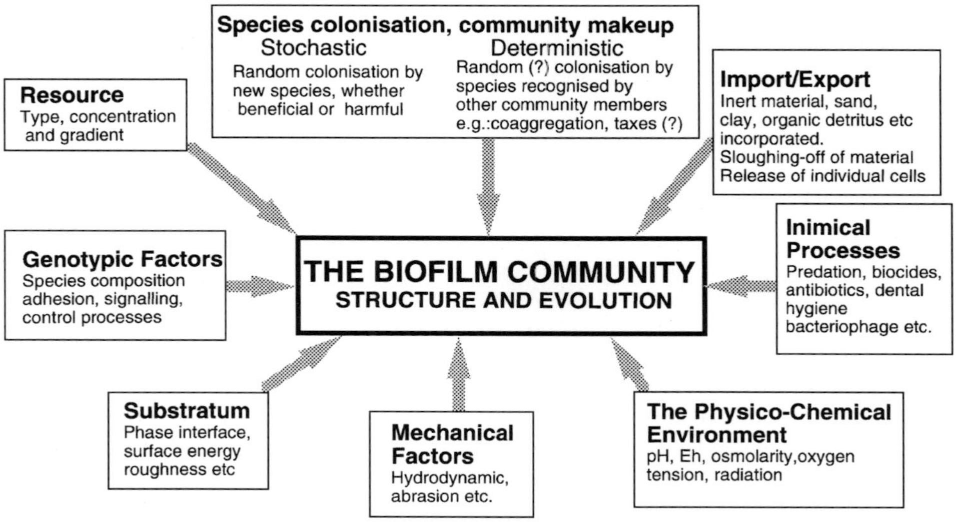

Figure 2 A possible scheme for community development in terms of the spatio-temporal stability of an ecosystem.

of the Dalapon. Mutation and selection are always ongoing in so called steady state cultures.

The author's own views on the relevance of spatio-temporal stability to microbial associations are illustrated in Figure 2, and in Figure 3 some of the many processes that will contribute to the structure and the community development of a biofilm are shown.

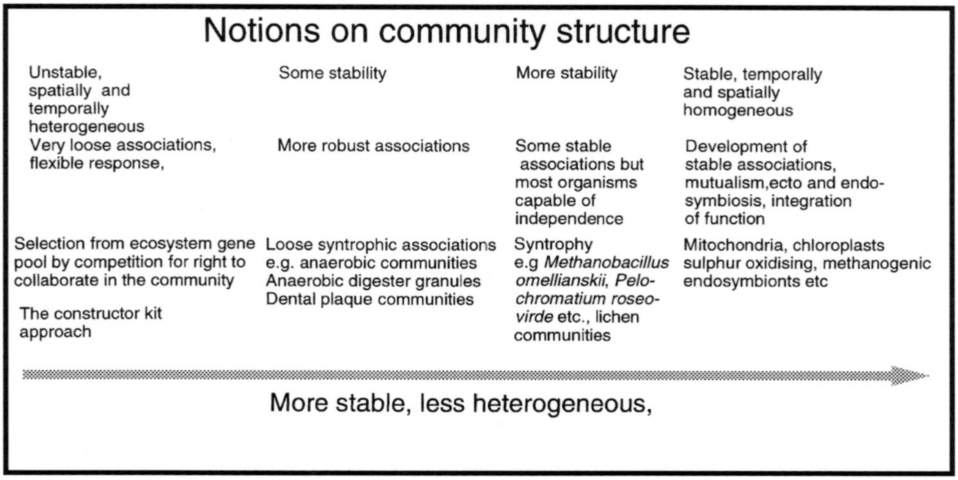

Figure 3 Factors that are involved in biofilm structure.

REFERENCES

Arcangeli J.P., Arvin E. (1995). Growth of an aerobic and an anoxic toluene degrading biofilm — a comparative study. *Water Sci Technol*, **32**, 125–132.

Asconcabrera M.A., Thomas D., Lebeault J.M. (1995). Activity of synchronised cells of a steady state biofilm recirculated reactor during xenobiotic biodegradation. *Appl Environ Microbiol*, **61**, 920–925.

Baillie G.S., Douglas L.J. (1998). Iron limited biofilms of *Candida albicans* and their susceptibility to amphotericin B. *Antimicrob Agents Chemother*, **42**, 2146–2149.

Bishop P.L., Zhang T.C., Fu Y.C. (1995). Effects of biofilm structure, microbial distributions and mass transport on biodegradation processes. *Water Sci Technol*, **31**, 143–152.

Blenkinsopp S.A., Lock M.A. (1994). The impact of storm-flow on river biofilm architecture. *J Phycol*, **30**, 807–818.

Bradshaw D.J., Homer K.A., Marsh P.D., Beighton D. (1994). Metabolic cooperation in oral microbial communities during growth. *Microbiology (Reading)*, **140**, 3407–3412.

Bradshaw D.J., Marsh P.D., Watson G.K., Allison C. (1998). Role of *Fusobacterium nucleatum* and coaggregation in anaerobe survival in planktonic and biofilm oral microbial communities during aeration. *Infect Immun*, **66**, 4729–4732.

Caldwell D.E., Korber D.R., Lawrence J.R. (1992a). Confocal laser microscopy and digital image analysis in microbial ecology. *Adv Microb Ecol*, **12**, 1–67.

Caldwell D.E., Korber D.R., Lawrence J.R. (1992b). Imaging of bacterial cells by fluorescence exclusion using scanning confocal laser microscopy. *J Microbiol Methods*, **15**, 249–261.

Caldwell D.E., Wolfaardt G.M., Korber D.R., Lawrence J.R. (1997). Do biofilm communities transcend Darwinism? *Adv Microb Ecol*, **15**, 105–191.

Costerton J.W., Stewart P.S., Greenberg E.P. (1999). Bacterial biofilms: a common cause of persistent infections. *Science*, **284**, 1318–1322.

Costerton J.W., Lewandowski Z., Caldwell D.E., Korber D.R., Lappin-Scott H.M. (1995). Microbial biofilms. *Ann Rev Microbiol*, **49**, 711–745.

Costerton J.W., Lewandowski Z., Caldwell D.E., Korber D.R., de Beer D., James G. (1994). Biofilms: the customized microniche. *J Bacteriol*, **176**, 2137–2142.

Davies D.G., Geesey G.C. (1995). Regulation of the alginate biosynthesis gene AlgC in *Pseudomonas aeruginosa* during biofilm development in continuous culture. *Appl Environ Microbiol*, **61**, 860–867.

Davies D.G., Chakrabarty A.M., Geesey G.G. (1993). Exopolysaccharide production in biofilms — substratum activation of alginate gene expression by *Pseudomonas aeruginosa. Appl Environ Microbiol*, **59**, 1181–1186.

Davies D.G., Parsek M.R., Pearson J.P., Iglewski B.H., Costerton J.W., Greenberg E.P. (1998). The involvement of cell-to-cell signals in the development of a bacterial biofilm. *Science*, **280**, 295–298.

de Beer D., Stoodley P. (1994). Effects of biofilm structures on oxygen distribution and mass transport. *Biotechnol Bioeng*, **43**, 1131–1138.

de Beer D., van den Heuvel J.C., Ottengraf S.P.P. (1993). Microelectrode measurements of activity distribution in nitrifying bacterial aggregates. *Appl Environ Microbiol*, **59**, 573–579.

de Beer D., Stoodley P., Lewandowski Z. (1996). Liquid flow and mass transport in heterogeneous biofilms. *Water Res*, **30**, 2761–2765.

Gibbons R.J., Nygaard M. (1970). Inter-bacterial aggregation of plaque bacteria. *Arch Oral Biol*, **15**, 1397–1400.

Gilbert P., Allison D.G. (1999). Dynamics in microbial communities: a Lamarkian perspective. In: Wimpenny J., Gilbert P., Walker J., Brading M., Bayston R. (eds) *Biofilms: the Good, the Bad and the Ugly*. Bioline, Cardiff, pp. 263–268.

Hermanowicz S.W. (1998). A model of two dimensional biofilm morphology. *Water Sci Technol*, 37, 219–222.

Keevil C.W., Walker J.T. (1992). Nomarski DIC microscopy and image analysis of biofilm. *Binary Comput Microbiol*, 4, 93–95.

Keevil C.W., Dowsett A.B., Rogers J. (1993). *Legionella* biofilms and their control. In: Denyer S.P., Gorman S.P., Sussman M. (eds) *Microbial Biofilms: Formation and Control*. Blackwell, London, pp. 201–215.

Kolenbrander P.E. (1988). Intergeneric coaggregation among human oral bacteria ecology of dental plaque. *Ann Rev Microbiol*, 42, 627–656.

Kolenbrander P.E. (1989). Surface recognition among oral bacteria: multigeneric coaggregations and their mediators. *Crit Rev Microbiol*, **17**, 137–159.

Korber D.R., James G.A., Costerton J.W. (1994). Evaluation of fleroxacin activity against established *Pseudomonas aeruginosa* biofilms. *Appl Environ Microbiol*, **60**, 1663–1669.

Kreft J.U., Booth G., Wimpenny J.W.T. (1998). BacSim, a simulator for individual based modeling of bacterial colony growth. *Microbiology (Reading)*, **144**, 3275–3287.

Kreft J.U., Picioreanu C., Wimpenny J., van Loosdrecht M. (1999). Individual based modeling of biofilms: why? In: Wimpenny J., Gilbert P., Walker J., Brading M., Bayston R. (eds) *Biofilms: the Good, the Bad and the Ugly*. Bioline, Cardiff, pp. 257–262.

Kwok W.K., Picioreanu C., Ong S.L., van Loosdrecht M.C.M., Ng W.J., Heijnen J.J. (1998). Influence of biomass production and detachment forces on biofilm structures in a biofilm airlift suspension reactor. *Biotechnol Bioeng*, **58**, 400–407.

Lee N.M., Welander T. (1994). Influence of predators on nitrification in aerobic biofilm processes. *Water Sci Technol*, **29**, 355–363.

Lewandowski Z., Altobelli S.A., Fukushima E. (1993). NMR and microelectrode studies of hydrodynamics and kinetics in biofilms. *Biotechnol Prog*, **9**, 40–45.

Marsh P.D., Bradshaw D.J. (1998). Microbial community aspects in dental plaque. In: Busscher H.J., Evans L.V. (eds) *Oral Biofilms and Plaque Control*. Harwood Academic Publishers, The Netherlands, pp. 43–55.

McGroarty J.A., Reid G. (1988). Detection of a lactobacillus that inhibits *Escherichia coli*. *Can J Microbiol*, **34**, 974–978.

Nyvad B., Fejerskov O. (1989). Structures of dental plaque and the plaque-enamel interface in human experimental caries. *Caries Res*, **23**, 151–158.

O'Toole G.A., Kotter R. (1998a). Flagellar and twitching motility are necessary for *Pseudomonas aeruginosa* biofilm development. *Mol Microbiol*, **30**, 295–305.

O'Toole G.A., Kolter R. (1998b). Initiation of biofilm formation in *Pseudomonas fluorescens* WCS365 proceeds via multiple, convergent signalling pathways: a genetic analysis. *Mol Microbiol*, **28**, 449–461.

Ohashi A., Desilva D.G.V., Mobarry B., Manem J.A., Stahl D.A., Rittmann B.E. (1995). Influence of substrate C/N ratio on the structure of multispecies biofilms consisting of nitrfiers and heterotrophs. *Water Sci Technol*, **32**, 75–84.

Okabe S., Hirata K., Watanabe Y. (1995). Dynamic changes in spatial microbial distribution in mixed population biofilms – experimental results and model simulation. *Water Sci Technol*, **32**, 67–74.

Picioreanu C., van Loosdrecht M.C.M., Heijnen J.J. (1998). Mathematical modeling of biofilm structure with a hybrid differential-discrete cellular automaton approach. *Biotechnol Bioeng*, **58**, 101–116.

Picioreanu C., van Loosdrecht M.C.M., Heijnen J.J. 1999). Discrete-differential modeling of biofilm structure. *Water Sci Technol*, **39**, 115–122.

Pratt L.A., Kolter R. (1998). Genetic analysis of *Escherichia coli* biofilm formation: roles of flagella, motility, chemotaxis and type 1 pili. *Mol Microbiol*, **30**, 285–293.

Rosan B., Correia F.F., DiRienzo J.M. (1998). Corncobs: a model for oral microbial biofilms. In: Busscher H.J., Evans L.V. (eds) *Oral Biofilms and Plaque Control*. Harwood Academic Publishers, The Netherlands, pp. 145–162.

Santegoeds C.M., Ferdelman T.G., Muyzer G., de Beer. D. (1998). Structural and functional dynamics of sulphate-reducing populations in bacterial biofilms. *Appl Environ Microbiol*, **64**, 3731–3739.

Schramm A., Larsen L.H., Revsbech N.P., Ramsing N.B., Amann R. (1996). Structure and function of a nitrifying biofilm as determined by in situ hybridization and the use of microelectrodes. *Appl Environ Microbiol*, **62**, 4641–4647.

Senior E., Bull A.T., Slater J.H. (1976). Enzyme evolution in a microbial community growing on the herbicide Dalapon. *Nature (Lond)*, **263**, 476–479.

Shotton D.M. (1989). Confocal scanning optical microscopy and its applications for biological specimens. *J Cell Sci*, **94**, 175–206.

Stoodley P., De beer D., Lewandowski Z. (1994). Liquid flow in biofilm systems. *Appl Environ Microbiol*, **60**, 2711–2716.

Tall B.D.W.H.N., George K.S., Gray R.T., Walch M. (1995). Bacterial succession within a biofilm in water supply lines of dental air-water syringes. *Can J Microbiol*, **41**, 647–634.

Tartakovsky B., Guiot S.R. (1997). Modeling and analysis of layered stationary anaerobic granular biofilms. *Biotechnol Bioeng*, **54**, 122–130.

Van Loosdrecht M.C.M., Eikelboom D., Gjaltema A., Mulder A., Tijhuis L., Heijnen J.J. (1995). Biofilm structures. *Water Sci Technol*, **32**, 235–243.

Villaverde S., Mirpuri R. Lewandowski Z., Jones W.A. (1997). Study of toluene degradation kinetics in a flat plate vapour phase bioreactor using oxygen microsensors. *Water Sci Technol*, **36**, 77–84.

Walker J.T., Mackerness C.W., Rogers J., Keevil C.W. (1995). In: Lappin-Scott H.M., Costerton J.W. (eds) *Microbial Biofilms*. Cambridge University Press, Cambridge, UK, pp. 196–204.

Wimpenny J.W.T. (1993). Microbial systems: patterns in time and space. *Adv Microb Ecol*, **12**, 469–522.

Wimpenny J.W.T., Colasanti R. (1997). A unifying hypothesis for the structure of microbial biofilms. *FEMS Microbiol Ecol*, **22**, 1–16.

Wolfaardt G.M., Lawrence J.R., Robarts R.D., Caldwell S.J. (1994). Multicellular organization in a degradative biofilm community. *Appl Environ Microbiol*, **60**, 434–446.

Zhang T.C., Fu Y.C., Bishop P.L. (1994). Competition in biofilms. *Water Sci Technol*, **29**, 263–270.

4 Microscopy Methods for Studying Biofilms

Iwona B. Beech, Rudi C. Tapper and Rolf J. Gubner

The importance of elucidating the dimensions and spatial arrangement of biofilms, including the distribution and composition of microorganisms within the biofilm matrix, is recognised as essential in understanding the function of biofilms, *i.e.*, their ability to act as a barrier against antimicrobial compounds such as biocides and antibiotics, regulation of interfacial processes (either promoting or inhibiting deterioration of materials), harbouring pathogenic microorganisms, scavenging pollutants (*e.g.* heavy metals), and facilitating transfer of genetic material. Biofilms formed under laboratory and intristic conditions in different environments including the clinical situation, on naturally occuring and man-made inanimate surfaces, as well as on biological surfaces, have been extensively studied using a wide range of microscopy techniques. Light microscopy (*e.g.* transmitted light-, epifluorescence-, differential interference contrast- and confocal scanning laser microscopy), electron microscopy (transmission-, scanning- and environmental scanning microscopy), and recently atomic force microscopy, have all been applied to investigate biofilm structure in various stages of its development. Each of these techniques offers a different degree of sensitivity and level of resolution, allowing imaging of the overall appearance and/or specific features of biofilms, *e.g.* microbial colonies and individual cells, the extracellular polymeric material (often termed glycocalyx or slime), and the presence of inorganic products within the biofilm. This communication presents an overview of conventional and state-of-the-art microscopy techniques used to investigate biofilms. A brief introduction to each method and examples of their use are provided.

KEY WORDS: biofilms, imaging, microscopy, structure

LIGHT MICROSCOPY IN STUDYING BIOFILMS

Various techniques of light microscopy have been employed to observe biofilms (Lawrence *et al.*, 1991; Keevil and Walker, 1992; Walker and Keevil, 1994 and references therein). Light microscopy uses glass lenses to bend and focus light rays in order to produce enlarged images of small subjects, such as bacterial cells. The resolution of a light microscope is determined by the numerical aperture of its lens system and by the wavelength of the light it employs, which in the case of a conventional light source is of the order of about 0.2 μm. The most common types of light microscopes are the bright-field-, dark-field-, phase-contrast-, and fluorescence microscopes. Each yields a distinctive image and may be used to observe different aspects of microbial morphology.

Bright and Dark Field Microscopy

Bright field microscopy is limited in its use, since most biological materials do not have inherent contrast, hence, most cells are fairly featureless under a bright-field microscope unless they are stained. Rather than employing both diffracted and undiffracted light, dark-field microscopy uses only the images formed by a hollow cone of a diffracted light illuminating the specimen. The hollow cone is at too wide an angle for light to enter the objective lens. Therefore, the field of view is dark (hence the name, dark-field) except for the light scattered by the object. Since this method relies on diffracted light only, the size of the object with respect to the wavelength is not relevant. The important factor is whether the object can scatter light. Particles smaller than 400 nm can be visualised by this method. Bakke and Olsson (1986) described biofilm thickness measurements applying bright-field microscopy. The authors used the vertical displacement of the sample required to focus from the biofilm-liquid interface to the substratum, measured by a stage micrometer. Biofilm thickness proved to be proportional, but not equal, to the measured vertical displacement, the proportionality factor being determined as the ratio of the refractive indexes of the liquid to the biofilm. It has, therefore, been proposed that the thickness of any transparent film might be determined by bright-field microscopy when the refractive index of the film is known. The method has been successfully adapted by Shieh and Mulcahy (1985), Bakke *et al.* (1990); Trulear and Characklis (1992) and further developed by Lauvvik and Bakke (1994) to automate biofilm thickness measurements by light microscopy based on variance analysis of optical images. This new *in situ* technique proved to be consistent with manually determined thickness measurements.

Fluorescent Microscopy

Traditional light microscopes produce images from light that passes through a specimen. Some of the light is absorbed depending on the density of the sample. Fluorescent microscopes work on the principle that an object is emitting light. Exposing a fluorescent specimen to ultraviolet, violet, or blue light results in the emission of the absorbed energy at a longer, specific, wavelength. The emitted light is used to form an image. Usually, specimens are stained with fluorochromes that fluoresce brightly upon exposure to light of a specific wavelength, but some microorganisms are autofluorescent (*e.g. algae* and *Pseudomonas fluorescens*). The green fluorescent protein (GFP) of the jellyfish *Aequorea victoria* has been found to fluorescence with fluorescein-like characteristics and is widely used as a strongly visible fluorescent reporter molecule which is species-independent and does not require cofactors or substrates (Herman, 1998). These findings are revolutionising cell biology as a gene coding can be constructed for a fusion of GFP with almost any other protein and the resulting fluorescent fusion should localise and behave similarly to the original protein. The method allows protein localisation to be visualised without having to inject cells or purify and label proteins. Stretton *et al.* (1998) describe the use of GFP to tag and investigate gene expression in marine bacteria. Since the study of biofouling control using fluorescent probes and image analysis tools is presented by Yu and McFeters (2000), the description of available probes that follows is limited to an overview of the microscopy methods and their applications.

Immunofluorescence is used to visualise specific bacterial genera or even strains within biofilms. It combines the specificity of antibodies with the high sensitivity of fluorescence. The antigen-antibody reaction is highly selective, hence, it can be applied for the identification, localisation and visualisation of cells within a biofilm. For example, Smith (1982) used this approach to identify sulphate-reducing bacteria (SRB). Specific labelling was applied to visualise and locate the waterborne pathogen *Legionella pneumophila* in biofilms grown on glass surfaces (Walker and Keevil, 1994).

Currently, fluorescent *in situ* hybridisation (FISH) is the predominant choice in biofilm research. FISH is very similar to immunofluorescence, except that it allows the direct visualisation and localisation of DNA and RNA sequences on chromosomes in biofilm cells. FISH is based on the hybridisation between target sequences of single-stranded DNA of chromosomes or nuclei with fluorescently labelled complementary sequences (cDNA). Amann *et al.* (1992) characterised the population architecture of sulphidogenic biofilms by fluorescent microscopy. For this purpose the authors amplified selectively a common region of the 16S rRNA sequence by the polymerase chain reaction. Sequences of the amplification products were used to design both general and specific hybridisation probes. Fluorescent versions of these probes were used to visualise specific SRB within developing and established biofilms. Acridine orange (AO) is an unspecific DNA-binding fluorochrome that has been widely used in enumeration and detection of bacteria (McFeters *et al.*, 1991). Verran *et al.* (1994) used AO staining in combination with image analysis to quantify the adhesion of micro-organisms to steel surfaces. Walker and Keevil (1994) used direct AO staining of microorganisms in a biofilm on copper.

Chan *et al.* (1996) introduced a non-toxic fluorochrome (4-p-dimethylaminostyryl-pyridinium) that allows imaging of viable cells using fluorescence microscopy methods. The effectiveness of SYTOX Green, a nucleic acid stain for measuring bacterial viability, was tested by Lebaron (1998) on starved populations of *Escherichia coli* and *Salmonella typhimurium*. The author concluded that the stain underestimated the fraction of the dead cells within starved populations containing cells with damaged nucleic acids or membranes. Mason *et al.* (1998) used a combination of the fluorescent nucleic acid binding dyes hexidium iodide (HI) and SYTO 13 to distinguish between Gram-negative and Gram-positive organisms. The SYTO dyes are lower-affinity nucleic acid stains that passively diffuse through cell membranes. These cell-permeant, visible light–excitable dyes can be used to stain RNA and DNA in both live and dead eukaryotic cells, as well as in Gram-positive and Gram-negative bacteria. Stevik *et al.* (1998) compared the fluorescent stain 4′,6-diamidino-2-phenylindole (DAPI) with the most probable number (MPN) method to quantify protozoa in infiltration systems. They concluded that the DAPI staining procedure gives a more realistic estimate of the number and distribution of protozoa in infiltration porous media than to the MPN method. Kalmbach *et al.* (1997) described the identification of the *in situ* dominating species from a drinking water biofilm applying this technique. With Gram-positive bacteria, the thick peptidoglycan layer of a cell wall presents a barrier for entry of horseradish peroxidase (HPR)-labelled probes (Bidnenko *et al.*, 1998). Therefore, such probes do not give any signal in FISH unless cells are first treated with enzymes which hydrolyse the peptidoglycan. The authors used this property of FISH to estimate the staining of cell walls of individual Gram-positive bacteria (Surman *et al.*, 1996).

Phase Contrast Microscopy

Phase contrast microscopy allows biological samples to be observed in their live and unstained state by taking advantage of the difference in refractive index between materials. This is a major advantage, since before the introduction of phase contrast microscopy, much of the information about cells was inferred from dead, stained cells. The principle involves the slowing down of the light waves, as they go through a material of a different refractive index. A phase ring in the condenser is used to illuminate the object. The light waves that are not refracted pass through another phase ring between the objective and the eyepiece and are shifted by 1/4 wavelength. The refracted waves do not pass through the phase ring and are not shifted. When these two waves passing through the object are summed, the phase-shifted light waves cancel each other, hence, the image is significantly darker. Thus the phase shifting of the wavelength enhances the contrast of refractive material. The Hoffman modulation phase contrast microscope is a modification of a bright-field microscope. It allows non-invasive imaging without the need for any prior staining or preparation of the specimens such as biofilms. Images can be obtained with high contrast resolution, resulting in the creation of a 3-D appearance, by conversion of opposite phase gradients to opposite intensities. One side of the object appears bright whilst the other is dark against a grey background (Hoffman, 1988). The images obtained with the Hoffman modulated phase contrast microscope should have good contrast without artefacts, such as halos, which are often observed in conventional phase contrast microscopy.

Differential Interference Contrast Microscopy

Differential interference contrast (DIC) microscopy is a different phase imaging technique using plane-polarised light based on refractive index (RI) differences in the sample, as opposed to light absorption differences in traditional bright-field microscopy. The DIC microscope has the advantage over the Hoffman microscope that more light passes through to the optics, allowing better imaging. However, DIC microscopes are more expensive compared to the Hoffman modulated microscopes. DIC microscopy is commonly used to visualise the structure of living, biological cells. Typical specimens viewed under the DIC microscope include highly transparent cells, which distort the phase of the propagating light wave. Rogers and Keevil (1992) applied gold immunolabelling and a fluorochrome for *Legionella pneumophila* to visualise the waterborne pathogen within aquatic biofilms developed on glass and polybutylene surfaces. The development of episcopic DIC enabled simultaneous visualisation of the total biofilm flora and gold-labelled legionellae. Walker and Keevil (1994) and Surman *et al.* (1996) used episcopic DIC in combination with a fluorescent set-up, combining the advantages of DIC with the use of specific fluorescent probes to visualise cells and their location within the biofilm. Walker *et al.* (1994) describe the detection of biofilms on corroded copper and polypropylene pipes to show the surface topography of microorganisms, biofilms and the surfaces of opaque substrata without artefacts. Reflected DIC and attenuated total refection Fourier transform infrared spectroscopy (ATR-FTIR) were used to obtain complementary data on the structural and chemical properties of a biofilm (Suci *et al.*, 1997). This non-destructive approach offers opportunities to examine the relationships between the population structure, distribution and chemistry of biofilms on a variety of substrata in the presence of a bulk aqueous phase under controlled hydrodynamic conditions. Using DIC, Walker and Keevil (1994) applied 5-cyano-2,3-diotyl tetrazolium chloride (CTC), which fluoresces when reduced by metabolically active bacteria, to confirm the viability of fresh water biofilms.

Confocal Scanning Laser Microscopy

Confocal scanning laser microscopy (CSLM), also referred to as laser scanning confocal microscopy (LSCM), is now established as a valuable tool for obtaining high resolution images and 3-D reconstructions of biofilms. The popularity of CSLM arises from its ability to produce blur-free, crisp images of thick specimens at various depths. Confocal imaging rejects the out-of-focus information by placing a pinhole in front of the detector, so that only the region of the specimen that is in focus is detected. Confocal imaging can only be performed with point-wise illumination and detection, which is a most important advantage of using CSLM. Scanning is achieved by either deflection of the laser beam or movement of the sample on a stage. As the laser scans across the specimen, the analogue light signal, detected by the photo-multiplier, is converted into a digital signal, contributing to a pixel-based image displayed on a computer monitor attached to the CSLM. The relative intensity of the fluorescent light, emitted from the laserhit point, corresponds to the intensity of the resulting pixel in the image (typically 8-bit greyscale). The plane of focus (z-plane) is selected by a computer-controlled fine-stepping motor, which moves the microscope stage up and down. Typical focus motors can adjust the focal plane in as little as 0.1 μm increments. Stacking 2-D optical sections collected in series can generate a 3-D reconstruction of a specimen. Image processing and analysis provides extensive and detailed information on microbial growth and behaviour under many conditions (Cadwell *et al.*, 1993). The application of CSLM with a high resolution and registration optical sectioning capability allows the visualisation of thick biofilms in three dimensions. The combination of CSLM, image analysis and fluorescent probes provides the basic tools for analysing changes in biofilm structure and chemistry, offering great potential for the analysis of the physico-chemical microenvironment surrounding microorganisms.

Attached bacterial cells, micro-colonies and biofilms have been studied using negative fluorescent staining in conjunction with CSLM (Cadwell *et al.*, 1992). The use of non-toxic fluorochromes allowed time-course investigation of micro-colony development. DeBeer *et al.* (1997) introduced the measurement of local diffusion coefficients in biofilms by micro-injection and CSLM. This new technique is based on the micro-injection of fluorescent dyes and quantitative analysis of the subsequent plume formation using CSLM. CSLM was also used to investigate the relationship between the presence of biofilms and the associated pepper-pot pitting of copper plumbing tubing (Walker *et al.*, 1998). Images of the biofilm and substrata at different locations within each sample were provided with minimum disruption and artefacts. The authors describe as a significant feature of this technique the elimination of out of focus information, thus, allowing a more analytical examination of the samples. CSLM has significantly contributed to the development of the current biofilm model proposed by Costerton *et al.* (1994), Lewandowski *et al.* (1995) and Lewandowski (2000).

SCANNING PROBE MICROSCOPY

Scanning probe microscopy (SPM) methods utilise a probe which is brought into close contact with the sample under study and then raster scanned over the sample in order to produce an image. There are several types of SPM. The first to be developed, scanning tunnelling microscopy (STM), makes use of a tungsten wire from which a stream of electrons "tunnels" across to the sample. In biological applications STM has been found to have a major drawback. Due to the nature of the method, the tip punctures and images the inside of as it advances and penetrates until it contacts the layer a few tens of nm thick which is in contact with the substrate (Ruppersberg *et al.*, 1989). Hence, the next SPM technique to appear, namely atomic force microscopy (AFM) is the more appropriate technique for imaging cells.

AFM utilises an atomically sharp tip (generally made of silicon nitride or silicon) to produce a topographical map of the surface, with maximum magnification of the order of ten million times. The tip is mounted on the underside of a cantilever, the flexing of which is generally detected by the deflection of a laser beam. The Piezolever™ (Park Scientific Instruments) works by measuring the stress-induced electrical resistance changes of an implanted conductive channel in the monolithic probe. This shortens the setup time of the instrument, since no alignment of a laser beam onto the cantilever is required. Surface features of cells imaged under fluid may be aquired at higher resolution using Cryo-AFM since cells are less deformable and thermal drift in the AFM is eliminated (Sheng and Shao, 1998).

Biologists have used AFM in combination with light microscopy (transmitted bright field, epifluorescence and surface interference) to gain both topographical and visual information from the sample (Vesenka *et al.*, 1995; Nagao and Dvorak, 1998). However, the diffraction limit of optical microscopy does not allow the observation of fine features. The final SPM technique, scanning near optical microscopy (SNOM) uses a fibre optic probe which allows the optical diffraction limit to be breached (Lewis *et al.*, 1995; Lieberman *et al.*, 1996). When the topographical information is simultaneously collected with fluorescence imaging of the cells, the combination is known as SNFM (scanning near-field fluorescence microscopy, Monobe *et al.*, 1998). The use of SNOM for the study of cell biology, which includes the investigation of *E. coli* containing green fluorescent protein (GFP) (Muramatsu *et al.*, 1996; Subramaniam *et al.*, 1997) and the use of the DNA dyes DAPI, Hoechst 33342 and ethidium bromide (Kirsch *et al.*, 1998) has been reviewed by Subramaniam *et al.* (1998). Ben-Ami *et al.* (1998) describe the first simultaneous imaging of bacteria using SNOM and AFM, also known as SNOAM (Tamiya *et al.*, 1997). Unstained bacteria were imaged and a comparison made with conventional AFM in both air and aqueous media. Certain internal (endospores) and external features (possibly the cell wall), not visible using AFM, were imaged using SNOM. Haydon *et al.* (1996) have described a confocal version of SNOM, terming it near-field confocal optical spectroscopy (NCOS).

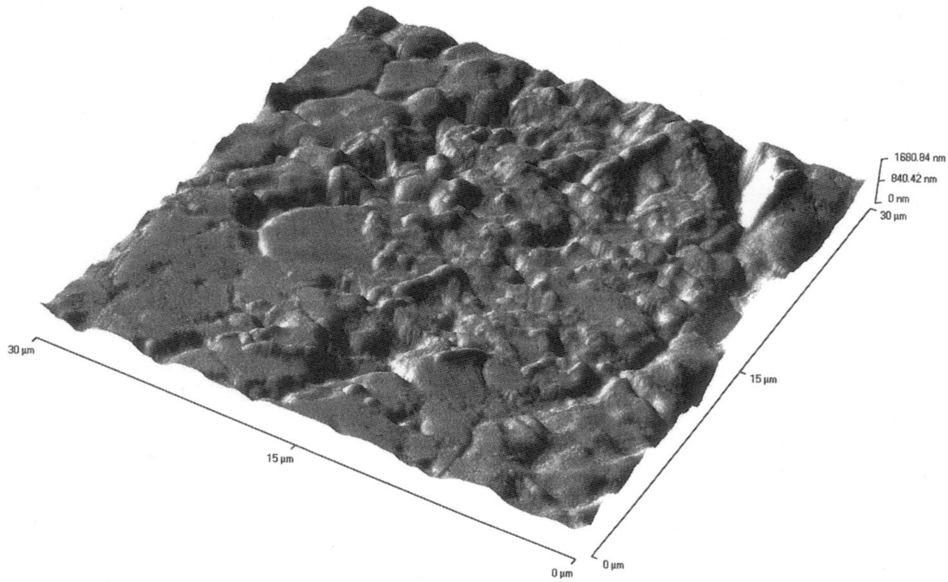

Figure 1 AFM image of a 30 day old biofilm formed under continuous flow conditions in a pure culture of marine *Pseudomonas sp.* NCIMB 2021 on an as-received surface of AISI 316 stainless steel; single bacterial cells are apparent.

AFM is often used in preference to methods such as scanning electron microscopy (SEM) as the technique has several major advantages. Since the sample need not be electrically conductive, no metallic coating of the specimen is required. Unlike the case with SEM, no dehydration of the sample is necessary and biofilms may be viewed in their hydrated state. This eliminates the shrinkage of biofilm associated with imaging using SEM, yielding a non-destructive technique. The resolution of AFM is higher than that of environmental SEM where hydrated images can also be obtained, and extracellular polymeric substances may not be imaged with clarity. The application of AFM for the study of biofilms on metal surfaces has been reviewed by Beech (1996). As an example, an AFM micrograph showing a biofilm formed in a continuous flow bioreactor by a pure culture of marine *Pseudomonas* spp. on an as-received 316 stainless steel surface is presented in Figure 1.

Modes of AFM Operation

There are several modes of AFM operation, each having been introduced to extend the variety of samples which can be imaged. The choice of mode is influenced by the sample, its topography, and the conditions of imaging, *e.g.* in air or in fluid for true hydrated imaging. Although imaging under liquid, using a so-called AFM wet-cell, reduces the danger of introducing artefacts, it generally offers inferior resolution compared with imaging in air. A useful starting point for the uninitiated reader and a source of information for the experienced user is provided by Howland and Benatar (1996), who describe the use of the various modes as well as AFM in general.

When operated in contact mode, the original mode of operation (Binnig *et al.*, 1986, 1987; Abraham *et al.*, 1988), the AFM can be used to image samples at the atomic level (Bachelot *et al.*, 1997). This mode of operation was improved by non-contact AFM (resonant or attractive mode) where a vibrating tip is oscillated at its resonant frequency to produce an image (Bachelot *et al.*, 1997). The lateral resolution is generally limited to a few nm due to the larger distance (> 2 nm) between sample and tip. However, the method allows imaging of more delicate samples. The third mode of operation is known as Tapping mode™ (Digital Instruments) AFM or TMAFM, and is a compromise between contact and non-contact AFM (Zhong *et al.*, 1993; Constant *et al.*, 1994; Hansma *et al.*, 1994). The cantilever is oscillated with a larger amplitude than in non-contact mode (several tens of nm) and a larger cantilever spring constant is used. Periodically, due to these differences, the tip can cross the large range force field and contact the surface under study. The large tip motion makes the mode insensitive to shear forces and their destructive influence. Shear force AFM was developed for use in SNOM. Since the optical fibre probe is very delicate, it is impossible to use with forces acting perpendicular to the surface. In shear mode, the probe vibrates parallel to the surface (Betzig *et al.*, 1992; Toledo-Crow *et al.*, 1992) with a vibration amplitude (typically a few nm) which is very sensitive to the tip-sample separation, so that a lateral resolution better than 10 nm may be obtained. A new mode of operation is now available, known as the magnetic a/c (MAC) mode of AFM imaging (Molecular Imaging), first reported by Zhong *et al.* (1993). This mode has a great advantage over TMAFM. In tapping mode, the oscillation of the tip results in mechanical excitations of the microscope as there is a damping effect due to the liquid present (Han *et al.*, 1996) disturbing the position of the laser beam (Florin *et al.*, 1993). This problem is overcome by directly driving the cantilever; a magnetic coating is applied and the cantilever deflected using a solenoid placed underneath the sample (Jarvis and Tokumoto, 1997). Using MAC mode there is no adhesion even to extremely sticky protein-coated surfaces, allowing imaging previously unobtainable with AFM (Lindsay *et al.*, 1998). MAC mode has allowed researchers to develop the use of carbon nanotube tips which break when used in TMAFM due to the large acoustic vibration (Li *et al.*, 1998). Carbon nanotube tips have a well defined tip unlike those made of silicon nitride or silicon, and allow higher resolution imaging. An added advantage is that the tips are more resistant to crashing into the sample, decreasing the cost of imaging.

Application of SPM to Biofilm Research

Adhesion

Adhesion is a very important process in biofilm development, and hence, the object of intensive studies. Morra and Cassinelli (1996) used AFM to investigate the effect of the surface upon adhesion of the bacterium *Staphylococcus epidermidis*, implicated in catheter-related urinary tract infections, and found that electron donor-electron acceptor interactions play a large part in the adhesion process. Gorman *et al.* (1997), studied the same bacterium and the influence of the conditioning film upon catheter material, measuring the surface roughness employing AFM. Hyde *et al.* (1997), also used AFM to correlate the effect of surface roughness and contact angle upon bacterial adherence and removal from fluorinated polymers, stainless steel, glass and polypropylene. Frank and Belfort (1997) measured the intermolecular forces between two layers of adsorbed EPS. This work studied the effect of seawater on EPS, providing information on the structure of the conditioning layer upon which biofilms grow. Baty *et al.* (1997) reported the use of AFM to image mussel adhesive proteins and study the mode of adhesion to polymers. Comparison of the images obtained with AFM contact and tapping modes allowed observation of the effect of hydration upon such a layer. It was found that dehydration had a pronounced effect upon the structure of the protein film on one polymer, but not on another. Bowen *et al.* (1998) reported the first use of a single, living, immobilised cell as a "cell probe" for the study of cell-surface adhesion in the presence of a liquid environment. Using different cells will allow measurement of key parameters in the fundamental study of cell adhesion, including the strength of cell-surface interactions, the time of development of adhesive contact, the influence of pH, ionic strength, the effect of substratum (type, roughness, preparation, coatings), the effect of cell life cycle and growth conditions, and the effect of weakening adhesion. The technique promises a new method of screening innovative antifouling materials and coatings. Tapping mode AFM was applied by McDonald *et al.* (1998) to visualise protein (fibronectin) binding to titanium implant surfaces, which is an important step in subsequent cell attachment. Steele *et al.* (1998) proved by a high-resolution AFM study of the surface of the Martian meteorite ALH84001 that the images of "biofilms" formed by alleged ultrananobacteria reported by a team at NASA were not artefacts created by the SEM sputter coating process. Whether the features observed by McKay *et al.* (1996) are indeed biological, remains to be resolved.

Assessment of antimicrobial action

AFM has been used by several researchers to investigate the effect and mode of action of antimicrobial agents on bacterial cells. Such studies include the action of penicillin on *Bacillus subtilis* (Kasas *et al.*, 1994), the effect of glutaraldehyde on aerobic marine biofilms formed on stainless steels (Tapper *et al.*, 1997), the influence of the antibiotic Cefodizime on *E. coli* (Braga and Ricci, 1998) and the action of Sterilox (superoxidised water) on *E. coli* and the SRB *Desulfovibrio indonensis* (Tapper *et al.*, 1998). Keresztes *et al.* (1998) studied the formation of metal sulfide layers on the surface of mild steel by the anaerobic SRB *Desulfovibrio desulfuricans* in the presence and absence of biocide.

Metal/microbe interactions

Bremer *et al.* (1992) used AFM to demonstrate the presence of bacterial biofilms on polished and unpolished copper surfaces. Steele *et al.* (1994) and Beech *et al.* (1996) compared and studied the corrosion of stainless steel in the presence of different types of bacterial biofilms. Extracellular polymeric material was visualised, as were micropits, with mixed bacterial cultures causing increased levels of corrosion compared with monocultures. Maurice *et al.* (1996) described an AFM study of the bacterial interaction with hydrous Fe (III) oxides, which are known to control the movement of metals and organic pollutants through soils. The authors discussed the use of AFM in soil research and the problems encountered. Grantham *et al.* (1997) reported the use of AFM in investigating the microbially catalysed dissolution of iron and aluminium oxyhydroxide mineral surface coatings to gain a better understanding of bacterial subsurface mobility. Washizu and Masuda (1997) applied AFM to studying the interaction between iron-oxidising bacteria (IOB) and corroding metal, and concluded that IOB tend to absorb to corrosion sites, and that they are activated by corrosion. While investigating the effect of biofilms on the deterioration of stainless steel, Steele (1996) imaged a fully hydrated biofilm formed by a fresh water bacterial consortium on the surface of AISI 316 stainless steel using a Nanoscope III AFM equipped with a wet cell, operated in contact mode. The biofilm was viewed under sterile saline. Such an arrangement facilitated the *in vivo* monitoring of the division of a single sessile cell over a period of 220 min.

The Future of SPM

Undoubtedly, SPM methods have developed considerably in a relatively short time. A new form of SPM termed magnetic resonance force microscopy (MRFM) is currently under development. MRFM could allow "non-destructive 3D imaging with Angstrom-scale resolution through the detection of single electronic or nuclear spins" (Noble, 1995). Such a device could be used for investigating proteins at a resolution better than that of traditional NMR, and for investigating subsurface structures of cells. Reading *et al.* (1998) describe the potential for another new type of probe. Photothermal measurements using infra red (IR) radiation with calorimetric analysis and scanning microscopy (CASM) could provide IR microscopy well below the diffraction limit of IR, ultimately in the range 20–30 nm. The probe could also be used for point heating and thus pyrolysis of the sample, analysing the evolved gases using mass spectroscopy (MS) or gas chromatography MS (GC-MS).

A final improvement to SPM is a change in the probe used in SNOM. Bergossi *et al.* (1997) describe a perturbative or apertureless SNOM probe made of tungsten, which allows a finer tip shape and hence a lateral resolution 10 times better than that obtained using an optic fibre.

ELECTRON MICROSCOPY IN BIOFILM IMAGING

Different forms of electron microscopy (EM) have been used to visualise biofilms associated with both non-biological and biological surfaces. Conventional scanning electron microscopy (SEM) requires fixation of samples in, for example, glutaraldehyde and/or osmium tetroxide, followed by dehydration using either critical point drying with liquid CO_2 or an alcohol series, and coating (sputtering) of biofilm with conductive metallic material, usually gold and/or palladium, or carbon. Conventional transmission electron microscopy (TEM) also requires fixation and dehydration procedures, followed by infiltration and polymerisation with epoxy resin. After ultrathin sectioning, embedded sections are stained with different electron opaque dyes, *e.g.* lead citrate and uranyl acetate. Such extensive sample treatment can cause considerable distortion of the specimen. The dehydration step can produce a significant shrinking effect due to the destruction of the highly hydrated extracellular polymeric (EPS) matrix (Richards and Turner, 1984; Fisher *et al.*, 1988). However, exopolymeric material, which is a mixture of macromolecules such as polysaccharides, proteins, lipids and nucleic acids, can be visualised in EM preparations applying polyanionic stains such as ruthenium red and alcian blue. The detection and preservation of the structure of the EPS can be improved using lectins and specific antibodies (Lambe *et al.*, 1994 and references therein; Sanford *et al.*, 1995). A sputter-cryo technique involving liquid nitrogen treatment of the biofilm, subsequent sputtering at –170°C or –180°C and viewing using a cold stage maintained at the same temperature, has also been reported as suitable for SEM preservation of the integrity of a biofilm slime matrix on pumice in an anaerobic filter treating molassess effluent (Richards and Turner, 1984) and of the EPS matrix of bacteria and fungi asssociated with clay minerals (Chenu and Jaunet, 1992). Conventional SEM and TEM imaging were widely employed to investigate biofilms of non-culturable animal gut bacteria *in situ* on epithelial or cuticular surfaces (Jolly *et al.*, 1993 and references therein).

Large numbers of studies describe SEM observations of pure and mixed culture bacterial biofilms on metals such as iron, copper and various alloys to determine the role of biofilms in corrosion of these materials (Coutinho *et al.*, 1993 and referenced therein; Percival *et al.*, 1998 and references therein). The importance of biofilms in wastewater treatment has also been investigated using electron microscopy (Zellner *et al.*, 1994; Sich and van Rijn, 1997). SEM and TEM examination of fungal and bacterial biofilms on wood helped in the understanding of the process of lignocellulose degradation (Daniel, 1994, and references therein). SEM was applied to monitor biodegradation of starch-plastic films in soil by mixed population biofilms harbouring fungi (Lopez-Llorca and Colom-Valiente, 1993) and polyhadroxyalkanoate (PHA) films in the presence of biofilms formed by bacteria and microalgae in marine and fresh water environments (Lopez-Llorca *et al.*, 1994). A typical image obtained using conventional SEM showing an heterogenous biofilm formed on a protective coating in a marine environment is presented in Figure 2.

The processes of biofilm involvement in bioleaching and biomineralisation have also been investigated applying EM techniques. SEM coupled with TEM and differential interference contrast microscopy have been employed to observe iron precipitation in a natural microbial biofilm (Brown *et al.*, 1998). The oxidation of pyrite by biofilms of *Thiobacillus ferrooxidans* has been studied using SEM combined with infrared spectroscopy (de Donato *et al.*, 1991).

Numerous accounts present the use of EM in studying biofilms on surgical implants, prosthetic devices, catheters and teeth or other solid oral structures to elucidate their role in infections (Speer *et al.*, 1988; Ganderton *et al.*, 1992; Zee *et al.*, 1997 and references therein). Modification of SEM and TEM preparation techniques has enabled visualisation of a glycocalyx in biofilms formed on bones by *Staphylococcus aureus*, thus demonstrating the importance of the EPS matrix in osteomyelitis (Evans *et al.*, 1998). Modern SEM and TEM offer high resolution imaging of cell/surface interactions such as reported in the study of biofilm formation on nasal turbinate tissue by *Pseudomonas aeruginosa* (Dowling and Wilson, 1998) and on mouse bladder lumenar cells and epithelial cells invaded by an uropathogenic strain of *E. coli* (Mulvey *et al.*, 1998).

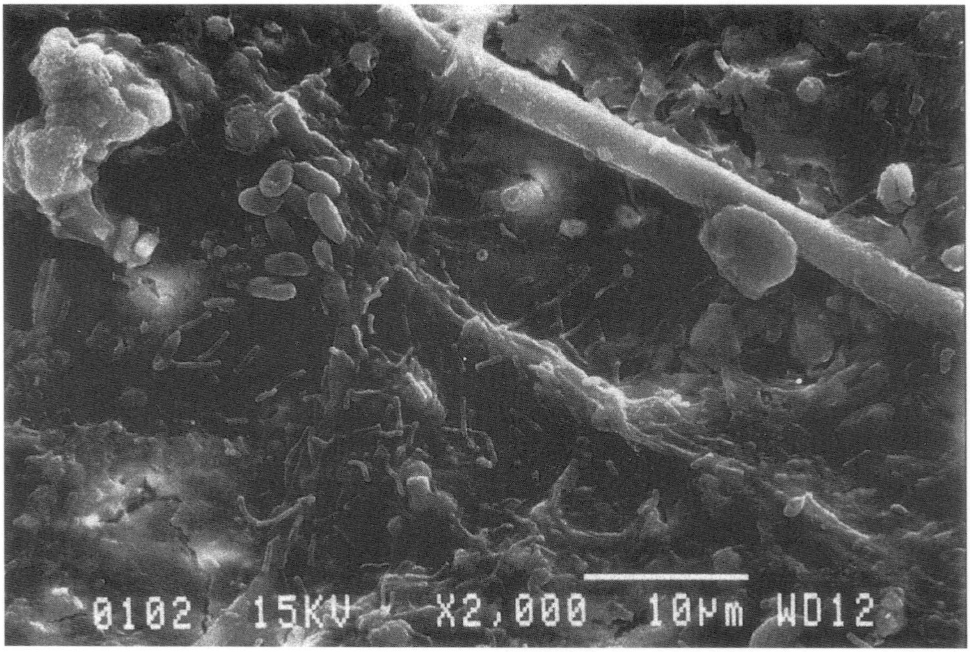

Figure 2 SEM photograph of a mixed population marine biofilm, harbouring bacteria and fungi, colonising the protective coating on a ship's ballast tank.

It is recognised that TEM and SEM can cause sample distortion and introduction of artefacts owing to the requirement for extensive specimen preparation prior to viewing (Little *et al.*, 1991; Sutton *et al.*, 1994). The loss of pure culture bacterial biofilms formed on glass beads as a result of sample preparation for SEM analysis was demonstrated by Chang and Rittman (1986). Great care should be taken when interpreting EM images of biofilm samples prepared in the conventional manner, to avoid erroneous conclusions with regard to the abundance and type of biofilm developed on a given substratum.

Environmental Scanning Electron Microscopy

Unlike traditional SEM which uses a high vacuum environment (10^{-7} torr) and either low energy secondary electrons or backscattered electrons to reproduce sample topography, the environmental scanning electron microscope (ESEM) developed in 1985 works with a specimen pressure chamber ten thousand times higher than that of the SEM. The pressure source in the chamber can be water vapour, air, argon, nitrogen or other gases. A vacuum gradient is used to maintain high vacuum conditions at the primary electron source, *i.e.* the gun system, whilst the chamber pressure can be varied up to aproximately 20 torr. The secondary electrons emitted from the specimen surface and accelerated toward the detector, collide with gas molecules, thus generating more free electrons and thereby providing more signal. Proper operating pressure controls the specimen surface charging. Elimination of charging allows examination

Figure 3 ESEM photograph of a fully hydrated biofilm formed on a carbon steel coupon exposed for 13 months in a marine environment; characteristic biofilm structures such as stacks and voids are clearly seen.

of unprepared, uncoated, nonconductive specimens in their natural environment (Baumgarten, 1990; Uwins, 1994; Li *et al.*, 1995). In addition, the elemental chemistry of specimens can be determined by combining ESEM observations with energy dispersive spectroscopy (EDS) analysis. ESEM observation of algal and fungal cells by the introduction of heavy metal stains, including potasium permanganate and osmium tetroxide to enhance backscattered signals (Collins *et al.*, 1993), have further proved that this technique is an effective tool for direct viewing of delicate biological samples. Use of water vapour as the imaging gas and for cooling the specimen enables viewing of hydrated samples. The latter conditions are particularly useful when observing biofilms. Results from SEM and ESEM studies of naturally occurring and laboratory generated biofilms were compared by Little *et al.* (1991) and Sutton *et al.* (1994). ESEM observations of biofilms on metal surfaces and their role in corrosion were reported by Wagner *et al.* (1992) and Beech *et al.* (1996). The advantages of the use of ESEM for the investigation of biofilm structure and morphology compared to other techniques were emphasised in both reports. ESEM has also been used investigate biofilm-influenced deterioration of materials such as iron, steel, copper and copper alloys, marine coatings and polymeric composites (Ray *et al.*, 1997). The study confirmed that ESEM is an excellent tool for demonstrating spatial relationships between biofilm and substrata and offered a unique insight into the role of microorganisms in the deterioration process. An ESEM image of a fully hydrated 13-month old marine biofilm formed on carbon steel is shown in Figure 3.

SUMMARY

Progress in refining microscopy techniques used for the study of biological material, *i.e.* the ability to obtain a three-dimentional image (*e.g.* CSLM), enhancing the resolving power to achieve the resolution at a molecular and even atomic scale (*e.g.* AFM) and eliminating sample pre-treatment (*e.g.* ESEM, AFM, CSLM), thus facilitating non-distructive, artefact-free observation of biofilms in their fully hydrated state, has resulted in the development of a current working model of a heterogenous biofilm (Costerton *et al.*, 1994; Lewandowsky, 2000). Coupling direct microscopy studies with the application of different fluorescent chemical, immuno- and molecular probes offer further insight into the relationship between biofilm structure and function (Yu and McFeters, 2000).

Comparison of DIC, ESEM, SEM, TEM, CSLM and AFM techniques for the examination of a naturally occuring mixed population biofilm consisting of bacteria, protozoa and ameoebae recovered from a water distribution system, formed on glass, titanium and silicone has been reported by Surman *et al.* (1996). It proved difficult to rank the methods as each offered a unique contribution to the understanding of biofilm structure and composition. Combining microscopy techniques such as SEM, TEM and AFM allowed visualisation of the morphology of *Pseudomonas putida* colonies forming biofilms at planar oil-water interfaces revealing bacterial flagella trapped within the biofilm and resolving bacterial surface features (Gunning *et al.*, 1996).

The microscopy techniques employed in biofilm research usually complement each other. Biofilms are heterogeneous, complex matrices composed of micro-colonies interspersed with channels allowing the movement of fluids (Lewandowsky, 2000). No technique can be said to be unequivocally better than another, as each of the methods adds a different dimension to the understanding of the spatial composition of biofilms. A combination of as many techniques as available is recommended to overcome the problems of artefacts and to provide the most accurate representation of the true biofilm structure and organisation.

REFERENCES

Abraham F.F., Batra I.P., Ciraci S. (1988). Effect of tip profile on atomic-force microscope images: a model study. *Phys Rev Letts*, **60**, 1314–1317.

Amann R.I., Stromley J., Devereux R., Key R., Stahl A. (1992). Molecular and microscopic identification of sulphate-reducing bacteria in multispecies biofilms. *Appl Environ Microbiol*, **58**, 614–623.

Bachelot R., Gleyzes P., Boccara A.C. (1997). Influence of both repulsive and attractive force fields in tapping mode atomic force microscopy. *Probe Microsc*, **1**, 89–97.

Bakke R., Olsson P.Q. (1986). Biofilm thickness measurements by light microscopy. *J Microb Methods*, **5**, 93–98.

Bakke, R., Salte K., Tengberg-Hansen H., Ingsy P. (1990). Xanthan degradation by biofilm in porous media. *Biofouling*, **2**, 311–321.

Baty A.M., Leavitt P.K., Siedlecki C.A., Tyler B.J., Suci P.A., Marchant R.E., Geesey G.G. (1997). Adsorption of adhesive proteins from the marine mussel, *Mytilus edulis*, on polymer films in the hydrated state using angle dependent x-ray photoelectron spectroscopy and atomic force microscopy. *Langmuir*, **13**, 5702–5710.

Baumgarten N. (1990). Introduction to the environmental scanning electron microscope. *Scanning*, **12**, I-36–I-37.

Beech I.B. (1996). The potential use of atomic force microscopy in studying biocorrosion. *Int Biodeterior Biodegr*, **37**, 141–149.

Beech I.B., Cheung C.W.S., Johnson D.B., Smith J.R. (1996). Comparative studies of bacterial biofilms on steel surfaces using techniques of atomic force microscopy and environmental scanning electron microscopy. *Biofouling*, **10**, 65–77.

Ben-ami N., Radko A., Ben-ami U., Lieberman K., Rothman Z., Rabin I., Lewis A. (1998). Near-field optical imaging of unstained bacteria: comparison with normal atomic force and far-field optical microscopy in air and aqueous media. *Ultramicroscopy*, **71**, 321–325.

Bergossi O., Bachelot R., Wioland H., Wurtz G., Laddada R., Adam P.M., Bijeon J.L., Royer P. (1997). Far field optical microscopy and spectroscopy with STM and AFM probes. *Acta Phys Pol*, A **93**, 393–398.

Betzig E., Finn P.L., Weiner J.S. (1992). Combined shear force and near-field scanning optical microscopy. *Appl Phys Letts*, **60**, 2484–2486.

Bidnenko E., Merchier C., Tremblay J., Tailliez P., Kulalauskas S. (1998) Estimation of the state of the bacterial cell wall by fluorescent *in situ* hybridisation. *Appl Environ Microbiol*, **64**, 3059–3062.

Binnig G., Quate C.F., Gerber Ch. (1986). Atomic force microscope. *Phys Rev Letts*, **56**, 930–933.

Binnig G., Gerber Ch., Stoll E., Albrecht T.R., Quate C.F. (1987). Atomic resolution with atomic force microscope. *Europhys Letts*, **3**, 1281.

Bowen W.R., Hilal N., Lovitt R.W., Wright C.J. (1998). Direct measurement of the force of adhesion of a single biological cell using an atomic force microscope. *Colloids Surf*, A **136**, 231–234.

Braga P.C., Ricci D. (1998). Atomic force microscopy: application to investigation of *Escherichia coli* morphology before and after exposure to Cefodizime. *Antimicrob Agents Chemother*, **42**, 18–22.

Bremer P.J., Geesey G.G., Drake B. (1992). Atomic force microscopy examination of the topography of a hydrated bacterial biofilm on a copper surface. *Curr Microbiol*, **24**, 223–230.

Brown A., Beveridge T.J., Keevil C.W., Sherriff B.L. (1998). Evaluation of microscopic techniques to observe iron precipitation in a natural microbial biofilm. *FEMS Microbiol Ecol*, **26**, 297–310.

Caldwell D.E., Korber D.R., Lawrence J.R. (1992). Imaging of bacterial cells by fluorescence exclusion using scanning confocal laser microscopy. *J Microbiol Methods*, **15**, 249–261.

Caldwell D.E., Korber D.R., Lawrence J.R. (1993). Analysis of biofilm formation using 2D vs 3D digital imaging. *J Appl Bacteriol Symp Suppl*, **74**, 52S–66S.

Chan E.C.S., Stranix B.R., Darling G.D., Noble P.B. (1996). A novel fluorochrome for microscopic observations of microbial morphology in wet mounts. *Can J Microbiol*, **42**, 875–879.

Chang H.T., Rittman B.E. (1986). Biofilm loss during sample preparation for scanning electron microscopy. *Water Res*, **20**, 1451–1456.

Chenu C., Jaunet A.M. (1992). Cryoscanning electron microscopy of microbial extracellular polysaccharides and their association with minerals. *Scanning*, **14**, 360–364.

Collins S.P., Pope R., Sceetz R.W., Ray I.R., Wagner P., Litle B. (1993). Advantages of environmental scanning electron microscopy in studies of microorganisms. *Microsc Res Tech*, **25**, 398–405.

Constant A., Putman J., Van der Werf K.O., De Grooth B.G., Van Hulst N.F., Greve J. (1994). Tapping mode atomic force microscopy in liquid. *Appl Phys Letts*, **64**, 2454–2456.

Costerton J.W., Lewandowski Z., DeBeer D., Caldwell D., Korber D., James G. (1994). Minireview: biofilms, the customized microniche. *J Bacteriol*, **176**, 2137–2142.

Coutinho C.M.L.M., Magalhaes F.C.M., Araujo-Jorge T.C. (1993). Scanning electron microscopy study of biofilm formation at different rates over metal surfaces using sulphate reducing bacteria. *Biofouling*, 7, 19–27.

Daniel G. (1994). Use of electron microscopy for aiding understanding of wood biodegradation. *FEMS Microbiol Rev*, **13**, 199–233.

De Beer D., Stoodley P., Lewandowski Z. (1997). Measurement of local diffusion coefficients in biofilms by microinjection and confocal microscopy. *Biotechnol Bioeng*, **53**, 151–158.

De Donato P., Mustin C., Berthelin J., Marion P. (1991). An infrared investigation of pellicular phases observed on pyrite by scanning electron microscopy, during its bacterial oxidation. *C R Acad Sci Paris*, **312**, 241–248.

Dowling R.B., Wilson R. (1998). Bacterial toxins which preturb ciliary function and respiratory epithelium. *J Appl Microbiol*, **85**, 138S-148S.

Evans R.P., Nelson C.L., Bowen W.R., Kleve M.G., Hickmon S.G. (1998). Visualization of bacterial glycocalyx with scanning electron microscopy. *Clin Orthop Relat Res*, **347**, 243–249.

Fisher W., Hanssel I., Paradies H.H. (1988). First results of microbial induced corrosion of copper pipes. In: Sequeira C.A.C., Tiller A.K. (eds) *Microbial Corrosion*. Elsevier Applied Science, London, pp. 300–327.

Florin E.L., Radmacher M., Fleck B., Gaub H.E. (1993). Atomic force microscope with magnetic force modulation. *Rev Sci Instrum*, **65**, 639–643.

Frank B.P., Belfort G. (1997). Intermolecular forces between extracellular polysaccharides measured using the atomic force microscope. *Langmuir*, **13**, 6234–6240.

Ganderton L., Chawla J., Winters C., Wimpenny J., Stickler D. (1992). Scanning electron microscopy of bacterial biofilms on indwelling bladder catheters. *Eur J Clin Microbiol Infect Dis*, **11**, 789–796.

Gorman S.P., Jones D.S., Mawhinney W.M., McGoven J.G., Adair C.G. (1997). Conditioning fluid influences on the surface properties of silicon and polyurethane peritineal catheters: implications for infection. *J Mater Sci: Mater Med*, **8**, 631–635.

Grantham M.C., Dove P.M., DiChristina T.J. (1997). Microbially catalysed dissolution of iron and aluminium oxyhydroxide mineral surface coatings. *Geochim Cosmochim Acta*, **61**, 4467–4477.

Gunning P.A., Kirby A.R., Parker M.L., Gunning A.P., Morris V.J. (1996). Comparative imaging of *Pseudomonas putida* bacterial biofilms by scanning electron microscopy and both DC contact and AC non-contact atomic force microscopy. *J Appl Bacteriol*, **81**, 276–282.

Han W., Lindsay S.M., Jing T. (1996). A magnetically-driven oscillating probe microscope for operation in liquids. Extracted from *Appl Phys Letts*, **23** Dec 1996, and published by the Lindsay Lab at http://green.la.asu.edu/pubs/APL122396/magneticprobes.html.

Hansma P.K., Cleveland J.P., Radmacher M., Walters D.A., Hillner P.E., Bezanilla M., Fritz M., Vie D., Hansma H.G., Prater C.B., Massie J., Fukunaga J., Gurley J., Elings V. (1994). Tapping mode atomic force microscopy in liquids. *Appl Phys Letts*, **64**, 1738–1740.

Haydon P.G., Marchese-Ragona S.P., Basarsky T.A., Szulczewski M., McCloskey M. (1996). Near-field confocal optical spectroscopy (NCOS): subdiffraction optical resolution for biological systems. *J Microsc*, **182**, 208–216.

Herman B. (1998). *Fluorescence Microscopy*. BIOS Scientific Publishers Ltd, Oxford, UK, pp. 64–68.

Hoffman R. (1988). Application of the modulation contrast microscope. *Int Lab*, July/August, 32–39.

Howland R., Benatar L. (1996). *A Practical Guide to Scanning Probe Microscopy*. Park Scientific Instruments.

Hyde F.W., Aulberg M., Smith K. (1997). Comparison of fluorinated polymers against stainless steel, glass and polypropylene in microbial biofilm adherence and removal. *J Ind Microbiol Biotechnol*, **19**, 142–149.

Jarvis S.P., Tokumoto H. (1997). Measurement and interpretation of forces in the atomic force microscope. *Probe Microsc*, 1, 65–79.

Jolley J.M., Lappin-Scott H.M., Anderson J.M., Clegg C.D. (1993). Scanning electron microscopy of the gut microflora of two earthworms: *Lumbricus terrestris* and *Octolasion cyaneum*. *Microb Ecol*, 26, 235–245.

Kalmbach S. Manz W., Szewzyk U. (1997). Isolation of the *in situ* dominating bacterial species from a drinking water biofilm. In: Wimpenny J., Handley P., Gilbert P., Lapin-Scott H., Jones M. (eds) *Biofilms: Community Interactions and Control*. BioLine, Cardiff, UK, pp. 183–191.

Kasas S., Fellay B., Cargnello R. (1994). Observation of the action of penicillin on *Bacillus subtilis* using atomic force microscopy: technique for the preparation of bacteria. *Surf Interf Anal*, 21, 400–401.

Keevil C.W., Walker J.T. (1992). Nomarski DIC microscopy and image analysis of biofilms. *Binary*, 4, 93–95.

Keresztes Z., Telegdi J., Beczner J., Kálmán E. (1998). The influence of biocides on the microbiologically influenced corrosion of mild steel and brass. *Electrochim Acta*, 43, 77–85.

Kirsch A.K., Subramaniam V., Striker G., Schnetter C., Arndt-Jovin D.J., Jovin T.M. (1998). Continuous wave two-photon scanning near-field optical microscopy. *Biophys J*, 75, 1513–1521.

Lambe Jr D.W., Jeffrey C., Ferguson K.P., Cooper M.D. (1994). Examination of the glycocalyx of four species of *Staphylococcus* by transmission electron microscopy and image analysis. *Microbios*, 78, 133–134.

Lauvvik T., Bakke R. (1994). Biofilm thickness measurements by variance analysis of optical images. *J Microb Methods*, 20, 219–224.

Lawrence J.R., Korber D.R., Hoyle B.D., Costerton J.W., Caldwell D.E. (1991). Optical sectioning of microbial biofilms. *J Bacteriol*, 174, 5732–5739.

Lebaron P., Catala P., Parthuisot N. (1998). Effectiveness of SYTOX green stian for bacterial viability assessment. *Appl Environ Microbial*, 64, 2697–2700.

Lewandowski Z. (2000). Structure and function of biofilms. In: Evans L.V. (ed) *Biofilms: Recent Advances in their Study and Control*. Harwood Academic Publishers, pp. 1–17.

Lewandowski Z., Stoodley P., Roe F. (1995). Internal mass transport in heterogeneous biofilms, recent advances. *Corrosion 95*, Paper No 222, NACE International, Houston, TX, USA.

Lewis A., Lieberman K., Ben-Ami N., Fish G., Khachatryan E., Ben-Ami U., Shalom S. (1995). New design and imaging concepts in NSOM. *Ultramicroscopy*, 61, 215–220.

Li J., Cassell A., Dai H. (1998). Application note: carbon nanotube tips for MAC mode AFM measurements in liquids. Molecular Imaging Corporation, Phoenix, AZ.

Li M.J., Rogers K., Rust C.A. (1995). Environmental scanning electron microscopes. *Adv Mat Proc*, 7, 24–25.

Lieberman K., Ben-Ami N., Lewis A. (1996) Fully integrated near-field optical, far-field optical and normal-force scanned probe microscope. *Rev Sci Instrum*, 67, 3567–3572.

Little B.J., Wagner P.A., Ray R.I., Pope R., Scheetz R. (1991). Biofilms: an ESEM evaluation of artefacts introduced during SEM preparation. *J Ind Microbiol*, 8, 213–222.

Lindsay S.M., Han W., Liu Y. (1998). Biological measurements and forces in MAC mode AFM. *Pico, Mol Imaging Newsletter*, 2, 1–2.

Lopez-Llorca V., Colom-Valiente M.F. (1993). Study of biodegradation of starch-plastic films in soil using scanning electron microscopy. *Micron*, 24, 457–463.

Lopez-Llorca V., Colom-Valiente M.F., Carcases M.J. (1994). Study of biofouling of polyhydroxy-alkanoate (PHA) films in water by scanning electron microscopy. *Micron*, 25, 45–51.

McDonald D.E., Markovic B., Allen M., Somasundaran P., Boskey A.L. (1998). Surface analysis of human plasma fibronectin adsorbed to commercially pure titanium materials. *J Biomed Mat Res*, 41, 120–130.

Mason D.J., Shanmuganathan S., Mortimer F.C., Gant V.A. (1998). A fluorescent Gram stain for flow cytometry and epifluorescense microscopy. *Appl Environ Microbiol*, **64**, 2681–2685.

Maurice P., Forsythe J., Hersman L., Sposito G. (1996). Application of atomic-force microscopy to studies of microbial interactions with hydrous Fe(III) oxides. *Chem Geol*, **132**, 33–43.

McFeters G.A., Singh A., Byun P.R., Williams S. (1991). Acridine orange staining reaction as an index of physiological activity in *Escheria coli*. *J Microbiol Methods*, **13**, 87–97.

McKay D.S., Gibson E.K., Thomaskeprta K.L., Vali H., Romanek C.S., Clemett S.J., Chillier X.D.F., Maechling C.R., Zare R.N. (1996). Search for past life on Mars – possible relic biogenic activity in martian meteorite ALH84001. *Science*, **273**, 924–930.

Monobe H., Koike A., Muramatsu H., Chiba N., Yamamoto N., Ataka T., Fujihira M. (1998). Scanning near-field fluorescence microscopy of a phase-separated hydrocarbon-fluorocarbon mixed monolayer. *Ultramicroscopy*, **71**, 287–293.

Morra M., Cassinelli C. (1996). *Staphylococcus epidermidis* adhesion to films deposited from hydroxyethylmethacrylate plasma. *J Biomed Mat Res*, **31**, 149–155.

Mulvey M.A., Lopez-Boado Y., Wilson C.L., Roth R., Parks W.C., Heuser J., Hultgren S.J. (1998). Induction and evasion of host defenses by type 1-piliated uropathogenic *Escherichia coli*. *Science*, **282**, 1494–1497.

Muramatsu H., Chiba A., Atika T., Iwabuchi S., Nagatani N., Tamiya E., Fujihira M. (1996). Scanning near-field optical/atomic force microscopy for fluorescence imaging and spectroscopy of biomaterials in air and liquid: observation of recombinant *Escherichia coli* with gene coding to green fluorescent protein. *Optical Rev*, **3**, 470–474.

Nagao E., Dvorak J.A. (1998). An integrated approach to the study of living cells by atomic force microscopy. *J Microsc (Oxf)*, **191**, 8–19.

Noble D. (1995). Magnetic resonance force microscopy. *Anal Chem*, November, 671A–673A.

Percival S.L., Knapp J.S., Edyvean R.G.J., Wales D.S. (1998). Biofilms, mains and stainless steel. *Wat Res*, 7, 2187–2201.

Ray R., Little B., Wagner P., Hart K. (1997). Environmental scanning microscopy investigation of biodeterioration. *Scanning*, **19**, 98–103.

Reading M., Hourston D.J., Song M., Pollock H.M., Hammiche A. (1998). Thermal analysis for the 21st century. *Am Lab*, **30**, 13.

Richards S.R., Turner R.J. (1984). A comparative study of techniques for the examination of biofilms by scanning electron microscopy. *Water Res*, **18**, 767–773.

Rogers J., Keevil C.W. (1992). Immunogold and fluorescein immunolabelling of *Legionella pneumophila* within an aquatic biofilm visualised by using episcopic differential interference contrast microscopy. *Appl Environ Microbiol*, **58**, 2326–2330.

Ruppersberg J.P., Horber J.K.H., Gerber Ch., Binnig G. (1989). Imaging of cell membraneous and cytoskeletal structures with a scanning tunneling microscope *FEBS Lett*, **257**, 460–464.

Sanford B.A., Thomas V.L., Mattingly S.J., Ramsay M.A., Miller M.M. (1995). Lectin-biotin assay for slime present in *in situ* biofilm produced by *Staphylococcus epidermis* using transmission electron microscopy (TEM). *J Ind Microbiol*, **15**, 156–161.

Sheng S., Shao Z. (1998). Biological cryo-atomic force microscopy: instrumentation and applications. *Jpn J Appl Phys*, **37**, 3828–3833.

Shieh W.K., Mulcahy L.T. (1985). Experimental determination of intrinsic kinetic coefficients for biological wastewater treatment systems. IAWPCR Specialised seminar, *Modelling of Biological Wastewater Treatment*, pp. 7–16.

Sich H., van Rijn J. (1997). Scanning electron microscopy of biofilm formation in denitrifying, fluidised bed bioreactors. *Water Res*, **31**, 733–742.

Smith A.D. (1982) Immunofluorescence of sulphate-reducing bacteria. *Arch Microbiol*, **133**, 118–121.

Speer A.G., Cotton P., Rode J., Seddon A.M., Neal C., Holton J., Costerton J.W. (1988). Biliary stent blockage with bacterial biofilms, a light and electron microscopy study. *Ann Int Med*, **108**, 546–553.
Steele A. (1996). The biodecontamination of stainless steel by bacterial biofilms. PhD thesis, University of Portsmouth, UK.
Steele A., Goddard D.T., Beech I.B. (1994). An atomic force microscopy study of the biodeterioration of stainless steel in the presence of bacterial biofilms. *Int Biodeterior Biodegr*, **34**, 35–46.
Steele A., Goddard D.T., Beech I.B., Tapper R.C., Stapleton D., Smith J.R. (1998). Atomic force microscopy imaging of fragments from the Martian meteorite ALH84001. *J Microsc (Oxf)*, **189**, 2–7.
Stevik T.K., Hanssen J.F., Jenssen P.D. (1998) A comparison between DAPI direct count (DCC) and most probable number method (MPN) to quantify protozoa in infiltration systems. *J Microbiol Methods*, **33**, 12–21.
Stretton S., Techkarnjanaruk S., McLennan A.M., Goodman A.E. (1998) Use of green fluorescent protein to tag and investigate gene expression in marine bacteria. *Appl Environ Microbiol*, **64**, 2554–2559.
Subramaniam V., Kirsch A.K., Rivera-Pomar R.V., Jovin T.M. (1997). Scanning near-field optical microscopy and microspectroscopy of green fluorescent protein in intact *Escherichia coli* bacteria. *J Fluorescence*, **7**, 381–385.
Subramaniam V., Kirsch A.K., Jovin T.M. (1998). Cell biological applications of scanning near-field optical microscopy (SNOM). *Cell Mol Biol*, **44**, 689–700.
Suci P., Siedlecki K.J., Palmer (Jr.) R.J., White D.C., Geesey G.G. (1997) Combined light microscopy and attenuated total refection Fourier transform infrared spectroscopy for integration of biofilm structure, distribution and chemistry at solid liquid interfaces. *Appl Environ Microbiol*, **63**, 4600–4603.
Surman S.B., Walker J.T., Goddard D.T., Morton L.H.G., Keevil C.W., Weaver W., Skinner A., Caldwell D., Kurtz J. (1996). Comparison of microscope techniques for the examination of biofilms. *J Microbiol Methods*, **25**, 57–70.
Sutton N.A., Hughes N., Handley P. (1994). A comparison of conventional SEM techniques, low temperature SEM and the Electroscan wet scanning electron microscope to study the structure of a biofilm of *Streptococcus crista* CR3. *J Appl Bacteriol*, **76**, 448–454.
Tamiya E., Iwabuchi S., Nagatani N., Murakami Y., Sakaguchi T., Yokoyama K. (1997) Simultaneous topographic and fluorescence imagings of recombinant bacterial cells containing a green fluorescent protein gene detected by a scanning near-field optical/atomic force microscope. *Anal Chem*, **69**, 3697–3701.
Tapper R.C. (1998). The use of biocides for the control of marine biofilms on stainless steel surfaces. PhD thesis, University of Portsmouth, UK.
Tapper R.C., Smith J.R., Beech I.B., Viera M.R., Guiamet P.S., Videla H.A., Swords C.L., Edyvean R.G.J. (1997). The effect of glutaraldehyde on the development of marine biofilms formed on surfaces of AISI 304 stainless steel. *Corrosion '97*, Paper No 205, NACE, Houston, TX, USA.
Toledo-Crow R., Yang P.C., Chen Y., Vaez-Iravani M. (1992). Near-field differential scanning optical microscope with atomic force regulation. *Appl Phys Letts*, **60**, 2957–2959.
Trulear M.G., Charaklis W.G. (1992). Dynamics in biofilm processes. *J Water Pollut Cont Fed*, **54**, 1288–1301.
Uwins P.J.R. (1994). Environmental scanning electron microscopy. *Mater Forum*, **18**, 51–75.
Verran J., Taylor R.L., Lees G.C. (1994). The use of image analysis to quantify microorganisms adherent to surfaces. *BINARY Bioline*, **6**, 55–57.
Vesenka J., Mosher C., Schaus S., Ambrosio L., Henderson E. (1995). Combining optical and atomic force microscopy for life sciences research. *Biotechniques*, **19**, 240–253.

Wagner P., Little B., Ray R., Jones-Meehan J. (1992). Investigation of microbiologically influenced corrosion using environmental scanning electron microscopy. *Corrosion '92*, Paper No 185, NACE, Houston, TX, USA.

Walker J.T., Keevil C.W. (1994). Study of microbial biofilms using light microscopy techniques. *Int Biodeterior Biodegr*, **34**, 223–236.

Walker J.T., Wagner D., Fisher W., Keevel C.W. (1994). Rapid detection of biofilms on corroded copper pipes. *Biofouling*, **8**, 55–63.

Walker J.T., Hanson K., Caldwell D., Keevil C.W. (1998). Scanning confocal laser microscopy study of biofilm induced corrosion on copper plumbing tubes. *Biofouling*, **12**, 333–444.

Washizu N., Masuda H. (1997). AFM observation of iron-oxidizing bacteria on surfaces of corroded metals. *J Jpn Inst Met*, **61**, 481–485.

Yu F.P., McFeters G.A. (2000). Study of biofouling control with fluoresent probes and image analysis. In: Evans L.V. (ed) *Biofilms: Recent Advances in their Study and Control*. Harwood Academic Publishers, pp. 401–418.

Zee K-Y., Samaranayake L.P., Attstrom R. (1997). Scanning electron microscopy of microbial colonization of rapid and slow dental plaque formers *in vivo*. *Arch Oral Biol*, **42**, 735–742.

Zellner G., Diekmann H., Austermann-Haun U., Seyfried C.F. (1994). Scanning electron microscopy of biofilm development in anaerobic fixed-bed reactors: influence of the inoculum. *Biotech Lett*, **16**, 315–320.

Zhong Q., Innis D., Kjoller K., Elings V.B. (1993). Fractured polymer silica fiber surface studied by tapping mode atomic-force microscopy. *Surf Sci Lett*, **290**, L688–692.

5 Gene Expression of Cells Attached to Surfaces

Amanda E. Goodman and Gill G. Geesey

Gene expression in biofilm microbial populations can now be assessed at the single cell level. Studies of genetically engineered bacterial populations attached to surfaces have revealed a variety of genes that are up-expressed when cells exist in biofilms. Real-time studies have shown that gene expression within a cell may be transient during residence on a surface. Within isogenic, surface-associated populations, gene expression can be heterogeneous, possibly reflecting microscale variations in environmental conditions. Since a significant proportion, if not the majority of microbial life forms spend some portion of their existence in association with surfaces, it is likely that many genes and cell functions yet to be discovered will be detected in microbial biofilm populations. The genetic capacity of microbial life cannot be realized and the emerging field of genomics cannot achieve full potential until a better understanding is gained of gene expression in cells on surfaces.

KEY WORDS: biofilm, reporter genes, green fluorescent protein

INTRODUCTION

Microbial biofilms represent a complex assemblage of individual cells that are associated with surfaces. Unlike microbial cells freely dispersed in an aqueous phase, biofilm cells associated with surfaces develop spatial relationships to each other that permit interactions approaching those of multicellular organisms. Because biofilm cells are fixed in space for at least short periods of time, their behavior can be evaluated on an individual cell basis. This provides the opportunity to determine intra-population variations as well as inter-population interactions in mixed species biofilm communities.

Spatial relationships between biofilm cells have been observed microscopically for decades (Mack *et al.*, 1975; Kudo *et al.*, 1987). Post-treatment of fixed populations with specific fluorescent-labeled antibodies has been used to reveal the locations and associations between cells of different microbial species. Fluorescently-labeled-oligonucleotide probes that hybridize *in situ* with specific sequences in the ribosomal RNA molecule in intact cells have been used to identify specific microbial populations and as indicators of overall cell activity in biofilm populations (Poulsen *et al.*, 1993; Møller *et al.*, 1996). While these microscopic approaches have been used to characterize activity and spatial relationships between biofilm cells at a single point in time, there is growing interest in following the development of spatial relationships and cell-cell interactions in real time (Caldwell *et al.*, 1992).

REPORTER GENES

Recent investigations have shown that communication, *via* chemical signals can occur between bacteria in biofilms. Reporter genes were used in this regard. Here the expression of a gene of interest is coupled to a promoterless gene (reporter gene) whose product is readily detectable by microscopy or other analytical instrumentation. Genes that have been used as reporters include *cat*, *xylE*, and *galK*. The most popular reporter gene in bacteria is *lacZ*, which codes for the enzyme β-galactosidase (β-gal). Chromogenic and fluorogenic substrates (chromophore and fluorophore-galactose conjugates which are colorless or non-fluorescent until cleavage) are used to detect *lacZ* expression and hence expression of the gene of interest. When used in conjunction with a fluorogenic substrate, this reporter gene can be used to detect and localize gene expression at the cellular level by fluorescence microscopy and flow cytometry. Quantification of the amount of enzyme produced can be achieved using fluorimetry or colorimetry.

Use of *lacZ* as a Reporter of Biofilm Activities

There is growing interest in evaluating the activity and, in particular, the expression of specific genes in individual biofilm cells non-destructively in real time. This has been made possible in recent years through the use of a combination of bioreactors, molecular techniques, and microscopic imaging systems (Davies *et al.*, 1993). An indirect approach that has been used for decades is assessment of specific activities in populations of cells by following the expression of genes encoding the enzymes associated with biochemical pathways linked to the activities. A more direct approach is to use reporter genes to assess particular gene activity in individual bacterial cells in biofilms.

Extracellular polymer production

The *lacZ* reporter has been used to follow the expression of *algC*, a "house-keeping" gene also involved in alginate biosynthesis in *Pseudomonas aeruginosa*. Since members of the genus *Pseudomonas* are naturally *lacZ*$^-$, the level of expression of this reporter gene does not have to be corrected for background expression. In *P. aeruginosa* strain 8830, *lacZ* was put under the control of the *algC* promoter in the *alg*C-*lacZ* transcriptional fusion plasmid pNZ63 (Davies *et al.*, 1993). When a comparison was made between a mature biofilm population growing on a Teflon substratum and a suspended cell population, *algC* expression was nearly 20-fold higher and alginate levels were over 2-fold higher in the former compared to the latter population. Cells shed from the biofilm into the bulk aqueous medium displayed *algC* expression levels in between those of the biofilm and original suspended cell populations.

The *lacZ* reporter has also been used to follow expression of *algD*, the gene encoding GDP-mannose dehydrogenase, which catalyzes the conversion of GDP-mannose to GDP-mannuronic acid in the alginate biosynthesis pathway in biofilms of *P. aeruginosa* (Hoyle *et al.*, 1993). *P. aeruginosa* 579 was transformed with plasmids pSDF13 and pSDF14 containing *lacZ* under the control of the *algD* promoter and conferring gentamicin resistance. β-Gal activity in extracts of suspended cell populations was significantly less than that of 1- and 4-day biofilm populations containing equivalent numbers of viable cells. Cells of 7-day biofilms displayed reporter activity that was not significantly different from that of suspended cells, suggesting that alginate production drops with biofilm maturation. The presence of NaCl appeared to depress but not completely eliminate *algD* expression. Hoyle *et al.* (1993) suggested that the decrease in *algD* expression in biofilm populations after day 4 was consistent with a decrease in production of mucoexopolysaccharide, based on the establishment of a plateau after day 1 in accumulation of neutral hexose in the suspended population as assayed by the method of Dubois *et al.* (1956). Electron transport activity, based on reduction of 2-(4-iodophenyl)-3-(4-nitrophenyl)-5-phenyltetrazolium chloride (INT) to INT-formazan was found to follow *algD* expression more closely than neutral hexose accumulation. The results were interpreted as strong evidence for enhanced but transient production of mucoexopolysaccharide by *P. aeruginosa* 579 following attachment to a surface.

Expression of the *P. aeruginosa algC* gene was also monitored during the initial phase of biofilm development in individual bacteria which had detached from an upstream biofilm and subsequently reattached to a glass substratum in the presence of the fluorogenic substrate methylumbelliferyl-β-D-galactoside (MUG) in a flow-through, flat plate, channel reactor with a viewing window that permitted fluorescent and phase contrast microscopic observation of the attached cells (Davies and Geesey, 1995). Using this approach it was found that the expression of *algC* was temporally related to surface attachment and colonization of individual bacterial cells. Although *algC* was down-expressed in the majority of the cells (>93%) that had been attached to the substratum for less than 15 min, during the subsequent 165 min period, *algC* expression in the attached cells increased from 26 to 50% of the total attached population, with 89% of the total attached population up-expressed for *algC* at the end of this observation period. Many of the attached cells displayed transient expression of *algC* and became detached from the substratum over the observation period. Over 70% of the cells that detached became down-expressed for *algC* just prior to detachment. These results suggest a relationship between the expression of certain genes and the ability of cells to remain associated with a surface.

The use of the *lacZ* reporter system in this application yielded new insight into biofilm bacterial cell behavior. The results demonstrated 1) variation in gene expression among cells of an isogenic population, 2) up-expression of *algC* in cells shortly after they become associated with a substratum, 3) transient expression of *algC* in cells during attachment to a substratum, and 4) bacteria that were down-expressed for *algC* while associated with the substratum showed a higher propensity to detach from the substratum than did bacteria which were up-expressed for *algC*.

Cell-cell signaling

Fuqua and colleagues used a *lacZ* reporter gene system to show that bacteria in biofilms produce cell-cell signaling molecules. In one study (McLean *et al.*, 1997) a community of freshwater bacteria, including strains of *Pseudomonas putida* and *Pseudomonas fluorescens*, growing as biofilms on rocks in a river, were found to produce acylated homoserine lactone (HSL) type chemicals. Stickler *et al.* (1998) found that pure cultures of *P. aeroginosa* biofilms colonizing model catheters produced acylated homoserine lactones, and showed that these compounds were produced in about 50% of biofilm-colonized catheters recovered from hospital patients. The three dimensional (3D) biofilm structure formed by *P. aeruginosa* cells has been found to rely on chemical communication occurring between cells. Davies *et al.* (1998) compared the 3D biofilm structures formed by a *lasI* mutant strain of *P. aeruginosa* and the wild-type. *LasI* directs synthesis of the cell-signaling molecule *N*-(3-oxododecanoyl)-L-homoserine lactone. Cells unable to produce *lasI* formed flat, non-structured biofilms that were sensitive to the detergent sodium dodecyl sulphate and produced as much extracellular polymer as the wild type. When exogenous HSL was supplied to the *las1* mutant strain, a thick, differentiated biofilm developed, similar to that produced by the wild type, which consisted of mushroom- and pillar-like structures attached to the substratum between liquid-filled spaces (Davies *et al.*, 1998). Another HSL molecule produced by *P. aeruginosa*, *N*-buytryl-L-homoserine lactone, did not appear to have any effect on the structure of the biofilm produced on a glass substratum (Davies *et al.*, 1998).

It should be noted that other biofilm-forming bacteria develop biofilm architectures completely different from that produced by *P. aeruginosa*. The structures of bacterial biofilms often depend on the nature of the substratum (reviewed in Dalton *et al.*, 1994; 1996; Lawrence *et al.*, 1995; Stretton *et al.*, 1998) as well as the nutrient concentration and composition of the aqueous phase (Lawrence *et al.*, 1995).

Use of the lacZ reporter gene to evaluate conditions of biofilm environment

Reporter genes have been used to probe the condition of the environment at the microscale. A *lacZ* reporter gene was used to show that individual *E. coli* cells (previously inactivated for their natural β-gal production) incorporated into a drinking water biofilm, expressed the anaerobically-induced *nirB* promoter in microcolonies when examined after 13 d of biofilm growth (Robinson *et al.*, 1995). This demonstrated that cells within the microcolonies were experiencing anaerobic conditions, whereas planktonic cells, which did not express the *nirB* promoter, were not.

Limitations of lacZ as a reporter of bacterial activity

The success of the *lacZ* gene in reporting gene expression in individual cells of a biofilm population depends on accessibility of the fluorogenic substrate to the cells, the uptake of the substrate by the cells, and the retention of sufficient quantities of the fluorescent product by the cells to elicit a detectable fluorescent signal. Many types of bacteria do not take up the substrate or retain the product in sufficient amount to produce a fluorescent cell, thereby precluding use of the *lacZ* reporter gene in studies of gene expression in individual cells. Furthermore, these substrates are costly to use in flow-through systems such as described above. In such situations, other reporter genes offer advantages over *lacZ*.

The use of *lacZ* to report expression of other genes in a bacterial cell is usually restricted to those strains that normally lack a functional copy of this gene or have had the normal gene deleted. Otherwise, the β-gal assay will report the combined activity of the reporter gene as well as that of the normal gene that is present.

Use of *lux* Genes as a Reporter of Biofilm Activities

Biofilm activities have also been evaluated using a *lux* gene cassette composed of 5 genes, *luxCDABE*. The *lux*AB genes encode a heterodimeric luciferase enzyme. The enzyme requires no fluorophore/enzyme substrate, and oxidizes a tetradecanol to a tetradecanoic acid using oxygen and reduced flavin mononucleotide ($FMNH_2$), yielding light as a byproduct (Meighen, 1991). The light is detected by an extremely sensitive photon-counting camera, producing spatially resolved quantitative images of photon flux at the level of resolution necessary to assay *lux* expression within single bacterial cells (Palmer *et al.*, 1996). Since most bacteria cannot make sufficient tetradecanol for prolonged light production, it must either be added as a supplement to the medium or the genes necessary for *in vivo* production (*luxCDE*), included in the cassette.

The *lux* reporter cassette has been inserted by transposon mutagenesis into the plasmid PUTK21 carrying genes for naphthalene catabolism to report *nahG* (salicylate hydrolase) gene expression in cells of *P. fluorescens* strain 5RL growing as a biofilm in a cell adhesion measurement module (CAMM) (Mittelman *et al.*, 1992). Luciferase-mediated light production was induced upon exposure to sodium salicylate, collected with a flexible liquid light cable and collimated beam probe, and detected as a photoelectric-induced current using a photomultiplier-digital readout system. Upon addition of sodium salicylate, induction of *nah*G based on light production from the *lux* reporter was similar in cells attached to glass and stainless steel. Light production was positively correlated with total attached cell densities. Such a light-based approach was used to relate substratum colonization rate to surface shear force.

Light emission from cells of the marine bacterium *Vibrio harveyi,* which carries *lux* genes naturally on the chromosome, was used to evaluate the efficacy of marine antifouling coatings on bacterial surface colonization (Mittelman *et al.*, 1993). Since light flux correlated positively with the surface densities of both viable and total direct counts, *lux* gene expression offered a simple, non-destructive, real time measure of the extent of bacterial surface colonization. Two copper-based coatings, Navy F-121 and International Paints BRA-640, were colonized less readily than a 15% dinitrophenol coating. The expression of *lux* genes in *V. harveyi* was found to be a useful indicator of antifouling efficacy under dynamic-flow conditions.

While the lux reporter system avoids the need for a fluorogenic enzyme substrate, its dependence on oxygen and sensitivity to oxygen concentration limits it use to environments of high, constant oxygen concentrations. This feature of the *lux* reporter has been exploited to measure oxygen mass transfer between the bulk liquid and bacterial cells growing on the surface of a hollow fiber reactor (Sheintuch *et al.*, 1992).

Since the respiratory activities of bacteria in biofilms generate strong oxygen gradients within the biofilm (Abrahamson *et al.*, 1996) gene expression is difficult to interpret from photon flux using the *lux* reporter system. The luciferase enzyme is also sensitive to other factors such as ATP and metal concentrations, as well as the ability of a cell to produce or regenerate FMNH. Any variation in the availability of these factors can complicate enzyme activity interpretation (Jacobs *et al.*, 1991).

Use of the *gfp* Gene to Report Biofilm Activities

Green Fluorescent Protein (GFP) from the jellyfish *Aequorea victoria* fluoresces upon transfer of energy from the Ca^{2+}-activated photoprotein, aequorin. The energy transfer is thought to proceed *via* direct interaction between these two proteins. Aequorin is a "precharged" quasi-stable enzyme peroxide intermediate formed by reaction of the coelenterate luciferase and luciferin with oxygen (Hastings, 1996). Apo-aequorin is, thus, coelenterate luciferase, which binds the substrate coelenterazine (luciferin) and reacts with oxygen to form aequorin, which is then stored until its further reaction is triggered by calcium. GFP emits green light (l_{max} = 510 nm) when excited with ultraviolet or blue light (l_{max} = 395 nm with a minor peak at 470 nm). GFP fluorescence can be monitored non-invasively by fluorescence microscopy and flow cytometry.

While full-length GFP is required for fluorescence, the minimal chromophore needed for light absorption is located within a hexapeptide at amino acid position 64 through 69. This region of the protein contains a ser^{65}-$dehydrotyr^{66}$-gly^{67} trimer which cyclizes to yield the chromophore. Mutant GFP proteins have been reported in which the excitation maximum is shifted from 395 nm to around 490 nm, and this causes increased intensity of protein fluorescence by changing the ser^{65} to thr (Heim *et al.*, 1995), or by various changes to amino acids at positions 64, 65, 68 or 69 (Delagrave *et al.*, 1995). Mutations both within the chromophore and at distal positions in the protein yield functional GFP mutants with altered fluorescence spectra. Red- and blue-shifted GFP mutants are available as reporters as are the filters needed to separate their fluorescence (Delagrave *et al.*, 1995).

The wild type *gfp* gene did not appear useful in prokaryotes, as the intensity of GFP fluorescence was so weak that cell populations of about 10^5–10^6 ml^{-1} were necessary for detection. Falkow and colleagues mutagenised the cloned *gfp* gene in *E. coli* and selected mutant proteins, GFPmut1-3, that produced high levels of fluorescence such that single bacterial cells were easily visualised (Cormack *et al.*, 1996). These mutant GFP proteins remained soluble in the bacterial cell, had their excitation maxima shifted to 481–501 nm (with negligible emission occurring when excited at 395 nm), and yielded about 100 times greater fluorescence compared to the wild type GFP. Cormack *et al.* (1996) suggested that the maximal fluorescence produced by these mutant GFPs results from simultaneous double mutations at amino acid positions 65 and 72. The expression of *gfp* does not adversely affect bacterial survival (Valdivia *et al.*, 1996) and requires no cofactors or addition of exogenous substrates or other factors. It is ideally suited, therefore, for use as a reporter gene, and constructs for use in bacteria have been developed (for example Matthysse *et al.*, 1996; Stretton *et al.*, 1998).

gfp as a reporter of chitinase gene activity

Using a vector construct designed for use with marine bacteria, Stretton *et al.* (1998) placed a *gfp* reporter gene under the control of a chitinase encoding gene, *chiA*, in the marine bacterium *Pseudoalteromonas* sp. S9. S9 *chiA-gfp* cells were grown on squid pen (a natural marine biodegradable polymer consisting of about 60% protein and 40% chitin by weight, Gooday, 1990) and found to colonize patches of the surface in small microcolonies. After 7 d, surface colonization still appeared to be patchy although microcolony volume had increased substantially. Visualization by laser scanning confocal microscopy showed that the *chiA* gene was strongly expressed in individual bacterial cells within microcolonies (Stretton *et al.*, 1998).

More recently, Baty and Geesey (unpublished results) have followed colonization of starved cells of *Pseudoalteromonas* sp. S9 on a thin film of pure chitin cast on an optically smooth silicon substrate. In the absence of other carbon, nitrogen and energy sources, cells colonized the substratum in a random manner. Following initial colonization, it was determined that a portion of the attached population synthesized chitinase enzyme(s), which permitted the utilization of the solid chitin film for attached cell growth and replication. Under these conditions, a biofilm formed over a 200-h period that consisted predominantly of a monolayer of evenly distributed cells across the surface. That not all cells of the isogenic population attached to the chitin surface produced the chitinase enzyme was demonstrated by incorporation of a *gfp* reporter gene under the control of a promoter for the chitinase-encoding gene, *chiA*. Through a combination of reflected differential interference contrast and epifluorescence microscopies, total attached cells and chitinase-producing cells could be located in the same field of view of the chitin surface. Although total cells were randomly distributed across the surface, chitinase-producing cells were clearly aggregated. Thus, reporter genes are useful in evaluating metabolic heterogeneity among cells within isogenic populations.

gfp as a reporter of contaminant biodegradation

Møller *et al.* (1998) developed a system to simulate the biodegradation of toluene and other related aromatic compounds by microbial biofilms. *P. putida* strains were constructed in which each of the promoters of the two operons, as well as appropriate activator genes, involved in the toluene degradation pathway were fused independently to a *gfp*-reporter and inserted into the chromosome. The *Pu* and *Pm* promoters drive the operons encoding genes for the oxidation of toluene to catechol and the subsequent transformation of catechol to Krebs cycle intermediates, respectively. In pure culture biofilms, growing in once through flow chambers supplied with benzyl alcohol as the carbon source, it was found that the *Pu* promoter was homogeneously expressed in all cells, and that the *Pm* promoter was strongly expressed in only a sub-population ($<0.01\%$) of cells. Twenty four h after addition of benzoate (a known inducer of the *Pm* promoter) strong expression of the *Pm* promoter in all cells was observed. This showed that biofilm cells degrading benzyl alcohol (or toluene) did not accumulate benzoate.

In mixed culture biofilms, similar results were found for the *Pu* promoter, although it showed heterogeneous expression after 24 h, with *P. putida* cells immediately surrounding an *Acinetobacter* microcolony displaying the highest *Pu* activity. It was found that the *Acinetobacter* sp. produced a diffusible "inducing agent" which spread through the biofilm with time so that in a 3-day old biofilm expression of *Pm* in the *P. putida* cells was relativley strong and homogenous. It was suggested that either the "inducing agent" leaching from the *Acinetobacter* cells (which also contain genes for degradation of benzyl alcohol) was benzoate or that the *Acinetobacter* cells caused a change in the way the *P. putida* cells were degrading benzyl alcohol such that the benzoate accumulated in these cells (Møller *et al.*, 1998).

gfp as a reporter of the dynamics of gene expression

Further variants of GFP have now been constructed by the addition of a peptide sequence to the C-terminal end of the molecule that renders it susceptible to protease degradation (Andersen *et al.*, 1998). These protein variants have a half-life of from 40 min to a few hours when synthesized in *E. coli* and *P. putida*, making them useful reporters of temporal gene expression.

Limitations of gfp as a reporter under low oxygen conditions

Because GFP requires molecular oxygen for fluorescence, it has not been used in anaerobic bacteria. Recent studies suggest, however, that only small amounts of oxygen are required for the fluorescence reaction in the cell (Gorby, Weaver, Brown, Romine, Neal, unpublished results). GFP fluorescence has been detected in individual cells of the facultatively anaerobic, dissimilatory iron reducing bacterium *Shewanella putrefaciens* MR-1, carrying a plasmid with a constitutively-expressed *gfp* gene in the presence of less than 0.7 mg l^{-1} dissolved oxygen in the bulk aqueous medium. GFP fluorescence has also been detected in actively-growing, individual cells of the dissimilatory sulfate-reducing bacterium *Desulfovibrio desulfuricans*, carrying a plasmid with a constitutively-expressed *gfp* gene (Neal, Techkarnjanaruk, Mead, unpublished results). Thus, it appears that the *gfp* gene can be transcribed and the GFP product induced to fluoresce in bacteria cultured under conditions of low oxygen concentration.

SUMMARY

Gene expression of cells attached to surfaces has received increased attention in the past few years. Gene fusions involving *lux*, *lacZ* and *gfp* have facilitated the detection of cells and changes in cell metabolism following attachment to a variety of surfaces. There is growing evidence that cells on surfaces, and in particular, those within biofilms, display unique physiologies different from those displayed by cells in suspension. There appears to be considerable heterogeneity in gene expression among isogenic population of cells on surfaces. The significance of this phenomenon may relate to survival and fitness of the population. What remains to be determined is how gene expression in one population affects gene expression and cell fitness in surrounding microbial populations within biofilms.

ACKNOWLEDGMENTS

The authors wish to acknowledge NSF grants No. OCE-9720151 and EEC-8907039, and U.S. Department of Energy grant No. ER-62719 and subcontract GOOO668 from Washington State University. Part of this work was supported by the Australian Research Council and The Flinders University of South Australia.

REFERENCES

Abrahamson M., Lewandowski Z., Geesey G., Skjak-Braek G., Strand W., Christensen B.E. (1996). Development of an artificial biofilm to study the effects of a single microcolony on mass transport. *J Microbiol Methods*, **26**, 161–169.

Andersen J.B., Sternberg C., Poulsen L.K., Bjørn S.P., Givskov M., Molin S. (1998). New unstable variants of green fluorescent protein for studies of transient gene expression in bacteria. *Appl Environ Microbiol*, **64**, 2240–2246.

Caldwell D.E., Korber D.R., Lawrence J.R. (1992). Confocal laser microscopy and digital image analysis in microbial ecology. *Adv Microb Ecol*, **12**, 1–67.

Cormack B.P., Valdivia R.H., Falkow S. (1996). FACS-optimized mutants of the green fluorescent protein (GFP). *Gene*, **173**, 33–38.

Dalton H.M., Goodman A.E., Marshall K.C. (1996). Diversity in surface colonization behavior in marine bacteria. *J Ind Microbiol*, **17**, 228–234.

Dalton H.M., Poulsen L.K., Halasz P., Angles M.L., Goodman A.E., Marshall K.C. (1994). Substratum-induced morphological changes in a marine bacterium and their relevance to biofilm structure. *J Bacteriol*, **176**, 6900–6906.

Davies D.G., Geesey G.G. (1995). Regulation of the alginate biosynthetic gene *algC* in *Pseudomonas aeruginosa* during biofilm development in continuous culture. *Appl Environ Microbiol*, **61**, 860–867.

Davies D.G., Chakrabarty A.M., Geesey G.G. (1993). Exopolysaccharide production in biofilms, substratum activation of alginate gene expression by *Pseudomonas aeruginosa*. *Appl Environ Microbiol*, **59**, 1181–1186.

Davies D.G., Parsek M.R., Pearson J.P., Iglewski B.H., Costerton J.W., Greenberg E.P. (1998). The involvement of cell-to-cell signals in the development of a bacterial biofilm. *Science*, **280**, 295–298.

Delagrave S., Hawtin R.E., Silva C.M., Yang M.M., Youvan D.C. (1995). Red-shifted excitation mutants of the green fluorescent protein. *Bio/Technology*, **13**, 151–154.

Dubois M., Gilles K.A., Hamilton J.K., Rebers P.A., Smith F. (1956). Colorimetric method for determination of sugars and related substances. *Anal Chem*, **28**, 350–356.

Gooday G.W. (1990). The ecology of chitin degradation. *Adv Microb Ecol*, **11**, 387–430.

Hastings W. (1996). Chemistries and colors of bioluminescent reactions, a review. *Gene*, **173**, 5–11.

Heim R., Cubitt A.B., Tsien R.Y. (1995). Improved green fluorescence. *Nature (Lond)*, **373**, 663–664.

Hoyle B.D., Williams L.J., Costerton J.W. (1993). Production of mucoid exopolysaccharide during development of *Pseudomonas aeruginosa* biofilms. *Infect Immun*, **61**, 777–780.

Jacobs M., Hill P.J., Stewart G.S.A.B. (1991). Highly bioluminescent *Bacillus subtilis* obtained through high-level expression of a *lux*AB fusion gene. *Mol Gen Genet*, **230**, 251–256.

Kudo H., Cheng K.-J., Costerton J.W. (1987). Interactions between *Treponema bryantii* and cellulolytic bacteria in the *in vivo* degradation of straw cellulose. *Can J Microbiol*, **33**, 244–248.

Lawrence J.R., Korber D.R., Wolfaardt G.M., Caldwell D.E. (1995). Behavioral strategies of surface-colonizing bacteria. *Adv Microb Ecol*, **15**, 1–75.

Mack W.N., Mack J.P., Ackerman A.O. (1975). Microbial film development in a trickling filter. *Microb Ecol*, **2**, 215–226.

Matthysse A.G., Stretton S., Dandie C., McClure N.C., Goodman A.E. (1996). Construction of GFP vectors for use in Gram-negative bacteria other than *Escherichia coli*. *FEMS Microbiol Lett*, **145**, 87–94.

McLean R.J.C., Whitely M., Stickler D.J., Fuqua W.C. (1997). Evidence of autoinducer activity in naturally occurring biofilms. *FEMS Microbiol Lett*, **154**, 259–263.

Meighen E.A. (1991). Bacterial bioluminescence, organization, regulation and application of the *lux* genes. *FASEB J*, **7**, 1016–1022.

Mittelman M.W., King J.M.H, Sayler G.S., White D.C. (1992). On-line detection of bacterial adhesion in a shear gradient with bioluminescence by *Pseudomonas fluorescens* (*lux*) strain. *J Microbiol Methods*, **15**, 53–60.

Mittelman M.W., Packard J., Arrage A.A., Bean S.L., Angell P., White D.C. (1993). Test system for determining antifouling coating efficacy using on-line detection of bioluminescence and fluorescence in a laminar-flow environment. *J Microbiol Methods*, **18**, 51–60.

Møller S., Pedersen A.R., Poulsen L.K., Arvin E., Molin S. (1996). Activity and three-dimensional distribution of toluene-degrading *Pseudomonas putida* in a multispecies biofilm assess by quantitative *in situ* hybridization and scanning confocal laser microscopy. *Appl Environ Microbiol*, **62**, 4632–4640.

Møller S., Sternberg C., Anderson J.B., Christensen B.B., Ramos J.L., Givskov M., Molin S. (1998). *In situ* gene expression in mixed-culture biofilms, evidence of metabolic interactions between community members. *Appl Environ Microbiol*, **64**, 721–732.

Palmer R. Jr., Phiefer C., Burlage R., Sayler G., White D. (1996) Single-cell bioluminescence and GFP in biofilm research. In: Hastings J.W., Kricka L.J., Stanley P.E. (eds) *Bioluminescence and Chemiluminescence, Molecular Reporting with Photons*. John Wiley and Sons, Chichester, UK, pp. 445–450.

Poulsen L.K., Ballard G., Stahl D.A. (1993). Use of rRNA fluorescence *in situ* hybridization for measuring the activity of single cells in young and established biofilms. *Appl Environ Microbiol*, **59**, 1354–1360.

Robinson P.J., Walker J.T., Keevil C.W., Cole J. (1995). Reporter genes and fluorescent probes for studying the colonisation of biofilms in drinking water supply line by enteric bacteria. *FEMS Microbiol Lett*, **129**, 183–188.

Sheintuch M., Vashitz O., Wolffberg A. (1992). Engineering applications of bioluminescence: modeling of mass transport in a hollow fiber and in a chemostat. *Chem Eng Sci*, **47**, 2615–2620.

Stickler D.J., Morris N.S., McLean R.J.C., Fuqua C. (1998). Biofilms on indwelling urethral catheters produce quorum-sensing signal molecules *in situ* and *in vitro*. *Appl Environ Microbiol*, **64**, 3486–3490.

Stretton S., Techkarnjanaruk S., McLennan A.M., Goodman A.E. (1998). Use of green fluorescent protein to tag and investigate gene expression in marine bacteria. *Appl Environ Microbiol*, **64**, 2554–2559.

Valdivia R.H., Hromocky A.E., Monack D., Ramakrishnan L., Falkow S. (1996). Applications for green fluorescent protein (GFP) in the study of host-pathogen interactions. *Gene*, **173**, 47–52.

6 Plasmid Transfer between Bacteria in Biofilms

Mark L. Angles and Amanda E. Goodman

Biofilms are environments of high microbial cell density where cell-cell contact is likely. Such conditions create a favourable niche for the spread of self-transmissible as well as mobilisable plasmids among members of the bacterial communities. Studies have demonstrated plasmid transfer among bacteria in a wide range of biofilm habitats, including the surfaces of stones in a river, the air-water interface, surfaces in soil and water microcosms, plant surfaces and insect as well as animal intestinal surfaces.

KEY WORDS: conjugative plasmids, biofilms, plasmid transfer

WHAT ARE PLASMIDS AND HOW DO THEY SPREAD THROUGH BACTERIAL POPULATIONS?

Plasmids are circular pieces of DNA found almost ubiquitously in bacteria. Plasmids often carry genes which are favourable for the survival of bacteria in adverse environments, where selection pressure would increase the likelihood of their transfer, for example catabolic genes and antibiotic and heavy metal resistance genes (Stotzky and Babich, 1986; Trevors *et al.*, 1987; Sayler *et al.*, 1990). Conjugation is the direct transfer of genetic information, in the form of plasmids, between bacterial cells and is dependent on cell-to-cell contact. Self-transmissible (conjugative) plasmids can be readily isolated from bacteria in most environments (Sizemore and Colwell, 1977; Stotzky and Babich, 1986; Hermansson *et al.*, 1987; Trevors *et al.*, 1987; Fredrickson *et al.*, 1988; Genthner *et al.*, 1988; Fry and Day, 1990; Sayler *et al.*, 1990; Dahlberg *et al.*, 1998a; 1998b). Conjugal transfer of plasmids occurs between closely related bacterial strains as well as between diverse genera (for reviews see Stotzky and Babich, 1986; Trevors *et al.*, 1987; Trevors and Oddie, 1986; Ippen-Ihler, 1989; Mazodier and Davies, 1991; Veal *et al.*, 1992). Cells into which plasmids have self-transferred (*i.e.* conjugated) are called transconjugants. There is evidence of conjugation occurring among bacteria in such diverse environments as the gastrointestinal tract of warm blooded animals, the human urinary and respiratory tracts, wounds, on plant surfaces, and in soil, water and sewage (Stotzky and Babich, 1986). As a result, plasmid transfer may contribute greatly to gene transfer in natural environments and to the adaptability of microorganisms to environmental stress.

Two major factors affecting conjugal plasmid transfer are the transfer efficiency and the host range of the plasmid (Veal *et al.*, 1992). In addition, plasmid transfer is dependent on donor and recipient cells becoming established in sufficient numbers and for an appropriate length of time in the environment. Other barriers to successful conjugation include plasmid incompatibility, the recipient's inability to replicate the introduced DNA, and degradation of the introduced plasmid by recipient restriction systems (Ippen-Ihler, 1989; Wilkins, 1990).

A feature common to all conjugative or self-transmissible plasmids is possession of *tra* genes which encode for plasmid self-transfer. One of the most studied conjugation systems is that of the F plasmid in *E. coli*, which carries 20 genes responsible for plasmid transfer (Ippen-Ihler and Minkley, 1986) on the one *tra* operon (Ippen-Ihler, 1989), comprising approximately 30 kb of the 100 kb plasmid. Other plasmid groups have the *tra* genes separated into discrete loci. Regardless of the organisation of the *tra* genes, self-transmissible plasmids are not less than 30 kb in size. Conjugative plasmids may also express surface exclusion proteins which block donor-to-donor transfer of the same plasmid, effectively stopping unproductive transfer between donors as well as limiting plasmid transfer to recipient cells to a single event.

The mechanism of conjugation is as follows: the pilus initiates contact with the recipient cell and forms a stable mating pair; the plasmid is nicked at the origin of transfer site, *oriT*, and a single strand of DNA is displaced in the 5' to 3' direction into the recipient cell; synthesis of a replacement strand in the donor and complementary strand in the recipient occurs. Following transfer the DNA is circularised in the recipient, while the donor maintains a copy of the plasmid.

Plasmid transfer must start at the *oriT* site but there is no limit to the length of DNA that can be transferred and transfer proceeds either until the mating pairs are physically separated, until there is a break in the DNA or the 3' end of the *oriT* site is reached. Transfer of the plasmid is usually considered to be unidirectional *i.e.* from host to recipient, with the *tra* genes usually transferred last (Ippen-Ihler, 1989). Other studies have shown, however, that retrotransfer of mobilisable plasmids (Perkins *et al.*, 1994) and chromosomal DNA (Mergeay *et al.*, 1987) can occur in the opposite direction *i.e.* from recipient to donor. The ability of some natural plasmids to retrotransfer other genetic elements indicates that this phenomenon may be more common than originally thought (Veal *et al.*, 1992). In addition, the F plasmid, and certain other plasmid systems, may recombine with the chromosome of the host by either expression of recombination (*rec*) genes or by site-specific recombination (Ippen-Ihler, 1989) which can result in part or all of the host chromosome being transferred with the plasmid (Veal *et al.*, 1992).

Certain plasmids which lack *tra* genes can be mobilised by conjugative plasmids resident in the same cell, or even separate cells. Mobilisable plasmids contain an *oriT* site which is recognised by a small number of proteins encoded by the mobilisation (*mob*) genes. The rest of the machinery necessary for transfer is supplied by a "helper" conjugative plasmid. Mobilisation ability is common, particularly in small plasmids which do not carry the *tra* genes (Veal *et al.*, 1992). Mobilisation is important to gene transfer in natural environments, as recipients in which the conjugative plasmid cannot become established may still acquire mobilisable plasmids (Ippen-Ihler, 1989).

Contact between donor and recipient cells is mediated *via* sex pili, produced on the surface of the donor cell, which interact with specific binding sites on the recipient cell surface. Pili can be divided into three functional groups; *viz.* thin flexible, thick flexible and rigid (Bradley, 1980; 1984). Conjugation studies have shown that plasmids encoding rigid pili (such as the IncP type) transfer more successfully at surfaces than in the aqueous phase (Bradley *et al.*, 1980; Bradley, 1984; Genthner *et al.*, 1988). It is still not clear whether pili act as a conduit for plasmid transfer or whether they are part of a mechanism by which donor and recipient cells achieve surface-to-surface contact. The very nature of the pilus, a hollow tube large enough for single stranded DNA to pass through, would support the premise that plasmids are translocated *via* pili. Indeed, one study has shown that plasmid transfer is possible without wall-to-wall contact between donors and recipients (Harrington and Rogerson, 1990), although transfer occurred at low frequencies such that this type of plasmid transfer may be an exceptional occurrence. Another model for plasmid transfer is that a pilus is necessary for initial cell contact after which pilus degradation (or retraction) brings donor and recipient cells closer together until membrane-to-membrane contact is achieved (Dreiseikelmann, 1994).

Abiotic factors, such as temperature, pH and nutrient availability, can affect plasmid transfer. Temperature effects on plasmid transfer have been studied extensively (Altherr and Kasweck, 1982; Singleton and Anson, 1983; Gauthier *et al.*, 1985; Bale *et al.*, 1988b; Richaume *et al.*, 1989; Rochelle *et al.*, 1989b; Sandt and Herson, 1991; Fernandez-Astorga *et al.*, 1992). Various naturally isolated plasmids have differing optimal transfer temperatures ranging from 10–20°C to 20–25°C for mercury resistant plasmids isolated from river epilithon (Fry and Day, 1990) to 30°C for plasmids isolated from marine strains (Gauthier *et al.*, 1985). The transfer frequency of a naturally occurring plasmid, encoding mercury resistance, was shown to increase with increasing nutrient concentrations (Fry and Day, 1990). Increased transfer of naturally occurring plasmids between *E. coli* strains was also shown on nutrient rich media, with no increase in plasmid transfer frequencies being detected in the absence of nutrients (Fernandez-Astorga *et al.*, 1992). The results of another study indicated that the pH optimum of plasmid transfer is dependent on the plasmid type, the effects of pH on donor or recipient cells or combinations of these factors (Singleton and Anson, 1983).

Plasmid Maintenance

The maintenance of conjugative plasmids in both donor and recipient populations is an important component of genetic exchange and the persistence of genetic elements in natural environments. Typically, natural bacterial plasmids are inherited with a high degree of stability (Nordström and Austin, 1989). Maintenance of plasmids in bacterial populations is dependent on plasmid replication and partitioning to daughter cells.

Plasmids reside in cells in either high (>5 plasmids per chromosome) or low copy number and this is dependent on the plasmid type. For plasmids which reside in the cell in a high copy number, Sherratt (1982) proposed a strategy based on random segregation where at least one plasmid will segregate to the majority of daughter cells. In the case of low copy number plasmids, replication and partitioning to daughter cells must be an active process if the plasmid is to be maintained in the population.

The replication and average copy number of plasmids within a host is strictly controlled by the plasmid through a negative feedback loop involving an origin of replication site, as well as *cop*, *inc* (involved in the initiation of replication) and *rep* (required for stable replication) genes (Couturier *et al.*, 1988). A decrease in plasmid copy number may cause inefficient partitioning to daughter cells resulting in plasmid loss from the population (Jones *et al.*, 1980).

Apart from plasmid copy number control, a combination of different systems is involved in the maintenance of plasmids in a population. Helper elements aid in the random distribution of plasmids whereas active elements work to ensure even plasmid distribution (Nordström and Austin, 1989). The two helper element systems, site-specific recombination and anticlumping, overcome the problems associated with the formation of plasmid multimers and noncovalent aggregates, particularly in low-copy number plasmids, which make partitioning to daughter cells impossible. It is probable that all natural plasmids make use of some type of site-specific recombination system (Nordström and Austin, 1989) which resolves multimeric plasmid forms into monomeric ones thus enabling proper partitioning.

Active element systems include true partition and killer systems. In true partition systems, plasmids are selectively moved to ensure that each daughter cell contains a copy of the plasmid and this system may act in much the same way as the segregation of chromosomes during mitosis in higher cells (Nordström and Austin, 1989). Genes responsible for the active partitioning of plasmids have been identified. For example, in the F plasmid this function is controlled by *sop* (stability of plasmid) genes (Mori *et al.*, 1986), and in plasmid pSC101 and IncFII plasmids control is determined by the *par* (partition) genes (Meacock and Cohen, 1980; Tucker *et al.*, 1984). Given the selective advantage involved, it is likely that all naturally occurring plasmids have an active partitioning system (Nordström and Austin, 1989).

Killer systems can affect plasmid maintenance by killing cells that have lost the plasmid thereby decreasing the level of plasmid free cells in a population (Nordström and Austin, 1989). Both *kor* (kill override) and *kil* (kill) loci in IncP1 plasmids (*ccd* genes in the F plasmid and *hok/sok* in IncFII plasmids) are expressed such that when the plasmid is lost the product of *kor* decays but the *kil* product persists in the cytoplasm thus killing the cell (Nordström and Austin, 1989). Similar systems have been found in other incompatibility groups (Miki *et al.*, 1984; Winans and Walker, 1985).

Factors such as growth under nutrient limited conditions and the difference in growth rates between host and plasmid-free cells can also affect plasmid maintenance in a population. Stable maintenance of RP1 (Melling *et al.*, 1977; Jones *et al.*, 1980) and ColE1-type plasmids (Jones *et al.*, 1980) in *E. coli* cells grown under nutrient limited conditions has been shown. Indeed, a study conducted in soil showed that numbers of plasmid-containing *E. coli* cells declined less rapidly than plasmidless cells (Devanas *et al.*, 1986). The results indicated that plasmid loss was primarily a function of the bacterial strain and the nutrient status of the soil rather than the plasmid type. Conversely, other studies have shown a decrease in plasmid maintenance under nutrient limitation; for example, pBR322 (Melling *et al.*, 1977; Jones *et al.*, 1980; Wouters *et al.*, 1980; Jones and Melling, 1984), TP120 (Godwin and Slater, 1979) and RSF2124 (Helling *et al.*, 1981) in *E. coli.* It is probable that plasmid loss under nutrient limited conditions is due to the high energy cost imposed on host cells from maintenance and replication of plasmids (Helling *et al.*, 1981; Sherratt, 1982).

Under laboratory conditions, evidence for the energy cost to host cells is the apparent decrease in their growth rates, resulting in the plasmidless cells being more competitive than the host cells. For example, loss of pBR322 from *E. coli* resulted in a plasmidless population that was more competitive than the host population under severe nutrient limitation (Wouters *et al.*, 1980); the presence of a small non-conjugative plasmid, RSF2124, decreased the growth rates of the host *E. coli* such that the proportion of plasmid containing cells declined (Helling *et al.*, 1981); and the formation of smaller plasmids, by deletion of drug resistance genes, resulted in increased growth rates in a host *E. coli* strain (Godwin and Slater, 1979). A study of the effects of plasmid size on growth rates in *E. coli* found that the majority of plasmids greater than 80 kb decreased growth rate, whereas smaller plasmids (<80 kb) did not (Zünd and Lebek, 1980).

The majority of studies on plasmid maintenance have involved using *E. coli* hosts grown in chemostat culture (Godwin and Slater, 1979; Jones *et al.*, 1980; Wouters *et al.*, 1980; Helling *et al.*, 1981; Jones and Melling, 1984; Caulcott *et al.*, 1987) and in immobilised cell cultures (Inloes *et al.*, 1983; de Taxis du Poët *et al.*, 1986; 1987; Nasri *et al.*, 1987; Sayadi *et al.*, 1987; 1989). These studies were aimed at gaining a better understanding of the ability of cells to maintain plasmids so as to optimise expression of recombinant genes carried by the plasmids for fermentation processes. While these investigations have provided important information on plasmid maintenance under laboratory conditions, they have provided little information on how plasmids might be maintained in natural environments or in indigenous bacteria. In terrestrial environments, plasmid maintenance has been studied in soil in *Pseudomonas fluorescens* (Van Elsas *et al.*, 1989) and in *E. coli* (Devanas *et al.*, 1986; Devanas and Stotzky, 1988) as well as in indigenous bacteria and *Pseudomonas* spp. in groundwater aquifer material (Jain *et al.*, 1987). In aqueous environments, plasmid maintenance has been studied in *Pseudomonas putida* strains in microcosms containing lake water (Sobecky *et al.*, 1992), in *Pseudomonas aeruginosa in situ* in tropical freshwater (Cruz-Cruz *et al.*, 1988), *in vitro* in enteric bacteria and *Pseudomonas* spp. in well water (Caldwell *et al.*, 1989; Griffiths *et al.*, 1990), in *E. coli* in river water (Flint, 1987) and in *E. coli* in artificial seawater (Byrd and Colwell, 1990; 1993). There have been relatively few studies, however, on the maintenance of indigenous plasmids in natural isolates (Jain *et al.*, 1987). In particular, little is known about plasmid maintenance in marine bacteria. Moreover, for aqueous environments generally there is a paucity of knowledge about how conjugal plasmids are maintained in complex microbial communities, such as mixed-species biofilms, where interactions may occur between different members of a community.

PLASMID TRANSFER IN BIOFILMS

Biofilms are ubiquitous in the environment and form whenever a solid surface is in contact with an aqueous phase. Biofilms consist of a high density of cells, within polymer matrices, lying in close proximity to one another and hence would allow stable cell-to-cell contact and plasmid transfer. The use of oxygen microelectrodes and scanning confocal laser microscopy, which can optically section biofilms at 0.2 μm vertical intervals without experiencing out-of-focus haze (Caldwell *et al.*, 1992), has revealed that biofilms are not necessarily continuous or planar in structure

(Lawrence *et al.*, 1991; Caldwell *et al.*, 1992; de Beer *et al.*, 1994). Rather, biofilms are complex arrangements of cell clusters, interstitial voids and channels such that diffusion from the aqueous phase through the biofilm, both vertically and horizontally, is facilitated (Caldwell *et al.*, 1992; Costerton *et al.*, 1994; de Beer *et al.*, 1994; Lewandowski, 2000). As biofilms represent the greatest activity and cell density in natural environments (Geesey *et al.*, 1978; Costerton *et al.*, 1995), it is highly likely that the greatest potential for interaction between microorganisms would also occur in biofilms. This may be particularly true with respect to gene transfer between bacteria in aqueous environments.

Aqueous Environments

Pioneering work on plasmid transfer in freshwater biofilms was conducted by the group of Fry and Day who demonstrated plasmid transfer in freshwater biofilms *in situ* (Bale *et al.*, 1987; 1988a; 1988b) and in laboratory microcosms (Rochelle *et al.*, 1989a; Hill *et al.*, 1994), with the latter allowing a level of control that is not possible with *in situ* studies (Barkay *et al.*, 1995). The group primarily investigated the transfer of natural mercury resistance plasmids *in situ* in biofilms attached to stones placed in a river.

These studies were significant in that they were performed in the absence of containment structures and they are the first direct demonstrations of controlled experiments *in situ.* The mercury-resistance plasmid, pQM1, has a narrow-host-range for some *Pseudomonas* spp. and was isolated from the indigenous epilithon (biofilm) population. In a preliminary set of experiments, transfer of pQM1 between *P. aeruginosa* strains was studied by placing recipients and donors together on filter membranes which were placed face down on the surface of either scrubbed, sterile river stones or unscrubbed stones (Bale *et al.*, 1987). Stones were either incubated in microcosms or placed in a river. Plasmid transfer frequencies were comparable in the two systems, *i.e.* microcosms or *in situ*. Transconjugants were detected in the aqueous phase of the microcosms but the results implied that most of the plasmid transfer had occurred on the stone or filter surfaces. Plasmid transfer was enhanced by heating unscrubbed stones and the authors suggested that this was a result of the release of nutrients from the heated epilithon. Both microcosm and *in situ* experiments revealed that plasmid transfer occurred at 6°C to 20°C but transfer frequencies were lower at the colder temperatures or when the natural epilithon community was present. Similarly, another set of experiments conducted on river stones *in situ* showed that a 2.6°C change in temperature yielded a 10-fold change in transfer frequency (Bale *et al.*, 1988b). *In situ* transfer of natural mercury resistance plasmids from the epilithon population to a known recipient was also shown (Bale *et al.*, 1988b).

Results that appeared to conflict with those in the first two studies were reported (Bale *et al.*, 1988a). In this study, donors and recipients were incorporated into the epilithon of separate river stones which were then placed together and immersed in a river. Transfer of pQM1 was detected on the stone surface in the presence of the indigenous population. In contrast to the previous experiments (Bale *et al.*, 1987; 1988b), no effect on plasmid transfer by either temperature or the presence of the epilithon community was evident. Rather, the donor to recipient ratio affected plasmid transfer, with a high ratio (490:1) giving the greatest plasmid transfer frequency.

A departure from *in situ* work was performed using river epilithon microcosms, by the same group (Rochelle *et al.*, 1989a; Hill *et al.*, 1994). In one set of experiments, a rotating slate disc microcosm was used to investigate the population dynamics and genetic interactions within a river epilithon (Rochelle *et al.*, 1989a). It was found that although donors and recipients were inoculated on opposite sides of the slate discs, transconjugants were detected quickly, indicating that the bacteria moved freely within the microcosms. In addition, the poor survival of strains and low numbers of transconjugants detected in the aqueous phase indicated that most of the plasmid transfer had occurred in the epilithon. Transfer frequencies in the microcosms were approximately 100 times lower than those obtained by plate mating. These results are similar to those found by other researchers who found a decrease in plasmid transfer frequencies among cells conjugated in broth and on plates compared to those existing under more natural environmental conditions (Bale *et al.*, 1987; O'Morchoe *et al.*, 1988).

Microcosms containing river stones were also used to show mobilisation of a catabolic plasmid, pD10, by plasmids isolated from the river epilithon (Hill *et al.*, 1994). As with the rotating disc microcosms (Rochelle *et al.*, 1989a), donors and recipients survived better in the epilithon community than in the aqueous phase over a three week period. Indeed, few transconjugants were isolated from the aqueous phase. Similar to the rotating disc microcosm, inoculation of donors and recipients in different parts of the microcosm revealed that donors and recipients were able to colonise other stones from which transconjugants could be isolated. More recently, transfer of the plasmid, pQKH6, was detected in biofilms of a percolating-filter sewage treatment system in the presence of invertebrate grazers and the natural indigenous microbial population (Ashelford *et al.*, 1995) at frequencies similar to those found in the river epilithon (Bale *et al.*, 1988b) and rotating disc microcosms (Rochelle *et al.*, 1989a).

In a study using a freshwater microcosm system, an increase in the incidence of transfer of the plasmid R68.45 between *P. aeruginosa* strains at the air/water interface was shown, with transconjugants being detected 100 times more frequently at the air/water interface than in the bulk liquid phase in laboratory microcosms (Jones *et al.*, 1991). This reflected the enrichment of the donor at this site and highlighted the need to take strain localisation into account when reporting transfer frequencies. No plasmid transfer to aquatic isolates was detected.

Christensen *et al.* (1998) investigated the establishment of a conjugative plasmid (pWWO, encoding toluene-degrading genes) in a mixed-species biofilm community actively degrading benzyl alcohol. In this flow chamber biofilm microcosm, the efficiency of pWWO transfer was low and establishment of the plasmid occurred by multiplication of transconjugants within the biofilm. Tagging the bacterial strains, as well as pWWO, with appropriate fluorescent markers enabled laser scanning confocal microscopy to be carried out to determine the structure of the mature biofilm. This showed that transconjugants did not establish new, or pure, microcolonies of their own, but rather were always associated with non-plasmid bearing recipient cells. The authors proposed that transconjugant multiplication took place on or within preexisting recipient microcolonies (Christensen *et al.*, 1998).

Apart from a limited number of studies, there is little information about plasmid transfer or the maintenance of plasmids in the marine environment (Goodman *et al.*,

1993; Dahlberg *et al.*, 1998a; 1998b; Sobecky *et al.*, 1998) and even less about plasmid transfer or maintenance in marine biofilms (Angles *et al.*, 1993; Angles, 1997). In general, the effects on plasmid transfer of differences in localisation of donor and recipient cells within mixed-species biofilms is still poorly understood. Further, it has been shown that the nature of the surface may be important in determining the structures of biofilms (Dalton *et al.*, 1994) and this in turn may affect plasmid transfer in this ecosystem. Whether this is the case generally is not known. Most of the studies of plasmid transfer in biofilms involve the addition of donor and recipient cells simultaneously. This situation may not be typical of natural environments where either donors or recipients may already be established within a biofilm and this may also have an effect on conjugation. In addition, the maintenance of plasmids within populations will affect both the occurrence of plasmid transfer as well as the extent to which genetic material will persist in the environment. The study of plasmid transfer and maintenance in mixed-species biofilms will lead to a better understanding of the mechanisms involved in the establishment of genes in biofilm populations.

Marine microcosms, used to study the effects of surface type on plasmid transfer in the aqueous phase and in mixed-species biofilms consisting of two species of marine bacteria (*Vibrio* and *Psychrobacter*), showed that plasmid transfer was significantly greater among cells in the biofilm than among cells in the aqueous phase. This was under non-flow conditions and in the absence of selection pressure for the plasmid. The hydrophobicity of the surface had no affect on plasmid transfer efficiency (Angles *et al.*, 1993). Further, marine microcosms were used under flow conditions, with no selection pressure for the plasmid, to study the effects of changing the order and length of time of strain colonisation on plasmid transfer and mixed-species biofilm stability (Angles, 1997). It was found that the order (*i.e.* whether donor or recipient colonised the substratum first, or whether colonisation was simultaneous) as well as the length of time of colonisation significantly affected plasmid transfer efficiency. This was highest when the donor colonised a pre-existing recipient biofilm which had been growing for 48 h, and was lowest under the reverse condition, *i.e.* when the recipient colonised a 48 h old pre-existing donor biofilm (Angles, 1997). Plasmid transfer was found to be correlated with plasmid maintenance in that plasmid maintenance in the donor was highest when the donor colonised a 48 h old pre-existing recipient biofilm (Angles, 1997). In the same study, a comparison of chitin (squid pen) and hydrophobic glass surfaces showed that plasmid transfer frequencies between the two bacterial species were comparable (Angles, 1997). The results indicated that the use of inert solid surfaces in the laboratory may be a suitable model for investigating plasmid transfer among biofilm-forming bacteria on natural biodegradable surfaces.

Sludge, Soil and Sediment Environments

These environments provide large surface areas on particulate matter for bacterial colonisation and biofilm development. Researchers have shown that plasmids isolated from wastewater are able to mobilise non-conjugative plasmids (McPherson and Gealt, 1986; Mancini *et al.*, 1987; McClure *et al.*, 1990). For example, a high number of plasmid containing bacteria from wastewater were shown to mobilise the non-conjugative plasmid, pBR325, in complex media and sterilised wastewater (McPherson and Gealt, 1986). Mobilisation of the plasmid pHSV106 from a laboratory *E. coli* strain by both laboratory and indigenous wastewater *E. coli* strains in a laboratory scale wastewater facility has also been shown (Mancini *et al.*, 1987). In this study, most of the transconjugants were isolated from sludge where flow was minimal; plasmid transfer was rare in areas where flow was more turbulent and appeared to be dependent on donor cell numbers as transconjugants were only detected when donor numbers were at least 10^7 cfu ml^{-1}.

The above studies were conducted in the absence of indigenous bacteria and protozoa, which may affect the survival of donor and recipient populations (McClure *et al.*, 1990). McClure *et al.* (1990) used a laboratory scale activated sludge unit to show acquisition of mobilising plasmids from the indigenous population by a *Pseudomonas putida* strain, harbouring the non-conjugative, mobilisable plasmid, pD10. Further, the activated sludge contained bacteria which could act as recipients of pD10 in plate matings. The fact that transfer was detected in the presence of a protozoan population further indicates that plasmid transfer is highly likely to occur in wastewater. Indeed, the presence of protozoa may increase the incidence of plasmid transfer by causing a localised increase in bacterial cell density through filter feeding and hence causing greater cell-cell contact for plasmid transfer (Otto *et al.*, 1997).

There has been considerable research into conjugal plasmid transfer in soils. It has been suggested that the probability of microbial movement within soils and hence, the occurrence of gene transfer, is lower than in ecosystems where the water phase is continuous. Exceptions are under conditions where the soil is saturated or in sediments (Stotzky, 1989). Other factors that may affect plasmid transfer in soil include the soil type and the presence of clays, the availability of nutrients, temperature, donor and recipient cell density, the persistence of donor cells, and the presence of the indigenous bacterial population (Klingmüller *et al.*, 1990). The same conditions are probably relevant to plasmid transfer within sediments, although the oxygen concentrations would be lower.

The majority of investigations of plasmid transfer in soil have involved the use of soil microcosms, often containing sterile soil. One study showed that the addition of either clay (15%) or organic matter (5%), the adjustment of the soil pH to 7.2, a soil moisture content of 8% and incubation temperature of 28°C were necessary for optimum transfer of a plasmid, pBLK-2, between *E. coli* and *Rhizobium fredii* (Richaume *et al.*, 1989). Similarly, Trevors and Oddie (1986) were only able to detect plasmid transfer between *E. coli* strains in sterile soil which had been supplemented with complex media; plasmid transfer frequencies were increased further in soil with a 100% water holding capacity. Rather than using *E. coli* to study plasmid transfer in soil, Raffii and Crawford (1988) were the first to show plasmid transfer between soil bacteria. Transfer of the conjugative plasmid, pIJ303, and mobilisation of the non-conjugative plasmid, pIJ702, by pIJ303, was shown to occur between *Streptomyces* strains in sterile soil, but at transfer frequencies lower than obtained on a complex medium. This may reflect the fact that *Streptomyces* strains grow sporadically in natural environments, being present as spores when nutrients are scarce. Plasmid transfer between another Gram-positive spore-forming bacterium, *Bacillus*, has also been shown in sterile soil and was increased approximately 100-times by the addition of clay and nutrients (Van Elsas *et al.*, 1987). Transfer was also higher in soil held at 27°C with a 60% water holding capacity than in soil at 15°C with a 20% water holding capacity. Plasmid transfer in non-sterile soil occurred only when clay, which provides greater surface area for microbial biofilms, was added (Van Elsas *et al.*, 1987). Addition of nutrients possibly increased the number of vegetative cells in the system leading to higher plasmid transfer frequencies. The use of pre-sterilised environment microcosms to assess plasmid transfer has limitations, however, particularly with respect to competition with indigenous populations and the survival ability of introduced donor strains. A decrease in donor numbers due to inhibition by indigenous bacteria cannot be predicted from a sterile system (Clewlow *et al.*, 1990). Nevertheless, it is often necessary to use sterile microcosms in order to obtain fundamental information on how interactions between donors and recipients may affect plasmid transfer before investigating plasmid transfer under more natural environmental conditions.

Evidence for plasmid transfer in the presence of indigenous soil populations has been reported. Transfer of the IncP plasmids, R68.45 and pJP4, carrying antibiotic and mercury resistance markers, respectively, between *Bradyrhizobium* spp. in non-sterile soil microcosms has been shown (Kinkle *et al.*, 1993), and plasmid transfer was only detected in soils amended with soya bean meal. More recently, plasmid transfer from *Alcaligenes eutrophus* to the indigenous population in non-sterile soil was reported (DiGiovanni *et al.*, 1996). Transfer of the plasmid, JMP134, which carried genes for mercury resistance and for the degradation of 2,4-dichlorophenoxyacetic acid (2,4-D) was shown in non-sterile soil microcosms that were amended with 2,4-D (DiGiovanni *et al.*, 1996). Significant numbers of transconjugants arose in the indigenous bacterial population over the first 3 weeks of the experiment. Of significance was the fact that the indigenous population had increased rates of 2,4-D degradation compared to microcosm populations in which no plasmid transfer occurred. The results indicated that utilisation of biodegradative genes is not dependent on survival of the introduced strain.

The rhizosphere of plants is an area where conjugation between bacteria may occur due to the increase in nutrients near and on root surfaces. Transfer of a broad-host-range plasmid, RP4, between introduced *Pseudomonas* strains in the rhizosphere of wheat was significantly stimulated by the presence of growing wheat roots (van Elsas *et al.*, 1988). In another study (Smit *et al.*, 1991), transfer of RP4 from *P. fluorescens* to indigenous soil bacteria was detected at 10^3 transconjugants g^{-1} of soil in the wheat rhizosphere. Transfer of the plasmid in the corresponding bulk soil was significantly lower, being slightly above the detection limit. No plasmid transfer was detected in soils of unplanted controls. Similarly, conjugation experiments on the rhizosphere of sugar beets revealed that while conjugation occurred in the soil mating mix, plasmid transfer was approximately 100 times higher on the peel surface (Lilley *et al.*, 1994). No transconjugants were detected *in vitro* in soil mating controls or *in situ* in soil more than 5 cm from the plants. From these results, it can be concluded that the rhizosphere has a stimulatory influence on plasmid transfer, and it is possible that plasmid transfer is occurring among bacteria forming thin-layered biofilms on plant root surfaces.

The presence of higher organisms may also affect plasmid transfer in soils. For example, bacterial cells may be transported through the soil by colonising the gut surface of macroinvertebrates. A genetically modified *P. fluorescens* strain, KTG, was shown to be physically retained in the gut of the woodlouse *Porcellio scaber* (Clegg *et al.*, 1996). The bacterium could be isolated from faeces for 6 days after ingestion by the woodlouse, in spite of a relatively short transit time of food through the gut (5 h) (Clegg *et al.*, 1994). The colonisation of gut surfaces in earthworms by bacteria has also been shown (Jolly *et al.*, 1993). In a study of the survival of *P. fluorescens* KTG in the earthworm *Octolasion cyaneum*, the bacterium was detected in casts (faeces) up to 15 days following a single exposure to the bacterium (Clegg *et al.*, 1995). Indeed, transfer of the plasmid, pJP4, from *P. fluorescens* to indigenous soil bacteria (Daane *et al.*, 1996) and between spatially separated *Alcaligenes eutrophus* donor and *P. fluorescens* recipient strains (Daane *et al.*, 1997) was shown to be facilitated by the presence of earthworms. In both studies, the depth of recovery of transconjugants in the soil was dependent on the burrowing behaviour of the earthworms (Daane *et al.*, 1996; 1997). Vilas-Bôas *et al.* (1998) found that a conjugative plasmid (containing an insecticidal crystal protein gene, *cry1Ac*) was transferred between *Bacillus thuringiensis* strains in soil microcosms having various plant covers and different organic contents, as well as in infected larvae of a lepidopteran insect.

Factors influencing plasmid transfer in soils may also hold for sediments, with the exception of the increased availability of water and, in most cases, the lower concentration of oxygen in sediments. Similar to sludge in wastewater treatment facilities, sediments represent a quiescent milieu which would favour plasmid transfer. Nevertheless, there are few studies of plasmid transfer in sediments. The extensive use of antibiotics in aquaculture would increase selection pressure for plasmid transfer, particularly in sediments, and may cause the acquisition of resistance genes by bacterial strains pathogenic to fish. This was shown by a study of the transfer of a broad-host-range, self-transmissible, antibiotic resistance plasmid from the fish pathogen, *Aeromonas salmonicida*, to the natural population in sediments (Sandaa and Enger, 1994). As predicted, placing plasmid transfer under selection pressure resulted in high plasmid transfer frequencies (3.6 donor^{-1}). DNA hybridisation revealed that only 46% of isolates carried the plasmid when no selection pressure was added, whereas 87% of isolates carried the plasmid when under an introduced selection pressure. Characterisation of transconjugants revealed that the presence of selection pressure, while increasing plasmid transfer frequency, decreased the level of diversity in the transconjugant population.

Transfer of plasmids carrying catabolic genes may be one way of increasing the bioremedial capacity of natural populations. Investigations on the transfer of a catabolic plasmid, pBRC60, carrying a transposon encoding genes for 3-chlorobenzoate degradation, in flowthrough freshwater mesocosms with selection pressure for the plasmid, showed that transfer of the plasmid and the transposable element to members of the natural bacterial populations was 10-fold higher in the sediments than in the aqueous phase (Fulthorpe and Wyndham, 1989; 1991). Conversely, Barkay *et al.* (1995) were unable to detect transfer of various IncP1 plasmids from a *Pseudomonas* donor to the indigenous population in estuarine microcosm sediments. The latter result was not surprising since plasmid transfer frequencies were low under laboratory conditions. As a decrease in plasmid transfer frequencies has been reported among cells existing under more natural environmental conditions compared to those conjugated under laboratory conditions, in broth and on plates (Bale *et al.*, 1987; Rochelle *et al.*, 1989a) this may have been a cause for plasmid transfer being below the level of detection in the microcosms of Barkay *et al.* (1995).

Plant and other Diverse Surface Environments

Experiments conducted to study conjugation on plants have shown plasmid transfer between *Erwinia chryanthemi* strains colonising maize plants after 5 h of mating (Lacy, 1978), between *E. coli* and *Pseudomonas* spp. colonising bean pods and other plant surfaces at frequencies as high as 10^{-1} transconjugants per donor *in planta* (Lacy and Leary, 1975), and intergenerically on the surface of pear blossoms between *E. herbicola* or *P. syringae* donors and an *E. amylovora* recipient at similarly high transfer frequencies (Lacy *et al.*, 1984). The study of plasmid transfer between *Bradyrhizobium* strains colonising bean pods, however, indicated that transfer had occurred in the soil, in the rhizosphere or on the root hairs as neither donor nor recipient strains could be isolated from the pods (Kinkle *et al.*, 1993). Manceau *et al.* (1986) showed that plasmid transfer occurred between *Xanthomonas campestris* and *E. herbicola* colonising hazelnut trees and that once populations had stabilised, transconjugants could be isolated from wounds in the trees throughout the seven months of the experiment. Transfer of the plasmid to indigenous epiphytic bacteria was also detected.

Lilley and Bailey (1997) found that transfer of conjugative plasmids, conferring mercury resistance and present in the indigenous microbiota, to *P. fluorescens* cells colonising sugar beet phytospheres occurred during a relatively short period coincident with midseason crop maturation. Normander *et al.* (1998) investigated transfer of a conjugative plasmid (encoding toluene degrading genes) between *P. putida* strains colonising the phylloplanes of bean plants. These authors found that metabolic activity of the bacteria did not affect conjugation efficiency, and that transconjugant cells were localised mainly at junctions between epidermal cells and in substomatal cavities. The high plasmid transfer rates suggested that aggregation of bacteria in microhabitats on the phylloplane stimulated plasmid transfer (Normander *et al.*, 1998).

The production of crown gall by *Agrobacterium tumefaciens* is encoded by a large tumour inducing (Ti) plasmid and infection of the host plant occurs in a fashion similar to bacterial conjugation (Farrand, 1992). Indeed, it has been shown that mobilisable plasmids are transmissible to plant cells when supplied with a complete complement of Ti *vir* genes (responsible for transfer of DNA into the plant cell) *in trans* (Buchanan-Wollaston *et al.*, 1987). Ti plasmid transfer between virulent and avirulent strains of agrobacteria when cultured on the surface of crown gall tumours has also been shown (Kerr *et al.*, 1977).

In an investigation on conjugation between bacteria on surfaces of diverse habitats, plasmid transfer was shown to occur between a bovine pathogenic strain of *E. coli* and a human *E. coli* strain on a hand towel contaminated with milk (Kruse and Sørum, 1994). Similarly, plasmid transfer was shown to occur between a porcine pathogenic *E. coli* strain and a human *E. coli* strain in minced meat and between a fish pathogen, *A. salmonicida*, and a human *E. coli* strain in fish placed on cutting boards (Kruse and Sørum, 1994). The study showed that plasmid transfer was possible between bacteria from diverse origins in the absence of selection pressure and highlights the ease with which pathogenic bacteria may acquire antibiotic resistance. The acquisition of multiple antibiotic resistance has been an increasing problem in the hospital environment where the transfer of antibiotic resistance between bacteria of different genera, mediated by transferable plasmids and transposons, is under strong selection pressure from intensive antibiotic use (for a review see DeFlaun and Levy, 1989).

Due to high densities of indigenous bacterial cells colonising gut surfaces, the gastrointestinal tract is another environment that is conducive to plasmid transfer by conjugation. For example, transfer of a broad-host-range, conjugative antibiotic resistance plasmid, pAMß1, occurred between a *Lactobacillus* sp. and an *Enterococcus* sp. in the digestive tract of infant mice, and between *Enterococcus* spp. and *Lactobacillus* in the digestive tract of adult mice administered with sub-therapeutic levels of antibiotic (McConnell *et al.*, 1991). Long-term administration of the antibiotic selected a modified form of the R plasmid which persisted in the bacteria long after antibiotic treatment had ceased, highlighting the undesirability of the long term administration of antibiotics. In another study, transfer of the broad-host-range plasmid, RP1, between *E. coli* donors and recipients was detected in the gastrointestinal tract of ex germ-free mice as well as *in vitro* in mouse gut extracts (Rang *et al.*, 1996).

CONCLUSIONS

The large number of self-transmissible and mobilisable genetic elements isolated from bacteria in natural environments indicates that conjugation may play a significant role in the dissemination of genes through bacterial communities. Although after plasmid transfer, maintenance of the plasmid in the new host cell is crucial to the persistence of genetic elements in microbial communities and will affect future plasmid transfer efficiency, maintenance is seldom studied in conjunction with plasmid transfer. The extent of plasmid transfer and maintenance in complex microbial communities is still poorly understood. In aqueous environments generally, the colonisation of surfaces by bacteria leading to high cell densities and biofilm development offers good opportunity for plasmid transfer among cells. Studies show that plasmid transfer between bacteria is indeed favoured in biofilms. Further research is required to understand better the contribution of plasmid transfer to the diversity of bacterial activities in the environment, particularly among complex microbial communities such as those comprising biofilms.

REFERENCES

Altherr M.R., Kasweck, K.L. (1982). *In situ* studies with membrane diffusion chambers of antibiotic resistance transfer in *Escherichia coli*. *Appl Environ Microbiol*, 44, 838–843.

Angles M.L. (1997). Gene transfer in marine biofilms. PhD Thesis, The University of New South Wales, Australia.

Angles M.L., Marshall K.C., Goodman A.E. (1993). Plasmid transfer between marine bacteria in the aqueous phase and biofilms in reactor microcosms. *Appl Environ Microbiol*, **59**, 843–850.

Ashelford K.E., Fry J.C. Learner, M.A. (1995). Plasmid transfer between strains of *Pseudomonas putida*, and their survival, within a pilot scale percolating-filter sewage treatment system. *FEMS Microbiol Ecol*, **18**, 15–26.

Bale M.J., Fry J.C., Day M.J. (1987). Plasmid transfer between strains of *Pseudomonas aeruginosa* on membrane filters attached to river stones. *J Gen Microbiol*, **133**, 3099–3107.

Bale M.J., Day M.J., Fry J.C. (1988a). Novel method for studying plasmid transfer in undisturbed river epilithon. *Appl Environ Microbiol*, 54, 2756–2758.

Bale M.J., Fry J.C., Day M.J. (1988b). Transfer and occurence of large mercury resistance plasmids in river epilithon. *Appl Environ Microbiol*, **54**, 972–978.

Barkay T., Kroer N., Rasmussen L.D., Sørensen S.J. (1995). Conjugal transfer at natural population densities in a microcosm simulating an estuarine environment. *FEMS Microbiol Ecol*, **16**, 43–54.

Bradley D.E. (1980). Morphological and serological relationships of conjugative pili. *Plasmid*, **4**, 155–169.

Bradley D.E. (1984). Characteristics and function of thick and thin conjugative pili determined by transfer-derepressed plasmids of incompatibility groups I_1, I_2, I_5, B, K and Z. *J Gen Microbiol*, **130**, 1489–1502.

Bradley D.E., Taylor D.E., Cohen D.R. (1980). Specification of surface mating systems among conjugative drug resistance plasmids in *Escherichia coli* K-12. *J Bacteriol*, **143**, 1466–1470.

Buchanan-Wollaston V., Passiatore J.E., Cannon F. (1987). The *mob* and *oriT* mobilization functions of a bacterial plasmid promote its transfer to plants. *Nature (Lond)*, **328**, 172–175.

Byrd J.J., Colwell R.R. (1990). Maintenance of plasmids pBR322 and pUC8 in nonculturable *Escherichia coli* in the marine environment. *Appl Environ Microbiol*, **56**, 2104–2107.

Byrd J.J., Colwell R.R. (1993). Long-term survival and plasmid maintenance of *Escherichia coli* in marine microcosms. *FEMS Microbiol Ecol*, **12**, 9–14.

Caldwell B.A., Ye C., Griffiths R.P., Moyer C.L., Morita R.Y. (1989). Plasmid expression and maintenance during long-term starvation-survival of bacteria in well water. *Appl Environ Microbiol*, **55**, 1860–1864.

Caldwell D.E., Korber D.R., Lawrence J.R. (1992). Confocal laser microscopy and digital image analysis in microbial ecology. *Adv Microb Ecol*, **12**, 1–67.

Caulcott C.A., Dunn A., Robertson H.A., Cooper N.S., Brown M.E., Rhodes P.M. (1987). Investigation of the effect of growth environment on the stability of low-copy-number plasmids in *Escherichia coli*. *J Gen Microbiol*, **133**, 1881–1889.

Christensen B.B., Sternberg C., Andersen J.B., Eberl L., Møller S., Givskov M., Molin S. (1998). Establishment of new genetic traits in a microbial biofilm community. *Appl Environ Microbiol* **64**, 2247–2255.

Clegg C.D., Anderson J.M., Lappin-Scott, H.M. (1996). Biophysical processes affecting the transit of a genetically-modified *Pseudomonas fluorescens* through the gut of the woodlouse *Porcellio scaber*. *Soil Biol Biochem*, **28**, 997–1004.

Clegg C.D., van Elsas J.D., Anderson J.M., Lappin-Scott H.M. (1994). Assessment of the role of a terrestrial isopod in the survival of a genetically modified pseudomonad and its detection using polymerase chain reaction. *FEMS Microbiol Ecol*, **15**, 161–168.

Clegg C.D., Anderson J.M., Lappin-Scott H.M., van Elsas J.D., Jolly J.M. (1995). Interaction of a genetically modified *Pseudomonas fluorescens* with the soil-feeding earthworm *Octolasion cyaneum* (Lumbricidae). *Soil Biol Biochem*, **27**, 1423–1429.

Clewlow L.J., Cresswell N., Wellington E.M.H. (1990). Mathematical model of plasmid transfer between strains of streptomycetes in soil microcosms. *Appl Environ Microbiol*, **56**, 3139–3145.

Costerton J.W., Lewandowski Z., Caldwell D.E., Korber D.R., Lappin-Scott H.M. (1995). Microbial biofilms. *Annu Rev Microbiol*, **49**, 711–745.

Costerton J.W., Lewandowski Z., DeBeer D., Caldwell D., Korber D., James G. (1994). Biofilms, the customized microniche. *J Bacteriol*, **176**, 2137–2142.

Couturier M., Bex F., Bergquist P.L., Maas W. (1988). Identification and classification of bacterial plasmids. *Microbiol Rev*, **52**, 375–395.

Cruz-Cruz N.E., Toranzos G.A., Ahearn D.G., Hazen T.C. (1988). *In situ* survival of plasmid-bearing and plasmidless *Pseudomonas aeruginosa* in pristine tropical waters. *Appl Environ Microbiol*, **54**, 2574–2577.

Daane L.L., Molina J.A.E., Sadowsky M.J. (1997). Plasmid transfer between spatially separated donor and recipient bacteria in earthworm-containing soil microcosms. *Appl Environ Microbiol*, **63**, 679–686.

Daane L.L., Molina J.A.E., Berry E.C., Sadowsky M.J. (1996). Influence of earthworm activity on gene transfer from *Pseudomonas fluorescens* to indigenous soil bacteria. *Appl Environ Microbiol*, **62**, 515–521.

Dahlberg C., Bergström M., Hermansson M. (1998a). *In situ* detection of high levels of horizontal plasmid transfer in marine bacterial communities. *Appl Environ Microbiol*, **64**, 2670–2675.

Dahlberg C., Bergström M., Andreasen M., Christenses B.B., Molin S., Hermansson M. (1998b). Interspecies bacterial conjugation by plasmids from marine environments visualized by *gfp* expression. *Mol Biol Evol*, **15**, 385–390.

Dalton H.M., Poulsen L.K., Halasz P., Angles M.L., Goodman A.E., Marshall K.C. (1994). Substratum-induced morphological changes in a marine bacterium and their relevance to biofilm structure. *J Bacteriol*, **176**, 6900–6906.

de Beer D., Stoodley P., Roe F., Lewandowski Z. (1994). Effects of biofilm structures on oxygen distribution and mass transport. *Biotechnol Bioeng*, **43**, 1131–1138.

de Taxis du Poët P., Dhulster P., Barbotin J.-N., Thomas D. (1986). Plasmid inheritability and biomass production: comparison between free and immobilized cell cultures of *Escherichia coli* BZ18(pTG201) without selection pressure. *J Bacteriol*, **165**, 871–877.

de Taxis du Poët P., Arcand Y., Bernier R. Jr., Barbotin J.-N., Thomas D. (1987). Plasmid stability in immobilized and free recombinant *Escherichia coli* JM105 (pKK223–200): importance of oxygen diffusion, growth rate, and plasmid copy number. *Appl Environ Microbiol*, **53**, 1548–1555.

DeFlaun M.F., Levy S.B. (1989). Genes and their varied hosts. In: Levy S.B., Miller R.V. (eds) *Gene Transfer in the Environment*. McGraw-Hill Publishing Company, New York, pp. 1–32.

Devanas M.A., Stotzky G. (1988). Survival of genetically engineered microbes in the environment: effect of host/vector relationship. *Dev Ind Microbiol*, **29**, 287–296.

Devanas M.A., Rafaeli-Eshkol D., Stotzky G. (1986). Survival of plasmid-containing strains of *Escherichia coli* in soil: effect of plasmid size and nutrients on survival of hosts and maintenance of plasmids. *Curr Microbiol*, **13**, 269–277.

DiGiovanni G.D., Neilson J.W., Pepper I.L., Sinclair N.A. (1996). Gene transfer of *Alcaligenes eutrophus* JMP134 plasmid pJP4 to indigenous soil recipients. *Appl Environ Microbiol*, **62**, 2521–2526.

Dreiseikelmann B. (1994). Translocation of DNA across bacterial membranes. *Microbiol Rev*, **58**, 293–316.

Farrand S.K. (1992). Conjugal transfer of bacterial genes on plants. In: Levy S.B., Miller R.V. (eds) *Gene Transfer in the Environment*. McGraw-Hill Publishing Company, New York, pp. 261–285.

Fernandez-Astorga A., Muela A., Cisterna R., Iriberri J., Barcina I. (1992). Biotic and abiotic factors affecting plasmid transfer in *Escherichia coli* strains. *Appl Environ Microbiol*, **58**, 392–398.

Flint K.P. (1987). The long-term survival of *Escherichia coli* in river water. *J Appl Bacteriol*, **63**, 261–270.

Fredrickson J.K., Hicks R.J., Li S.W., Brockman F.J. (1988). Plasmid incidence in bacteria from deep subsurface sediments. *Appl Environ Microbiol*, **54**, 2916–2923.

Fry J.C., Day M.J. (1990). Plasmid transfer in the epilithon. In: Fry J.C., Day M.J. (eds) *Bacterial Genetics in Natural Environments*. Chapman and Hall, London, pp. 55–80.

Fulthorpe R.R., Wyndham R.C. (1989). Survival and activity of a 3-chlorobenzoate catabolic genotype in a natural system. *Appl Environ Microbiol*, **55**, 1584–1590.

Fulthorpe R.R., Wyndham R.C. (1991). Transfer and expression of the catabolic plasmid pBRC60 in wild bacterial recipients in a freshwater ecosystem. *Appl Environ Microbiol*, **57**, 1546–1553.

Gauthier M.J., Cauvin F., Breittmayer J.-P. (1985). Influence of salts and temperature on the transfer of mercury resistance from a marine pseudomonad to *Escherichia coli*. *Appl Environ Microbiol*, **50**, 38–40.

Geesey G.G., Mutch R., Costerton J.W., Green R.B. (1978). Sessile bacteria: an important component of the microbial population in a small mountain stream. *Limnol Oceanogr*, **23**, 1214–1223.

Genthner F.J., Chatterjee P., Barkay T., Bourquin A.W. (1988). Capacity of aquatic bacteria to act as recipients of plasmid DNA. *Appl Environ Microbiol*, **54**, 115–117.

Godwin D., Slater J.H. (1979). The influence of the growth environment on the stability of a drug resistance plasmid in *Escherichia coli* K12. *J Gen Microbiol*, **111**, 201–210.

Goodman A.E., Hild E., Marshall K.C., Hermansson M. (1993). Plasmid transfer between marine bacteria under oligotrophic conditions. *Appl Environ Microbiol*, **59**, 1035–1040.

Griffiths R.P., Moyer C.L., Caldwell B.A., Ye C., Morita, R.Y. (1990). Long-term starvation-induced loss of antibiotic resistance in bacteria. *Microb Ecol*, **19**, 251–257.

Harrington L.C., Rogerson A.C. (1990). The F pilus of *Escherichia coli* appears to support stable DNA transfer in the absence of wall-to-wall contact between cells. *J Bacteriol*, **172**, 7263–7264.

Helling R.B., Kinney T., Adams J. (1981). The maintenance of plasmid-containing organisms in populations of *Escherichia coli*. *J Gen Microbiol*, **123**, 129–141.

Hermansson M., Jones G.W., Kjelleberg S. (1987). Frequency of antibiotic and heavy metal resistance, pigmentation, and plasmids in bacteria of the marine air-water interface. *Appl Environ Microbiol*, **53**, 2338–2342.

Hill K.E., Fry J.C., Weightman A.J. (1994). Gene transfer in the aquatic environment: persistence and mobilization of the catabolic recombinant plasmid pD10 in the epilithon. *Microbiology (Reading)*, **140**, 1555–1563.

Inloes D.S., Smith W.J., Taylor D.P., Cohen S.N., Michaels A.S., Robertson C.R. (1983). Hollow-fibre membrane bioreactors using immobilized *E. coli* for protein synthesis. *Biotechnol Bioeng*, **25**, 2653–2681.

Ippen-Ihler K. (1989). Bacterial conjugation. In: Levy S.B., Miller R.V. (eds) *Gene Transfer in the Environment*. McGraw-Hill Publishing Company, New York pp. 33–72.

Ippen-Ihler K.A., Minkley E.G. (1986). The conjugation system of F, the fertility factor of *Escherichia coli*. *Annu Rev Genet*, **20**, 593–624.

Jain R.K., Sayler G.S., Wilson J.T., Houston L., Pacia D. (1987). Maintenance and stability of introduced genotypes in groundwater aquifer material. *Appl Environ Microbiol*, **53**, 996–1002.

Jolly J.M., Lappin-Scott H.M., Anderson J.M., Clegg C.D. (1993). Scanning electron microscopy of the gut microflora of two earthworms: *Lumbricus terrestris* and *Octolasion cyaneum*. *Microb Ecol*, **26**, 235–245.

Jones G.W., Baines L., Genthner F.J. (1991). Heterotrophic bacteria of the freshwater neuston and their ability to act as plasmid recipients under nutrient deprived conditions. *Microb Ecol*, **22**, 15–25.

Jones I.M., Primrose S.B., Robinson A., Ellwood D.C. (1980). Maintenance of some ColE1-type plasmids in chemostat culture. *Mol Gen Genet*, **180**, 579–584.

Jones S.A., Melling J. (1984). Persistence of pBR322-related plasmids in *Escherichia coli* grown in chemostat cultures. *FEMS Microbiol Lett*, **22**, 239–243.

Kerr A., Manigualt P., Tempé J. (1977). Transfer of virulence *in vivo* and *in vitro* in *Agrobacterium*. *Nature (London)*, **265**, 560–561.

Kinkle B.K., Sadowsky M.J., Schmidt E.L., Koskinen W.C. (1993). Plasmids pJP4 and R68.45 can be transferred between populations of Bradyrhizobia in nonsterile soil. *Appl Environ Microbiol*, **59**, 1762–1766.

Klingmüller W., Dally A., Fentner C., Steinlein M. (1990). Plasmid transfer between soil bacteria. In: Fry J.C., Day M.J. (eds) *Bacterial Genetics in Natural Environments*. Chapman and Hall, London, pp. 133–151.

Kruse H., Sørum H. (1994). Transfer of multiple drug resistance plasmids between bacteria of diverse origins in natural environments. *Appl Environ Microbiol*, **60**, 4015–4021.

Lacy G.H. (1978). Genetic studies with plasmid RP1 in *Erwinia chrysanthemi* strains pathogenic on maize. *Phytopathology*, **68**, 1323–1330.

Lacy G.H., Leary J.V. (1975). Transfer of antibiotic resistance factor RP1 into *Pseudomonas glycinea* and *Pseudomonas phaseolicola in vitro* and *in planta*. *J Gen Microbiol*, **88**, 49–57.

Lacy G.H., Stromberg V.K., Cannon N.P. (1984). *Erwinia amylovora* mutants and *in planta*-derived transconjugants resistant to oxytetracycline. *Can J Microbiol*, **6**, 33–39.

Lawrence J.R., Korber D.R., Hoyl B.D., Costerton, J.W., Caldwell D.E. (1991). Optical sectioning of microbial biofilms. *J Bacteriol*, **173**, 6558–6567.

Lewandowski Z. (2000). Structure and function of biofilms. In: Evans L.V. (ed) *Biofilms: Recent Advances in their Study and Control*. Harwood Academic Publishers, pp. 1–17.

Lilley A.K., Bailey M.J. (1997). The acquisition of indigenous plasmids by a genetically marked pseudomonad population colonizing the sugar beet phytosphere is related to local environmental conditions. *Appl Environ Microbiol* **63**, 1577–1583.

Lilley A.K., Fry J.C., Day M.J., Bailey M.J. (1994). *In situ* transfer of an exogenously isolated plasmid between *Pseudomonas* spp. in sugar beet rhizosphere. *Microbiology (Reading)*, **140**, 27–33.

Manceau C., Gardan L., Devaux M. (1986). Dynamics of RP4 plasmid transfer between *Xanthomonas campestris* pv. *corylina* and *Erwinia herbicola* in hazelnut tissues, *in planta*. *Can J Microbiol*, **32**, 835–841.

Mancini P., Fertels S., Nave D., Gealt M.A. (1987). Mobilization of plasmid pHSV106 from *Escherichia coli* HB101 in a laboratory-scale waste treatment facility. *Appl Environ Microbiol*, **53**, 665–671.

Mazodier P., Davies J. (1991). Gene transfer between distantly related bacteria. *Annu Rev Genet*, **25**, 147–171.

McClure N.C., Fry J.C., Weightman A.J. (1990). Gene transfer in activated sludge. In: Fry J.C., Day M.J. (eds) *Bacterial Genetics in Natural Environments*. Chapman and Hall, London, pp. 111–129.

McConnell M.A., Mercer A.A., Tannock G.W. (1991). Transfer of plasmid pAMß1 between members of the normal microflora inhabiting the murine digestive tract and modification of the plasmid in a *Lactobacillus reuteri* host. *Microb Ecol Health Dis*, **4**, 343–355.

McPherson P., and Gealt M.A. (1986). Isolation of indigenous wastewater bacterial strains capable of mobilizing plasmid pBR325. *Appl Environ Microbiol*, **51**, 904–909.

Meacock P.A., Cohen S.N. (1980). Partitioning of bacterial plasmids during cell division: a cis-acting locus that accomplishes stable plasmid inheritance. *Cell*, **20**, 529–542.

Melling J., Ellwood D.C., Robinson A. (1977). Survival of R-factor carrying *Escherichia coli* in mixed cultures in the chemostat. *FEMS Microbiol Lett*, **2**, 87–89.

Mergeay M., Lejeune P., Sadouk A., Gerits J., Fabry L. (1987). Shuttle transfer (or retrotransfer) of chromosomal markers mediated by plasmid pULB113. *Mol Gen*, 61–70.

Miki T., Chang Z.-T., Horiuchi T. (1984). Control of cell division by sex factor F in *Escherichia coli*. II. Identification of genes for inhibitor protein and trigger protein on the 42.84–43.6F segment. *J Mol Biol*, **174**, 627–646.

Mori H., Kondo A., Ohshima A., Ogura T., Hiraga S. (1986). Structure and function of the F plasmid genes essential for partitioning. *J Mol Biol*, **192**, 1–15.

Nasri M., Sayadi S., Barbotin J.-N., Dhulster P., Thomas D. (1987). Influence of immobilization on the stability of pTG201 recombinant plasmid in some strains of *Escherichia coli*. *Appl Environ Microbiol*, **53**, 740–744.

Nordström K., Austin S.J. (1989). Mechanisms that contribute to the stable segregation of plasmids. *Annu Rev Genet*, **23**, 37–69.

Normander B., Christensen B.B., Molin S., Kroer N. (1998). Effect of bacterial distribution and activity on conjugal gene transfer on the phylloplane of the bush bean (*Phaseolus vulgaris*). *Appl Environ Microbiol* **64**, 1902–1909.

O'Morchoe S.B., Ogunseitan O., Sayler G.S., Miller R.V. (1988). Conjugal transfer of R68.45 and FP5 between *Pseudomonas aeruginosa* strains in a freshwater environment. *Appl Environ Microbiol*, **54**, 1923–1929.

Otto K., Weichart D., Kjelleberg S. (1997). Plasmid transfer between *Vibrio* strains during predation by the heterotrophic microflagellate *Cafeteria roenbergensis*. *Appl Environ Microbiol*, **63**, 749–752.

Perkins C.D., Davidson A.M., Day M.J., Fry J.C. (1994). Retrotransfer kinetics of R300B by pQKH6, a conjugative plasmid from the river epilithon. *FEMS Microbiol Ecol*, **15**, 33–44.

Rafii F., Crawford D.L. (1988). Transfer of conjugative plasmids and mobilization of a nonconjugative plasmid between *Streptomyces* strains on agar and in soil. *Appl Environ Microbiol*, **54**, 1334–1340.

Rang C.A., Kennan R.M., Midtvedt T., Chao L., Conway P.L. 1996. Transfer of plasmid RP1 *in vivo* in germ free mice and *in vitro* in gut extracts and laboratory media. *FEMS Microb Ecol*, **19**, 133–140.

Richaume A., Angle J.S., Sadowsky M.J. (1989). Influence of soil variables on *in situ* plasmid transfer from *Escherichia coli* to *Rhizobium fredii*. *Appl Environ Microbiol*, **55**, 1730–1734.

Rochelle P.A., Fry J.C., Day M.J. (1989a). Plasmid transfer between *Pseudomonas* spp.within epilithic films in a rotating disc microcosm. *FEMS Microbiol Ecol*, **62**, 127–136.

Rochelle P.A., Fry J.C., Day M.J. (1989b). Factors affecting conjugal transfer of plasmids encoding mercury resistance from pure cultures and mixed natural suspensions of epilithic bacteria. *J Gen Microbiol*, **135**, 409–424.

Sandaa R.-A., Enger O. (1994). Transfer in marine sediments of the naturally occurring plasmid pRAS1 encoding multiple antibiotic resistance. *Appl Environ Microbiol*, **60**, 4234–4238.

Sandt C.H., Herson D.S. (1991). Mobilization of the genetically engineered plasmid pHSV106 from *Escherichia coli* HB101(pHSV106) to *Enterobacter cloacae* in drinking water. *Appl Environ Microbiol*, **57**, 194–200.

Sayadi S., Nasri M., Barbotin J.N., Thomas D. (1989). Effect of environmental growth conditions on plasmid stability, plasmid copy number, and catechol 2,3-dioxygenase activity in free and immobilized *Escherichia coli* cells. *Biotechnol Bioeng*, **33**, 801–808.

Sayadi S., Nasri M., Berry F., Barbotin J.N., Thomas D. (1987). Effect of temperature on the stability of plasmid pTG201 and productivity of *xylE* gene product in recombinant *Escherichia coli*: development of a two-stage chemostat with free and immobilized cells. *J Gen Microbiol*, **133**, 1901–1908.

Sayler G.S., Hooper S.W., Layton A.C., King J.M.H. (1990). Catabolic plasmids of environmental and ecological significance. *Microb Ecol*, **19**, 1–20.

Sherratt D.J. (1982). The maintenance and propagation of plasmid genes in bacterial populations. *J Gen Microbiol*, **128**, 655–661.

Singleton P., Anson A.E. (1983). Effect of pH on conjugal transfer at low temperatures. *Appl Environ Microbiol*, **46**, 291–292.

Sizemore R.K., Colwell R.R. (1977). Plasmids carried by antibiotic-resistant marine bacteria. *Antimicrob Agents Chemother*, 12, 373–382.

Smit E., van Elsas J.D., van Veen J.A., de Vos W.M. (1991). Detection of plasmid transfer from *Pseudomonas fluorescens* to indigenous bacteria in soil by using bacteriophage fR2f for donor counterselection. *Appl Environ Microbiol*, **57**, 3482–3488.

Sobecky P.A., Schell M.A., Moran M.A., Hodson R.E. (1992). Adaptation of model genetically engineered microorganisms to lake water: growth rate enhancements and plasmid loss. *Appl Environ Microbiol*, **58**, 3630–3637.

Sobecky P.A., Mincer T.J., Chang M.C., Toukdarian A., Helinski D.R. (1998). Isolation of broad-host-range replicons from marine sediment bacteria. *Appl Environ Microbiol*, **64**, 2822–2830.

Stotzky G. (1989). Gene transfer among bacteria in soil. In: Levy S.B., Miller R.V. (eds) *Gene Transfer in the Environment*. McGraw-Hill Publishing Company, New York, pp. 165–222.

Stotzky G., Babich H. (1986). Survival of, and genetic transfer by, genetically engineered bacteria in natural environments. *Adv Appl Microbiol*, **31**, 93–138.

Trevors J.T., Oddie K.M. (1986). R-plasmid transfer in soil and water. *Can J Microbiol*, **32**, 610–613.

Trevors J.T., Barkay T., Bourquin A.W. (1987). Gene transfer among bacteria in soil and aquatic environments: a review. *Can J Microbiol*, **33**, 191–198.

Tucker W.T., Miller C.A., Cohen S.N. (1984). Strutural and functional analysis of the *par* region of the pSC101 plasmid. *Cell*, **38**, 191–201.

Van Elsas J.D., Govaert J.M., van Veen J.A. (1987). Transfer of plasmid pFT30 between bacilli in soil as influenced by bacterial population dynamics and soil conditions. *Soil Biol Biochem*, **19**, 639–647.

Van Elsas J.D., Trevors J.T., Starodub M.E. (1988). Bacterial conjugation between pseudomonads in the rhizosphere of wheat. *FEMS Microbiol Ecol*, **53**, 299–306.

Van Elsas J.D., Trevors J.T., van Overbeek L.S., Starodub M.E. (1989). Survival of *Pseudomonas fluorescens* containing plasmids RP4 or pRK2501 and plasmid stability after introduction into two soils of different texture. *Can J Microbiol*, **35**, 951–959.

Veal D.A., Stokes H.W., Daggard G. (1992). Genetic exchange in natural communities. *Adv Microb Ecol*, **12**, 383–430.

Vilas-Bôas G.F.L.T., Vilas-Bôas L.A., Lereclus D., Arantes O.M.N. (1998). *Bacillus thuringiensis* conjugation under environmental conditions. *FEMS Microbiol Ecol*, **25**, 369–374.

Wilkins B.M. (1990). Factors influencing the dissemination of DNA by bacterial conjugation. In: Fry J.C., Day M.J. (eds) *Bacterial Genetics in Natural Environments*. Chapman and Hall, London, pp. 22–30.

Winans S.C., Walker G.C. (1985). Identification of pKM101-encoded loci specifying potentially lethal gene products. *J Bacteriol*, **161**, 417–424.

Wouters J.T.M., Driehuis F.L., Polaczek P.J., van Oppenraay M.-L.H.A., van Andel J.G. (1980). Persistence of the pBR322 plasmid in *Escherichia coli* K12 grown in chemostat cultures. *Antonie van Leeuwenhoek*, **46**, 353–362.

Zünd P., Lebek G. (1980). Generation time-prolonging R plasmids: correlation between increases in the generation time of *Escherichia coli* caused by R plasmids and their molecular size. *Plasmid*, **3**, 65–69.

7 Bacterial Interactions with Marine Fouling Organisms

Carola Holmström and Staffan Kjelleberg

The development of biofouling communities at surfaces in the marine environment is a complex process with various interactions occurring between the different organisms involved. Bacteria being among the first colonisers have been shown to be important on both living and non-living surfaces. They have been demonstrated to produce secondary metabolites which can either stimulate or inhibit the settlement of invertebrate larvae and algal spores. Such bacterial interactions with higher organisms commonly occur in the marine environment and the bacterially-produced compounds may influence the settlement of larvae and spores either by a contact-dependent mode of action or by producing signalling molecules which can communicate with organisms in the surrounding environment. Moreover, specific adaptations between bacteria and host organisms have been demonstrated in the marine environment. These adaptations could be based on occasional interactions with host organisms or represent a symbiotic relationship which has developed as a result of production of secondary metabolites by both the bacteria and the host organisms.

KEY WORDS: biofouling, bacterial interactions, invertebrate larvae, algal spores, marine bacteria

INTRODUCTION

The attachment and settlement of marine organisms and the development of biofouling communities on surfaces in the marine environment have been described as "a race of various marine organisms to the surfaces" (Clare *et al.*, 1992) in which bacteria are among the first colonisers (Characklis, 1981; Wahl, 1989; Henschel and Cook, 1990). The competition between organisms for space at surfaces is intensive and multiple factors have been implicated in regulating the settlement as well as the metamorphosis of marine organisms and hence the development of biofouling. For example, biological factors such as predation and grazing from fish, gastropods and other organisms influence the establishment of a fouling community. Furthermore, variations in climate and sea temperature as well as geographic variations and seasonal and environmental factors have been reported to effect the development of fouling communities (Wahl, 1989; Pawlik *et al.*, 1991; Richmond and Seed, 1991; Pawlik, 1992; Rodriguez *et al.*, 1993). The age of settling larvae, the availability of food and the recognition of other members of their own species (gregarious behavior) also effect the ability of free-living larvae to settle (Knight-Jones and Stevenson, 1950; Knight-Jones, 1951; Knight-Jones and Crisp, 1953; Crisp and Meadows, 1962; 1963; Crisp, 1974; Buss, 1981; Characklis, 1981; Larman *et al.*, 1982; Bertness and Grosholz, 1985; Crisp *et al.*, 1985; Wahl, 1989; Henschel and Cook, 1990; Svane and Havenhand, 1993). In addition, a series of interactions between bacteria and fouling organisms take place in the marine environment; many of these events are highly specific and bacteria and their exoproducts have been found to offer several means of specificity and strongly select the recruited fauna and flora (Maki *et al.*, 1990; 1994; Mary *et al.*, 1993; Holmström and Kjelleberg, 1994; 1999; Wieczorek *et al.*, 1995; Holmström *et al.*, 1996; Wieczorek and Todd, 1997). Bacterial components such as exopolysaccharides, proteins and fatty acids have been isolated and demonstrated to either increase or decrease the settlement of larvae and spores of sessile organisms (Kirchman *et al.*, 1982b; Maki and Mitchell, 1985; Maki *et al.*, 1990; Szewzyk *et al.*, 1991; Goto *et al.*, 1992; Holmström *et al.*, 1992; 1997; Kitamura *et al.*, 1993; Leitz and Wagne, 1993; Altena and Butler, 1994).

This chapter will review several aspects of the complex development of biofouling communities, with specific reference to bacterial interactions with invertebrate larvae, and algal spores. The aim is to provide evidence for the importance of bacteria in the development of biofouling communities. Furthermore, host bacterial interactions and chemically mediated responses that occur in biofouling communities will be discussed and evidence for the specificity in interactions between marine fouling organisms will be presented. These interactions include secondary metabolites produced by the host organisms which can be used as non-specific chemical defenses as well as interfering in a specific fashion in different stages of bacterial colonisation processes (Wahl *et al.*, 1994; Steinberg *et al.,* 1997; Maximilien *et al.*, 1998).

INTERACTIONS BETWEEN BACTERIA AND INVERTEBRATE LARVAE

The settlement of marine organisms is influenced by passive and active processes in which both microorganisms and their extracellular components play an important role (Butmann, 1987; Butmann *et al.*, 1988; Black and Moran, 1991). Both living and non-living surfaces submerged in the sea are quickly colonised by bacteria, followed by the attachment and settlement of higher organisms (Bakus *et al.*, 1986; Wahl, 1989; De Nys *et al.*, 1994; Wahl *et al.*, 1994; Hay and Fenical, 1996). In contrast to living surfaces which may produce metabolites which protect them against fouling, non-living surfaces are dependent on artificial defenses or being colonised by an antifouling producing organism.

Bacteria are ubiquitous on all surfaces in marine environments. This suggests that studies on bacteria and their numerous coexistences with different hosts are of great relevance in order to understand the development of biofouling communities. Several studies have demonstrated the importance of bacterially produced extracellular components in regulating the settlement of invertebrate larvae (Mihm *et al.*, 1981; Kirchman *et al.*, 1982a; Holmström *et al.*, 1992; Mary *et al.*, 1993; Johnson and Sutton, 1994; Maki *et al.*, 1994; Satuito *et al.*, 1996). Some of these studies have shown how bacteria can enhance larval settlement while other studies have demonstrated their inhibition of larval settlement. Bacteria may therefore be suitable for screening studies aimed at obtaining novel active components. The isolation of bacterial strains which stimulate the settlement of invertebrate larvae may be greatly relevant for farming of commercially important organisms such as abalone and oysters. In addition, bacterial strains that inhibit fouling organisms will be directly applicable in the development of novel environmentally friendly antifouling paints.

Inhibition of Settlement of Invertebrate Larvae

To date several studies which include screening for settlement inhibitory bacterial strains have been performed. However, while several of these studies resulted in the isolation of settlement inhibiting bacteria, detailed chemical characterization of the inhibitory compounds was generally not performed. An ubiquinone compound produced by a strain *Alteromonas* sp. isolated from a marine sponge, *Halichondria okadai,* was found to be inhibitory against barnacle larvae (Kon-ya *et al.*, 1995). Fatty acid compounds have been isolated in various studies both from bacteria (Burke, 1983; Leitz and Wagne, 1993) and higher organisms (Burke, 1983; Pawlik, 1986; Kitamura *et al.*, 1993; Mizobuchi *et al.*, 1993) and have been shown to affect the settlement of invertebrate larvae (Burke, 1983; Pawlik, 1986; Kitamura *et al.*, 1993; Leitz and Wagne, 1993; Mizobuchi *et al.*, 1993). Goto *et al.* (1992), demonstrated a stronger antilarval effect by mixing fatty acids as compared to using the same amount of each fatty acid singly, suggestive of a synergistic effect of the active metabolites. A phospholipid isolated from a sponge, *Crella incrustans*, was found to inhibit settlement of larvae of both the tunicate, *Ciona moluccensis* and the bryozoan, *Bugula neritina,* but was found to be inactive against the blue mussel *Mytilus edulis* (Altena and Butler, 1994). Such target organism specificity has also been demonstrated for the bacterium *Halomonas marina* (*Pseudomonas marina* = *Deleya marina*) which was shown to stimulate the settlement of spirorbid polychaete larvae but was found to inhibit the settlement of bryozoan larvae (Maki *et al.*, 1989) and barnacle larvae (Maki *et al.*, 1992). As stated above, fatty acid compounds have been demonstrated to have a broad range of activities. Such a variety of effects displayed by one group of compounds emphasizes the importance of performing broad range bioassays in order to properly describe the activity of metabolites of interest.

A well studied antifouling producing bacterium is *Pseudoalteromonas tunicata* strain D2. This strain was originally isolated from an adult tunicate collected at a depth of 10 m off the Swedish westcoast (Holmström *et al.*, 1992). The strain is dark green pigmented and has been demonstrated to inhibit the settlement of common fouling organisms such as invertebrate larvae, algal spores and diatoms, and to inhibit the growth of bacteria and fungi (Holmström *et al.*, 1992; 1996; 1997; 1998; Holmström and Kjelleberg, 1994; 1999; Egan, 1995; James *et al.*, 1996). It has been shown that *P. tunicata* D2 cells produce different components that specifically inhibit different classes of fouling organisms. The production by the cells of settlement inhibitory metabolites seems to coincide with the production of the pigment(s) produced by this bacterium but it has clearly been demonstrated that the pigment(s) are not identical to the anti-larval, -spore and -bacterial molecules. However, the production of the antifouling components is closely related to the synthesis of the pigment(s). This has been demonstrated by generating mutants with different pigmentation and studying their effects against fouling organisms (unpublished data). Furthermore, D2 cells grown on rich bacteriological growth media such as LB20 (Luria broth medium) lose both the expression of the pigment and the antifouling activity. The antilarval component is a small (< 500 Da in size), polar, heat stable component which is extracellularly released (Holmström *et al.*, 1992). The antibacterial protein which is 190 kDa in size consists of at least two subunits (James *et al.,* 1996). By separating the subunits the activity against bacteria is lost. D2 cells display autoinhibition but the cells seem to become more resistant against the antibacterial protein when reaching stationary phase, which may offer an explanation for the survival of D2 cells. Furthermore, this protein is active against a broad range of different bacteria, including both Gram-negative and Gram-positive bacteria isolated from marine waters and medical and terrestial environments (James *et al.*, 1996). In addition, *P. tunicata* was the only isolate among a variety of different bacterial strains tested which expressed autoinhibition. This function may suggest that the antibacterial protein is not a commonly produced bacterial compound. The antispore component is suggested to be a relatively heat stable peptide which is around 3 kDa in size (Egan, 1995), while the antifungal effect appears to be due to at least one fatty acid molecule (unpublished data). The active metabolite against the diatom *Amphora* sp. has not yet been identified.

Stimulation of Settlement of Invertebrate Larvae

Studies of the effects by bacteria on the settlement of invertebrate larvae, as cited above, have given more emphasis to the detection of metabolites with inhibitory activity rather than those exhibiting settlement inducing cues. However, both stimulatory and inhibitory signals are present at surfaces in marine waters (Busscher and Weerkamp, 1987; Woodin, 1991). These cues derive from bacteria as well as from eukaryotes. Although this review mainly focuses on bacterial/higher organism interactions, a few examples of settlement enhancing metabolites that are proposed to be produced by eukaryotic organisms will be provided because these metabolites may derive from the bacteria colonising the higher organisms rather than from the host.

Exopolysaccharides (EPS) have been demonstrated to enhance the settlement of larvae by a contact-dependent recognition of specific moieties. A well studied bacterial/larval interaction is that between the spirorbid polychaete, *Janua brasiliens,* and the bacterium *H. marina*. It was shown that a polymer containing glucose moieties produced by the bacterium, is recognized by a lectin on the larval surface and that binding between the two organisms only occurs if both the larval and the bacterial components are present (Kirchman *et al.*, 1982a; 1982b; Maki and Mitchell, 1985). In contrast to this very specific binding, a study performed by the authors of this review demonstrated that *Pseudoalteromonas* sp. S9 (previously *Pseudomonas* S9) which produces EPS during stationary phase and starvation conditions causes increased settlement of larvae of the ascidian *Ciona tunicata* compared to a S9 mutant strain which does not release the EPS (Szewzyk *et al.*, 1991). Two different modes of larval attachment were proposed; an active settlement by use of the adhesive organ and a passive settlement due to trapping of the larvae in the exopolymer material. Both the actively and the passively attached larvae developed into juveniles and thereby contributed to the biofouling community.

Studies have also demonstrated that surface wettability exerts some control over the molecular rearrangements of adsorbed macromolecules and bacterial surface components, causing different domains to be exposed (Mihm *et al.*, 1981; Maki *et al.*, 1988; 1989; 1990; O'Connor and Richardson, 1998). Maki and coworkers have studied extensively the exopolymer producing bacterium *H. marina.* The exopolymer was collected and coated on different surfaces prior to exposure to barnacle larvae. Maki *et al.* (1990) made the observation that polystyrene surfaces coated with the exopolymer were inhibitory against *Balanus amphitrite* larvae while tissue culture polystyrene and glass surfaces resulted in a stimulatory effect compared to the unfilmed surfaces. There might be two explanations for these results. Either the different substrata allow for different domains to be exposed when the exopolymer is adsorbed, or at least two components are involved in regulating the settlement process. The latter case would involve a stimulatory component that only attaches to a hydrophilic substratum, and an inhibitory component that only attaches to a hydrophobic substratum.

Several studies of the effect by naturally preconditioned surfaces against macrofouling have been performed. However, these focus more on the effects of the age of the mixed bacterial biofilm rather than on specific mechanisms of biofilm development (Maki *et al.*, 1990; Todd *et al.*, 1993; Neal and Yule, 1994; Todd and Keough, 1994; Keough and Raimondi, 1995; Wieczorek *et al.*, 1995; Wieczorek and Todd, 1997). Due to the changes occurring within a biofilm it may be difficult to draw general conclusions from mixed biofilm experiments. For example, the production of EPS by different bacterial strains is age dependent (Wrangstadh *et al.*, 1986; 1990). Some strains produce EPS during starvation conditions while other strains produce EPS during growth. The concentration and composition of other excreted metabolites such as peptides, proteins and fatty acids which may help the bacterial cells in establishing themselves in the biofilm community have also been demonstrated to be dependent on age and nutrient availability.

Although artificial inducer components will not be discussed in this review, naturally produced analogs of dihydroxyphenylalanine (L-DOPA) and γ-aminobutyric acid (GABA) and a few well studied host-larvae interactions mediated by these analogs will be mentioned. These examples highlight methodological problems inherent to the isolation of active components. For example, a natural small peptide (GABA analog) associated with coralline red algae was found to induce settlement and metamorphosis of the abalone *Haliotis rufescens* (Morse, 1984; 1985; 1990; Morse, 1991). The postulated model is that the abalone larvae have receptors located on the cell surface which recognize the signal molecule, triggering the transition from the larval to the juvenile stage (Morse, 1985; Morse, 1991; Trapido-Rosenthal and Morse, 1985; 1986). Yet, the biological relevance of GABA compounds in natural system has been questioned (Mountfort and Pybus, 1992; Kaspar and Mountfort, 1995). GABA can both be produced and degraded by bacteria and Mountfort and Pybus (1992) suggested that the induction of abalone larvae to settle may be due to bacterially produced components. Kaspar and Mountfort (1995) performed a study where they examined whether GABA could be detected on crustose coralline algal surfaces. However, they did not detect the concentrations of GABA needed to trigger the settlement response of abalone larvae, and it was concluded that GABA does not play a significant role in larval settlement in the natural habitat.

In the oyster *Crassostrea gigas,* settlement and metamorphosis have been demonstrated to be induced by L-DOPA (Coon and Bonar, 1985; Coon *et al.*, 1990). A model for the settlement of *C. gigas* consisting of two control pathways has been proposed, *viz.* a behavioral pathway and an adrenergic morphogenetic pathway (Bonar *et al.*, 1990; Coon *et al.*, 1990). The behavioral pathway is induced by suitable environmental cues such as bacterially produced components but it could also be triggered by ammonia (NH_3) and other weak bases (Bonar *et al.*, 1990). These authors also found that a wide variety of bacteria release ammonia in mid to late log phase of growth (Bonar *et al.*, 1990). The authors suggested that ammonia and L-DOPA utilise separate mechanisms in triggering the behavioral pathway.

With L-DOPA compounds, as with GABA analogs and other metabolites, conflicting results with respect to whether natural or artificial inducers were obtained in the sampling procedure can also be found in the literature (Jensen *et al.*, 1990; Mountfort and Pybus, 1992; Kaspar and Mountfort, 1995). The natural inducer of settlement and metamorphosis of the tube worm *Phragmatopoma* has been suggested to be free fatty acids (Pawlik, 1990). However, Jensen *et al.* (1990) postulated the inducer to be a protein (rich in residues of DOPA) and argued that the fatty acids that were previously suggested to be inducers were the result of biological contamination.

Host-associated Bacterial Biofilms and Signalling Activity in Microbial Habitats

Bacterial biofilms have been demonstrated to exert a variety of specific effects on their host organisms (Johnson *et al.*, 1991; Johnson and Sutton, 1994; Littler and Littler, 1995; Holmström *et al.*, 1996; Kushmaro *et al.*, 1996). Such bacterial interactions can be beneficial to the host (Gil-Turnes *et al.*, 1989) but negative effects have also been reported (Gil-Turnes *et al.*, 1989; Littler and Littler, 1995; Kushmaro *et al.*, 1996). It has been hypothesised that host organisms lacking defense mechanisms against fouling may be colonised by an antifouling-producing bacterial biofilm which may help the host in the defense against biofouling. For example, in a study which focused on bacteria isolated from living surfaces it was demonstrated that 74% (n = 23) of isolates from different seaweeds and 30% (n = 23) of the bacterial strains isolated from marine animals inhibited barnacle larval settlement (Holmström *et al.*, 1996). Also, Gil-Turnes *et al.* (1989) demonstrated that a strain of *Alteromonas* sp. isolated from embryos of the shrimp *Palaemon macrodactylus* chemically defends the crustacean embryos from a pathogenic fungus. With respect to negative effects, significant degradation of host tissue or mortality of the hosts can occur (Littler and Littler, 1995; Kushmaro *et al.*, 1996; Kushmaro and Rosenberg, 1998). Such is the case for coral bleaching which is caused by the marine bacterium *Vibrio* AK-1 (Kushmaro *et al.*, 1996; Kushmaro and Rosenberg, 1998).

A high diversity of intercellular signalling systems exist in bacteria. Many Gram-negative bacteria have regulatory systems which are switched on by small molecules and allow the bacteria to react rapidly to environmental changes (Kjelleberg *et al.*, 1997). Acylated homoserine lactones (AHLs) are such self-produced signals and function to control gene regulation for expression of a broad range of phenotypes under conditions of high cell density (Fuqua *et al.*, 1996; Swift *et al.*, 1996). This system is termed quorom sensing. The presence of AHLs in naturally occuring biofilms has been demonstrated and it has been suggested that AHLs might play a role in biofilm differentiation and development (Davies *et al.*, 1998).

An example of the control of bacterial biofilm communities exerted by animate surfaces is provided by the red alga *Delisea pulchra* (Maximilien *et al.*, 1998). *D. pulchra* has been found to produce small halogenated compounds, furanones, which have been demonstrated to interfere with bacterial growth and with the bacterial signalling systems that are mediated by AHLs (Givskov *et al.*, 1996; Gram *et al.*, 1996; Kjelleberg *et al.*, 1997; Steinberg *et al.*, 1997). In addition, the compounds released by *D. pulchra* prevent settlement of common fouling organisms (de Nys *et al.*, 1994).

Bacterial abundance on the surface of *D. pulchra* was demonstrated to differ significantly along the plant thallus, ranging from approximately 10^6 cells cm^{-2} at the apical tip of the plant to more than 10^7 cells cm^{-2} at the base (Maximilien *et al.*, 1998). Furthermore, the ratio of Gram+/Gram– bacteria throughout a 12 months period (at 3 different sampling times) was consistently around 50% at the tips compared to a much lower proportion of Gram positive bacteria on the basal parts of the plant (unpublished data). This variation in bacterial abundance and composition along the thallus of the plant, and hence control by the plant of its surface associated bacteria has not been reported for other seaweeds. Studies demonstrated a higher production of furanones at the tip of *D. pulchra* compared to other parts, and these differences were found to correlate with changes in the bacterial colonisation pattern. The halogenated metabolites were subsequently demonstrated to interfere with many AHL regulatory system in several Gram-negative bacteria such as swarming motility, bioluminescence and exoenzyme synthesis (Givskov *et al.*, 1996; Gram *et al.*, 1996; Kjelleberg *et al.*, 1997). A large number of AHL-dependent phenotypes may facilitate bacterial colonisation on or in higher organisms and bacteria employing this regulatory system possibly have an advantage in the colonisation of surfaces. Given that the host can interfere with different bacterial colonisation traits it is suggested that *D. pulchra* manipulates the composition of the associated bacterial population. It may be hypothesised that eukaryote metabolites that act on specific bacterial regulatory systems may constitute a common chemical regulation of host-bacterial interactions.

INTERACTIONS BETWEEN BACTERIA AND ALGAE

Algal communities may constitute a major contribution to marine biofouling. A limiting factor for the development of algal communities is the quantity of light but also, as with marine animals, space availability. The attachment and settlement of different algal spores may be sequential, with spores of green and brown algae attaching first followed by spores of red algae (Evans, 1981). Most green and brown algae have motile spores as a result of hair like flagella (Fletcher and Callow, 1992). It has been proposed that motility and the hairlike flagella mediate site selection and initial adherence (Fletcher and Callow, 1992). Red algae on the other hand have non-motile spores (Lin *et al.*, 1975) and the dispersal of spores is a passive process where their sinking rate is determined by their size and density (Okuda and Neushul, 1981).

The development of algal communities reflects a sequence of events for which it would appear that bacterial films play an important role (Tosteson and Corpe, 1975; Dillon *et al.*, 1989; Holmström *et al.*, 1996). The extent and specificity of bacterial/spore interactions in the marine environment are not known, however, mainly because detailed studies of these interactions are rare. In a settlement study where three different target organisms, barnacle larvae, algal spores and diatoms, were tested for their responses to exposure to different marine bacterial isolates the results showed that 16 of the 93 isolates prevented growth of spores of the green alga *Ulva lactuca*, while 9 of the 93 strains prevented settlement of barnacle larvae (Holmström *et al.*, 1996). This may indicate that algal spore settlement is not as selective in the recognition of cues as is the case for invertebrate larvae. Furthermore, only 4 bacterial isolates were found to be inhibitory against both larvae and spores, which suggest that different compounds are needed for successful interference in the settlement process of the two target organisms. The specificity of components that regulate spore attachment and settlement has been demonstrated with the bacterium *Pseudoalteromonas tunicata* strain D2. As mentioned earlier, this bacterium produces different components that are active against different target organisms. The antialgal component would appear to be a peptide around 3kDa in size (Egan, 1995), which has been shown not to interfere in larval settlement or result in the inhibition of bacterial growth and colonisation.

The attachment of spores consists of two different phases; an initial attachment and a permanent attachment (Fletcher and Callow, 1992). With respect to the permanent stage it has been demonstrated that all algal spores produce an adhesive material (probably a polysaccharide-protein complex) which strengthens spore attachment to the surface. Furthermore, algal spores have been shown to display contact-dependent recognition of specific signals (Tosteson and Corpe, 1975; Klut *et al.*, 1983). Moreover, a chemotactic response between bacteria and the target organism has been demonstrated by Lovejoy *et al.* (1998). They observed a high concentration of a *Pseudoalteromonas* strain in the vicinity of harmful algal bloom species. Paradoxically, extracellular component(s) produced by the bacterial strain were found to have algicidal effects by way of lyzing the spores (Lovejoy *et al.*, 1998). Algicidal components produced by bacteria have also been reported in other studies. In some of these studies, the active bacterial components were demonstrated to be species specific in their mode of action (Imai *et al.*, 1995; Yoshinaga *et al.*, 1997), and lysis of algae by bacteria may be mediated by extracellularly produced components or cell-to-cell contact (Kato *et al.*, 1998). It is proposed that bacterial species composition and production of extracellular components may play a significant role in the control of harmful algal blooms.

Many members of the genus *Pseudoalteromonas* (previously designated *Alteromonas*) have been found to produce components that demonstrate activity against growth and settlement of other bacteria as well as higher organisms (Bein, 1954; Leitz and Wagne, 1993; Uchida *et al.*, 1995; Holmström *et al.*, 1998). *Pseudoalteromonas* strains have been isolated from marine environments around the world and often in association with living surfaces (Weiner *et al.*, 1988; Simidu *et al.*, 1990; Akagawa-Matsushita *et al.*, 1992; Ivanova *et al.*, 1996; 1998; Yoshikawa *et al.*, 1997). Very little is known about the means by which *Pseudoalteromonas* species affect their target organisms. However, compounds produced by these species have been demonstrated to interfere with the growth or the viability of other Gram-negative and/or Gram-positive bacteria and in addition may display an autoinhibition mode of action (Gauthier and Flatau, 1976; Gauthier, 1976, 1977; Gauthier and Breittmayer, 1979; Shiozawa *et al.*, 1993; McCarthy *et al.*, 1994; James *et al.*, 1996; Ivanova *et al,* 1998). Studies addressing interactions between *Pseudoalteromonas* strains and higher organisms frequently report the presence of a bacterially derived inhibitory activity against the settlement of invertebrate larvae and algal spores to surfaces. In this respect, *Pseudoalteromonas tunicata* is one the most well-studied antifouling producing strains (Holmström *et al.*, 1992; 1998; Holmström and Kjelleberg, 1994; Gatenholm *et al.*, 1995; James *et al.*, 1996). As mentioned above this strain produces inhibitory compounds against invertebrate larvae, algae spores, bacteria, fungi and diatoms. The different compounds produced by *P. tunicata* D2 clearly display specific activities against different classes of organisms, highlighting the importance and complexity of bacterial/host interactions occurring in the marine environment.

SUMMARY

The development and composition of a biofouling community have been found, in several studies, to be strongly influenced by bacteria and bacterially produced components. Not only do bacteria and their metabolites interfere in the settlement and metamorphosis of higher organisms in the marine environment, they can also sense and produce chemical signals which interfere with different processes and interactions in biofouling communities. This broad host-range activity expressed by some marine bacteria has elicted an interest in the search for novel antifouling components. An example of a bacterial strain which has been demonstrated to produce many biologically active compounds is *P. tunicata* D2. This marine bacterial isolate produces at least five extracellular components and each expresses an inhibitory activity against a specific group of target biofouling organisms. Many bacteria have also been shown to employ small extracellular signalling molecules to mediate different bacterial colonisation traits such as swarming and exoprotease production. A red alga, *Delisea pulchra,* produces secondary metabolites that inhibit the settlement of higher organisms and also interfere with different bacterial colonisation traits that are driven by extracellular signalling molecules. These observations suggest that the alga itself may be able to control bacterial abundance and composition on its surface. Given that bacteria may actively respond to or "communicate" with environmental stimuli through signal-driven regulatory systems and given that these systems can be interfered with in a non-toxic fashion, additional exciting research areas for an improved understanding of the complex development of marine biofouling communities can be suggested.

ACKNOWLEDGEMENTS

The authors gratefully acknowledge valuable discussions with Sally James, Suhelen Egan, Ashley Franks, Rocky de Nys and Peter Steinberg. Studies in the authors' laboratory were supported by the Australian Research Council, a Vice Chancellor's post doctoral research fellowship at the University of New South Wales to CH and by the Centre for Marine Biofouling and Bio-Innovation at UNSW.

REFERENCES

Akagawa-Matsushita M., Matsuo M., Koga Y., Yamasato K. (1992). *Alteromonas atlantica* sp. nov., *Alteromonas carrageenovora* sp. nov., bacteria that decompose algal polysaccharides. *Int J Syst Bacteriol*, **42**, 621–627.

Altena I., Butler A.J. (1994). Antifoulants from marine invertebrates. In: Kjelleberg S., Steinberg P. (eds) *Proc Workshop Biofouling: Problems and Solutions*. Sydney, Australia, pp. 80–86.

Bakus G.J., Targett N.M., Schulte B. (1986). Chemical ecology of marine organisms: an overview. *J Chem Ecol*, **12**, 951–987.

Bein S.J. (1954). A study of certain chromogenic bacteria isolated from "red tide" water with a description of a new species. *Bull Mar Sci Gulf Caribb*, **4**, 110–119.

Bertness M.D., Grosholz E. (1985). Population dynamics of the ribbed mussel, *Geukeusia demissa*: the costs and benefits of an aggregated distribution. *Oecologia*, **67**, 192–204.

Black K.P., Moran P.J. (1991). Influence of hydrodynamics on the passive dispersal and initial recriutment of larvae of *Acanthaster planci* (Echinodermata: Asteroidea) on the Great Barrier Reef. *Mar Ecol Prog Ser*, **69**, 55–65.

Bonar D.B., Coon S.L., Walch M., Weiner R.M., Fitt W. (1990). Control of oyster settlement and metamorphosis by endogenous and exogenus chemical cues. *Bull Mar Sci*, **46**, 484–498.

Burke R.D. (1983). The induction of metamorphosis of marine invertebrate larvae: stimulus and response. *Can J Zool*, **61**, 1701–1719.

Buss L.W. (1981). Group living, competition, and the evolution of cooperation in a sessile invertebrate. *Science*, **213**, 1012–1014.

Busscher H.J., Weerkamp A.H. (1987). Specific and non-specific interactions in bacterial adhesion to solid substrata. *FEMS Microbiol Rev*, **46**, 165–173.

Butmann C.A. (1987). Larval settlement of soft-sediment invertebrates: the spatial scales of pattern explained by active habitat selection and the emerging role of hydrodynamical processes. *Oceanogr Mar Biol Annu Rev*, **25**, 113–165.

Butmann C.A., Grassle J.P., Webb C.M. (1988). Substrate choices made by marine larvae settling in still water and in a flume flow. *Nature*, **333**, 771–773.

Characklis W.G. (1981). Bioengineering report. Fouling biofilm development: a process analysis. *Biotechnol Bioeng*, **23**, 1923–1960.

Clare A.S., Rittschof D., Gerhart D.J., Maki J.S. (1992). Molecular approaches to nontoxic antifouling. *Invertebr Reprod Dev*, **22**, 67–76.

Coon S.L., Bonar D.B. (1985). Induction of settlement and metamorphosis of the pacific oyster, *Crassostrea gigas*, by L-DOPA and Catecholamines. *J Exp Mar Biol Ecol*, **94**, 211–221.

Coon S.L., Fitt W.K., Bonar D.B. (1990). Competence and delay of metamorphosis in the pacific oyster, *Crassostrea gigas*, by L-DOPA and Catecholamines. *Mar Biol*, **106**, 379–387.

Crisp D.J. (1974). Factors influencing the settlement of marine invertebrate larvae. In Grant P.T., Mackie A. (eds) *Chemoreception in Marine Organisms*. Academic Press, New York, pp. 177–265.

Crisp D.J., Meadows P.S. (1962). The chemical basis of gregariousness in cirripedes. *Proc Roy Soc*, **156**, 499–520.

Crisp D.J., Meadows P.S. (1963). Adsorbed layers: the stimulus to settlement in barnacles. *Proc Roy Soc Lond*, **158**, 363–387.

Crisp D.J., Walker G., Young G.A., Yule A.B. (1985). Adhesion and substrate choice in mussels and barnacles. *J Colloid Interface Sci*, **104**, 40–50.

Davies D.G., Parsek M.R., Pearson J.P., Iglewski B.H., Costerton J.W., Greenberg E.P. (1998). The involvement of cell-to-cell signals in the development of a bacterial biofilm. *Science*, **280**, 295–297.

de Nys R., Steinberg P.D., Willemsen P., Dworjanyn S.A., Gabelish C.L., King R.J. (1994). Broad spectrum effects of secondary metabolites from the red alga *Delisea pulchra* in antifouling assays. *Biofouling*, **8**, 250–271.

Dillon P.S., Maki J.S., Mitchell R. (1989). Adhesion of *Entermorpha* swarmers to microbial films. *Microb Ecol*, **17**, 39–47.

Egan S. (1995). Inhibition of algal spores germination by marine bacteria. Honors Thesis, University of New South Wales, Australia.

Evans L.V. (1981). Marine algae and fouling; a review with particular reference to ship fouling. *Bot Mar*, **24**, 167–171.

Fletcher R.L., Callow M. (1992). The settlement, attachment and establishment of marine algal spores. *Br Phycol J*, **27**, 303–329.

Fuqua C., Winans C., Greenberg E.P. (1996). Census and consensus in bacterial ecosystems: the LuxR-LuxI family of quorom-sensing transcriptional regulators. *Annu Rev Microbiol*, **50**, 727–751.

Gatenholm P., Holmström C., Maki J.S., Kjelleberg S. (1995). Toward biological antifouling surface coatings-marine bacteria immobilized in hydrogel inhibit barnacle larvae. *Biofouling*, **8**, 293–301.

Gauthier M.J. (1976). *Alteromonas rubra* sp. nov. and a new marine antibiotic-producing bacterium. *Int J Syst Bacteriol*, **26**, 459–466.

Gauthier M.J. (1977). *Alteromonas citrea,* a new Gram-negative, yellow-pigmented species from seawater. *Int J Syst Bacteriol*, **27**, 349–354.

Gauthier M.J., Flatau G.N. (1976). Antibacterial activity of marine violet-pigmented *Alteromonas* with special reference to the production of brominated compounds. *Can J Microbiol*, **22**, 1612–1619.

Gauthier M.J., Breittmayer V.A. (1979). A new antibiotic-producing bacterium from seawater: *Alteromonas aurantia* sp. nov. *Int J Syst Bacteriol*, **29**, 366–372.

Gil-Turnes M.S., Hay M.E., Fenical W. (1989). Symbiotic marine bacteria defend crustacean embryos from a pathogenic fungus. *Science*, **240**, 116–118.

Givskov M., De Nys R., Manefield M., Gram L., Maximilien R., Eberl L., Molin S., Steinberg P., Kjelleberg S. (1996). Eukaryotic interference with homoserine lactone mediated procaryotic signalling. *J Bacteriol*, **178**, 6618–6622.

Goto R., Kado R., Muramoto K., Kamiya H. (1992). Fatty acids as antifoulants in a marine sponge. *Biofouling*, **6**, 61–68.

Gram L., de Nys R., Maximilien R., Givskov M., Steinberg P.D., Kjelleberg S. (1996). Inhibitory effects of secondary metabolites from the red alga *Delisea pulchra* on swarming motility of *Proteus mirabilis*. *Appl Environ Microbiol*, **62**, 4284–4287.

Hay M.E., Fenical W. (1996). Chemical ecology and marine biodiversity: insights and products from the sea. *Oceanography*, **9**, 10–20.

Henschel J.R., Cook P.A. (1990). The development of a marine fouling community in relation to the primary film of microorganisms. *Biofouling*, **2**, 1–11.

Holmström C., Kjelleberg S. (1994). The effect of external biological factors on settlement of marine invertebrate and new antifouling technology. *Biofouling*, **8**, 147–160.

Holmström C., Kjelleberg S. (1999). Factors influencing the settlement of macrofoulers. In: Fingerman M., Nagabhushanam R., Thompson M. (eds) *Recent Advances in Marine Biotechnology*, Volume 3. Science Publishers Incorporated, New Delhi, pp. 173–201.

Holmström C., Rittschof D., Kjelleberg S. (1992). Inhibition of settlement by larvae of *Balanus amphitrite* and *Ciona intestinalis* by a surface-colonizing marine bacterium. *Appl Environ Microbiol*, **58**, 2111–2115.

Holmström C., James S., Egan S., Kjelleberg S. (1996). Inhibition of common fouling organisms by marine bacterial isolates with special reference to the role of pigmented bacteria. *Biofouling*, **10**, 251–259.

Holmström C., James S., Egan S., Kjelleberg S. (1997). Regulation of activity and settlement of marine organisms by bacterial extracellular components. In Brady R.F. (ed) *Proc U.S-PacificRim Workshop Emerging Nonmetalic Materials for the Marine Environment*. Honolulu, Hawaii.

Holmström C., James S., Neilan B.A., White D.C., Kjelleberg S. (1998). *Pseudoalteromonas tunicata* sp. nov., an antifouling producing bacterium. *Int J Syst Bacteriol*, **48**, 1205–1212.

Imai I., Ishida Y., Sakaguchi K., Hata Y. (1995). Algicidal marine bacteria isolated from Northern Hiroshima bay, Japan. *Fish Sci Tokyo*, **61**, 628–636.

Ivanova E.P., Kiprianova E.A., Mikhailov V.V., Levanova G.F., Garagulya A.D., Gorshikova N.M., Yumoto N., Yoshikawa S. (1996). Characterization and identification of marine *Alteromonas nigrifaciens* strains and emendation of the description. *Int J Syst Bacteriol*, **46**, 223–228.

Ivanova E.P., Kiprianova E.A., Mikhailov V.V., Levanova G.F., Garagulya A.D., Gorshkova N.M., Vysotskii M.V., Nicolau D.V., Yumoto N., Taguchi T., Yoshikawa S. (1998). Phenotypic diversity of *Pseudoalteromonas citrea* from different marine habitats and emendation of the description. *Int J Syst Bacteriol*, **48**, 247–256.

James S., Holmström C., Kjelleberg S. (1996). Purification and characterization of a novel antibacterial protein from the marine bacterium D2. *Appl Environ Microbiol*, **62**, 2783–2788.

Jensen R.A., Morse D.E., Petty R.L., Hooker N. (1990). Artificial induction of larval metamorphosis by free fatty acids. *Mar Ecol Prog Ser*, **67**, 55–71.

Johnson C.R., Sutton D.C. (1994). Bacteria on the surface of crustose coralline algae induce metamorphosis of the crown-of-thorns starfish *Acanthaster planci*. *Mar Biofoul*, **120**, 305–310.

Johnson C.R., Sutton D.C., Olson R.R., Giddins R. (1991). Settlement of crown-of-thorns starfish: role of bacteria on surfaces of coralline algae and a hypothesis for deepwater recruitment. *Mar Ecol Prog Ser*, **71**, 143–162.

Kaspar H.F., Mountfort D.O. (1995). Microbial production and degradation of γ-aminobutyric acid (GABA) in the abalone larval settlement. *FEMS Microbial Ecol*, **17**, 205–211.

Kato J., Amie J., Murata Y., Kuroda A., Mitsutani A., Ohtake H. (1998). Development of a genetic transformation system for an alga-lysing bacterium. *Appl Environ Microbiol*, **64**, 2061–2064.

Keough M.J., Raimondi P.T. (1995). Responses of settling invertebrate larvae to bioorganic fims: effects of different types of film. *J Exp Mar Biol Ecol*, **185**, 235–253.

Kirchman D., Graham S., Reish D., Mitchell R. (1982a). Bacteria induce settlement and metamorphosis of *Janua* (*Dexiospira*) *brasiliensis* Grube (Polychaeta: Spirorbidae). *J Exp Mar Biol Ecol*, **56**, 153–163.

Kirchman D., Graham S., Reish D., Mitchell R. (1982b). Lectins may mediate in the settlement and metamorphosis of *Janua* (Dexiospira) *brasiliensis* Grube (Polychaeta: Spirorbidae). *Mar Biol Lett*, **3**, 131–142.

Kitamura H., Kitahara H., Koh B. (1993). The induction of larval settlement and metamorphosis of two sea urchins, *Pseudocentrotus depressus* and *Anthocidaris crassispina*, by free fatty acids extracted from coralline red alga *Corallina pilulifera*. *Mar Biol*, **115**, 387–392.

Kjelleberg S., Steinberg P., Givskov M., Gram L., Manefield M., de Nys R. (1997). Do marine natural products interfere with prokaryotic AHL regulatory systems? *Aquat Microb Ecol*, **13**, 85–93.

Klut M.E., Bisalputra T., Antia N.J. (1983). Agglutination of the chlorophycean flagellate *Adunaliella tertiolecta* by treatment with lectins or divalent cations at alkaline pH. *J Phycol*, **19**, 112–115.

Knight-Jones E.W. (1951). Gregariousness and some other aspects of the settling behaviour of *Spirorbis*. *J Mar Biol Assoc UK*, **30**, 201–222.

Knight-Jones E.W., Stevenson J.P. (1950). Gregariousness during settlement in the barnacle *Elminius modestus* Darwin. *J Mar Biol Assoc UK*, **29**, 281–297.
Knight-Jones E.W., Crisp D.J. (1953). Gregariousness in barnacles in relation to the fouling of ships and to antifouling research. *Nature (Lond)*, **171**, 1109–1110.
Kon-ya K., Shimidzu N., Otaki N., Yokoyama A., Adachi K., Miki W. (1995). Inhibitory effect of bacterial ubiquinones on the settling of barnacle. *Balanus amphitrite*. *Experientia*, **51**, 153–155.
Kushmaro A., Rosenberg E. (1998). Effect of temperature. *Mar Ecol Prog Ser*, **171**, 131–137.
Kushmaro A., Loya Y., Fine M., Rosenberg E. (1996). Bacterial infection and coral bleaching. *Nature (Lond)*, **380**, 396.
Larman V.N., Gabbot P.A., East J. (1982). Physico-chemical properties of the settlement factor proteins from the barnacle *Balanus balanoides*. *Comp Biochem Physiol*, **72**, 329–338.
Leitz T., Wagne R.T. (1993). The marine bacterium *Alteromonas espejiana* induces metamorphosis of the hydroid *Hydractinia echinata*. *Mar Biol*, **115**, 173–178.
Lin H.P., Sommerfeld M.R., Swafford J.R. (1975). Light and electron observations on motile cells of *Porphyridium purpureum* (Rhodophyta). *J Phycol*, **11**, 452–457.
Littler M.M., Littler D.S. (1995). Impact of CLOD pathogen on Pacific coral reef. *Science*, **267**, 1356–1360.
Lovejoy C., Bowman J.P., Hallegraeff G.M. (1998). Algicidal effects of a novel marine *Pseudoalteromonas* isolate (Class *Proteobacteria*, Gamma subdivision) on harmful algal bloom species of the genera *Chattonella, Gymnodinium*, and *Heterosigma*. *Appl Environ Microbiol*, **64**, 2806–2813.
Maki J.S., Mitchell R. (1985). The involvement of lectins in the settlement and metamorphosis of marine invertebrate larvae. *Bull Mar Sci*, **37**, 675–683.
Maki J.S., Rittschof D., Mitchell R. (1992). Inhibition of larval barnacle attachment to bacterial films: an investigation of physical properties. *Microb Ecol*, **23**, 97–106.
Maki J.S., Rittschof D., Costlow J.D., Mitchell R. (1988). Inhibition of attachment of larval barnacles, *Balanus amphitrite*, by bacterial surface films. *Mar Biol*, **97**, 199–206.
Maki J.S., Yule A.B., Rittschof D., Mitchell R. (1994). The effect of bacterial films on the temporary adhesion and permanent fixation of cypris larvae, *Balanus amphitrite* Darwin. *Biofouling*, **8**, 121–131.
Maki J.S., Rittschof D., Schmidt A.R., Snyder A.G., Mitchell R. (1989). Factors controlling attachment of bryozoan larvae. A comparison of bacterial films and unfilmed surfaces. *Biol Bull*, **177**, 295–302.
Maki J.S., Rittschof D., Samuelsson M.-O., Szewzyk U., Yule A.B., Kjelleberg S., Costlow J.D., Mitchell R. (1990). Effect of marine bacteria and their exopolymers on the attachment of barnacle cypris larvae. *Bull Mar Sci*, **46**, 499–511.
Mary A.S., Mary V. SR., Rittschof D., Nagabhushanam R. (1993). Bacterial-barnacle interaction: potential using juncellins and antibiotica to alter structure of bacterial communities. *J Chem Ecol*, **19**, 2155–2167.
Maximilien R., de Nys R., Holmström C., Gram L., Givskov M., Crass K., Kjelleberg S., Steinberg P.D. (1998). Chemical mediation of bacterial surface colonisation by secondary metabolites from the red alga *Delisea pulchra*. *Aquat Microb Ecol*, **15**, 233–246.
McCarthy S.A., Johnson R.M., Kakimoto D. (1994). Characterization of an antibiotic produced by *Alteromonas luteoviolacea* Gauthier 1982, 85 isolated from Kinko Bay, Japan. *J Appl Bacteriol*, **77**, 426–432.
Mihm J.W., Banta W.C., Loeb G.I. (1981). Effects of adsorbed organic and primary fouling films on bryozoan settlement. *J Exp Mar Biol Ecol*, **54**, 167–179.
Mizobuchi S., Shimidzu N., Katsuoka M., Adachi K., Miki W. (1993). Antifouling substances against the mussel in an octocoral *Dendronephthya* sp. *Nippon Suisan Gakkaishi*, **59**, 1195–1199.

Morse A.N.C. (1991). How do planktonic larvae know where to settle? *Am Sci*, **79**, 154–167.
Morse D.E. (1984). Biochemical control of larval recruitment and marine fouling. In: Costlow J.D., Tipper R.C. (eds) *Marine Biodeterioration: an Interdisciplinary Study*. Annapolis, Maryland: Naval Institute Press, USA, pp. 134–140.
Morse D.E. (1985). Neurotransmitter-mimetic inducers of larval settlement and metamorphosis. *Bull Mar Sci*, **37**, 697–706.
Morse D.E. (1990). Recent progress in larval settlement and metamorphosis: closing the gaps between molecular biology and ecology. *Bull Mar Sci*, **46**, 465–483.
Mountfort D.O., Pybus V. (1992). Regulatory influences on the production of gamma-aminobutyric acid by a marine *Pseudomonas*. *Appl Environ Microbiol*, **58**, 237–242.
Neal A.L., Yule A.B. (1994). The interactions between *Elminius modestus* Darwin cyprids and biofilms of *Deleya marina* NCMB 1877. *J Exp Mar Ecol*, **176**, 127–139.
O'Connor N.J., Richardson D.L. (1998). Attachment of barnacle (*Balanus amphitrite* Darwin) larvae: responses to bacterial films and extracellular materials. *J Exp Mar Biol Ecol*, **226**, 115–129.
Okuda T., Neushul M. (1981). Sedimentation studies on red algal spores. *J Phycol*, **17**, 113–118.
Pawlik J.R. (1986). Chemical induction of larval settlement and metamorphosis in the reef-building tube worm *Pragmatopoma california* (Sabellariidae: Polychaeta). *Mar Biol*, **91**, 59–68.
Pawlik J.R. (1990). Natural and artificial induction of metamorphosis of *Phragmatopoma lapidosa california* (Polychaeta: Sabellariidae), with a critical look at the effects of bioactive compounds on marine invertebrate larvae. *Bull Mar Sci*, **46**, 512–536.
Pawlik J.R. (1992). Chemical ecology of the settlement of benthic marine invertebrates. *Oceanogr Mar Biol Annu Rev*, **30**, 273–335.
Pawlik J.R., Butmann C.A., Starczak V.R. (1991). Hydrodynamic facilitation of gregarious settlement of a reef-building tube worm. *Science*, **251**, 421–424.
Richmond M.D., Seed R. (1991). A review of marine macrofouling communities with special reference to animal fouling. *Biofouling*, **3**, 151–168.
Rodriguez S.R., Ojeda F.P., Inestrosa N.C. (1993). Settlement of benthic marine invertebrates. *Mar Ecol Prog Ser*, **97**, 193–207.
Satuito C.G., Shimizu K., Natoyama K., Yamazaki M., Fusetani N. (1996). Age-related settlement success by cyprids of the barnacle *Balanus amphitrite*, with special reference to consumptiom of cyprid storage protein. *Mar Biol*, **127**, 125–130.
Shiozawa H., Kagasaki T., Kinoshita T., Haruyana H., Domon H., Utsui Y., Kodama K., Takahashi S. (1993). Thiomarinol, a new hybrid antimicrobial antibiotic produced by amarine bacterium. *J Antibiot (Tokyo)*, **46**, 1834–1842.
Simidu U., Kita-Tsukamoto K., Yasumoto T., Yotsu M. (1990). Taxonomy of four marine bacterial strains that produce tetrodotoxin. *Int J Syst Bacteriol*, **40**, 331–336.
Steinberg P.D., Schneider R., Kjelleberg S. (1997). Chemical defenses of seaweed against microbial colonization. *Biodegradation*, **8**, 211–220.
Svane I., Havenhand J.N. (1993). Spawning and dispersal in *Ciona intestinalis* (L). *Mar Ecol*, **14**, 53–66.
Swift S., Throup J.P., Williams P., George P.C., Salmond P.C., Stewart G.S.A.B. (1996). Quorom sensing: a population-density component in the determination of bacterial phenotype. *Trends Biochem Sci*, **21**, 214–219.
Szewzyk U., Holmström C., Wrangstadh M., Samuelsson M.-O., Maki J.S., Kjelleberg S. (1991). Relevance of the exopolysaccharide of marine *Pseudomonas* sp. Strain S9 for the attachment of *Ciona intestinalis* larvae. *Mar Ecol Prog Ser*, **75**, 259–265.
Todd C.D., Keough M.J. (1994). Larval settlement in hard substratum epifaunal assemblages: a manipulative field study of the effcts of substratum filming and the presence of incumbents. *J Exp Mar Biol Ecol*, **181**, 159–187.

Todd J.S., Zimmerman R.C., Crews P., Randall S.A. (1993). The antifouling activity of natural and synthetic phenolic acid sulfate esters. *Phytochemistry*, **34**, 401–404.

Tosteson T.R., Corpe W.A. (1975). Enhancement of adhesion of the marine *Chlorella vulgaris* to glass. *Can J Microbiol*, **21**, 1025–1031.

Trapido-Rosenthal H.G., Morse D.E. (1985). L- Diamino acids facilitate GABA induction of larval metamorphosis in a gastropod mollusc (*Haliotis rufescens*). *J Comp Physiol*, **155**, 403–414.

Trapido-Rosenthal H.G., Morse D.E. (1986). Regulation of receptor-mediated settlement and metamorphosis in larvae of a gastropod mollusc (*Haliotis rufescens*). *Bull Mar Sci*, **39**, 383–392.

Uchida M., Nakayama A., Abe S. (1995). Distribition and characterization of bacteria capable ofdecomposing brown algae fronds in waters associated with *Laminaria* vegetation. *Fish Sci (Tokyo)*, **61**, 117–120.

Wahl M. (1989). Marine epibiosis. 1. Fouling and antifouling: some basic aspects. *Mar Ecol Prog Ser*, **58**, 175–189.

Wahl M., Jensen P.R., Fenical W. (1994). Chemical control of bacterial epibiosis on ascidians. *Mar Ecol Prog Ser*, **110**, 45–57.

Weiner R.M., Coyne V.E., Brayton P.P., Raiken S.F. (1988). *Alteromonas colwelliana* sp. nov. and an isolate from oyster habitats. *Int J Syst Bacteriol*, **38**, 240–244.

Wieczorek S.K., Todd C.D. (1997). Inhibition and facilitation of bryozoan and ascidian settlement by natural multi-species biofilms: effects of film age and the roles of active and passive larval attachment. *Mar Biol*, **128**, 463–473.

Wieczorek S.K., Clare A.S., Todd C.D. (1995). Inhibitory and facilitatory effects of microbial films on settlement of *Balanus amphitrite* larvae. *Mar Ecol Prog Ser*, **119**, 221–228.

Woodin S.A. (1991). Recruitment of infauna: positive or negative cues? *Am Zool*, **31**, 797–807.

Wrangstadh M., Conway P.L., Kjelleberg S. (1986). The production and release of an extracellular polysaccharide during starvation of a marine *Pseudomonas* sp. and the effect thereof on adhesion. *Arch Microbiol*, **145**, 220–227.

Wrangstadh M., Szewzyk J., Östling J., Kjelleberg S. (1990). Starvation specific formation of a peripheral exopolysaccharide by a marine *Pseudomonas* sp S9. *Appl Environ Microbiol*, **56**, 2065–2072.

Yoshikawa K., Takadera T., Adachi K., Nishijima M., Sano H. (1997). Korormicin, a novel antibiotic specifically active against marine Gram-negative bacteria, produced by a marine bacteria. *J Antibiot Tokyo*, **50**, 949–953.

Yoshinaga I., Kawai T., Ishida Y. (1997). Analysis of algicidal ranges of the bacteria killing the marine dinoflagellate *Gymnodinium mikimotoi* isolated from Tanabe Bay Wakayama Pref., Japan. *Fish Sci (Tokyo)*, **63**, 94–98.

8 Biofilm Infections on Implant Surfaces

Roger Bayston

Implantable devices, made from biocompatible biomaterials, are used to replace or restore a structure or function which is absent or nonfunctional due to trauma, disease or congenital defect. All are at risk of infection, although the rates of infection vary considerably with the siting of the device, whether it is used for constant infusion of fluids (*e.g.* an intravenous catheter) or is totally implanted (*e.g.* a prosthetic hip joint), the age of the patient and other risk factors. Most are caused by coagulase negative staphylococci and other organisms which form biofilms on the implant surface. The two main features of biomaterials-related infection are chronicity, due to evasion of the immune/inflammatory responses, and the need for surgical removal as part of treatment. Prevention may include the use of prophylactic antibiotics but some devices are not amenable to this approach and modified biomaterials offer the best option in these cases. It is likely that increasing numbers of implants will be used and the problem of infection, already a serious resource burden, will increase accordingly.

KEY WORDS: biofilms, biomaterials, infection, treatment, prevention

AETIOLOGY AND RESOURCE BURDEN OF BIOFILM INFECTIONS

INTRODUCTION

Implantable devices are used in medicine to replace a structure or restore a function which has been damaged by disease or injury or is congenitally absent. The first fully implantable device in general use was the shunt used for control of hydrocephalus, a condition caused by accumulation of fluid in the brain, in the late 1950's. Nowadays, devices in common use include large joint replacements, cardiac valves and pacemakers, intravenous catheters and vascular grafts. Various biomaterials are in use, including steel and titanium alloys, polymethylmethacrylate, high density polyethylene, woven polymers such as polytetrafluoroethane and terephthalate, and silicone elastomer. This last biomaterial is used in the majority of implantable devices.

While biomaterials, by definition, are highly biocompatible, they are all prone to infection by bacteria or fungi, although the incidence of infection varies with the type of biomaterial and particularly with its application (Table 1). The organisms causing the infections differ between devices. The most common causative organism *is Staphylococcus epidermidis*, which predominates in most implant infections, but in

Table 1 Incidence of infection in implantable devices.

Implant	*Site*	*Infection rate (%)*
Prosthetic hip/knee joint	Hip or knee	1–3
Vascular grafts	Thoracic aorta Abdominal aorta/groin	1 3–6
CSF shunts	Brain to abdomen	3–20
Central venous catheters	Major venous system	3–10
CAPD catheters	Abdominal wall	1.3 per patient/year
Voice prostheses	Trachea/oesophagus	*c.* 100*

*This figure represents prostheses becoming colonised with *C. albicans*; not all require removal as a consequence

Table 2 Causative microorganisms of biofilm infections in implants.

Implant	*Causative microorganisms*
Prosthetic hip/knee joint	*S. epidermidis*, *S. aureus*; peptococci, streptococci, Gram-negative bacteria, *P. acnes*
Vascular grafts	*S. epidermidis*, *S. aureus*, Gram-negative bacteria
CSF shunts	*S. epidermidis*, *S. aureus*, coryneforms, Gram-negative bacteria
Central venous catheters	*S. epidermidis*, *S. aureus*, Gram-negative bacteria, *Candida* spp.
CAPD catheters	*S. epidermidis*, *S. aureus*, Gram-negative bacteria, *Candida* spp.
Voice prostheses	*Candida* spp., *S. aureus*

some cases other organisms such as *Staphylococcus aureus*, *Pseudomonas aeruginosa* or *Candida albicans* (Table 2) are also important. The period of risk for infection also varies with the device. Examples of implant infections serve to illustrate the aetiological principles.

Aetiology

Large joint replacements

Prosthetic replacement for the hip was introduced in the late 1960's, and similar replacements are now available for the knee, elbow, shoulder and other joints. They are made from steel or titanium alloy, often with a polyethylene component and are often bedded into the bone using polymethylmethacrylate (PMMA) cement (Figure 1). Approximately 95,000 are inserted each year in the United Kingdom. The rate of infection is between 1% and 3% of operations, depending on the condition of the patient. Most infections are caused by bacteria which gain access to the implant during

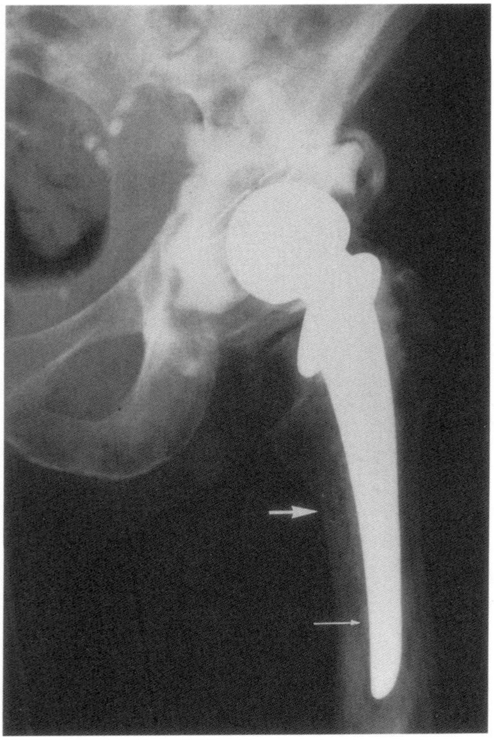

Figure 1 An X-ray of a prosthetic hip replacement. The arrows show bone resorption and remodelling due to infection.

surgical insertion and which originate on the skin of either the patient or the theatre staff. The relatively low infection rate is due to strenuous efforts to minimise this perioperative contamination, including rigorous aseptic technique, ultraclean air and prophylactic antibiotics (Lidwell *et al.*, 1984). Gram-positive bacteria such as *S. epidermidis* or *S. aureus* are believed to comprise the majority of causative organisms, although this might have to be revised in the light of recent findings. In one study, removed implants were sonicated and the sonicate cultured aerobically and anaerobically. Of 19 infected implants, 10 grew *Propionibacterium acnes*, an anaerobic Gram-positive bacillus of skin origin (Tunney *et al.*, 1997). A minority of infections are caused by bacteria which reach the bloodstream from infections in the respiratory, urinary or alimentary tracts at any time after surgery, and these late-onset infections, often due to Gram-negative bacteria such as *Escherichia coli* or *Haemophilus influenzae*, may appear years afterwards (Atkins & Bowler, 1998). Some infections are rapid in onset and give rise to severe symptoms and local suppuration, and diagnosis does not present much difficulty. However, others, particularly those due to *S epidermidis*, cannot easily be distinguished from other conditions such as loosening of the prosthesis due to ongoing bone disease or mechanical problems. Eventually most infected joint prostheses have to be removed in order to eradicate the infection, and this is followed

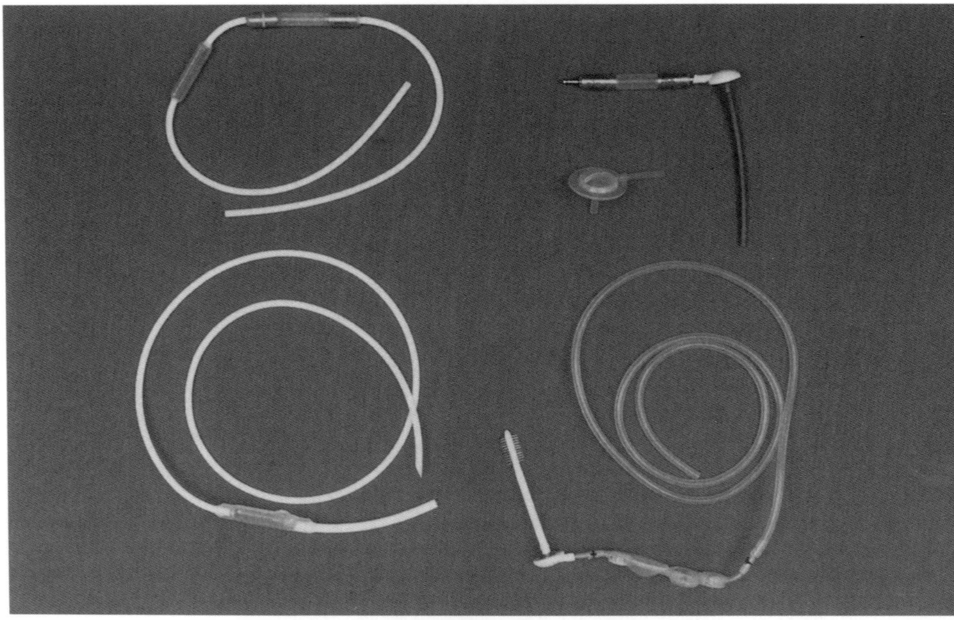

Figure 2 A collage of CSF shunts used in the control of hydrocephalus. Although some have metal parts, all have silicone elastomer as their main biomaterial and a valve to control the rate and direction of flow. Despite their different designs, all are functionally similar.

by a period of antibiotic treatment before insertion of a new prosthesis. Occasionally this is not possible due to the bone damage caused by the infection. Sometimes the infection relapses to involve the new prosthesis and the process may be repeated.

Hydrocephalus shunts

Hydrocephalus is due to accumulation of cerebrospinal fluid (CSF) in the cerebral ventricles, with consequent damage to the surrounding brain. It can be caused by any process, such as haemorrhage, meningitis, trauma or congenital defect, which impedes drainage of the CSF. Most cases occur in children. The treatment in common use is insertion of a valved tube from the cerebral ventricles to the heart or the peritoneal cavity to drain away the excess fluid. The device, made from silicone elastomer, and sometimes including steel components (Figure 2), is intended to be in place for life. Approximately 4000 are used each year in the United Kingdom. The rate of infection varies between 2 and 5% in adults and older children to 20% or more per operation in infants under 6 months of age (Pople *et al.*, 1992). CSF shunts are unusual in that almost all infections are introduced during surgery, although some do not become clinically apparent for months or years (Bayston, 1989). Almost all bacteria are derived from the patient's skin or mucous membranes at surgery. Over 75% of infections are due to *S. epidermidis* (or other coagulase negative staphylococci), with a minority caused by *S. aureus*, coryneforms and Gram-negative bacilli. As with hip prostheses, infected CSF shunts must be removed in order to eradicate the infection, followed by a period of external drainage of CSF (which has a high risk of secondary infection) while a course of antibiotics is administered (Bayston *et al.*, 1995). A new shunt is then inserted, but relapse sometimes occurs.

Vascular grafts

Narrowing or weakening of the major blood vessels is often due to ageing, compounded by factors such as diabetes or smoking. The affected segment of vessel is resected and replaced with a tubular graft of knitted or woven material. This must be sealed to prevent leakage, and this is usually done by the manufacturer, using modified gelatin. The graft later becomes incorporated, that is the endothelial lining of the vessel to which it connects grows into it. Infection occurs in about 5% of cases, either due to bacteria entering the graft area during surgery, or from intravenous catheterisation in the immediate post-operative period, or in a minority of cases from bacteria entering the bloodstream from another site such as the alimentary, urinary or respiratory tracts long after surgery. Grafts to the abdominal aorta or femoral artery have the highest rate of infection. Most infections are due to *S. epidermidis*, and these may not manifest themselves for months or years (Bandyk *et al.*, 1984), and are difficult to diagnose because of the frequent absence of specific features. However, the real threat is from infections due to *S. aureus* or *P. aeruginosa* (Geary *et al.*, 1990). These usually present early after surgery, with generalised illness (fever, signs of toxaemia), but can result in catastrophic haemorrage to the point of exsanguination. This is because the bacteria are capable of producing potent proteases such as elastase and collagenase which cause the anastomosis between the vessel and the graft to break down. Up to 47% of cases are fatal even when appropriate treatment is instituted, and up to 60% of survivors require limb amputation due to interruption of blood supply.

Continuous ambulatory peritoneal dialysis (CAPD)

For the treatment of renal failure, CAPD offers distinct advantages over haemodialysis in that the patient can self-administer the dialysis while carrying on a relatively normal existence in the community. A silicone catheter is inserted surgically into the abdominal wall to give access to the peritoneal cavity, and several litres of dialysis fluid are run in daily, allowed to remain for a few hours, and run out again to be discarded. Though the patient's quality of life could be said to be much better than for haemodialysis, there is a risk of peritonitis of about 1.3 patient episodes per year (Keane *et al.*, 1996), and infections around the catheter (exit site infections) also occur, although less frequently. These latter present with local pain, inflammation and suppuration. CAPD peritonitis is characterised by cloudy dialysate, abdominal pain and sometimes fever. A wide range of causative organisms has been reported, but the majority of cases of peritonitis are due to *S. epidermidis*, with about 10% each of *S. aureus* and Gram-negative bacilli such as *P. aeruginosa*. Peritonitis often relapses after treatment, and after several attempts to eradicate the infection the catheter is removed and the patient reverts to haemodialysis, with consequent disruption of lifestyle. Most laboratory research has concentrated on *S. epidermidis*, with the assumption that relapses are associated with biofilm inside the catheter. However, there is some evidence which calls this approach into question. A recent study (Bayston *et al.*, 1999) has found firstly *that S. epidermidis* infection rarely leads to catheter removal, this being associated with *S. aureus* or *P. aeruginosa* infection, and secondly that relapses are probably related to infection of the terephthalate fabric cuff around part of the catheter. This cuff is intended to prevent loosening of the catheter in its track and ingress of organisms by this route. However it provides an ideal environment for formation of a complex biofilm which is often not eradicated by conventional treatment. Most of these cuff infections are caused by *S. aureus* or Gram-negative bacilli such as *P. aeruginosa*. The biofilm of *S. epidermidis* inside the catheter is usually eradicated eventually by repeated courses of high-dose intraperitoneal antibiotics. Exit site infections are a more clinically obvious form of extraperitoneal infection and again are usually due to *S. aureus*, often requiring catheter removal. *C. albicans* occasionally causes CAPD peritonitis, and this is a life-threatening infection which can progress to generalised fungaemia.

Central venous catheters (CVCs)

Catheterisation of the central, rather than peripheral venous system is used in cases where access is needed for more than 1 or 2 d, and catheters can be in place for months. Two main clinical specialities account for the majority of CVCs. Patients who, either through trauma, disease or surgical resection are deprived of segments of small bowel (or its function) cannot absorb sufficient nutrients by the oral route, and some or all of their nutritional needs are met by intravenous infusion of amino acids, lipids, glucose and minerals. These must be administered in solutions of high concentrations which would cause necrosis of peripheral veins, but if infused into the central venous system they are diluted so rapidly that no harm results. A similar situation applies to oncology, where cytotoxic drugs, if given by a peripheral vein, would cause necrosis but can be given relatively safely by the central route. The catheters in common use are the Hickman and the Broviac, each of which may have more than one lumen. Made from silicone elastomer, they have a terephthalate cuff similar to that of the CAPD catheter. Infection is a major problem, with approximately 5% of the 20 million or so of the vascular catheters used annually in the United States becoming infected (Jansen, 1997). Again, most of the causative organisms, predominantly *S. epidermidis*, originate from the patient's skin (Sherertz, 1997). In the first 8 days of use, most of the biofilm in infected catheters is on the outer surface, but those becoming infected later show biofilm growth on the luminal suface (Raad *et al.*, 1993). Ideally the catheter should be removed in order to ensure eradication of infection, though often successful treatment can be achieved using intra-catheter antibiotics (Wang *et al.*, 1984) or by the "antibiotic lock technique" (Krzywda *et al.*, 1995). The application of a very small electric field has been shown to enhance the antibacterial effect of antibiotics so that biofilm bacteria can be eradicated by therapeutically achievable concentrations (Costerton *et al.*, 1994). Direct currents of very small densities ($<100\ \mu A\ cm^{-2}$) appear to be effective, and this technology is being developed further.

Infection in a central venous catheter causes bacteraemia, with continuous release of bacteria into the blood, resulting in fever and secondary infections including endocarditis, which can be fatal.

Voice prostheses

Loss of the ability to vocalise is the inevitable consequence of removal of the larynx due to malignancy. Twenty years ago, a device was developed which allows speech using a valved connector between the oesophagus and cervical tracheal remnant. Air is admitted to the oesophagus and is used to form voice sounds. The original, developed by Blom and Singer (Singer and Blom, 1980), has been followed by several others including the Groningen device. Most are made from silicone elastomer. Though the prostheses are intended to be in place for up to a year before being replaced, they are often removed early because of either leakage of liquids through the valve or failure to vocalise. This is often caused by colonisation of the prosthesis by microorganisms, and particularly *Candida* spp. (van Weissenbruch *et al.*, 1997). While most voice prostheses eventually become colonised by *Candida* spp. (Neu *et al.*, 1994), this does not always result in removal, though *Candida* spp. are capable of physically obstructing the valve mechanism, preventing closure and causing leakage of food and saliva. The most likely source of the organisms is the mouth, throat or skin.

CLINICAL IMPORTANCE

The incidence of infection in many implantable devices is low, but the number of such devices used each year means that the absolute number of infected cases is high. An example of this is prosthetic hip replacement, where the infection rate is approximately 2% if high risk cases are included, but as approximately 80,000 are inserted annually in the United Kingdom, at least 1600 infections result. Each of these involves expenditure of hospital resources during the months of investigation which it often takes to reach a diagnosis, followed by 4–6 weeks of antibiotic treatment, removal of the prosthesis, 2 more weeks of antibiotics and, where possible, insertion of a new prosthesis followed by convalescence (Saravolatz, 1993). Costs vary greatly but while the hospital costs of insertion of the first prosthesis are approximately £4500, the costs of an infection are often at least £15000 to £18000. The discomfort and distress to the patient is not quantifiable, but loss of earnings or even of employment, and expenditure by relatives on hospital visits can be considerable. Similarly, it is difficult to put an exact figure on the increased reliance on, for example, social services and rehabilitation. However, by a simple calculation, it can be shown that the hospital cost alone of infection annually in the United Kingdom approaches £30 million. When viewed similarly, hydrocephalus shunt infections produce about 450 infections each year at a total hospital cost of £6.5 million. Most of these are in young children (Pople *et al.*, 1992), and each episode of infection increases the risk of further physical and mental deterioration, again leading to enormous burdens on carers and social and educational services. The impact of sudden death from anastomotic breakdown or overwhelming systemic sepsis in patients with vascular graft infection is great, but that of the of amputation of part or all of one or both lower limbs in a high proportion of the survivors is arguably greater. Similar considerations apply to central venous catheter infection (Pittet *et al.*, 1994). It will be clear therefore that each type of implant must be assessed differently in order to determine the medical, psychosocial and financial impact of infection. Most of the scale of this impact is due to difficulties in diagnosis and in treatment, factors inherent in biomaterials-related infection because of their pathogenesis.

PATHOGENESIS OF MEDICAL BIOFILM INFECTIONS

Biofilm formation on medical implants almost always involves a single species, and mixed infections are rare. In this respect they differ considerably from biofilms in many industrial or natural systems, which are often a complex community of bacteria and other organisms. The exceptions are implants such as urinary catheters and voice prostheses, which do not require a surgical procedure for insertion, and which lie in a system (urinary tract, oesophagus/trachea) which has a mixed microbial flora. However, several basic principles are common to both types.

The most important primary event in biofilm formation, and therefore infection, is adherence of the causative organism to the implant. Many laboratory studies of bacterial adherence to biomaterials have been carried out, and certain conclusions can be drawn. Firstly, most organisms have been found to adhere more avidly to highly hydrophobic materials (Hogt *et al.*, 1983), but there are important species differences, and the antecedent cultural conditions and growth phase of strains are also critical (Fletcher, 1977). Also, test methods often dictate the result and great care must be taken in their design.

Another criticism of many of the microbiological laboratory studies on adherence to biomaterials is based on their lack of consideration of a conditioning film, and such studies are of limited value in predicting events *in vivo*. When a device is implanted into the human body, its surfaces rapidly become coated with glycoproteins derived from plasma and tissue fluids, so that the presenting surface to the bacteria is usually the conditioning film, and adherence to naked biomaterial rarely takes place. One exception is the hydrocephalus shunt, which if implanted into a patient with a normal CSF protein level (about 400 mg l^{-1}) will take much longer to be conditioned. Another is the lumen of the central venous catheter, unless it is used for aspiration of blood samples or administration of blood products. Conditioning films consist of complexes which include fibrinogen, fibronectin, laminin, albumen, collagen and sometimes platelets and other host-derived material. Data from aquatic environments indicate that the constituents of the conditioning film are not static, and considerable turnover can be expected *in vivo*. Examination of the effects of conditioning films on microbial adherence *in vitro* have shown that they may be clinically important. Albumen has been found to decrease adherence of *S. epidermidis*, but fibronectin enhances adherence of *S. aureus* (Maxe *et al.*, 1986; Herrmann *et al.*, 1988). However, it appears that, at least *in vitro*, both the concentration of fibronectin and differences between staphylococcal strains might affect the result, and there are some reports of fibronectin in these circumstances inhibiting adherence (Pascual *et al.*, 1986; Espersen *et al.*, 1990;

Brokke *et al.*, 1991; Brydon *et al.*, 1996). Plasma protein conditioning films were also found to reduce adherence of *S. epidermidis*, particularly at low shear stress (*i.e.* <15 dyn cm^{-2}), because of diminished non-specific adherence due to charge and hydrophobicity (Vacheethanasee *et al.*, 1998).

After adherence has occurred, slow multiplication takes place to produce plaques which eventually become multilayered. During this phase, the cells usually produce exopolymers which serve to support the developing biofilm by facilitating cell-cell adherence. As the biofilm thickens, the metabolism of the cells in the deeper layers slows and changes in response to nutritional deprivation (Gilbert *et al.*, 1990) and the susceptibility of the bacterial cells to antimicrobial agents decreases substantially (Brown *et al.*, 1988; Evans *et al.*, 1990). Deeper cells undergo phenotypic modification resulting in significant down-regulation of their electron transport sytem (Proctor and von Humbolt, 1998). This causes a general slowing of metabolism with little synthesis of cell wall material, and greatly decreased proton motive force at cell membrane level. When retrieved from explanted devices such cells grow very slowly to produce very small colonies (Small Colony Variants, SCV) which exhibit a spectrum of auxotrophy. These phenotypes in the biofilm do not express, or utilise, the target sites for antibiotic action (Bayston and Wood, 1997). An example is the insusceptibility to β-lactam antibiotics such as cephalosporins, which inhibit bacterial cell wall synthesis, but which are therapeutically irrelevant if the bacteria are not synthesising cell wall material. Similarly, the ineffectiveness of aminoglycosides is because of their requirement for bacteria-derived energy in order to transport them into the cell.

From the foregoing, it appears that exopolymer production is essential for the construction of biofilms, which will create the conditions in which this phenotypic switching can occur and in which he auxotrophic phenotypes can be maintained. The exopolymers produced by the various bacteria which can cause biomaterials-related infection are generally not well-characterised, although they are closely analogous to those found in environmental and industrial settings. The exopolymer of *P. aeruginosa* has been identified as alginate (Pier *et al.*, 1983), but that of *S. epidermidis* is a β-1,6-linked glycosaminoglycan (Mack *et al.*, 1996b). *S. epidermidis* exopolymer has now been named polysaccharide intercellular adhesin (PIA) (Mack *et al.*, 1996a). Recently, PIA synthesis has been found to be encoded by the *ica*ABCD operon, and considerable strides in understanding of biosynthesis and its genetic control have been made (Gerke *et al.*, 1998). In addition, the pheromones which control cell density in Gram-negative microbial communities by "quorum sensing" (Pesci and Iglewski, 1997) are now being considered as future therapeutic targets which might be subject to blocking by nonsense analogues. A similar system involving octapeptides has now been discovered in Gram-positive bacteria (Ji *et al.*, 1995), and it is possible that interference with quorum sensing might either turn off virulence factors or inhibit biofilm production, or both.

TREATMENT OF BIOFILM INFECTIONS

One of the key features of medical biofilm infections is the need to remove the device in order to eradicate the infection. This is because of the effect of the biofilm mode of growth on the susceptibility of the bacteria to antimicrobials, although in some cases, such as CSF shunts, there are additional pharmacological problems. Also, most biomaterials-related infections are caused by *S. epidermidis*, and this group of organisms exhibits multiresistance to antibiotics even when grown in conventional mode. Consequently, treatment of implant infections is often very difficult and treatment failures or relapses contribute significantly to their morbidity (Bayston, 1989). An example is the CSF shunt for hydrocephalus. The biofilm is formed inside the tubing of the shunt, and is therefore not accessible to antibiotics other than *via* the CSF. The infection gives rise to a feeble inflammatory response in the cerebral ventricles, and in these circumstances most drugs given orally or intravenously fail to penetrate into the CSF. If therapeutic concentrations of drugs are to be achieved in the CSF, they must be introduced directly into the ventricular system using the shunt as access. However, many antibiotics are toxic when so used, and this limits the choice. If, as is usually the case, the infection is caused by *S. epidermidis*, the choice is likely to be limited even further by multiresistance. If the role of phenotypic modification in the biofilm is then considered in increasing the concentration of antibiotic required for eradication to several hundred times the human toxic level, a radical approach involving removal of the device and the biofilm is seen to be almost always necessary. Even then, infection will often persist in the ventricular system. In order to eradicate it, powerful antibiotics such as vancomycin must be injected into the CSF daily for at least a week (Bayston *et al.*, 1995). In addition, intravenous antibiotics, usually rifampicin, must be given to enhance the effect of the vancomycin. When the infection has been eradicated, a new shunt is inserted. If eradication has not been absolute, a relapse of the infection is likely and the whole process must be repeated. While the pharmacological problems are not present in infections of prosthetic hip replacements, the other factors still dictate a surgical approach. The usual method is to administer intravenous antibiotics for at least 4 weeks to reduce the biofilm and associated tissue infection to a minumum, then to remove the prosthesis and after a further week or so of antibiotics to insert a new hip (Saravolatz, 1993). Antibiotics are continued for several weeks afterwards. Again, there is a risk of relapse. However, hope for a non-surgical approach based on a greater understanding of the pathogenesis is suggested by the work of Zimmerli *et al.* (1998), who have used a combination of rifampicin and ciprofloxacin in successful non-surgical eradication of infection in hip replacement. However, the infections were diagnosed very promptly after surgery (<1 month) and the drugs had to be given for 3–6 months. It remains to be seen whether the success of such an approach can be confirmed and whether the regimen can be improved. In the case of vascular graft infections, if the patient survives without haemorrhage, long term antibiotic treatment is necessary with removal of the infected graft and, after a period of bypass with its further risks of thrombosis, haemorrhage and secondary infection, insertion of a new graft. During this period, the blood supply to the lower limbs can be compromised, and amputation may be required (Bunt, 1994). Voice prostheses soon become leaky and dysfunctional when colonised by *C. albicans*, and have to be replaced.

PREVENTION OF BIOFILM INFECTIONS

Where the period of risk is equal to the life of the implant, its management and care become crucial to prevention of infection. An example is CAPD, where connections and disconnections of the dialysate system to the catheter are carried out by the patient in a community setting. Failure to comply with a strict regime of aseptic technique and hygiene increases the risk of peritonitis and exit site infection. The use of long term prophylactic antibiotics is not indicated as they are associated with development or selection of resistant organisms. However, the carriage of *S. aureus* in the nose of CAPD patients is associated with an increased risk of exit site infection and peritonitis, and its eradication lowers this risk (Wanten *et al.*, 1996). Systems specially designed to reduce infection, such as the "Y set" (Port *et al.*, 1992) also contribute.

Infection by *C. albicans* in voice prostheses has been addressed by daily oral application of nystatin, an anticandida drug, and this has significantly reduced candida infection (Leder and Erskine, 1997). However, recently it has been shown that dietary lactobacilli and *Streptococcus thermophilus* produce surfactants which can *in vitro* reduce significantly candida biofilm on voice prostheses as well as *E. coli* on urinary catheters (van der Mei and Busscher, 1999), and there is also evidence that drinking or eating foods such as live yoghurt can significantly prolong the life of the voice prosthesis by delaying candida colonisation.

The incidence of infection in prosthetic joint surgery has been reduced substantially in recent years, and this has been associated in part (though not necessarily directly causally) by the development of strict aseptic regimens and the use of filtered laminar air flows in the operating theatre (Lidwell *et al.*, 1984). However, the administration of a broad spectrum antibiotic immediately pre-operatively also reduces the incidence of infection, with or without the use of special theatre conditions. Though the incidence of infection in prosthetic joint surgery is now very low, and may be too low for statistically valid trials to be undertaken, each case of infection is so devastating that further measures are constantly being sought. Antibiotic powder can be mixed into the uncrosslinked polymethylmethacrylate bone cement in theatre immediately before use, and this has been shown to be released from the cement after crosslinking (Bayston and Milner, 1982). It has been shown to be beneficial when used in cases of already infected hip prostheses, considerably reducing the risk of relapse in the new hip but its use in all primary prosthetic joint operations is controversial (Strachan, 1995). The role of prophylactic antibiotics in reducing the incidence of surgery-associated infection of vascular grafts is debatable (Schmitt, 1990), although on theoretical grounds this should be effective if a satisfactory regimen can be devised.

The use of intravenous prophylactic antibiotics at insertion of central venous catheters has been claimed to reduce the risk of infection, but controlled trails do not generally support this (McKee, 1985; Ranson *et al.*, 1990) and while a great deal can be achieved by specially trained teams of hospital personnel, other measures are needed. A variety of processes for coating the surfaces of intravenous catheters with antimicrobials such as antibiotics (Raad *et al.*, 1996), antiseptics (Maki *et al.*, 1997) or silver (Heard *et al.*, 1998) have been described, but there are problems attached to their clinical use. Coatings have two major problems, *viz.* they are rapidly eluted from the surface by blood or infusate, and they are easily obliterated by conditioning film. Another approach has been taken in the case of CSF shunts, where prophylactic antibiotics do not help, even when administered intraventricularly (Brown *et al.*, 1994) and the infection rate, especially in young children, is high. Rather than coat the shunt catheters with antimicrobials, a process has been developed whereby the whole of the silicone material from which the shunt is made can be impregnated with antimicrobials, thus protecting both inner and outer surfaces. Stringent *in vitro* tests have shown that protective activity can be expected for up to 2 months (Bayston and Lambert, 1997), and the shunts are now available commercially. The advantage of impregnation over coating is that, in the former, the conditioning film makes no difference to the activity of the antimicrobials, and the shunts can be expected to protect against infection in the highest risk group, *i.e.* children who survive premature birth and cerebral haemorrhage and who often have high CSF protein levels. Similarly, impregnation of the gelatin sealant of vascular grafts with rifampicin promises to be effective in treatment as a new graft can be placed in the infected site with much less risk of re-infection (Strachan, 1995).

CONCLUSIONS AND FUTURE STRATEGIES

Many implantable devices are used in either the very young or the elderly, and both these groups are likely to expand in size. In addition, more sick and premature children are likely to be rescued due to improved emergency and intensive care facilities, and elderly people will live much longer and require more vascular grafts and prosthetic joint replacements. In this sense the problem of biomaterials-related infection is likely to increase. Options for treatment should be broadened to include more successful non-surgical therapy, but this could be offset by an increase in bacterial resistance to vital antibiotics. Fortunately there is now worldwide concern over resistance, and radical revisions in thinking on antibiotic use are being undertaken. Changes in clinical practice can in some instances reduce the need for antibiotics, and there may in the near future be a place for antimicrobial strategies involving substances other than antibiotics. However, there is a clear need for antibiotics, wisely used. Prevention is extremely important and every effort should be made to develop more effective strategies which are firmly based on a clear knowledge of aetiology.

REFERENCES

Atkins B.L., Bowler I.C.J.W. (1998). The diagnosis of large joint sepsis. *J Hosp Infect*, **40**, 263–274.

Bandyk D.F., Berni G.A., Theile B.L., Towne J.B. (1984). Aortofemoral graft infection due to *Staphylococcus epidermidis*. *Arch Surg*, **119**, 102–107.

Bayston R. (1989) *Hydrocephalus Shunt Infections*. Chapman and Hall Medical, London, pp. 160.

Bayston R., Milner R.D.G. (1982). The sustained release of antimicrobial drugs from bone cement. *J Bone Jt Surg J Br Vol*, **64B**, 460–464.

Bayston R., Lambert E. (1997). Duration of activity of cerebrospinal fluid shunt catheters impregnated with antimicrobials to prevent bacterial catheter-related infection. *J Neurosurg*, **87**, 247–251.

Bayston R., Wood H. (1997). Small colony variants: are they anything to do with biofilms? In: Wimpenny J., Handley P., Gilbert P., Lappin-Scott H., Jones M. (eds) *Biofilms: Community Interactions and Control*. Bioline Publications, Cardiff, pp. 161–165.

Bayston R., Andrews M., Rigg K., Shelton A. (1999). Recurrent infection and catheter loss in patients on continuous ambulatory peritoneal dialysis. *Perit Dial Int*, **19**, 550–555.

Bayston R., de Louvois J., Brown E.M., Hedges A.J., Johnston R.A., Lees P. (1995). Treatment of infections associated with shunting for hydrocephalus. Working Party on Use of Antibiotics in Neurosurgery of the British Society for Antimicrobial Chemotherapy. *Br J Hosp Med*, **53**, 368–373.

Brokke P., Dankert J., Carballo J., Feijen J. (1991). Adherence of coagulase-negative staphylococci onto polyethylene catheters *in vitro* and *in vivo*: a study on the influence of various plasma proteins. *J Biomater Appl*, **5**, 204–266.

Brown M.R.W., Allison D.G., Gilbert P. (1988). Resistance of bacterial biofilms to antibiotics: a growth rate related effect. *J Antimicrob Chemother*, **22**, 777–789.

Brown E.M., de Louvois J., Bayston R., Hedges A.J., Johnston R.A., Lees P. (1994). Antimicrobial prophylaxis in neurosurgery and after head injury: British Society for Antimicrobial Chemotherapy Working Party Report on Use of Antibiotics in Neurosurgery. *Lancet*, **344**, 1547–1551.

Brydon H.L., Bayston R., Hayward R., Harkness W. (1996). Reduced bacterial adhesion to hydrocephalus shunt catheters mediated by cerebrospinal fluid proteins. *J Neurol Neurosurg Psychiatr*, **60**, 671–675.

Bunt T.J. (1994). Treatment options for graft infections. In: Bunt T.J. (ed) *Vascular Graft Infections*. Futura Publishing Company Incorporated, New York, pp. 175–210.

Costerton J.W., Ellis B., Lam K., Johnson F., Khoury A.E. (1994). Mechanism of electrical enhancement of efficacy of antibiotics in killing biofilm bacteria. *Antimicrob Agents Chemother*, **38**, 2803–2809.

Espersen F., Wilkinson B.J., Gahrn-Hansen B., Rosdahl V.T., Clemmensen I. (1990). Attachment of staphylococci to silicone catheters *in vitro*. *Acta Path Microbiol Immunol Scand*, **98**, 471–478.

Evans D.J., Brown M.R.W., Allison D.G., Gilbert P. (1990). Susceptibility of bacterial biofilms to tobramycin: role of specific growth rate and phase in division cycle. *J Antimicrob Chemother*, **25**, 585–591.

Fletcher M. (1977). The effects of culture concentration and age, time, and temperature on bacterial attachment to polystyrene. *Can J Microbiol*, **23**, 1–6.

Geary K.J., Tomkiewicz Z.J., Harrison H.N., Fiore W.M., Geary J.E., Green R.M., De Weese J.A., Ouriel K. (1990). Differential effects of a Gram-negative and Gram-positive infection on autogenous and prosthetic grafts. *J Vasc Surg*, **11**, 339–347.

Gerke C., Kraft A., Sussmuth R., Schweitzer O., Götz F. (1998). Characterisation of the N-acetylglucosaminyltransferase activity involved in the biosynthesis of the *Staphylococcus epidermidis* polysaccharide intercellular adhesin. *J Biol Chem*, **273**, 18586–18593.

Gilbert P., Collier P.J., Brown M.R.W. (1990). Influence of growth rate on susceptibility to antimicrobial agents: biofilms, cell cycle, dormancy and stringent response. *Antimicrob Agents Chemother*, **34**, 1865–1868.

Heard S.O., Wagle M., Vijayakumar E., McClean S., Brueggemann A., Napolitano L.M., Edwards P., O'Connell F., Puyuna J.C., Doern G.V. (1998). Influence of triple lumen central venous catheters coated with chlorhexidine and silver sulfadiazine on the incidence of catheter-related bacteraemia. *Arch Int Med*, **158**, 81–87.

Herrman M., Vaudaux P.E., Pittet D., Auckenthaler R., Lew P.D., Schumacher-Perdreau F., Peters G., Waldvogel F.A. (1988). Fibronectin, fibrinogen, and laminin act as mediators of adherence of clinical staphylococcal isolates to foreign material. *J Infect Dis*, **158**, 693–701.

Hogt A.H., Dankert J., de Vries J.A., Feijen J. (1983). Adhesion of coagulase-negative staphylococci to biomaterials. *J Gen Microbiol*, **129**, 2959–2968.

Jansen B. (1997). Current approaches to the prevention of catheter-related infections. In: Seifert H., Jansen B., Farr B.M. (eds) *Catheter-Related Infections*. Marcel Dekker Incorporated, New York, pp. 411–446.

Ji G., Beavis R.C., Novick R.P. (1995). Cell density control of staphylococcal virulence mediated by an octapeptide pheromone. *Proc Natl Acad Sci USA*, **92**, 12055–12059.

Keane W.F., Alexander S.R., Baillie G.R., Boeschoten E., Gokal R., Golper T.A., Holmes C.J., Huang C-C., Kawaguchi Y., Piriano B., Riella M., Schaefer F., Vas S. (1996). Peritoneal dialysis-related peritonitis treatment recommendations: 1996 update. *Perit Dial Int*, **16**, 557–573.

Krzywda E.A., Andris D.A., Edmiston C.E., Quebbeman E.J. (1995). Treatment of Hickman catheter sepsis using antibiotic lock technique. *Infect Control Hosp Epidemiol*, **16**, 596–598.

Leder S.B., Erskine M.C. (1997). Voice restoration after laryngectomy: experience with the Blom-Singer extended-wear indwelling tracheoesophageal voice prosthesis. *Head and Neck*, **19**, 92–97.

Lidwell O.M., Lowbury E.J.L., Whyte W., Blowers R., Stanley S.J., Lowe D. (1984). Infection and sepsis after operations for total hip or knee-joint replacement: influence of ultraclean air, prophylactic antibiotics and other factors. *J Hyg*, **93**, 505–529.

Mack D., Haeder M., Siemssen N., Laufs R. (1996a). Association of biofilm production of coagulase-negative staphylococci with expression of a specific polysaccharide intercellular adhesin. *J Infect Dis*, **174**, 881–884.

Mack D., Fischer W., Krokotsk A., Leopold K., Hartmann R., Egge H., Laufs R. (1996b). The intercellular adhesin involved in biofilm accumulation of *Staphylococcus epidermidis* is a linear β-1,6-linked glucosaminoglycan: purification and structural analysis. *J Bacteriol*, **178**, 175–183.

Maki D.G., Stoltz S.M., Wheeler S., Mermel L.A. (1997). Prevention of central venous catheter-related bloodstream infection by use of an antiseptic-impregnated catheter. A randomized controlled trial. *Ann Intern Med*, **127**, 257–266.

Maxe I., Rydén C., Wadström T., Rubin K. (1986). Specific attachment of Staphylococcus aureus adherence. *Infect Immun*, **54**, 695–704.

McKee R. (1985). Does antibiotic prophylaxis at the time of catheter insertion reduce the incidence of catheter-related sepsis in intravenous nutrition? *J Hosp Infect*, **6**, 419–425.

Neu T.R., Verkerke G.J., Herrmann I.F., Schutte H.K., van der Mei H.C., Busscher H.J. (1994). Microflora on explanted silicone rubber voice prostheses: taxonomy, hydrophobicity and electrophoretic mobility. *J Appl Bacteriol*, **76**, 521–523.

Pascual A., Fleer A., Westerdaal N.A.L., Verhoef J. (1986). Modulation of adherence of coagulase-negative staphylococci to Teflon catheters *in vitro*. *Eur J Clin Microbiol*, **5**, 515–522.

Pesci E.C., Iglewski B.H. (1997). The chain of command in *Pseudomonas* quorum sensing. *Trends Microbiol*, **5**, 132–134.

Pier G.B., Matthews W.J., Eardley D.B. (1983). Immunochemical characterisation of the mucoid exopolysaccharide of *Pseudomonas aeruginosa*. *J Infect Dis*, **147**, 494–503.

Pittet D., Tarar D., Wenzel R.P. (1994). Nosocomial bloodstream infection in critically ill patients: excess length of stay, extra costs, and attributable mortality. *J Am Med Assoc*, **271**, 1598–1601.

Pople I.K., Bayston R., Hayward R.D. (1992). Infection of cerebrospinal fluid shunts in infants: a study of aetiological factors. *J Neurosurg*, **77**, 29–36.

Port F.K., Held P.J., Knolfe K.D., Turenne M.N., Wolfe R.A. (1992). Risk of peritonitis and technique failure by CAPD connection technique: a national study. *Kidney Int*, **42**, 867–974.

Raad I., Darouiche R., Hachem R., Mansouri M., Bodey G.P. (1996). The broad spectrum activity and efficacy of catheters coated with minocycline and rifampicin. *J Infect Dis*, **173**, 418–424.

Raad I., Costerton W., Sabharwal U., Sacilowski M., Anaissie E., Bodey G.P. (1993). Ultrastructural analysis of indwelling vascular catheters: a quantitative relationship between luminal colonisation and duration of placement. *J Infect Dis*, **168**, 400–407.

Ranson M.R., Oppenheim B.A., Jackson A., Kamthan A.G., Scarffe J.H. (1990). Double blind placebo controlled study of vancomycin prophylaxis for central venous catheter insertion in cancer patients. *J Hosp Infect*, **15**, 95–102.

Saravolatz L.D. (1993). Infection in implantable prosthetic devices. In: Wenzel R.P. (ed) *Prevention and Control of Nosocomial Infections*, 2nd Edition. Williams and Wilkins, Baltimore, pp. 683–707.

Schmitt D.D. (1990). Antibiotic usage in the prevention and treatment of graft infection. *Semin Vasc Surg*, **3**, 77–80.

Sherertz R.J. (1997). Pathogenesis of Vascular Catheter-Related Infections. In: Seifert H., Jansen B., Farr B.M. (eds) *Catheter-Related Infections*. Marcel Dekker Incorporated, New York, pp. 1–29.

Singer M.I., Blom E.D. (1980). An endoscopic technique for restoration of voice after laryngectomy. *Ann Otol Rhinol Laryngol*, **89**, 529–533.

Strachan C.J.L. (1995). The prevention of orthopaedic implant and vascular graft infections. *J Hosp Infect*, **30**, 54–63.

Tunney M.M., Patrick S., Gorman S.P., Nixon J.R., Anderson N., Hanna D., Rammage G., McKenne J.P. (1997). Isolation of anaerobes from orthopaedic implants. *Rev Med Microbiol*, **8**(S1), S92–S93.

Vacheethasanee K., Temenoff J.S., Higashi J., Gary A., Anderson J.M., Bayston R., Marchant R.E. (1998). Bacterial surface properties of clinically isolated *Staphylococcus epidermidis* strains determine adhesion on polyethylene. *J Biomed Mater Res*, **42**, 425–432.

van der Mei H.C., Busscher H.J. (1999). Microbial and host factors influencing adhesion to prosthetic devices. Abstract. In: *21st Int Congr Chemotherapy*, Birmingham.

van Weissenbruch R., Albers F.J.W., Bouckaert S., Nelis H.J., Criel G., Remon J.P., Sulter A.M. (1997). Deterioration of the Provox silicone rubber tracheoesophageal voice prosthesis: microbial aspects and structural changes. *Acta Otolaryngol*, **117**, 452–458.

Wang E.E.L., Prober C.G., Ford-Jones L., Gold R. (1984). The management of central intravenous catheter infections. *Ped Infect Dis*, **3**, 110–113.

Wanten G.J.A., van Oost P., Schneeberger P.M., Koolen M.I. (1996). Nasal carriage and peritonitis by Staphylococcus aureus in patients on continuous ambulatory peritoneal dialysis: a prospective study. *Perit Dial Int*, **16**, 352–356.

Zimmerli W., Widmer A.F., Blatter M., Frei R., Ochsner P.E. (1998). Role of rifampin for treatment of orthopedic implant related staphylococcal infections – a randomised controlled trial. *J Am Med Assoc*, **279**, 1537–1541.

9 Animal Models for the Study of Bacterial Biofilms

Merle E. Olson, Douglas W. Morck, Howard Ceri, Ronald R. Read and Andre G. Buret

Bacterial biofilms on medical devices and within tissue are constituted of bacterial and host cells, soluble bacterial and host products. Animal models are necessary in order to fully investigate the complex host-pathogen interactions which occur within this type of infection. Animal models are used to evaluate biocompatability of implanted medical devices, the prophylactic efficacy of antimicrobial agents and coatings, and the ability of chemotherapeutic agents to eliminate biofilm bacteria. Species such as rats, rabbits, pigs or sheep are commonly employed as they are of an appropriate size, and have well characterised physiological, pathological and immunological responses. Sheep and pigs are routinely used as they are sufficiently large to accommodate medical devices which are used in humans. Medical implants or materials are placed within tissue or organs of animals in such a manner that models true clinical situations. Implants may also be provocatively challenged with pathogens that are frequently associated with biofilm infections such as *Staphylococcus aureus*, *Staphylococcus epidermidis* or *Pseudomonas aeruginosa*. By using animal models, antimicrobials can be evaluated to determine their *in vivo* ability to resist bacterial colonization, and therefore their utility in reducing device-associated infections. In addition, chemotherapeutic agents or treatment strategies for the control of biofilm infections have been developed using such models. Animal models have also been developed for certain tissue biofilm infections including chronic pneumonia, biliary and bladder stones, prostatitis, vegetative endocarditis, keratitis and osteomyelitis. This chapter discusses a number of models which have generated improved and novel strategies in the management of biofilm infections.

KEY WORDS: *Staphylococcus aureus*, *Staphylococcus epidermidis, Pseudomonas aeruginosa*, antibiotic, biocompatability, biomaterial, bladder, endocarditis, CAPD, catheter, inflammation, infection, keratitis, osteomyelitis, pneumonia, pig, prostatitis, rabbit, rat, sheep

INTRODUCTION

Animal models are an integral component of biomedical research in general, and their use in the study of bacterial biofilms is of central importance. The significance of infections associated with bacterial biofilms on medical devices and within tissue has been recognised over the past two decades. Device-associated infections are some of the most common types of bacterial infection and are the most difficult to treat (Vaudaux *et al.*, 1994). Such infections are usually initiated by autochthonous micro-organisms from the skin (*Staphylococcus aureus, Staphylococcus epidermidis*), the bowel

(*Escherichia coli*) or the environment (*Pseudomonas aeruginosa*). When otherwise not associated with an implant, such organisms are usually readily eliminated by host defences or chemotherapeutic agents. It has been shown in clinical disease and in animal models that implantation of a foreign body increases the susceptibility to bacterial infection (Vaudaux *et al.*, 1994). Animal models have improved understanding of complex interactions between adherent bacteria on a medical device and the surrounding tissue. In turn, this will help develop therapeutic agents and/or medical devices less prone to infections. Implant infections frequently result when bacteria become protected from host elimination through association with the implant. A true representation of such associations can only be studied by using animal models. Ultimately, the information gained in studies employing animal models will help establish a rational basis for initiation of human clinical trials. This chapter is not an all-inclusive survey of animal models for biofilm infections, but discusses effective and commonly used models that have provided valuable information regarding the pathogenesis and treatment of tissue and device-associated infections.

LEGAL AND REGULATORY ISSUES

Europe, Canada, the U.S., Japan and Australia have all enacted laws or strict regulatory mechanisms which provide protection to animals used in research and testing (Olfert *et al.*, 1993). Good Laboratory Practices Regulations also require that ethical and scientific reviews of animal experiments are conducted. Although the use of animals may be necessary, the study protocols must be developed to ensure that the minimum number of animals are used and that they are not exposed to pain and excessive stress. In any infectious disease study a clearly defined endpoint is necessary as animals should not be permitted to die of the infection. Carefully designed studies that have been reviewed by an Animal Care and Use Committee (ACUC) will ensure optimal results and minimal or no adverse effects to the animal (Olfert *et al.*, 1993).

ANIMAL MODELS FOR STUDYING BIOFILM-HOST INTERACTIONS

Bacteria colonizing the surface of an implant are protected from the humoral and cellular immune response of the host (Vaudaux *et al.*, 1994; Zimmerli *et al.*, 1982). The glycocalyx of biofilm bacteria has been shown to impair recognition and elimination of the bacteria by host antibodies or activated phagocytes.

Implant Materials

Numerous publications describe animal models where foreign material is implanted within the tissue or body cavity of mice, rats or rabbits. Rabbits are commonly used as their body size is more adaptable to larger pieces of material than can be implanted in mice or rats. Typically pieces or sheets of raw material (1–4 cm^2) are surgically placed within the peritoneal cavity or subcutaneous tissues. When raw materials are used, it is essential that they are soaked and washed in saline before implantation to remove possible irritating contaminants. Materials must be sterilized by autoclaving, ethylene oxide exposure or a chemical treatment prior to implantation. In order to gain a better understanding of the colonization kinetics on a given material, bacterial biofilms are frequently allowed to form on implants prior to insertion. Most commonly, devices or biomaterials are exposed to logarithmically growing bacteria in a nutrient broth for 2 to 18 h. The implant is then washed in saline to remove planktonic organisms and surgically implanted. Alternatively, bacteria may be injected onto the surface of the material at various time intervals following surgical implantation.

Animal models are particularly useful for evaluating host responses to implant materials whether colonized with bacteria or not. These interactions cannot be readily studied in human patients as the implant and associated tissue must be removed in order to determine biocompatability and the presence and location of the bacteria. When a sterile implant is placed in an animal there is some degree of foreign body reaction. Sterile implants induce various degrees of fibroblast proliferation and deposition of extracellular proteins like fibrin and collagen (Buret *et al.*, 1991). Most biocompatible implant materials have few inflammatory cells (macrophages, polymorphonuclear leukocytes, lymphocytes) associated with this fibrous response. This host reaction, by itself, may interfere with the function of the implant, if it obstructs luminal flow or inhibits proper integration within the tissue. Colonization with a bacterial biofilm considerably exacerbates the host response to the implant. Indeed, when the foreign body is colonized by bacteria, the fibrous capsule is greatly increased in thickness, and large numbers of activated phagocytes are distributed throughout its layers (Buret *et al.*, 1991; Ward *et al.*, 1992). These cells release or cause the release of pro-inflammatory mediators like proteases, tumour necrosis factor, interleukins, prostaglandins, complement and coagulation factors. Typical host reactions toward sterile and bacteria-colonized implant material in a rabbit model are illustrated in Figure 1. Very large differences have been reported in the host biocompatability of different sterile and bacteria-colonized biomaterials. The cause for such differences is often difficult to ascertain.

Implanted Growth Chambers

Bacteria growing in pure culture in the laboratory have been shown to differ considerably from those growing within a host. In order to study the characteristics of *in vivo* grown bacteria, implantable chambers have been developed (Lambert *et al.*, 1990). In this model, sheets of implant material previously colonized with the bacteria of interest are placed in a chamber that has millipore filters on either end. This chamber is then surgically placed in the peritoneum or subcutaneous tissue of an animal such as a rat, rabbit or pig (Lambert *et al.*, 1990; Ward *et al.*, 1992; McDermid *et al.*, 1993a). The bacteria receive nutrients from the host but remain localised to the implant surface as they are restricted to the lumen of the implant. This model has the advantage of enabling the recovery of *in vivo* grown bacteria without contamination by host cellular material. As bacterial products are able to leave the chamber, the host immune response to these bacteria, as well as their soluble products, can be evaluated. Such chambers are very valuable for the study of the phenotypic plasticity of biofilm bacteria associated with implanted medical devices.

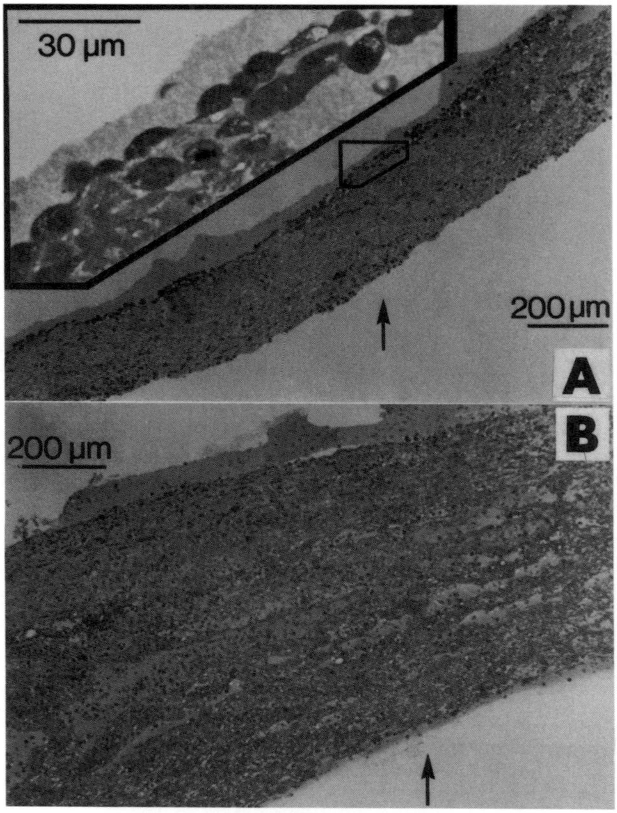

Figure 1 Light micrographs of biofilms from uncolonized (A) and colonized (B) silicone implants at 4 d post implantation. The implant capsule recovered from a control animal is significantly thinner than the colonized implant and is covered by a layer of macrophages (insert). There are a higher numbers of neutrophils in the colonized implant. Arrows = the bottom of the biofilm which coated the implant. (Reproduced from Buret *et al.*, 1991, with permission.)

ANIMAL MODELS OF DEVICE-ASSOCIATED INFECTIONS

Vascular Catheters

Vascular catheters are readily colonized by bacteria causing localized infection and inflammation. Critically ill and cancer patients who frequently have multiple catheters and are immunosuppressed are particularly susceptible to catheter-associated infections. These catheter associated infections can lead to septicaemia and the dissemination of pathogens to other organs such as the kidney, lung and heart. The bacteraemia may lead to organ failure and death. It is therefore extremely important to understand the pathogenesis of vascular catheter associated infections so that improved control strategies may be developed.

Colonization of catheters (*e.g.* vascular, peritoneal) can be evaluated by tunnelling the catheters subcutaneously on the backs of rabbits or pigs (McDermid *et al.*, 1993b). Several catheters are placed perpendicular to the spine using a trochar that can accommodate the catheter within its lumen. The trochar with its catheter is inserted into the subcutaneous space, and the catheter is delivered using a lumenal stylette as the trochar is withdrawn. In this way it is implanted with minimal trauma which is representative of most clinical situations. The skin exit sites are then inoculated with *S. aureus* or *S. epidermidis* and the movement of bacteria along the catheter is evaluated, as well as the local tissue reaction. Catheters can also be implanted without a skin exit site. The bacterial challenge is injected into the tissue surrounding the catheter or intravenously to determine if a particular material is susceptible to colonization by local contaminating pathogens or by haematogenous bacteria.

Ultimately vascular catheters are required to be placed within a vessel. It has been shown that when a catheter is removed from a vessel, the majority of the biofilm is stripped from the surface of the catheter (Maki, 1994). It is therefore extremely important when examining catheter biofilms and the local vascular reaction to split the vessel and examine the catheter and vessel *in situ*. Clearly, this requires the use of a laboratory animal model. Most vessels of small laboratory animals are too small to permit the placement of vascular catheters typically used in human medicine. The jugular vein of rabbits may be used to evaluate smaller vascular catheters. When studying bacterial colonization and host reactions to intravascular catheters, sheep are ideal as they have large vessels and do not self mutilate the catheter exit sites. The sheep external jugular vein catheter model has permitted evaluation of vascular catheter materials, catheter coatings and catheter placement strategies (Olson *et al.*, 1992). The sheep model is particularly valuable in examining the consequences of septicaemia-associated colonized vascular catheters. Using the sheep model, it has been shown that vascular catheters can lead to septicaemia, embolic pneumonia and vegetative endocarditis (Olson *et al.*, 1992).

Urinary Catheters

Foley catheters are used to drain urine from the bladder and to facilitate repair of injuries to the urethra. They are extensively used following surgery and in critically ill patients. Catheter-associated urinary tract infection (UTI) is the most common nosocomial infection and there is an accumulative risk of infection for each day of catheterization (Kaye and Hessen, 1994). Urinary tract infections in critically ill patients can lead to significant morbidity and morality. Urinary catheters are readily colonized, usually by faecal pathogens, which may cause ascending bladder infections. Catheter-associated infections usually result from meatal mircroorganisms that attach and migrate along the exterior of the urinary catheter. Alternatively pathogens may contaminate the drainage bag and ascend into the bladder through the catheter lumen.

An animal model has been developed to evaluate Foley catheters and urine drainage systems that resist colonization, and to evaluate chemotherapeutic agents that can be employed to prevent and treat catheter-associated urinary tract infections (Olson *et al.*, 1989b; Nickel *et al.*, 1991; Morck *et al.*, 1994). The diameter of the rabbit urethra is unusually large for its body size and a 8 to 12 French Foley catheter

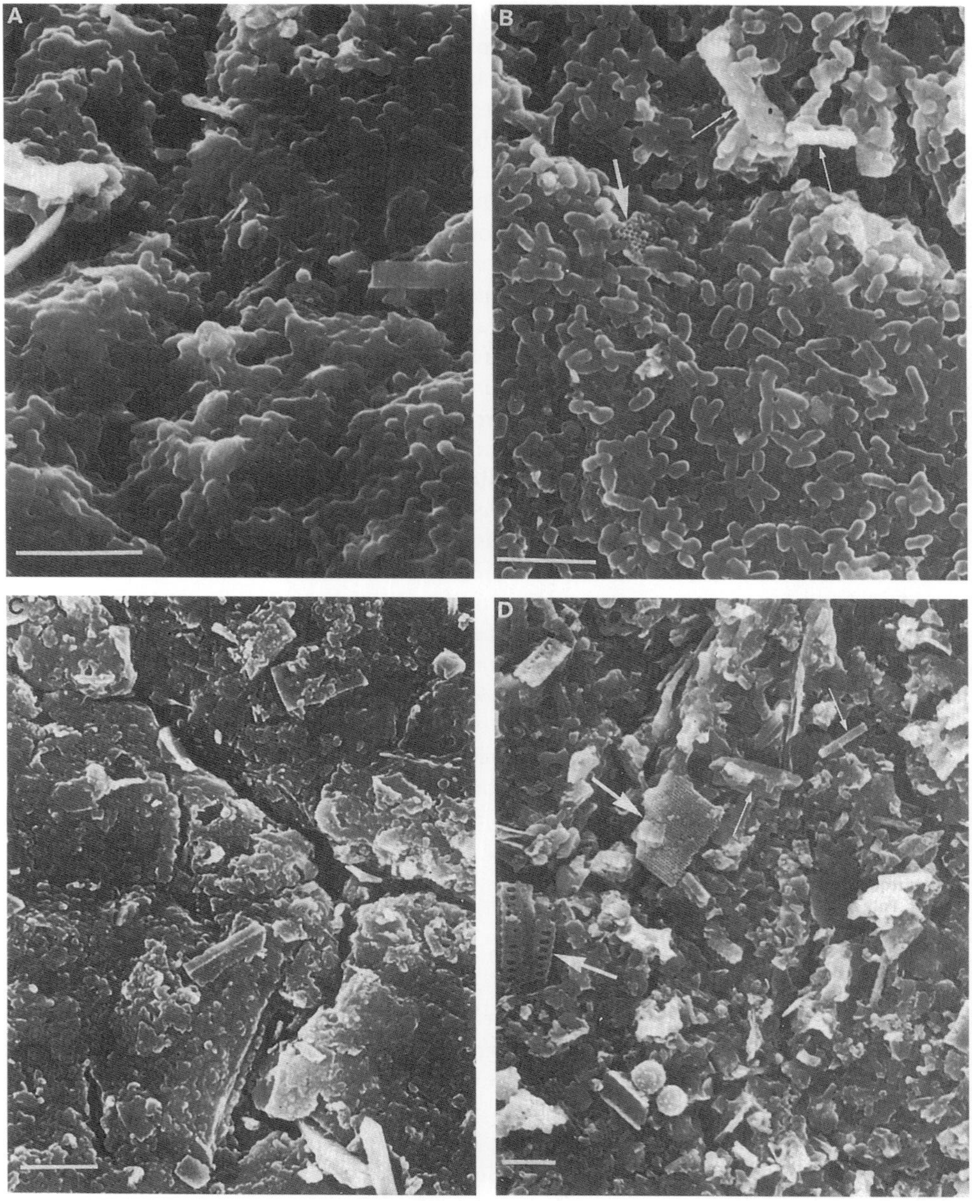

Figure 2 Scanning electron micrographs of intraluminal surface of urethral catheters which show bacteria embedded in thick biofilm of untreated rabbit (A), and in rabbit treated with 100 mg kg^{-1} amdinocillin (B). Biofilm shows signs of deterioration. Cracks expose underlying catherter surface and diatom (large arrow) and crystals (small arrows). Biofilm remains, but has deteriorated, in rabbit treated with 200 mg kg^{-1} amdinocillin (C). In rabbit treated with 400 mg kg^{-1} amdinocillin (D), biofilm and bacteria are absent, exposing underlying urine phosphate crystals (small arrows) and diatoms (large arrows). Bar = 5 μm. (Reproduced from Olson *et al.*, 1989b, with permission.)

can be used to catheterize a male rabbit. To promote urine flow intravenous fluids and nutrients are provided. Animals must be lightly restrained in a special cage during the study to prevent mutilation of the catheterized sites. The meatus or drainage system is inoculated with a uropathogen to provocatively challenge the system. Unprotected catheters and their associated draining systems will become colonized within 72 h. This model has been used to demonstrate the kinetics of biofilm formation on the urinary catheters following meatal challenge (Olson *et al.*, 1989b). Figure 2 illustrates the typical bacterial biofilm formed on the surface of a Foley catheter in a rabbit with a catheter-associated urinary tract infection. This model has been valuable in developing biocompatible catheters and coatings that resist colonization, as well as devising drainage bags that are resistant to environmental contamination (Khoury *et al.*, 1989; Nickel *et al.*, 1991).

The same animal model system has been used to evaluate chemotherapeutic treatment regimes for controlling catheter-associated urinary tract infections (Olson *et al.*, 1989b; Morck *et al.*, 1994). After a catheter-associated UTI is induced in the rabbit, specific treatment regimes are initiated. Figure 2 illustrates the effective elimination of bacterial biofilm from the surface of a Foley catheter. This model has demonstrated that certain chemotherapeutic agents are superior to others *in vivo*, in spite of equivalent, favourable minimum inhibitory concentration (MIC) data (Morck *et al.*, 1994). The efficacy may be associated with the specific metabolism of biofilm bacteria, the ability of the antibiotic to penetrate the biofilms or pharmacokinetics which enable the antibiotic to persist within the tissue for an extended period. This model is able to evaluate all of these parameters. Table 1 illustrates that efficacy in elimination of bacteria from the catheter surface and tissue closely correlates with the ability of the antibiotic to eliminate the bacterial biofilm. These data are achieved by measuring the minimum biofilm eradication concentration (MBEC) using the Calgary biofilm device (Ceri *et al.*, 1999). In contrast, the MIC does not correlate with the efficacy of the antibacterial agent. The MBEC model has been a powerful tool to evaluate and select antibacterial chemotherapeutic agents.

Table 1 Efficacy in elimination of bacteria from the catheter surface and tissue. Comparison of MIC and MBEC as measured by the Calgary biofilm device.

Treatment	*MIC* (μg ml^{-1})	*MBC* (μg ml^{-1})	*MBEC* (μg ml^{-1})	*Catheter surface bacteria* (cfu cm^{-2})	*Bladder mucosa bacteria* (cfu g^{-1})
Fleroxacin	0.16	0.32	2.0	0.00	0.00
Ampicillin	2.0	4.0	>1024	0.00	26
Trimethoprim-sulphamethoxasole	1.0	>128	>1024	46773	24547
Gentamicin	0.5	1.0	512	0.00	26
No Antibiotic	NA	NA	NA	40738	16218

MIC = minimum inhibitory concentration on planktonic bacteria (μg ml^{-1})
MBC = minimum bactericidal concentration on planktonic bacteria (μg ml^{-1})
MBEC = minimum biofilm eradication concentration from the Calgary biofilm device on sessile bacteria (μg ml^{-1})
NA = not applicable

Biliary Stents

Biliary stents are used to drain the bile from the obstructed biliary tree. Such obstructions are associated with stones or biliary malignancies. Stents are normally placed endoscopically into the common bile duct to drain bile into the duodenum. A bacterial biofilm forms (usually within 4 months) within these stents (Leung *et al.*, 1988), and the biofilm and biliary sludge result in stent blockage and cholangitis.

As domestic cats have bile that is similar to humans, a feline animal model was developed to study the pathogenesis of biliary stent blockage and to develop stenting systems that resist obstruction (Sung *et al.*, 1990; Libby *et al.*, 1994). This cat model involves surgically placing the a polyethylene stent (1.0 mm external diameter, 0.86 mm internal diameter) in the common bile duct through an enterotomy incision or through an incision into the duct itself. Stents become colonized by ascending bacteria from the duodenum or descending microorganisms from a transient bacterobilia from the biliary tract. The model is well suited to the study of biofilm obstruction of biliary stents as obstruction readily occurs once the biofilm is established. Using this model, it has been demonstrated that the stent is colonized by duodenal bacteria and that colonization may be prevented by placing the stent entirely within the bile duct (Sung *et al.*, 1990). Certain prophylactic antibiotics (ciprofloxacin) also prevent biofilm from forming, hence avoiding obstructions (Libby *et al.*, 1994).

CAPD Catheters

Continuous ambulatory peritoneal dialysis (CAPD) has become a favourable alternative to haemodialysis because it is less costly and provides patients with the freedom of not being tied to a hospital haemodialysis unit. A major drawback of CAPD is associated with peritonitis. Peritoneal infections result from contaminated dialysate, from bacterial colonization of the catheters or drainage systems, and/or tissue infections around the catheter. These infections can be life threatening and usually require high levels of antibiotics, and frequently require removal of the catheter.

Rabbit and pig models have been developed to study the pathogenesis of these infections, and to develop catheters and drainage systems that resist colonization (Read *et al.*, 1989; McDermid *et al.*, 1993a; 1993b; Ababio *et al.*, 1995). The rabbit model uses a Tenchoff catheter that is placed in the peritoneum and exits through a dorsal skin incision. Inoculating the skin exit site with environmental pathogens (*S. epidermidis, S. aureus*, and *P. aeruginosa*) provocatively challenges the system. This is an extremely useful model, but it has certain drawbacks that are not encountered in the pig model. Pig skin closely resembles human skin and the pig can be made uremic by a partial nephrectomy. In the pig model, one kidney is totally removed and approximately 3/4 of the other kidney is removed using large tissue staples. This produces a mildly uremic animal that closely represents humans undergoing CAPD (McDermid *et al.*, 1993a). Animals made uremic were more susceptible to infections and their immune response to challenge organisms is markedly different from that seen in immunocompetent animals. The catheter exit sites may be challenged in a similar manner as in the rabbit model. The pig model has been extremely useful for investigating the pathogenesis of CAPD catheter infections and is recommended for the development of new catheter systems that may resist infection.

Intrauterine Devices

Women using intrauterine contraceptive devices (IUCD) are at risk of developing uterine infections and pelvic inflammatory disease (Toivonen, 1993). An animal model was developed to study the pathophysiology of these infections. The model involves placement of a human IUD into the uterine horns of rabbits (Jacques *et al.*, 1986). The rabbit is an ideal model for studying IUCD as the right uterus is totally separate from the left. Each uterine horn has its own cervix that enters the vagina thus allowing the establishment of a control within the same animal. Following implantation, autochthonous bacterial biofilms readily form on the surface of the IUCD. Using this model it was shown that IUCDs placed surgically within the vagina were not colonized by bacteria. However, when the IUCD was passed through the cervix it became colonized. This research has led to the speculation that the retrieval tails acted to wick bacteria from the vagina through the cervical barrier, and that bacterial colonization of the device may contribute to the contraceptive activity of the IUD (Jacques *et al.*, 1986).

ANIMAL MODELS OF BIOFILM TISSUE INFECTIONS

Bacterial biofilms are important in many organ infections and particularly chronic infections. The pathogenesis of these infections involves both the bacterial pathogen and the host. Initially the bacterium attaches to the tissue surface through a host receptor or a non-specific mechanism. The bacteria produce a glycocalyx which prevents elimination by host products and cells (*e.g.* immunoglobulins, phagocytes). The bacteria can then multiply on the surface or invade and disseminate within the tissue. Bacterial microcolonies release toxins and the host phagocytes release factors which damage the surrounding tissue. Specific models using various pathogens are required to aid in understanding the pathogenesis of the disease process.

Pneumonia

Bacterial pneumonic diseases are common in both humans and animals and lead to significant mortalities. The mammalian lung, however, has a great capacity to eliminate pathogens from the respiratory system. This is achieved by highly efficient activity of the ciliated respiratory epithelium in the respiratory tree, and phagocyte-mediated clearance. Biofilms play a major role in the development of chronic pneumonia by establishing microcolonies in alveolar spaces or within pulmonary phagocytes (macrophages, neutrophils). Animal models have been established to help understand the pathogenesis of the disease and develop strategies for prevention and treatment.

Lung infections are not easily established in laboratory rodents and often require a mechanism of promoting the disease (Pennington, 1985). Rats and guinea pigs are the best models for studying pulmonary infections as they are sufficiently large to inoculate the airway with the pathogen through the oral cavity or following a tracheotomy. Mice can be used but need to be challenged by intranasal or aerosol challenge. Inoculation of up to 10^8 organisms (*e.g. P. aeruginosa*, *Streptococcus pneumoniae*, *Haemophilus influenzae*) into the lung of a healthy animal usually leads to the rapid elimination of the organism. Larger numbers of bacteria may cause acute or severe pneumonia, inflammation, septicaemia and toxaemia. Animals usually die acutely from bacterial toxins and the inflammatory reaction induced by the bacteria, and not from the colonization with bacteria as biofilm microcolonies. Chronic pneumonia can be induced in rodents by inoculation of bacteria as a microcolony (*e.g.* agar bead *P. aeruginosa* model), by combining the pathogen with an organism which produces an exopolysaccharide (*e.g. S. aureus*), by challenging immunosuppressed animals or by inoculating the lung with an intracellular pathogen (*Legionella pneumophilia, Mycobacterium tuberculosis*). The agar bead model involves production of a bacterial microcolony by enrobing *P. aeruginosa* within agar beads which are injected through a catheter intrabronchially into the lungs of rats, either oro-tracheally or through a tracheotomy (Cash *et al.*, 1979). Chronic streptococcal pneumonia is produced by inoculation of a mixed culture of *S. pneumoniae* and *S. aureus* into the lung of rats (Smith *et al.*, 1990). *Legionella* pneumonia can be induced in guinea pigs by inoculation of 10^6 *L. pneumophilia* cells into the lung through a tracheotomy incision (Wright *et al.*, 1993). These models have been vital in the examination of both host and bacterial factors involved in chronic bacterial pneumonia.

Endocarditis

Endocarditis is an infection of the inner lining of the heart (the endocardium) and usually involves the heart valves. Bacterial endocarditis is a serious complication of septicaemia and vascular foreign body infections. Endocarditis results when intravascular bacteria colonize a normal or damaged heart valve. Following an injury to a heart valve, bacteria, platelets and fibrin readily adhere to the surface and the bacteria begin to multiply. The bacteria form a microcolony within the protective biofilm and the host inflammatory response causes further tissue damage and fibrosis. Both the rabbit and the rat have been extensively used to study the pathogenesis and treatment of this disease (Durack *et al.*, 1973; Santaro and Levison, 1978). Bacterial colonization (*Streptococcus faecalis*, *S. aureus*) is promoted by damaging the heart valve with an intravascular polyethylene catheter or wire. Once the heart valve is colonized, the bacteria are protected from elimination from the host by antibiotics, because of the glycocalyx and the extensive fibrosis. Heart valves can be colonized for weeks before animals become clinically ill, thereby making the model useful for the study of pathogenic mechanisms and therapeutic measures. The vegetative endocarditis model has been used extensively to screen potential antibiotics for the treatment of chronic bacterial infections (Ades and Craig, 1998).

Osteomyelitis

When bacteria (usually *S. aureus*) invade and colonize bone, a serious chronic infection, known as osteomyelitis ensues. In this infection, bacterial numbers are low but the host inflammatory reaction and bacterial toxins induce severe tissue damage resulting in both bone loss and soft tissue destruction. Animal models of osteomyelitis have been described in the dog, rabbit, dog, chicken and rat (Mayberry-Carson *et al.*, 1984; Power *et al.*, 1990). The rabbit model has a significant shortcoming as this species has a very large medullary cavity and thin cortex for its body size. For studying the role of biofilms in osteomyelitis a rat model has been extensively employed (Power *et al.*, 1990). In this model a burr hole is drilled into the proximal tibia and *S. aureus* inoculated with a sclerosing agent (arachidonic acid) into the medullary cavity. The burr hole is then sealed with bone wax. This leads to a chronic infection within 3 weeks and the infection persists over 12 weeks. The infection is characterised by bone deformation, inflammation with large numbers of polymorphonuclear leukocytes, and microcolonies of glycocalyx-encased cocci sequestered within bone. This model closely mimics the human infection and has permitted investigations into the pathogenesis of disease and the development new methods of treatment (Power *et al.*, 1990).

Prostatitis

In males, chronic bacterial prostatitis is the most common cause of relapsing urinary tract infection (Nickel *et al.*, 1990). The pathogenesis of the disease is poorly understood and treatment requires prolonged antimicrobial therapy. Treatment failures are common. A reliable and reproducible animal model using a uropathogenic strain of *E. coli* as the challenge organism was developed to study the pathogenesis and treatment of bacterial prostatitis (Nickel *et al.*, 1990). In order to induce the disease, the organism is inoculated into the ventral prostate using a urethral catheter. Although only approximately 10^4 organisms enter the prostatic tissue, infection is established in up to 100% of the challenged animals. Clearly, the prostate is an organ vulnerable to bacterial infections. The bacteria divide and invade the prostate tissue, inducing a severe acute inflammatory reaction which frequently becomes chronic (Figure 3). In this model, chronic prostatitis develops and the low numbers of bacteria encased within a glycocalyx which induce extensive fibrosis, scarring and accumulation of immune complexes within tissue. The low numbers of bacteria within the prostate are not readily eliminated, in spite of the development of a strong specific immune response (Nickel *et al.*, 1990).

Stone Formation

Under certain conditions, bacterial colonization may lead to the formation of stones. When a urease-producing bacterium like *Proteus mirabilis* colonizes the urinary tract urine urea is broken down to ammonia. This results in the production of alkaline urine and the formation of struvite stones ($MgNH_4PO_46H_2O$). An animal model has been developed to investigate the role of bacterial biofilms in the formation of these stones (Olson *et al.*, 1989a). In this model a foreign body (zinc disc) is surgically placed in the bladder of rats and 1 week later *P. mirabilis* is inoculated into the bladder using a urethral catheter. Struvite is rapidly deposited on the foreign body. This model allowed the demonstration of the presence of bacterial biofilms on the surface of the struvite crystal and within the crystal structure itself (Olson *et al.*, 1989a).

Brown pigment stones (also know as biliary infection stones) are formed in a similar fashion within the gall bladder and biliary tree. These crystals are formed when β-glucuronidase producing bacteria (*E. coli*, *Enterobacter* spp.) hydrolyse bilary bilirubin and cause precipitation of calcium bilirubinate. Furthermore, the surgical placement of a foreign body in the gall bladder of cats has lead to the formation of stones composed of bilirubinate and cholesterol (Sung *et al.*, 1991). This model has

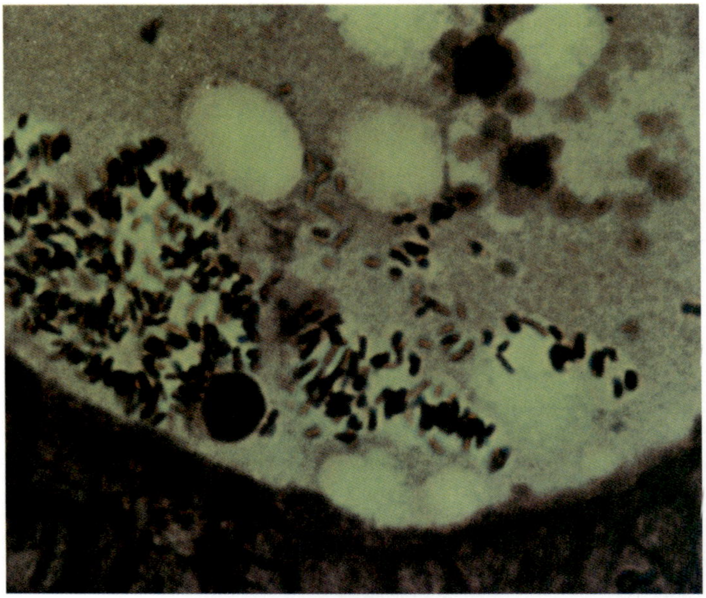

Figure 3 Microcolony of biofilm bacteria found on the mucosal surface of the acini of a rat prostate stained with methylene blue – basic fuschin, showing polymorphonuclear leukocytes at the infection area.

been useful to investigate the pathogenesis of brown pigment stone formation, and to study methods of treatment and prevention.

Corneal Infections

Corneal infections are predominantly due to either trauma or in the presence of a foreign body such as a contact lens (Baker and Schein, 1994). Microbial keratitis, which is initiated by a defect in the corneal epithelium, can lead to pain, photophobia, and decreased visual acuity. If the foreign body is not removed and the infection is allowed to progress, permanent damage to the eye can occur. Induction of microbial keratitis is induced by causing superficial trauma followed by inoculation of pathogens such as *Staphylococcus* spp., *Streptococcus* spp., *Neisseria* spp., *Serratia* spp. or *Pseudomonas* spp. The rabbit model has been shown to be well suited to studying the pathogenesis of corneal infections (Schultz *et al.*, 1997). A corneal abrasion is produced by scratching the corneal epithelial surface with a 26 gauge needle. The eye is then inoculated with the pathogen. A contact lens may be placed over the defect. The eyelids are sutured together to prevent photophobia and to hold the contact lens in place. Topical anesthetics should be applied to the eye and systemic analgesics are recommended. Edema, conjunctivitis and exudate are produced after 24 h. Large numbers of mononuclear cells and polymorphonuclear leukocytes migrate to the area of injury and infection.

CONCLUSION

The selection of animal models is based upon an understanding of the biology and the diseases of different laboratory animals. Because the host response is so important in device-associated infections, animal models may be the only method enabling a full understanding of the disease process in its entire complexity to be achieved. This knowledge may in turn lead to the development of new implant materials and chemotherapeutic agents.

REFERENCES

Ababio G.O., Rogars J.A., Morck D.W., Olson M.E. (1995). Effectiveness of sustained release ciprofloxacin microspheres against device-associated *Pseudomonas aeruginosa* biofilm infections in the rabbit peritoneal model. *J Med Microbiol*, **43**, 368–376.

Ades D.R., Craig W.A. (1998). Pharmacodynamics of fluoroquinolones in experimental models of endocarditis. *Clin Infect Dis*, **27**, 47–50.

Baker A.S., Schein O.D. (1994). Ocular infections. In: Bisno A.L., Waldvogel F.A. (eds) *Infections Associated with Indwelling Medical Devices*. ASM Press, Washington, pp. 111–134.

Buret A.G., Ward K.H., Olson M.E., Costerton J.W. (1991). An *in vivo* model to study the pathobiology of infectious biofilms on bacterial surfaces. *J Biomed Mater Res*, **25**, 865–874.

Cash H.A., Woods D.E., McCullough B., Johansson W.G., Bass J.A. (1979). A rat model of chronic respiratory infection with *Pseudomonas aeruginosa*. *Am Rev Respir Dis*, **119**, 453–459.

Ceri H., Olson M.E., Stremick C., Read R.R., Morck D.W., Buret A.G. (1999). The Calgary biofilm device: new technology for rapid determination of antibiotic susceptibilities of bacterial biofilms. *J Clin Microbiol*, **37**, 1771–1776.

Durack D.T., Beeson P.B., Petersdorf R.G. (1973). Experimental bacterial endocarditis. III. Production and progress of disease in rabbits. *Br J Exp Pathol*, **54**, 142–151.

Jacques M., Olson M.E., Costerton J.W. (1986). Microbial colonization of tailed and tailless intrauterine contraceptive devises: Influences of the mode of insertion in the rabbit. *J Obstet Gynaecol*, **154**, 648–655.

Kaye D., Hessen M.T. (1994). Infections associated with foreign bodies in the urinary tract. In: Bisno A.L., Waldvogel F.A. (eds) *Infections Associated with Indwelling Medical Devices*. ASM Press, Washington, pp. 291–307.

Khoury A.E., Olson M.E., Nickel J.C., Costerton J.W. (1989). Evaluation of the retrograde contamination guard in a bacteriologically challenged rabbit model. *Br J Urol*, **63**, 384–388.

Lambert P.A., Shorrock J.J., Aichson E.J., Domingue P.A.G., Power M.E., Costerton J.W. (1990). Effect of *in vivo* growth conditions upon expression of surface protein antigens in *Enterobacter faecalis*. FEMS *Microbiol Immunol*, **64**, 51–54.

Leung J.W.C., Ling T.K.W., Kung J.L.S., Vallence-Owen J. (1988). The role of bacteria in the blockage of biliary stents. *Gastroenterol Endosc*, **34**, 19–22.

Libby E., Morck D.W., Olson M.E., Leung J.W.C. (1994). Ciprofloxacin prevents stent blockage in an animal model. *Gastroenterology*, **106**, A346.

Maki D.G. (1994). Infections caused by intravascular devices used for infusion therapy: pathogenesis, prevention and management. In: Bisno A.L., Waldvogel F.A. (eds) *Infections Associated with Indwelling Medical Devices*. ASM Press, Washington, pp. 155–212.

Mayberrry-Carson K.J., Tober-Meyer B.K., Smith J.K., Lambe D.W., Costerton J.W. (1984). Bacterial adherence and glycocalyx formation in osteomyelitis induced with *Staphyloccocus aureus*. *Infect Immun*, 43, 825–833.

McDermid K.P., Morck D.W., Olson M.E., Dasgupta M.K., Costerton J.W. (1993a). Effect of growth conditions on expression and antigenicity of *Staphylococcus epidermidis* RP62A cell envelope proteins. *Infect Immun*, 61, 1743–1749.

McDermid K.P., Morck D.W., Olson M.E., Boyd N.D., Khoury A.E., Dasgupta M.K., Costerton J.W. (1993b). A porcine model of *Staphylococcus epidermidis* catheter associated infection. *J Infect Dis*, 168, 897–903.

Morck D.W., Lam K., McKay S.G., Olson M.E., Prosser B., Ellis B.D., Cleeland R., Costerton J.W. (1994). Comparative evaluation of fleroxacin, ampicillin, trimethoprim-sulfadethoxazole and gentamicin as a treatment of catheter associated urinary tract infection in a rabbit model. *Antimicrob Agents Chemother*, 4, S21–S27.

Nickel J.C., Olson M.E., Barabas A., Benediksson H., Dasgupta M.K., Costerton J.W. (1990). Pathogenesis of chronic bacterial prostatis in an animal model. *Br J Urol*, 66, 47–54.

Nickel J.C., Grant S.K., Lam K., Olson M.E., Costerton J.W. (1991). Evaluation in a bacteriologically-stressed animal model of a new closed catheter drainage system incorporating a microbicidal outlet tube. *Urology*, 38, 280–289.

Olfert E.D., Cross B.M., McWilliam A.A. (1993). Responsibility for the care and use of experimental animals. In: Olfert E.D., Cross B.M., McWilliam A.A. (eds) *Guide to the Care and Use of Experimental Animals* (2nd Edition), Volume 1. Canadian Council on Animal Care, Ottawa, pp. 1–13.

Olson M.E., Nickel J.C., Costerton J.W. (1989a). Animal model of human disease: infection-induced struvite urolithiasis in rats. *Am J Pathol*, 135, 581–583.

Olson M.E., Lam K., Bodey J.P., King G.E., Costerton J.W. (1992). Evaluation of strategies for central venous catheter replacement. *Crit Care Med*, 20, 797–804.

Olson M.E., Nickel J.C., Khoury A.E., Morck D.W., Cleeland R., Costerton J.W. (1989b). Amdinocillin treatment of catheter-associated bacteruria in rabbits. *J Infect Dis*, 159, 1065–1072.

Pennington J.E. (1985). Animal models of pneumonia for evaluation of antimicrobial therapy. *J Antimicrob Chemother*, 16, 1–6.

Power M.E., Olson M.E., Domingue P.A.G., Costerton J.W. (1990). A rat model of *Staphylococcus aureus* chronic osteomyelitis that provides a suitable system for studying the human infection. *J Med Microbiol*, 33, 189–198.

Read R.R., Eberwein P., Dasgupta M.K., Costerton J.W. (1989). Peritonitis in peritoneal dialysis: bacterial colonization by biofilm spread along the catheter surface. *Kidney Int*, 35, 614–621.

Santaro J., Levison M.E. (1978). Rat model of experimental endocarditis caused by *Staphylococcus aureus*. *J Infect Dis*, 141, 331–337.

Schultz C.L., Morck D.W., McKay S.G., Olson M.E., Buret A.G. (1997). Lipopolysaccharide induced acute red eye and corneal ulcers. *Exp Eye Res*, 64, 3–9.

Smith G.M., Boon R.J., Beale A.S. (1990). Influence of claolanic acid on the activity of amoxicillin against an experimental *Streptococcus-Staphylococcus aureus* mixed respiratory infection. *Antimicrob Agents Chemother*, 34, 21–214.

Sung J.Y., Olson M.E., Leung J.W.C., Lundberg M.S., Costerton J.W. (1990). The sphincter of Oddi is the boundary of bacterial colonization of the feline biliary and gastrointestinal tract. *Microb Ecol Health Dis*, 3, 199–207.

Sung J.Y., Leung J.W.C., Olson M.E., Lundberg M.S., Costerton J.W. (1991). Demonstration of transient bacterobilia by foreign body implantation in the feline biliary tract. *Dig Dis Sci*, 36, 943–948.

Toivonen J.T. (1993). Intrauterine contraceptive device and pelvic inflammatory disease. *Ann Med*, 25, 171–173.

Ward K.H., Olson M.E., Lam K., Costerton J.W. (1992). Mechanism of persistent infection associated with peritoneal implants *J Med Microbiol*, **36**, 406–413.

Wright J.B., Rechnitzer C., Kharazmi A., Sorensen B.A., Bertelsen T.B. (1993). Guinea pig alveolar macrophage function during the course of sublethal *Legionella pneumophila* infection. *Immunol Infect Dis*, **3**, 181–187.

Vaudaux P.E., Lew D.P., Waldvogel F.A. (1994). Host factors predisposing to and influencing therapy of foreign body infections. In: Bisno A.L., Waldvogel F.A. (eds) *Infections Associated with Indwelling Medical Devices*. ASM Press, Washington, pp. 1–33.

Zimmerli W., Waldvogel F.A., Vaudaux P., Nydegger U.E. (1982). Pathogenesis of foreign body infection: description and characteristics of an animal model. *J Infect Dis*, **146**, 487–497.

10 Antimicrobial Resistance of Biofilms

David G. Allison, Tomas Maira-Litran and Peter Gilbert

It is generally accepted that the majority of bacteria in nature have a marked tendency to interact and grow in close association with surfaces, forming biofilms through binding and inclusion within exopolymeric matrices. The exopolymers not only immobilise the cells on the colonised surface but also allow different species to interact. These interactions give the biofilm community metabolic and physiological capabilities which are not possible for individual, unattached cells. Notable amongst the unique properties of biofilms is their high level of resistance towards antibiotics, biocides and disinfectants. Such resistance has been attributed to a variety of mechanisms, and while some aspects of resistance are poorly understood at present, a number of dominant processes have been identified. The most frequent attribution of resistance relates to the properties of the exopolymer matrix (glycocalyx). Although the glycocalyx does not represent a significant barrier to diffusion in its own right it can restrict access to the depths of the biofilm by quenching chemically-reactive agents or through binding highly charged antimicrobials. In addition extracellular enzymes, such as ß-lactamases and formaldehyde lyase, bound within the matrix are able to augment the diffusion limitation through destruction of susceptible compounds. The close proximity of cells also makes the aquisition of nutrients and oxygen problematic for the deeper lying cells. Thus spatial gradients of growth rate will develop within the community structure, associated with different growth-limiting nutrients at points remote from the accessible surface. During exposure to antimicrobials slower-growing cells at the core of the biofilm will generally outsurvive the faster metabolising cells found at the periphery. Slow growing cells will, in addition, express dormant, starvation-phenotypes which often overexpress non-specific defences such as shock proteins, multi-drug efflux pumps (*acrAB*) as well as further exopolymer synthesis. It is also now believed that attachment of microorganisms to surfaces causes the expression of biofilm-specific phenotypes, possibly regulated through quorum sensing mechanisms, which may contribute to biofilm resistance. However, none of these processes alone can provide a complete explanation for the observed levels of resistance *in situ*. It is more likely that they will collectively delay eradication of the treated population and provide a window of time during which other selection events may occur.

KEY WORDS: biofilms, antimicrobials, growth rate, diffusion limitation, exopolymers, attachment-physiologies

INTRODUCTION

There is growing concern within the scientific community that the widespread, indiscriminate use of antibiotics has led to the development and emergence of antibiotic resistant bacterial strains (Dixon, 1998). Similarly, the widespread use, and dissemination within the environment, of chemical antimicrobials is leading to a reduction in their effectiveness. Moreover, the prospect of resistance to one type of antimicrobial agent, such as a biocide, leading to cross-resistance to a separate, unrelated agent such as an antibiotic, has serious potential (Levy, 1998). This, coupled to increasing demands by society on the control of bacteria in an ever-widening sphere of applications has heightened the need to understand the mechanisms associated with antimicrobial resistance and tolerance development.

Microbial biofilms are notable in their recalcitrance towards treatment with antibiotics, biocides and/or disinfectants that would adequately control the same organism growing in planktonic mode (Allison and Gilbert, 1995). Indeed, given the close proximity of cells within the exopolymeric matrix of a biofilm, biofilms might represent a nidus for both the generation and rapid spread of antimicrobial resistance. Many medical biofilm-related problems are associated with the surfaces of implanted medical devices such as prostheses, endocardial pacemakers and catheters, as well as with soft-tissue surfaces such as the nasopharynx and gut epithelia and hard surfaces such as bone (Costerton *et al.*, 1987). Equally, serious infection can arise from the failure to disinfect adequately the organisms attached to medical equipment such as fibre-optic endoscopes and washers (Spach *et al.*, 1993; Griffiths *et al.*, 1997). Biofilm infections, associated with indwelling medical devices are often chronic and act as the reservoir of bacteraemia. Whilst the latter respond readily to antibiotic treatment dictated by the results of conventional sensitivity testing (Thomson *et al.*, 1995), the biofilms from which they derive display a greatly enhanced resistance and often fail to respond to even the most aggressive antibiotic prescribing (Kunim and Steel, 1985; Nickel *et al.*, 1985; Gristina *et al.*, 1987; Costerton *et al.*, 1993). If the device is not surgically removed prior to antibiotic treatment then the infection will generally recur.

While biofilms have been widely implicated in medicine as a source of chronic infections (Costerton *et al.*, 1987), problems related to microbial biofilm are not unique to the biomedical field. In many industrial settings, biofouling and biocorrosion have been related to the presence of biofilms causing physical blockage of pipework and heat exchangers (Characklis, 1990; Little *et al.*, 1990). They may also increase the frictional resistance to fluid flow on ship hulls and in water conduits and promote the corrosion of metallic substrata. To these problems can be added contamination and spoilage in the food industry (Holah *et al.*, 1994; Eginton *et al.*, 1998; Holah and Gibson, 2000). The resistance of biofilms is not restricted to antibiotics but is also demonstrable with respect to a wide range of chemical biocides. These include isothiazolones (Costerton and Lashen, 1984), quaternary ammonium compounds (Costerton and Lashen, 1984; Evans *et al.*, 1990b), halogens and halogen-release agents (Favero *et al.*, 1983). Failure of available antimicrobials to adequately control microbial biofilms, together with an increasing dependence of modern medicine upon the implantation of devices, has stimulated the search for antimicrobials which have activity directed primarily towards the biofilm phenotype (Gilbert and Brown, 1995). Current control strategies involve the design of antimicrobial agents that are specifically targeted at cells growing within the biofilm. These include molecules with high diffusion-reaction ratios and agents targeted at slow or non-growing cells. Such approaches have, to date, only met with limited success and the need to develop efficient, low cost hygienic cleansing systems remains as urgent as ever. To aid the search for novel antimicrobial targets there is a need to not only develop knowledge of biofilm physiology, but also to examine the various mechanisms associated with resistance of biofilms towards antimicrobial agents.

The failure of microorganisms to succumb to antimicrobial treatment may arise through (i) an inherent insusceptibility to the agents employed, (ii) the acquisition of resistance, by previously susceptible strains, either by genetic mutation or by transfer of genetic material from another species or genus, and (iii) the emergence of pre-existing but unexpressed resistance phenotypes. The extent to which such adaption towards less susceptible phenotypes is influenced by growth as a biofilm is currently a matter for debate. This article considers the current understanding of resistance mechanisms associated with microbial biofilms gained as a result of attachment-associated events and growth of the cells to become microcolonies entrapped within extracellular polymers. Such mechanisms include diffusional resistance of the extracellular matrix augmented by chemical/enzymic modification of the agent (reaction-diffusion limitation), physiological changes due to slow growth rate and starvation responses, and the induction of attachment-specific, drug-resistant physiologies. It is unlikely that any single mechanism will account for the general observation of resistance, rather, these mechanisms are likely to be compounded in biofilms to create insusceptibility and an environment well suited for the emergence of tolerant genotypes.

BIOFILM PHYSIOLOGY

Biofilms may be considered as functional consortia of microbial cells enveloped within sometimes extensive matrices of extracellular polymers (the glycocalyx) and the concentrated products of their own metabolism. The latter include ions and nutrients sequestered from the environment and extracellular enzymes such as lyases, proteases, and ß-lactamases. In the majority of natural habitats, such consortia are made up of a variety of species and genera, but in biomedical situations, particularly those associated with soft tissue infection or infections of indwelling devices, monocultures are more usual.

The structural organisation of the glycocalyx, which forms the intercellular matrix, varies according to the prevailing physico-chemical environment. Thus, in situations of high shear (*i.e.* the tooth surface during mastication, and the gastro-intestinal tract), the biofilm population is organised within stratified compacts of exopolymeric material delimiting the boundries between different component species (Marsh, 1995; Newman and Barber, 1995). Under low to moderate shear, with little or no nutrient being accessed from the colonised surface, biofilms take on the appearance of attached floccules. These anchor the microcolony to the substratum and maximise diffusive interactions with a nutrient-bearing environment (Costerton *et al.*, 1994). If nutrients are derived from the substratum rather than from the bathing fluids, then diffuse layers of biofilm cells which completely coat the available surface are favoured, such as in the colonisation of the nasopharynx. With the exception of cells that are located at the outer limits of the biofilm community, the access and availability of nutrients and the elimination of metabolic by-products are restricted to a greater extent than they would be for the same cells growing as individuals in liquid culture. Thus cells deep within the biofilm matrix have available to them only those materials from the bathing fluids that have failed to be sequestered by the more out-lying cells. Conversely, these microorganisms have greater access to the secreted metabolic products of their neighbours. This leads to spatial organisation of species within mixed-species biofilm communities, with associated cross-feeding, and the development of functional interspecies dependences. In both mono- and multi-species biofilms, nutrient and gaseous gradients generated by metabolism will cause nutrient availability and the growth rate of the enveloped cells to vary with location relative to the substratum and biofilm/liquid phase interface. In well-mixed, suspension cultures all members of the community experience a common environment at any particular time. Single phenotypes therefore dominate such cultures which might, as a consequence, demonstrate a singular response towards antimicrobial treatment. At any given time within biofilm communities, however, a plethora of phenotypes is represented for each component species, the breadth of which reflects the extent of chemical heterogeneity within the film (Gilbert *et al.*, 1990). The outcome of any antimicrobial treatment will therefore reflect the susceptibility of the most resistant phenotype represented within it. As biofilms mature and exopolymer deposition increases the magnitude of the nutrient and gaseous gradients will become increased and the net growth rate of the community will become further reduced, possibly with the onset of dormancy and the triggering of stringent response genes (Zambrano and Kolter, 1995).

RESISTANCE MECHANISMS

The Exopolymeric Matrix

A common feature of all bacterial biofilms is the presence of an extracellular polymer matrix (Cooksey, 1992). Direct light and electron microscopy show this to be an ordered array of fine fibres providing a relatively thick, hydrated coating around the cells. Following attachment, cells initiate the production and accumulation of extracellular polymers which eventually surround and envelop the developing microcolony (Allison and Sutherland, 1987). Accumulation of extracellular matrices can occur within hours of initial adhesion. Although such matrices are often referred to as the 'glycocalyx', they do not provide a common defined structure. Whilst the predominant components of the exopolymeric matrix are gelled and highly hydrated exopolysaccharides found in an ordered cofiguration (Sutherland, 1997), other macromolecules such as nucleic acids, proteins, globular glycoproteins and lipids and ions may also be present (Sutherland, 1995). Whether the matrix polymers differ from those associated with planktonically grown cells, and also whether the matrix polymers, which bind cells to other cells, differ from 'foot-print polymers' which cement the primary colonisers to the substratum is uncertain at present (Sutherland, 1997; Allison, 1998). Nevertheless, matrix polymers determine the physical properties of the biofilm. Bacterial polysaccharides are chemically heterogeneous and contain non-carbohydrate substituents, many of which are species specific and are usually negatively charged (Sutherland, 1985). They may also be neutral, or more rarely positively charged depending upon the component of the repeat units. Furthermore, whilst polysaccharides are generally hydrophilic, in nature they may also possess some hydrophobic properties (Neu and Marshall, 1990; Sutherland, 1997). Indeed, many polymers are heterogeneous with respect to lipohilicity and hydrophobicity (Sutherland, 1997). In mixed-species biofilms each component species will produce a different set of polymers, and these will merge to give heterogeneous regions of polymers within the matrix (Cooksey, 1992). The physicochemical properties of the blended exopolymers will differ significantly from those of purified components and will also be sustantially affected by the ionic strength of the surrounding medium and the nature of the cationic species (Allison and Matthews, 1992).

Regulation of exopolymer synthesis

The synthesis of matrix polymers appears to be regulated by a variety of factors, of which surface-attachment appears to be of particular importance. Thus, Davies *et al.* (1993) showed exopolysaccharide (alginate) production to be derepressed in biofilm cells compared to planktonic cells, and Evans *et al.* (1994) showed exopolysaccharide production to be increased at low growth rates and to be substantially higher for biofilm than planktonic cells. The latter effect would provide for increased exopolymer production within the slow growing core of a thick micro-colony/biofilm. This would alter the distribution and density of cells throughout the matrix, and confer some structural organisation upon the community, resulting in customised micro-niches at various points within the biofilm (Costerton *et al.*, 1994). In this respect it is notable that with the exception of the alginates the majority of the extracellular polysaccharides are soluble to some degree in water. Biofilm structure is therefore dynamic, with solubilisation of polymers at the periphery being compensated by increased polymer production within the depths. Recently, it has also been suggested that in some Gram-negative organisms the production of exopolysaccharides, such as alginate, may be under the control of signal substances in the form of *N*-acyl homoserine lactone (HSL) (Davies *et al.*, 1998). These are global regulators of transcriptional activation in bacteria which may also act as regulators of biofilm specific physiology (Williams *et al.*, 1992; Gambello *et al.*, 1993; Cooper *et al.*, 1995; Heys *et al.*, 1997). They are responsible for cell-cell signalling in Gram-negative bacteria, and are implicated in cell density-mediated events. In biofilms, signal substances such as HSL would become concentrated within the geometric centre of the micro-colonies/biofilm, thereby increasing exopolymer production. The extent and nature of exopolymer production is also dependent upon physiological factors such as the relative availability of carbon and nitrogen (Sutherland, 1985).

The relationship between the exopolymeric matrix and resistance towards antimicrobial agents may be considered in terms of diffusion limitation, modified exopolymer properties and reaction-diffusion-limitation.

Diffusion Limitation

The presence of a charged, hydrated exopolymer matrix surrounding individual cells and microcolonies will influence profoundly the access of molecules and ions, including protons, to the cell wall and membranes. Restricted diffusion of agents from the surrounding medium may occur through a combination of ionic-interaction and molecular-sieving events. In this respect, the polymers of the extracellular matrix act as an ion-exchange resin where strongly charged molecules are actively removed from solution as they pass through. It is not surprising therefore that many groups of workers have suggested that the glycocalyx might physically prevent the access of antimicrobials to cell surfaces, and that the recalcitrance of biofilms is simply a matter of exclusion (Slack and Nichols, 1981; 1982; Costerton *et al.*, 1987; Suci *et al.*, 1994). Accordingly, if the antimicrobial agents are strongly charged (*i.e.* tobramycin) or chemically highly reactive (*i.e.* halogens/peroxygens), they will be quenched within the matrix during diffusion. Such universal explanations have been refuted (Gordon *et al.*, 1988; Nichols *et al.*, 1988; 1989) since reductions in the diffusion coefficients of antibiotics such as tobramycin and cefsulodin within biofilms or microcolonies are insufficient to account for the observed change in susceptibility. In this context Gristina *et al.* (1987) found no difference in the susceptibility towards antibiotics of slime producing and non-slime producing strains of *Staphylococcus epidermidis*. However, Evans *et al.* (1991) in a similar study assessed the susceptibility to the quinolone antibiotic ciprofloxcacin of mucoid and non-mucoid strains of *Pseudomonas aeruginosa* grown as biofilms and demonstrated that whilst the possession of a mucoid phenotype may be associated with decreases in susceptibility, reductions in the diffusion coefficient across polymeric matrices relative to liquid media are insufficient to account for it, since at equilibrium the concentrations at the cell surface and in the surrounding bulk aqueous phase will be equal. Diffusion limitation studies have generally focussed on antibiotics rather than biocides and upon medically relevant biofilm populations rather than biofouling situations. The dimensions of biofilms *in vivo* are in the order of tens of micrometres, whilst for industrial biofilms they may be in the order of tens of centimetres thick (Nichols *et al.*, 1989). Whilst thickness will not affect diffusion properties *per se*, it will affect the mass-flux of charged antimicrobials within the depths of industrial biofilms (Nichols, 1993). Similar conclusions have been reached by Stewart (1996), albeit using a theoretical model to investigate antibiotic penetration into microbial biofilms. On the basis of the data available in the literature, the extent of retardation of antibiotic diffusion due to sorption does not appear to be sufficient to account for reduced biofilm susceptibility.

Reaction-Diffusion-Limitation

In addition to their potential action as a diffusion barrier, exopolymer and cellular materials at the the periphery of a biofilm may react chemically with, and neutralise, the treatment agent and thereby further reduce its availability. Such effects will be most pronounced with biocides such as (i) iodine and iodine-polyvinylpyrollidone complexes (Favero *et al.*, 1983), and (ii) chlorine and peroxygens (Huang *et al.*, 1995) that react directly in a consumptive manner with the exopolymer and cellular materials but may also relate to the deposition of drug-inactivating enzymes such as ß-lactamases (Giwercman *et al.*, 1991), formaldehyde lyase and dehydrogenase (Sondossi *et al.*, 1985) which would cause the degradation of ß-lactam antibiotics and formaldehyde respectively. Such effects would be in addition to losses in the activity of highly charged drug molecules, such as the glycopeptides, caused through irreversible binding to the matrix (Hoyle *et al.*, 1992). Macrolide antibiotics, which are also positively charged, but also very hydrophobic, are relatively unaffected by the presence of the exopolymers (Ichimiya *et al.*, 1994). Poor penetration through anionic matrices might therefore be a phenomenon restricted to the more hydrophilic, positively charged agents. Clearly, whether or not the exopolymeric matrix constitutes a physical barrier to antimicrobial penetration depends upon the nature of the agent, the binding capacity of the polymeric matrix towards it, the levels of agent used therapeutically (Nichols, 1993), the distribution of biomass and local hydrodynamics (DeBeer *et al.*, 1994) and the rate of turnover of the microcolony relative to the antibiotic diffusion rate (Kumon *et al.*, 1994). For antibiotics such as tobramycin and cefsoludin such effects are likely to be minimal (Nichols *et al.*, 1988; 1989), but would be high for positively charged antibiotics such as the aminoglycosides, which will bind to polyanionic matrix polymers (Nichols *et al.*, 1988).

In such a fashion the resistance of *Klebsiella pneumoniae* and *P. aeruginosa* biofilms towards monochloramine treatment has been explained as depletion of the biocide within the interior of the biofilm through reaction-diffusion interactions on the periphery (Huang *et al.*, 1995). These authors grew biofilms of the two organisms together on stainless steel surfaces using a continuous-flow annular reactor. Biofilms were treated with 2 mg l^{-1} of monochloramine for 2h and stained using a fluorogenic redox indicator that could differentiate respiring from non-respiring cells. Epifluorescent micrographs of frozen, cross-sections taken at regular time intervals revealed gradients of respiratory activity within the biofilms in response to monochloramine treatment. Cells near the biofilm-bulk fluid interface lost respiratory activity early in the treatment whereas residual respiratory activity persisted near the substratum or in the centre of small, viable cell-clusters even after 2 h treatment. A further study, using both oxidising and non-oxidising biocides (Stewart *et al.*, 1998), used an artificial biofilm construct of alginate-entrapped *Enterococcus faecalis* to demonstrate a similar lack of penetration and action against the entrapped cells by chlorine, glutaraldehyde, isothiazolone and quaternary ammonium biocides. In a real-life situation, however, the volume and reactive-capacity of a biofilm would be insufficient to deplete the bulk availability of biocide, and interactive sites within the matrix polymers would become saturated with adsorbed/reacted biocide. The net effect would therefore be to delay, rather than prevent, the inhibitory process (Huang *et al.*, 1995). Provided that exposure to biocide was brief or that the biocide was not greatly in excess of requirements, reaction diffusion limitation could allow the survival of cells at the base of the biofilm which would fluorish once the biocide was removed.

As indicated above, the reaction-diffusion-limitation properties of the glycocalyx would be enhanced if it contained extracellular enzymes that were capable of degrading the diffusing substrate. A catalytic (*e.g.* enzyme) reaction, provided that turn-over was sufficiently rapid, could lead to severe antibiotic penetration failure (Stewart, 1996). In this respect hydrolytic enzymes such as ß-lactamases are induced/de-repressed in adherent populations and in those exposed to sub-lethal concentrations of imipenem and/or pipericillin (Giwercman *et al.*, 1991; Lambert *et al.*, 1993). These enzymes become trapped and concentrated within the biofilm matrix and further impede the action of susceptible antibiotics. Similarly, inactivation of formaldeyde by the enzymes formaldeyde lyase and formaldehyde dehydrogenase (Sondossi *et al.*, 1985) has been observed within biofilms of *P. aeruginosa*.

Modification of Exopolymer Properties

The presence of adsorbed ions within the biofilm matrix polymers will affect its net charge and thereby its antibiotic-exclusion properties. In this respect, Hoyle *et al.* (1992) have shown that the capacity of bacterial exopolysaccharides to bind tobramycin is less important, in terms of reduced susceptibility, than is the reduction in diffusivity of the matrix imposed by Ca^{2+} condensation of the polymer matrix. Such effects are dependent upon the nature of the adsorbed species and are not seen, for example, with Mg^{2+}. Such reduction can be explained on the basis of non-covalent associations occcurring between the different polysaccharides, forming a three-dimensional 'weak-gel' network, cross-linked by Ca^{2+}. In this manner, tobramycin is bound to the negatively charged polysaccharide gel and is prevented from reaching the cells (Allison and Matthews, 1992). These observations are important since body fluids at the site of an infection normally show relatively high concentrations of soluble salts. However, modification of the physico-chemical properties of the glycocalyx through ion adsorption is insufficient to account totally for biofilm resistance, since biofilm bacteria are found to be resistant to biocides even when growing in purified, deionised waters.

Modulation of the Growth Environment

It is well established that the susceptibility of bacterial cells towards antibiotics, biocides and preservatives is significantly affected by their nutrient status and growth rate, as well as by their temperature and pH and prior exposure to sub-effective concentrations of antimicrobials (Brown and Williams, 1985; Brown *et al.*, 1990; Williams, 1988). Such responses include changes in a variety of cellular components including membrane fatty acids, phospholipids and envelope proteins, and the production of extracellular enzymes and polysaccharides. Adherent microcolonies form functional consortia and influence their micro-environment through the localised concentration of enzymes and metabolic products, and the relative exclusion of gases such as oxygen (Costerton *et al.*, 1987). In this fashion, the exopolymeric matrix performs a homeostatic function, minimising the consequences of fluctuations in the surrounding macro-environment. A major consequence of such homeostasis is that cells deep within the biofilm are exposed to concentrations of substrates, hydrogen ions and also oxidation otentials that are substantially different from those experienced by cells on the periphery and by cells growing planktonically in the same medium. Growth rates will therefore be reduced within biofilms through the imposition of nutrient deficiences, which may or may not reflect the composition of the bulk aqueous phase.

Since mature biofilms are composed of multilayers of bacteria embedded in an exoplysaccharide matrix, diffusion and transport of nutrients through the biofilm becomes an important consideration. In nutrient rich environments, oxygen and nutrients are rapidly utilised by aerobic bacteria at the biofilm:liquid interface, thereby diminishing the availability of such nutrients in the depths of the biofilm. This leads to the formation of anaerobic and anoxic zones (Marshall, 1992). In this manner, nutrient gradients are likely to be established within thick biofilms and will generate cell populations which are very heterogeneous with respect to growth rate. As a consequence, growth rates are likely to decrease with the depth of the cells within the biofilm. At any particular time a plethora of phenotypes is represented within the community which reflects the chemical heterogeneity of the biofilm. Whilst this has long been considered to be the case for biofilm communities (Brown *et al.*, 1990; Gilbert *et al.*, 1990) it was visualised by Wentland *et al.* (1996) who used acridine orange staining of *K. pneumoniae* biofilms to show that the regions of fastest growth occurred in the outer 30 µm of the biofilm, closest to the bulk liquid. Dispersed regions of slow growth, tending towards the substratum, were also noted. A major contributor to the observed resistance of biofilms is therefore associated with physiological gradients of growth rate and nutritional status.

A distinction can be made between effects related to the nature of the least available nutrient (nutrient limitation/depletion) and the cellular growth rate. Within the depths of a biofilm growth rates will generally be suppressed relative to planktonic cells growing in the same environment. In this respect Ashby *et al.* (1994) used biofilm:planktonic ratios of isoeffective concentration (growth inhibition and bactericidal activity), determined for a wide range of antibiotics against cells grown either in broth or on urinary catheter discs, to indicate the extent of biofilm resistance. They noted that such ratios followed closely those generated between non-growing and actively growing cultures. With the exception of ciprofloxacin, antibiotic agents that were most effective against non-growing cultures (*i.e.* imipenem, meropenem) were also the most active against these biofilms. Other workers have used perfused biofilm fermenters (Gilbert *et al.*, 1989) to control directly and study the effects of growth rate within biofilms. Using control populations of planktonic cells grown in a chemostat, the separate contributions of growth rate and association within a biofilm can be evaluated. Decreased susceptibility of *S. epidermidis* to tobramycin (Duguid *et al.*, 1992a) and of *Escherichia coli* to tobramycin (Evans *et al.*, 1990a) and cetrimide (Evans *et al.*, 1990b) could be explained largely in terms of growth rate. Cells resuspended from growth rate controlled biofilms and planktonic cells of the same growth rate possessed almost identical susceptibilities to these agents. When intact biofilms were treated susceptibility however decreased somewhat from that of planktonic and resuspended biofilm cells, indicating some benefit to the cells of organisation within an exopolymeric matrix.

Stewart (1994) developed a mathematical model which incorporated the concepts of metabolism-driven oxygen gradients and growth rate dependent killing, to examine the susceptibility of *S. epidermidis* biofilms to various antibiotics. The model accurately predicted that susceptibility would be reduced in thicker biofilms due to oxygen limitation. Oxygen gradients within the biofilm may also directly influence the activity of some antibacterials (Shepherd *et al.*, 1988; Zabinski *et al.*, 1995). Since nutrient and gaseous gradients will increase in extent as biofilms thicken and mature, growth rate effects on susceptibility, such as these, will become particularly marked in aged biofilms (Anwar *et al.*, 1989; 1990). Such changes probably contribute to reports that aged biofilms are more recalcitrant to antibiotic and biocide treatment than younger ones (Anwar *et al.*, 1989).

The existence of physiological gradients accross biofilms cannot explain the long term survival of biofilm populations in the presence of inimical concentrations of antimicrobial. Such gradients depend on the growth and metabolism of cells at the periphery consuming nutrients before they permeate to the more deeply placed cells.

The peripheral cells will have growth rates and nutrient profiles that are not very dissimilar from those of planktonic cells. They will therefore be relatively sensitive to the treatments imposed and will quickly cease to metabolise. This will increase the availability of nutrients to the underlying survivors which will step up their metabolism and growth rate, adopt a more susceptible phenotype and die. Indeed, the growth rate of cells within this layer might even exceed that of untreated peripheral cells through mobilisation of nutrients from the dead biomass. This phenomenon would occur throughout the biofilm, proceeding inwards from the outside, until the biofilm was completely killed. Should the supply of antimicrobial agent cease then the biofilm could re-establish almost as fast as it was desytroyed because of the relative abundance of nutrients. Growth rate as an explanation of resistance cannot therefore account solely for the observed resistance of biofilms.

While reaction diffusion limitation of the access agent and the existence of physiological gradients within biofilms provide an explanation for the reduced susceptibility of biofilms, they do not explain their long term tolerance of antimicrobial agents. For this to occur, the biofilm population must adapt to a resistant phenotype during the 'time-window' of opportunity provided by this buffering effect.

Antibiotic Resistance Phenotypes

None of the potential mechanisms of resistance described above can account for the extreme tolerance towards antibiotics and biocides observed in the field and in medicine. Each of these can delay the onset of a lethal effect but cannot, unless turnover of the matrix polymers and cells is very rapid, account for survival during chronic long term treatments. Long-term survival of a biofilm community must relate to the adoption of more resistant phenotypes during the delayed action of the treatment agent. One effect of a delayed action with respect to the underlying cells within a biofilm is that they will be exposed, for an extended period, to the presence of sub-inhibitory levels of treatment agent. These might act as inducers/transcriptional activators of a more tolerant phenotype. It is worthy note in this light that the expression of the multi-drug resistance operons such as *mar* and efflux pumps such as *acrAB*, have been shown to be upregulated by exposure to sub-effective concentrations of antibiotics such as tetracycline and chloramphenicol (George and Levy, 1983; Ma *et al.*, 1993), and xenobiotics such as salicylate. *Mar* is chromosomal, variously induced and represented within a wide range of Gram-negative bacteria. Induction of operons such as *mar* during the delayed onset of the action of inducer antibiotics directed at biofilms is a tempting explanation of the biofilms long-term resistance. The importance of *mar* would be far greater, however, if it were induced by growth as a biofilm *per se*. Ciprofloxacin exposure will not induce the expression of *mar* or *acrAB* in *E. coli* but such expression will confer limited protection against this agent. *E. coli* biofilms, in which the *mar* locus was either constitutively expressed or deleted, were perfused for 48h with various concentrations of ciprofloxacin. These experiments, whilst demonstrating reduced susceptibility in the *mar* constitutive strain showed little or no difference between wild-type and *mar*-deleted strains (Maira *et al.*, 1998). Similar experiments using biofilms constructed from strains in which the efflux pump *acrAB* was either deleted or constitutively expressed (Maira-Litran,

1998) showed the *acrAB* deletion did not significantly affect susceptibility over that of the wild-type strain. Clearly neither *mar* nor *acrAB* is induced by sub-lethal treatment of biofilms with other than inducer substances. On the other hand, constitutive expression of *acrAB* protected the biofilm against low concentrations of ciprofloxacin and studies conducted in continuous culture with a *lacZ* reporter gene fused to *marO* showed, *mar* expression to be inversely related to specific growth rate (Maira *et al.*, 1998). Hence, following exposure of biofilms to sub-lethal levels of ß-lactams, tetracyclines and salicylates *mar* expression will be greatest within the depths of the biofilm, where growth rates are suppressed, and might account of the long-term survival of the community. Similar systems under the regulation of different inducer agents might extend this explanation of biofilm tolerance to include other treatment agents.

Attachment Specific Phenotypes

The possibility remains that bacteria are able to sense the presence of a surface, to which they become attached, and, as a consequence, transcriptionally activate genes/operons to confer an attachment-specific phenotype which has a modified susceptibility towards antimicrobials. Variations of the 'bottle-effect', whereby the metabolic activity of microorganisms is stimulated by their attachment to surfaces, have been reported in the literature since the early 1940's (Zobell, 1943; Fletcher, 1984; 1986). Only recently, however, have attempts been made to define the genetic and physiological bases of such phenomena. Dagostino *et al.* (1991) utilised transposon mutagenesis to randomly insert into the chromosome of *E. coli* a marker gene which lacked its own promoter element. They then isolated mutant cell lines which expressed the gene when attached to a polystyrene surface, but not when grown on agar or in liquid media. The isolation of mutant cells such as these, with reporter genes that respond to attachment onto surfaces, has not diminished the level of debate on the cause of surface-induced metabolic stimulation. This may reflect derepression or induction of specific operons/genes or it may be a physico-chemical manifestation (*i.e.* localised concentration of nutrients, viscosity changes, pH effects) of the proximity of the surface (van Loosdrecht *et al.*, 1990).

Evidence in favour of physico-chemical effects includes work on the regulation of lateral flagella gene transcription in *Vibrio parahaemolyticus*. This organism has been shown to produce a single polar flagellum in liquid, and numerous lateral, unsheathed flagella on solid culture media (Belas *et al.*, 1984; 1986; McCarter *et al.*, 1988). Changes in flagellation in this instance reflect an increased viscosity at the surface which restricts the movement of the polar flagellum and regulates the *laf* genes. In a similar vein, Lee and Falkow (1990) recognised that reduced oxygen tension, as experienced by cells enveloped within a biofilm or in association with a surface, causes the triggering of expression of *Salmonella* invasins.

Other examples of surface-induced behaviour defy such physico-chemical explanation. Thus, nitrilotriacetate does not adsorb to surfaces but its breakdown is enhanced when the degradative organisms are attached to inert surfaces (McFeters *et al.*, 1990). This suggests increased expression of the degradative enzymes by attached cells. In a similar fashion, gliding bacteria do not synthesise extracellular polymers when they are grown in suspension culture (Humphrey *et al.*, 1979; Abbanat *et al.*, 1988), but do so rapidly after they become irreversibly bound to a surface.

In balance, it now seems probable that bacteria can sense the close proximity of surfaces, and, through cell density transcriptional activation (Cooper *et al.*, 1995), the close proximity of other cells. In this light, Davies and Geesey (1995) have shown, using a *lacZ* reporter gene fused to *algC* of *P. aeruginosa*, upregulation of alginate production within minutes of attachment to a surface. The expression of *algC* generally ceases when large amounts of alginate are accumulated (Davies *et al.*, 1993). The ability of biofilm bacteria to modulate exoplymer production as a consequence of changes in population density through HSL mediated quorum sensing is a phenomenon that is only beginning to be understood (Allison *et al.*, 1998; Davies *et al.*, 1998). Under conditions of high cell density, where the concentration of signalling molecules will be high due to being trapped within the exopolymeric matrix, the possibility exists that additional exopolymer deposition will occur, mediated through mechanisms of cell-cell communication. Such exopolymers will not only strengthen attachment of cells to the substratum (Heys *et al.*, 1997) but also provide an additional exoplymeric barrier to antimicrobial agents.

Using novel spectrophotometric methods which allowed for simultaneous monitoring of the growth of planktonic and biofilm bacteria within the wells of a microtitre plate, Das *et al.* (1998) showed that the susceptibility of *P. aeruginosa* and *Staphylococcus aureus* to a range of different biocides changed rapidly after cellular attachment and biofilm formation. In some instances 3–5-fold decreases in susceptibility occurred immediately on attachment and could occur in the presence of biocide concentrations which exceeded the minimum inhibitory concentration for planktonic populations. These observed changes in susceptibility occurred before there had been any significant growth of the microorganisms on the surface and before exopolymers had accumulated. In a similar study, Fujiwara *et al.* (1998) investigated the immediate effect of adherence on antimicrobial susceptibility of *P. aeruginosa*, *Serratia marcescens* and *Proteus mirablis*. Their results demonstrated that after incubation for 1 h, involving adherence of bacteria to the surface of a plastic tissue culture plate, the minimal bactericidal concentrations for adherent bacteria were markedly elevated before biofilm formation had occurred. The magnitude of the decreases in susceptibility observed after bacterial attachment, but before biofilm formation, are, however, in general far less marked than those observed in mature biofilms. This indicates that the attachment-specific phenotype *per se* cannot explain the high levels of resistence seen in biofilms.

CONCLUSIONS

Resistance of microbial biofilms to a wide variety of antimicrobial agents is clearly associated with the organisation of cells within an extensive exoplymer matrix. The mechanisms associated with such resistance remain, however, unclear particularly when related to the long-term survival of biofilms in the presence of harsh treatment agents. Firstly, the dense organisation of the biofilm population within an exopolymeric matrix sets up a reaction-diffusion-limitation of the access of agent from its point of application to the deeper lying cells. These deeper lying cells will out-survive those on the surface and, if the bulk of the treatment agent is depleted or the exposure transient, will multiply and divide. Nutrients, as well as biocides and antibiotics, will suffer from reaction-diffusion-limitation of their availability to individual cells within biofilms. This will lead to the establishment of spatial gradients of growth rate within the community structure. Different growth-limiting nutrients will also prevail at different points in the biofilm. This will provide for a plethora of phenotypes within the biofilm, each reflecting the physico-chemical microenvironment of individual cells and their proximity to neighbours. Faster growing, more susceptible cells will generally lie on the periphery of the biofilm with slow growing recalcitrant cells being more deeply placed. In both instances, at the fringes of action, selection pressures will enrich the populations with the least susceptible genotype. It is possible under such circumstances for repeated chronic exposure to sub-lethal treatments to select for a resistant population that shows cross-resistance to other forms of antimicrobial. Whilst neither of these mechanisms can provide a complete explanation for recalcitrance, together they will delay eradication of the treated population and allow other selection, regulation events to occur.

These mechanisms of resistance acquisition are not dissimilar to *in vitro* 'training' experiments whereby microorganisms that survive a low concentration of an antimicrobial agent are gradually exposed to increasing concentrations of the agent in a stepwise manner. In this fashion cells can be 'trained' to acquire resistance (Brown *et al.*, 1969; Gilleland *et al.*, 1989). This process has often been dismissed in the past as being 'artificial' on the grounds that it could not possibly occur in nature. However, biofilms exposed to antimicrobial agents in a pulsed fashion will, assuming incomplete eradication of the cells, produce a front of killing within the biofilm which contains a mixture of live and dead cells. The live cells will subsequently be selected as being more resistant and at the next passage of antimicrobial treatment will move closer to the biofilm periphery. Such resistance problems will be compounded by the inappropriate use of different antimicrobial agents.

A third explanation for the resistance of biofilm communities lies with their expression of biofilm-specific phenotypes that are so different to those of planktonic cells that the agents developed against the latter fail to operate. Whilst such phenotypes are known to be expressed and might be regulated through quorum sensing mechanisms, they to do not appear to contribute greatly to the susceptibility pattern of individual biofilm cells. Such processes do, however, offer the possibility of novel agents that might prevent the formation of dense, polymer-encased communities and thereby circumvent the problem of killing intact, mature biofilms.

REFERENCES

Abbanat D.R., Godchaux W., Leadbetter E.R. (1988). Surface-induced synthesis of new sulphonolipids in the gliding bacterium *Cytophaga johnsonae. Arch Microbiol*, **149**, 358–364.

Allison D.G. (1998). Exopolysaccharide production in bacterial biofilms. *Biofilm*, **3**, BF98002.

Allison D.G., Sutherland I.W. (1987). The role of exopolysaccharides in adhesion of freshwater bacteria. *J Gen Microbiol*, **133**, 1319–1327.

Allison D.G., Matthews M.J. (1992). Effect of polysaccharide interactions on antibiotic susceptibility of *Pseudomonas aeruginosa. J Appl Bacteriol*, **73**, 484–488.

Allison D.G., Gilbert P. (1995). Modification by surface association of antimicrobial susceptibility of bacterial populations. *J Ind Microbiol*, **15**, 311–317.

Allison D.G., Ruiz B., SanJose C., Jaspe A., Gilbert P. (1998). Extracellular products as mediators of the formation and detachment of *Pseudomonas fluorescens* biofilms. *FEMS Microbiol Lett*, **167**, 179–184.

Anwar H., Costerton J.W. (1990). Enhanced activity of combination of tobramycin and piperacillin for eradication of sessile biofilm cells of *Pseudomonas aeruginosa*. *Antimicrob Agents Chemother*, **34**, 1666–7161.

Anwar H., Dasgupta M., Lam K., Costerton J.W. (1989). Tobramycin resistance of mucoid *Pseudomonas aeruginosa* biofilm grown under iron limitation. *J Antimicrob Chemother*, **24**, 647–565.

Ashby M.J., Neale J.E., Knott S.J., Critchley I.A. (1994). Effect of antibiotics on non-growing cells and biofilms of *Escherichia coli*. *J Antimicrob Chemother*, **33**, 443–452.

Belas R., Simon M., Silverman M. (1986). Regulation of lateral flagellar gene transcription in *Vibrio parahaemolyticus*. *J Bacteriol*, **167**, 210–218.

Belas R., Mileham A., Simon M., Silverman M. (1984). Transponson mutagenesis of marine *Vibrio* spp. *J Bacteriol*, **158**, 890–896.

Brown M.R.W., Williams P. (1985). The influence of environment on envelope properties affecting survival of bacteria in infections. *Annu Rev Microbiol*, **39**, 527–556.

Brown M.R.W., Watkins W.M., Foster J.H. (1969). Step-wise resistance to polymyxin and other agents by *Pseudomonas aeruginsa*. *J Gen Microbiol*, **55**, 17–18.

Brown M.R.W., Collier P.J., Gilbert P. (1990). Influence of growth rate on the susceptibility to antimicrobial agents: modification of the cell envelope and batch and continuous culture. *Antimicrob Agents Chemother*, **34**, 1623–1628.

Characklis W.G. (1990). Microbial fouling. In: Characklis W.G., Marshall K.C. (eds) *Biofilms*. Wiley, New York, pp. 523–584.

Cooksey K.E. (1992). Extracellular polymers in biofilms. In: Melo L.F., Bott T.R., Fletcher M., Capdeville B. (eds) *Biofilms: Science and Technology*. Kluwer Academic Press, Dordrecht, pp. 137–47.

Cooper M., Batchelor S.M., Prosser J.I. (1995). Is cell density signalling applicable to biofilms? In: Wimpenny J., Handley P., Gilbert P., Lappin-Scott H. (eds) *The Life and Death of Biofilm*. Bioline Press, Cardiff, pp. 93–97.

Costerton J.W., Lashen E.S. (1984). Influence of biofilm on the efficacy of biocides on corrosion-causing bacteria. *Mater Perf*, **23**, 34–37.

Costerton J.W., Khoury A.E., Ward K.H., Anwar H. (1993). Practical measures to control device-related bacterial infections. *Int J Artif Organs*, **16**, 765–770.

Costerton J.W., Lewandowski Z., DeBeer D., Caldwell D., Kober D., James G. (1994). Biofilms, the customised microniche. *J Bacteriol*, **176**, 2137–2142.

Costerton J.W., Cheng K.J., Geesey G.G., Ladd T.I., Nickel J.C., Dasgupta M., Marrie T.J. (1987). Bacterial biofilms in nature and disease. *Annu Rev Microbiol*, **41**, 435–464.

Dagostino L., Googman A.E., Marshall K.C. (1991). Physiological responses induced in bacteria adhering to surfaces. *Biofouling*, **4**, 113–119.

Das J.R., Bhakoo M., Jones M.V., Gilbert P. (1998). Changes in the the biocide susceptibility of *Staphylococcus epidermidis* and *Escherichia coli* cells associated with rapid attachment to plastic surfaces. *J Appl Microbiol*, **84**, 852–858.

Davies D.G., Geesey G.G. (1995). Regulation of the alginate biosynthesis gene algC in *Pseudomonas aeruginosa* during biofilm development in continuous culture. *Appl Environ Microbiol*, **61**, 860–867.

Davies D.G., Chakrabarty A.M., Geesey G.G. (1993). Exopolysaccharide production in biofilms: substratum activation of alginate gene expression by *Pseudomonas aeruginosa*. *Appl Environ Microbiol*, **59**, 1181–1186.

Davies D.G. Parsek M.R., Pearson J.P., Iglewski B.H., Costerton J.W., Greenberg E.P. (1998). The involvement of cell-to-cell signals in the development of a bacterial biofilm. *Science*, **280**, 295–298.

DeBeer D., Srinivasan R., Stewart P.S. (1994). Direct measurement of chlorine penetration into biofilms during disinfection. *Appl Environ Microbiol*, **60**, 4339–4344.

Dixon B. (1998). New and resurgent infections — new and future prospects. *Sci Prog*, **81**, 273–285.

Duguid I.G., Evans E., Brown M.R.W., Gilbert P. (1992a). Growth-rate-independent killing by ciprofloxacin of biofilm-derived *Staphylococcus epidermidis*: evidence for cell-cycle dependency. *J Antimicrob Chemother*, **30**, 791–802.

Duguid I.G., Evans E., Brown M.R.W., Gilbert P. (1992b). Effect of biofilm culture upon the susceptibility of *Staphylococcus epidermidis* to tobramycin. *J Antimicrob Chemother*, **30**, 803–810.

Eginton P.J., Holah, J,. Allison D.G., Handley P.S., Gilbert P. (1998). Changes in the strength of attachment of microorganisms to surfaces following treatment with disinfectants and cleansing agents. *Letters in Applied Microbiology*, 27, 101–106.

Evans E., Brown M.R.W., Gilbert P. (1994). Iron cheletor, exopolysaccharide and protease production of *Staphylococcus epidermidis*: a comparative study of the effects of specific growth rate in biofilm and planktonic culture. *Microbiology*, **140**, 153–157.

Evans D.J., Brown M.R.W., Allison D.G., Gilbert P. (1990a). Susceptibility of bacterial biofilms to tobramycin: role of specific growth rate and phase in the division cycle. *J Antimicrob Chemother*, **25**, 585–591.

Evans D.J., Brown M.R.W., Allison D.G., Gilbert P. (1990b). Growth rate and resistance of Gram-negative biofilms towards Cetrimide USP. *J Antimicrob Chemother*, **26**, 473–478.

Evans D.J., Brown M.R.W., Allison D.G., Gilbert P. (1991). Susceptibility of *Escherichia coli* and *Pseudomonas aeruginosa* biofilms to ciprofloxacin: effect of specific growth rate. *J Antimicrob Chemother*, **27**, 177–184.

Favero M.S., Bond W.W., Peterson N.J., Cook E.H. (1983). Scanning electron microscopic observations of bacteria resistant to iodophor solutions. In: *Proc Int Symp Povidone*. University of Kentucky, Lexington, USA, pp. 158–166.

Fletcher M. (1984). Comparative physiology of attached and free-living bacteria. In: Marshall K.C. (ed) *Microbial Adhesion and Aggregation*. Springer Verlag, Berlin, pp. 223–232.

Fletcher M. (1986). Measurement of glucose utilisation by *Pseudomonas fluorescens* that are free living and that are attached to surfaces. *Appl Environ Microbiol*, **52**, 672–676.

Fujiwara S., Miyake Y., Usui T., Suginaka H. (1998). Effect of adherence on antimicrobial susceptibility of *Pseudomonas aeruginosa*, *Serratia marcescens* and *Proteus mirablis*. *Hiroshima J Med Sci*, 47, 1–5.

Gambello M.J., Kaye S., Inglewski B.H. (1993). *LasR* of *Pseudomonas aeruginosa* is a transcriptional activator of the line protease gene (*apr*) and an enhancer of exotoxin A expression. *Infect Immun*, **61**, 1180–1184.

George A.M., Levy S.B. (1983). Amplifiable resistance to tetracycline, chloramphenicol, and other antibiotics in *Escherichia coli*: involvement of a non-plasmid-determined efflux of tetracycline. *J Bacteriol*, **155**, 531–540.

Gilbert P., Brown M.R.W. (1995). Screening for novel antimicrobial activity/compounds in the pharmaceutical industry. In: Brown M.R.W., Gilbert P. (eds) *Microbial Quality Assurance: A Guide Towards Relevance and Reproducibility of Inocula*. CRC Press, Boca Raton, USA, pp. 247–260.

Gilbert P., Collier P.J., Brown M.R.W. (1990). Influence of growth rate on susceptibility to antimicrobial agents: biofilms, cell cycle and dormancy. *Antimicrob Agents Chemother*, **34**, 1865–1868.

Gilbert P., Allison D.G., Evans D.J., Handley P.S. & Brown M.R.W. (1989). Growth rate control of adherent bacterial populations. *Appl Environ Microbiol*, **55**, 1308–1311.

Gilleland L.B., Gilleland H.E., Gibson J.A., Champlin F.R. (1989). Adaptive resistance to aminoglycoside antibiotics in *Pseudomonas aeruginosa*. *J Med Microbiol*, **29**, 41–50.

Giwercman B., Jensen E.T., Hoiby N., Kharazmi A., Costerton J.W. (1991). Induction of ß-lactamase production in *Pseudomonas aeruginosa* biofilms. *Antimicrob Agents Chemother*, **35**, 1008–1010.

Gordon C.A., Hodges N.A., Marriot C. (1988). Antibiotic interaction and diffusion through alginate and exopolysaccharide of cystic fibrosis derived *Pseudomonas aeruginosa*. *J Antimicrob Chemother*, **22**, 667–674.

Griffiths P.A., Babb J.R., Bradley C.R., Fraise A.P. (1997). Glutaraldehyde-resistant *Mycobacterium chelonae* from endoscope washer disinfectors. *J Appl Microbiol*, **82**, 519–526.

Gristina A.G., Hobgood C.D., Webb L.X., Myrvik Q.N. (1987). Adhesive colonisation of biomaterials and antibiotic resistance. *Biomaterials*, **8**, 423–426.

Heys S.J.D., Gilbert P., Allison D.G. (1997). Homoserine lactones and bacterial biofilms. In: Wimpenny J., Handley P., Gilbert P., Lappin-Scott H., Jones M. (eds) *Biofilms: Community Interactions and Control*. Bioline Press, Cardiff, pp. 103–112.

Holah J.T., Gibson H. (2000). Food industry biofilms. In: Evans L.V. (ed) *Biofilms: Recent Advances in their Study and Control*. Harwood Academic Publishers, pp. 211–235.

Holah J.T., Bloomfield S.F., Walker A.J., Spenceley H. (1994). Control of biofilms in the food industry. In: Wimpenny J.T., Nichols W.W., Stickler D., Lappin Scott H. (eds) *Bacterial Biofilms and their Control in Medicine and Industry*. Bioline Press, Cardiff, pp. 163–168.

Holmes C.F., Evans R. (1986). Biofilm and foreign body infection — the significance to CAPD peritonitis. *Perit Dial Bull*, **6**, 168–177.

Hoyle B.D., Wong C.K., Costerton J.W. (1992). Disparate efficacy of tobramycin on Ca(2+)-, Mg(2+)-, and HEPES-treated *Pseudomonas aeruginosa* biofilms. *Can J Microbiol*, **38**, 1214–1218.

Huang C.T., Yu F.P., McFeters G.A., Stewart P.S. (1995). Nonuniform spatial patterns of respiratory activity within biofilms during disinfection. *Appl Environ Microbiol*, **61**, 2252–2256.

Humphrey B.A., Dixon M.R., Marshall K.C. (1979). Physiological and *in situ* observations on the adhesion of gliding bacteria to surfaces. *Arch Microbiol*, **120**, 231–238.

Ichimiya T., Yamaski T., Nasu M. (1994). *In vitro* effects of antimicrobial agents on *Pseudomonas aeruginosa* biofilm formation. *J Antimicrob Chemother*, **34**, 331–341.

Kumon H., Tomochika K-I., Matunaga T., Ogawa M., Ohmori H. (1994). A sandwich cup method for the penetration assay of antimicrobial agents through *Pseudomonas* exopolysaccharides. *Microbiol Immunol*, **38**, 615–619.

Kunim C.M., Steel C. (1985). Culture of the surfaces of urinary catheters to sample urethral flora and study the effect of antimicrobial therapy. *J Clin Microbiol*, **21**, 902–908.

Lambert P.A., Giwercman B., Hoiby N. (1993). Chemotherapy of *Pseudomonas aeruginosa* in cystic fibrosis. In: Wimpenny J., Nichols W., Stickler D., Lappin-Scott H. (eds) *Bacterial Biofilms and their Contriol in Medicine and Industry*. Bioline Press, Cardiff, pp. 151–153.

Lee C.A., Falkow S. (1990). The ability of *Salmonella* to enter mammalian cells is affected by bacterial growth state. *Proc Nat Acad Sci USA*, **87**, 4304–4308.

Levy S.B. (1992). Active efflux mechanisms for antimicrobial resistance. *Antimicrob Agents Chemother*, **36**, 695–703.

Levy S.B. (1998). The challenge of antibiotic resistance. *Sci Am*, **278**, 46–53.

Little B.J., Wagner P.A., Characklis W.G., Lee W. (1990). Microbial corrosion. In: Characklis W.G., Marshall K.C (eds) *Biofilms*. Wiley, New York, pp. 635–670.

Ma D., Cook D.N., Alberti M., Pong N.G., Nikaido H., Hearst J.E. (1993). Molecular cloning and characterization of *acrAB* and *acrE* genes of *Escherichia coli*. *J Bacteriol*, **175**, 6299–6313.

Maira-Litran T. (1998). An investigation into the potential of the *mar* operon to moderate the antibiotic resistance of biofilms. PhD Thesis, University of Manchester, Manchester, UK.
Maira T., Levy S.B., Allison D.G., Gilbert P. (1998). Influence of growth environment and biofilm formation on the expression of *mar* in *Escherichia coli* (abstract). Ninety eighth Gen Meet Am Soc Microbiol, May, 1998. ASM Press, Washington DC, pp. V–89.
Marsh P. (1995). Dental plaque. In: Lappin-Scott H.M., Costerton J.W. (eds) *Microbial Biofilms*. Cambridge University Press, Cambridge, pp. 282–300.
Marshall K.C. (1992). Biofilms: an overview of bacterial adhesion, activity and control at surfaces. *ASM News*, **58**, 202–207.
McCarter L., Hilmen M., Silverman M. (1988). Flagellar dynamometer controls swarmer cell differentiation of *Vibrio parahaemolyticus*. *Cell*, **54**, 345–351.
McFeters G.A., Egil T., Wilberg E., Adler A., Schneider R., Snozzy M., Geiger W. (1990). Activity and adaptation of nitrilo(NTA)-degrading bacteria: Field and laboratory studies. *Water Res*, **24**, 875–881.
Neu T.R., Marshall K.C. (1990). Bacterial polymers: physicochemical aspects of their interactions at surfaces. *J Biomat Appl*, **5**, 107–133.
Newman H.N., Barber P.M. (1995). Dental plaque structure *in vivo*. In: Wimpenny J., Handley P., Gilbert P., Lappin-Scott H. (eds) *The Life and Death of Biofilm*. Bioline Press, Cardiff, pp. 27–32.
Nichols W.W. (1993). Biofilm permeability to antibacterial agents. In: Wimpenny J., Nichols W.W., Stickler D., Lappin-Scott H. (eds) *Bacterial Biofilms and their Control in Medicine and Industry*. Bioline Press, Cardiff, pp. 141–149.
Nichols W.W., Dorrington S.M., Slack M.P.E., Walmsley H.L. (1988). Inhibition of tobramycin diffusion by binding to alginate. *Antimicrob Agents Chemother*, **32**, 518–523.
Nichols W.W., Evans M.J., Slack M.P.E., Walmsley H.L. (1989). The penetration of antibiotics into aggregates of mucoid and non-mucoid *Pseudomonas aeruginosa*. *J Gen Microbiol*, **135**, 1291–1303.
Nickel J.C., Ruseska I., Wright J.B., Costerton J.W. (1985). Tobramycin resistance of *Pseudomonas aeruginosa* cells growing as a biofilm on urinary catheter materials. *Antimicrob Agents Chemother*, **27**, 619–624.
Shepherd J.E., Waigh R.D., Gilbert P. (1988). Antimicrobial action of 2-bromo-2-nitropropan-1,3-diol (Bronopol) against *Escherichia coli*. *Antimicrob Agents Chemother*, **32**, 1693–1698.
Slack M.P.E., Nichols W.W. (1981). The penetration of antibiotics through sodium alginate and through the exopolysaccharide of a mucoid strain of *Pseudomonas aeruginosa*. *Lancet*, **11**, 502–503.
Slack M.P.E., Nichols W.W. (1982). Antibiotic penetration through bacterial capsules and exopolysaccharides. *J Antimicrob Chemother*, **10**, 368–372.
Sondossi M., Rossmore H.W., Wireman J.W. (1985). Observation of resistance and cross-resistance to formaldehyde and a formaldeyde condensate biocide in *Pseudomonas aeruginosa*. *Int Biodeterior*, **21**, 105–106.
Spach D.H., Siverstein F.E., Stamm W.E. (1993). Transmission of infection by gastrointestinal endoscopy and bronchoscopy. *Ann Intern Med*, **118**, 117–128.
Stewart P.S. (1994). Biofilm accumulation model that predicts antibiotic resistance of *Pseudomonas aeruginosa* biofilms. *Antimicrob Agents Chemother*, **38**, 1052–1058.
Stewart P.S. (1996). Theoretical aspects of antibiotic diffusion into microbial biofilms. *Antimicrob Agents Chemother*, **40**, 2517–2522.
Stewart P.S., Grab L., Diemer J.A. (1998). Analysis of biocide transport limitation in an artificial biofilm system. *J Appl Microbiol*, **85**, 495–500.
Suci P.A., Mittelman M.W., Yu F.U., Geesey G.G. (1994). Investigation of ciprofloxacin penetration into *Pseudomonas aeruginosa* biofilms. *Antimicrob Agents Chemother*, **38**, 2125–2133.

Sutherland I.W. (1985). Biosynthesis and composition of Gram-negative bacterial extracellular and wall polysaccharides. *Annu Rev Microbiol*, **39**, 243–270.

Sutherland I.W. (1995). Biofilm specific polysaccharides: do they exist? In: Wimpenny J., Handley P., Gilbert P., Lappin-Scott H. (eds) *Life and Death of Biofilm*. Bioline Press, Cardiff, pp. 103–107.

Sutherland I.W. (1997). Microbial biofilm exopolysaccharides — superglues or velcro? In: Wimpenny J., Handley P., Gilbert P., Lappin-Scott H., Jones M. (eds) *Biofilms: Community Interactions and Control*. Bioline Press, Cardiff, pp. 33–39.

Thomson K.S., Bakken J.S., Sanders C.C. (1995). Antimicrobial susceptibility testing in the Clinic. In: Brown M.R.W., Gilbert P. (eds) *Microbial Quality Assurance: A Guide Towards Relevance and Reproducibility*. CRC Press, Boca Raton, USA, pp. 275–289.

van Loosdrecht M.C.M., Lyklema J., Norde W., Zehnder A.J.B. (1990). Influence of interfaces on microbial activity. *Microbiol Rev*, **54**, 75–87.

Wentlamd E.J., Stewart P.S., Huang C.T., McFeters G.A. (1996). Spatial variations in growth rate within *Klebsiella pneumoniae* colonies and biofilm. *Biotechnol Prog*, **12**, 316–321.

Williams P. (1988). Role of the cell envelope in bacterial adaption to growth *in vivo* in infections. *Biochimie*, **70**, 987–1011.

Williams P., Bainton N.J., Swift S., Chhabra S.R., Winson M.K., Stewart G.S.A.B., Salmond G.P.C., Bycroft B.W. (1992). Small molecule-mediated density dependent control of gene expression in prokaryotes: bioluminescence and the biosynthesis of carbapenem antibiotics. *FEMS Microbiol Letts*, **100**, 161–168.

Zabinski R.A., Walker K.J., Larsson A.J., Moody J.A., Kaatz G.W., Rotschafer J.C. (1995). Effect of aerobic and anaerobic environments on antistaphylococcal activities of five fluoroquinolones. *Antimicrob Agents Chemother*, **39**, 507–512.

Zambrano M.M., Kolter R. (1995). Changes in bacterial cell properties on going from exponential growth to stationary phase. In: Brown M.R.W., Gilbert P. (eds) *Microbial Quality Assurance: A Guide Towards Relevance and Reproducibility*. CRC Press, Boca Raton, USA, pp. 21–30.

Zobell C.E. (1943). The effect of solid surfaces upon bacterial activity. *J Bacteriol*, **46**, 39–56.

11 Biofilms in the Oral Cavity: Impact of Surface Characteristics

M. Quirynen, M. Brecx and D. van Steenberghe

The oral cavity has two particularities which are of interest for the study of biofilms, *viz.* it is an open growth system and it possesses non-shedding surfaces (*e.g.* teeth, dentures). The bacterial accumulation on teeth or prosthetic surfaces is commonly called dental plaque. Besides intra- and inter-individual differences in *de novo* plaque formation, surface free energy (sfe) and especially surface roughness (sr) have been found to have a significant impact on bacterial build-up, especially supragingivally and to a lesser extent in the subgingival environment. An increase in sr above the threshold R_a value of 0.2 μm or an increase in substratum sfe will favour bacterial adhesion, especially if the microbiota primarily consists of species with a high sfe. These non-invasive observations in the oral cavity can contribute to the understanding of biofilm formation in other environments.

KEY WORDS: dental biofilm, periodontal disease, dental plaque growth, surface characteristics, surface free energy, surface roughness

INTRODUCTION

Throughout life, all interface surfaces in the human body are exposed to colonisation by a wide range of microorganisms. In general, microbiota live in harmony with the host. Constant renewal of the epithelial surfaces by shedding prevents the accumulation of large masses of microorganisms. In the oral cavity, teeth or prosthetic devices provide non-shedding surfaces which allow extensive bacterial deposits to form. Such an accumulation of microbes on a hard surface is commonly called "dental plaque" because of its yellowish colour reminiscent of the mucosal plaques caused by syphilis. The accumulation and/or metabolic products of bacteria have been associated with dental caries, gingivitis, periodontitis, peri-implant infections and stomatitis.

From an ecological view point, the oral cavity and the oro-pharynx as a whole, should be considered as an "open growth system". There is an uninterrupted ingestion and removal of microorganisms and their nutrients. A dynamic equilibrium exists between the adhesion forces of microorganisms and a variety of removal forces such as swallowing, friction by food intake, the tongue and oral hygiene implements as well as the wash-out effect of the salivary and crevicular fluid outflow. The latter is an inflammatory exudate that streams out of the periodontal pocket (*i.e.* the gap between tooth root and surrounding gingiva after destruction of the alveolar bone due to chronic periodontal infection). Most organisms can only survive in the mouth when they adhere to non-shedding surfaces.

This chapter aims to review dental plaque formation in the oral cavity with special reference to the impact on the rate of microbial build-up of both individual dependent variables and the surface characteristics of the intra-oral hard substrata. Modification of these variables might facilitate the prevention of oral diseases and be of clinical relevance.

THE COMPLEXITY OF THE ORAL CAVITY

The Diversity of the Oral Microbiota

After birth the oral cavity is inhabited by a complex array of bacteria. Indeed, within hours after birth colonisation of the sterile oral cavity by low numbers of mainly facultative and aerobic bacteria occurs (Socransky and Manganiello, 1971). From the second day on, anaerobic bacteria can be detected in the infant's edentulous mouth (Rotimi and Duerden, 1981; Evaldson *et al.*, 1982). Whether these organisms are already part of the developing indigenous flora, or just transient, is not known. After tooth eruption a more complex oral flora becomes established. It is estimated that more than 400 different species are capable of colonising the adult mouth and any individual may harbour 150 or more different species (Moore and Moore, 1994; Socransky and Haffajee, 1997). In 1 mm^3 of dental plaque, weighing approximately 1 mg (one single tooth can harbour 10 mg), more than 10^8 bacteria are present (Schroeder and De Boever, 1970). The number of bacteria in a single periodontal pocket can exceed 10^9 colony forming units (c.f.u.).

Although the presence of pathogenic bacteria is a *conditio sine qua non*, the etiology of both caries and periodontitis is multi-factorial, involving host susceptibility. During the last decades, bacterial "specificity" in the etiology of infections, caries and periodontitis, has become largely accepted. For caries the pathogenic species (Loesche, 1976; Tanzer, 1992) are mainly Gram-positive facultative organisms (*e.g.* mutans streptococci such as *S. mutans* and *S. sobrinus*, and *Lactobacillus* species). Periodontal infections (Socransky and Haffajee, 1992; Wolff *et al.*, 1994; Brecx, 1997; Socransky *et al.*, 1998) on the other hand involve a mainly Gram-negative anaerobic flora (*e.g. Porphyromonas gingivalis*, *Prevotella intermedia*, *Bacteroides forsythus*, *Fusobacterium nucleatum*, *Campylobacter rectus* and spirochetes) and some Gram-negative facultative and microaerophilic species (*e.g. Eikenella corrodens* and *Actinobacillus actinomycetemcomitans*). The recognition of such bacterial specificity has led to a more microbiological approach to both the prevention and treatment of the above mentioned infections. This has been translated at the clinical level by microbiological monitoring of the treatment, the use of antimicrobials, the search for microbial virulence factors or parameters which increase the host susceptibility, and the recognition of the possible transmission of oral pathogens between relatives. However, regular removal of the complete biofilm from all non-shedding intra-oral surfaces remains the most predictable approach to preventing the initiation and/or the recurrence of both diseases (Axelsson and Lindhe,

1981; Axelsson *et al.*, 1991). This cumbersome approach explains the increased interest in factors which could interfere with biofilm formation.

The Diversity of Intra-oral Surfaces for Bacterial Adhesion

Teeth and permucosal implants are unique because of the ectodermal interruption they imply. Thus the external environment has direct contact with bodily tissues through a seal of epithelial/connective tissue. However, when dental plaque accumulation occurs a swelling of the gingival margin due to the inflammatory reaction will result in the creation of a small crevice between the gingiva and the tooth crown which favours an anaerobic environment. With time, such an inflammatory process will create, at least in susceptible patients, a periodontal pocket because of the tissue breakdown (Figure 1). This pocket forms a new niche which favours bacterial survival because of anaerobioses and shielding from removal forces. Moreover, these gingival pockets with their subgingival area offer a range of new possibilities for bacteria to survive in the oral environment. They are rather like a solitary swimming-pool (for review see Quirynen and Bollen, 1995), filled with crevicular fluid. Bacteria can swim/float around or adhere to the root cementum (Nyvad & Fejerskov, 1987) or its collagen appendages (Naito & Gibbons, 1988), or invade the dentine tubules (Adriaens *et al.*, 1988). Bacteria can also adhere to and/or invade the junctional epithelium (Liakoni *et al.*, 1987) with its large intercellular spaces which line the pocket to the other side.

Besides hard surfaces, the oral cavity also offers different soft tissues with distinct ecological determinants for bacterial adhesion, *e.g.* the gingiva around the teeth, the alveolar mucosa lining the cheeks and the floor of the mouth, the dorsum of the tongue roughened by the presence of papillae and the tonsils. Although all these soft tissues have a high desquamation rate, they are colonised by very large numbers of bacteria.

Thus bacteria that normally reside in the oral cavity (*i.e.* the indigenous microbiota) can select their habitat from different ecosystems. Most periodontopathogens (with the exception of spirochetes) are able to colonise all the above mentioned niches, even though they are more or less exclusively Gram-negative anaerobic species. Even in the edentulous mouth of infants or of denture wearers the proportions of periodontopathogens [with the exception of *A. actinomycetemcomitans* and *P. gingivalis* (Könönen *et al.*, 1992; Danser, 1996)] can become very high. Thus the role of teeth as "port d' entrée" for these bacteria seems negligible. This hypothesis is supported by the observation that successful periodontal therapy (including periodontal pocket elimination) had only a limited effect on the detection frequency of pathogenic species on the buccal mucosa and/or in the saliva (Danser *et al.*, 1996). The intra-oral distribution of cariogenic species, on the other hand, is clearly associated with solid surfaces. For that reason *Streptococcus mutans* is often called an obligate periphyte (Tappuni and Challacombe, 1993). This tropism is supported by the observations of Carlsson and co-workers (Carlsson *et al.*, 1969, 1970) who studied the life history of infection by mutans streptococci. In infants they could only recover these species from the time the deciduous teeth erupted (Carlsson *et al.*, 1970). In a longitudinal observation of adults with severe dental caries, the cariogenic species fell below detection level after full-mouth extraction but reappeared within a few days after artificial denture insertion (Carlsson *et al.*, 1969). If these patients withdrew their

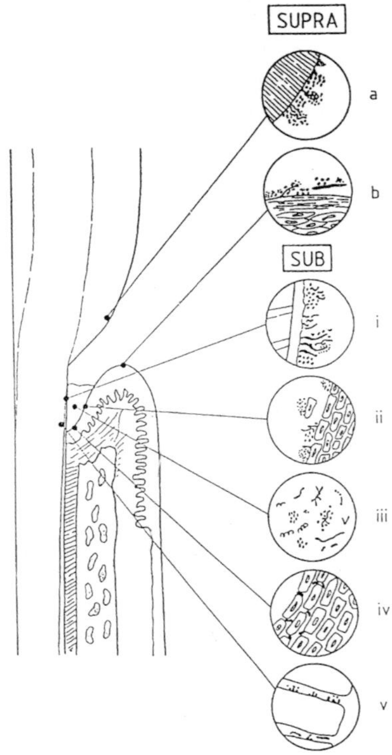

Figure 1 Ecological differences in the supra- and subgingival environment which are of importance when bacterial adhesion is considered. Supragingivally bacteria can adhere to the enamel surface of the tooth (a) or, to a lesser extent, to the desquamating oral epithelium (b). Subgingivally more niches are available for bacterial survival, *viz.* (i) adhesion to the root cementum or dentin, (ii) adhesion to the desquamating pocket epithelium. (iii) swimming/ floating in the creviculair fluid, (iv) invasion of the soft tissue, (v) invasion into the hard tissue *via* the dentine tubules. (Adapted from Quirynen and Bollen, 1995, and reproduced with permission.)

dentures for a few days, the bacteria disappeared again. Based on these reports and on their own observations Caufield and Gibbons (1979) assumed that most of the *S. mutans* cells in the saliva or on the tongue are derived from the biofilm present on teeth, prostheses or both.

The Principle of Transmission and Translocation

Recent studies, applying the method of bacterial fingerprinting, have clearly proved that periodontal pathogens are transmissible within members of a family (Zambon, 1996). The term bacterial transmission between subjects should be used with caution. Indeed, transmission should not be confused with contagion (the term contagious referring to the likelihood of a microorganism being transmitted from an infected to an uninfected host and creating disease).

Intra-oral translocation of bacteria moving from one niche to another has also been proven even if it has received less attention. This is surprising since transmission of these bacteria, from one *locus* to another, can jeopardise the outcome of periodontal therapy. The introduction of two phase type oral implants provided a favourable experimental set-up to study intra-oral bacterial translocation. Indeed, when the transmucosal part of the implant (the abutment) is inserted on top of the endosseous part, a bacteriologically "virgin" surface is available. Since such implant abutments can be replaced without any discomfort to the patient these "artificial" surfaces offer an excellent model to study the build-up of a biofilm and the intra-oral translocation of bacteria (Quirynen *et al.*, 1996a). These abutments are also useful for the study of the influence of surface characteristics (*e.g.* roughness, free energy) on initial supra- and subgingival colonisation (Quirynen *et al.*, 1994).

The translocation of bacteria to these newly installed abutments was examined in partially edentulous patients by means of differential phase contrast microscopy (Quirynen *et al.*, 1996a). This allows the investigator to follow the relative proportion of spirochetes which, intra-orally, almost exclusively colonise periodontal pockets and have been associated with periodontal breakdown (Listgarten and Heldén, 1978). A few weeks after abutment insertion, the proportion of spirochetes around the initially sterile abutments was found to be comparable to that around the teeth, at least if the implants and the teeth were present in the same jaw. This observation not only suggested intra-oral translocation of bacteria between both abutment types, but also indicated that this transmission preferentially occurs within a jaw, maybe through oral hygiene utensils (Quirynen *et al.*, 1996a). Papaioannou *et al.* (1996), using DNA probes, also reported a nearly comparable subgingival microbiota (*P. gingivalis*, *P. intermedia*, *B. forsythus*) around teeth and implants from partially edentulous patients with different types of periodontal infections. The similarity became more evident when the artificial abutments were located in deeper pockets (*i.e.* an environment favourable for periodontal pathogens). The vehicle for the intra-oral transportation of the pathogenic species is still not known. Since all species seem to survive in the saliva, this medium probably plays an important role. Whether translocation of periodontopathogens into a periodontal pocket occurs by salivary flow is arguable, since the continuous outflow of crevicular fluid from the pocket (0.5 to 2.4 ml d^{-1}) makes the spontaneous entrance of saliva nearly impossible. Oral hygiene aids (which can penetrate the pocket up to 3 mm) or dental instruments (especially pocket probes and curettes) however, could contribute to this transmission.

THE ULTRASTRUCTURE OF *DE NOVO* PLAQUE FORMATION

Many papers report studies on the initial adhesion to, as well as the bacterial colonisation of different removable artificial substrata inserted in the oral cavity. These substrata can be used for clinical observations (see later) or for ultrastructural or confocal microscopy which allow analysis of the first steps in bacterial adhesion.

Immediately following professional cleaning, a thin saliva-derived layer called the acquired pellicle covers the tooth surface. This pellicle consists of numerous components including mucins, glycoproteins, proline-rich proteins, histidine-rich proteins, enzymes such as α-amylase and other molecules. Today the term "acquired pellicle" is less frequently used because it is misleading. Indeed, it may imply that

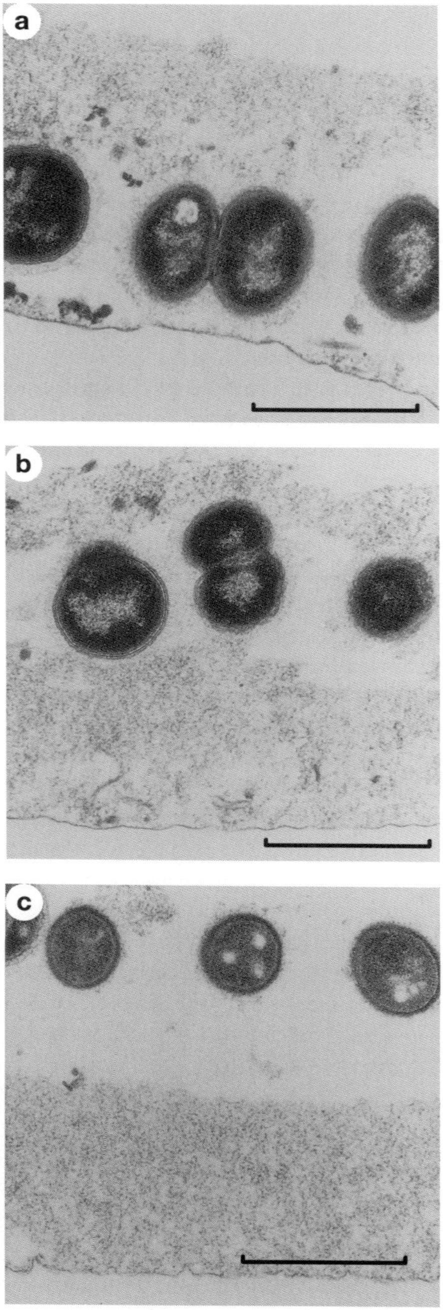

Figure 2 Thin sections of 4 h biofilms formed on plastic films. Groups of cocci are found close to the tooth surface (a), in the middle of the biofilm (b) or at its surface (c). The horizontal line at the bottom of the illustrations represents the interface between the plastic film and the dental plaque. (Photograph from M. Brecx.) Bar = 1 μm.

bacteria only colonise the tooth surfaces when this pellicle has been in place for some hours. However, it has been proven that bacteria are part of the very early deposit (Figure 2), within seconds of prophylaxis (Rönström *et al.*, 1977). The term "dental biofilm" is thus more appropriate because it includes both the acquired pellicle and the dental plaque. Streptococci, the principal early colonisers, bind to acidic proline-rich-proteins (Hsu *et al.*, 1994) and other receptors like α-amylase (Scannapieco *et al.*, 1995) and sialic acid (Hsu *et al.*, 1994) in the dental biofilm. These streptogocci also participate in intrageneric coaggregation (Kolenbrander *et al.*, 1990; Hsu *et al.*, 1994), which offers an extra advantage in allowing them to bind to the nascent monolayer of already bound streptococci (Nyvad & Kilian 1990; Skopek *et al.*, 1993). In addition, actinomyces, which are also primary colonisers, bind to the pellicle (Gibbons *et al.*, 1988) and to the streptococci (Kolenbrander and London, 1992). Each streptococcal and actinomyces strain binds specific salivary molecules. Thus from a common pool of salivary molecules, each strain of early coloniser may be coated with distinct molecules. Identical cells, coated with a specific salivary molecule may agglutinate, which would lead to a microconcentration and juxtapositioning of a particular strain. Alternatively, growth of a particular accreted strain can lead to a micro-colony coated with specific salivary molecules. Such events alter the diversity of salivary molecules exposed to later colonisers. Both streptococci and actinomyces are facultatively anaerobic, and doubling times for microbial populations during the first 4 h of plaque development are less than 1 h (Table 1, Weiger *et al.*, 1995). Consequently, these two groups of primary colonisers are thought to prepare the environment for later colonisers that have more fastidious growth requirements (Weiger *et al.*, 1995). Other bacteria like fusobacteria (that coaggregate with all other human oral bacteria (Kolenbrander *et al.*, 1995; Whittaker *et al.*, 1996), veillonellae, capnocytophagae and prevotellae bind to streptococci and/or actinomyces (Kolenbrander and London, 1992). Each new accreted cell becomes itself a nascent surface and therefore might act as a coaggregation bridge to the next potentially-accreting cell type that passes by. Most cooaggregations among strains of different genera are mediated by lectin-like adhesins and can be inhibited by lactose and other galactosides. The diversity of oral bacteria

Table 1 Comparison (literature review) of *in vivo* microbial generation time during human dental plaque formation.

Author	*Method*	*Generation time (median in h, and time interval)*				
Socransky *et al.*, 1977	CFU			4.6 (8–24 h)	2.6 (4–24 h)	11.4 (24–48 h)
Brecx *et al.*, 1983	TEM		1.0 (4–8 h)	4.4 (8–24 h)	2.6 (4–24h)	
Nyvad and Killian, 1987	CFU		1.3 (4–8 h)	8.2 (8–24 h)	4.4 (4–24 h)	
	CFU		2.4 (4–12 h)	5.3 (12–24 h)		
Mikkelsen, 1993	CFU	1.0 (2–6 h)	1.9 (4–12 h)	3.8 (6–24 h)	2.8 (4–24 h)	5.0 (24–48 h)
	CFU			5.1 (12–24 h)		
Brecx *et al.*, 1994	SEM	1.7 (1–4 h)	0.9 (4–8 h)	8.0 (8–24 h)	6.0 (4–24 h)	
Weiger *et al.*, 1995	CFU	0.8 (1–4 h)	2.2 (4–8 h)	7.5 (8–24 h)	4.6 (4–24 h)	14.8 (24–72 h)
Weiger *et al.*, 1995	BC_{vf}	0.8 (1–4 h)	2.2 (4–8 h)	7.3 (8–24 h)	4.5 (4–24 h)	12.5 (24–72 h)

CFU = colony forming units; TEM = transmission electron microscopy; SEM = scanning electron microscopy; BC_{vf} = total bacterials counts estimated by vital microscopic counts. (Table adapted from Weiger *et al.*, 1995.)

and their multiple adherence mechanisms makes dental plaque a dynamic biofilm (Whittaker *et al.*, 1996). Oral hygiene disrupts the bacterial community, which immediately reforms with readily available nutrients both from food intake and host-derived salivary and serous molecules.

A study on the growth dynamics of the biofilm formation on teeth reported some important changes in the growth rate within the first 24 h (Liljemark *et al.*, 1997). During the first 2–8 h the adherent pioneer streptococci saturated the salivary pellicular binding sites and thereby covered 3 to 30% of the adhesio enamel surface (Liljemark and Bloomquist, 1996). Instead of the expected steady growth during the next 20 h (a 4 to 6 h generation time was suggested by Loesche, 1988), a short period of rapid growth was observed. As bacterial densities approached approximately 2 to 6 million cells mm^{-2} on the enamel surface, a marked increase in growth rate (about 3 to 4 doublings) could be observed, resulting in 32 million bacteria mm^{-2}. This growth period was subject-, surface-, tooth location-, and time-independent but appeared to be cell density-dependent. Such a cell density-dependent growth suggests that the induction of growth was caused by bacteria. A better understanding and improved control of this active period of bacterial division might have a major impact on the prevention of dental diseases.

THE IMPACT OF HARD TISSUE VARIABLES ON DENTAL PLAQUE FORMATION

Clinical and Ultrastructural Features of Supragingival Plaque Formation

Clinically, early undisturbed plaque on teeth follows an exponential growth curve when measured planimetrically (Quirynen *et al.*, 1989). During the first 24 h after perfect tooth cleaning (including scaling and polishing), the increase in the amount of plaque is negligible (< 3% coverage of the vestibular tooth surface, which is clinically nearly undetectable). After 1 d the term biofilm is fully deserved because organisation takes place within it. Microorganisms, packed closely together, form a palisade (Figure 3) while others start to develop pleomorphism (Figure 4). Each crack is filled with one type of microorganism (Figure 5). The thickness of the plaque increases slowly with time, amounting to 20–30 μm after 3 d (Figure 6). From day 2 to day 4, the plaque growth rate is much faster but, from then on, there is a tendency for growth to slow down. After 96 h, on average 30% of the total tooth crown area will be covered with plaque (Figure 7). Although plaque does not increase substantially with time after the fourth day, several reports have shown that its bacterial composition will further change, with a shift towards a more anaerobic and a more Gram-negative flora, including an influx of fusobacteria, filaments, spiral forms and spirochetes (Theilade *et al.*, 1966; Syed and Loesche, 1978).

The slow start of the plaque growth curve can be explained partly by the fact that a single colony of bacteria has to reach a certain size before it can be clinically detected. Brecx *et al.* (1983) illustrated that the early increase in plaque mass originates largely from the proliferation of bacteria already present, and only to a limited extent from new adhering species, an observation which is also suggestive of an exponential

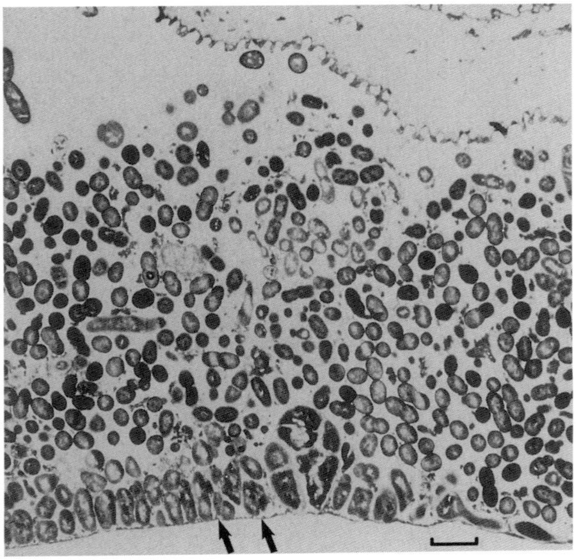

Figure 3 Thin section of 24 h dental biofilm on plastic film which is covered by numerous layers of bacteria. Close to the surface of the film a compact zone of palisading organisms is seen (arrows). In the external part of the deposit (top right) an epithelial cell is present. (Reproduced from Brecx *et al.*, 1983, with permission.) Bar = 1 μm.

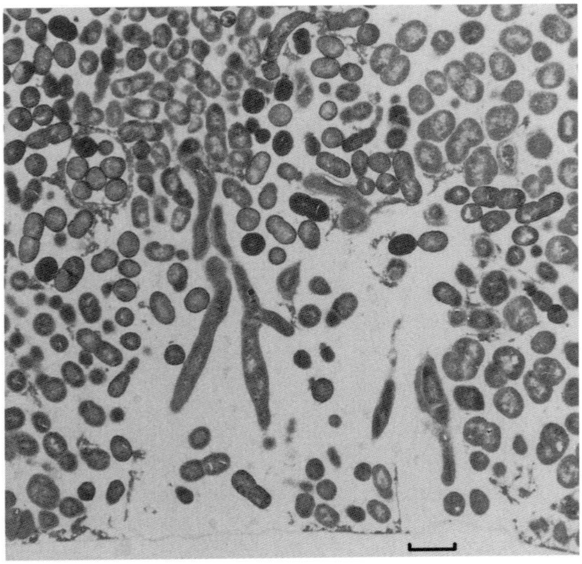

Figure 4 Thin section of 24 h dental biofilm on plastic film containing numerous bacteria comprising coci, rods and filaments, one of which is branching. (Reproduced from Brecx *et al.*, 1983, with permission.) Bar = 1 μm.

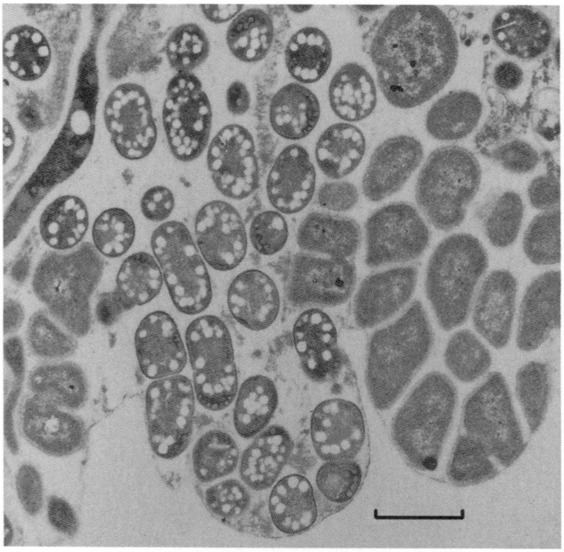

Figure 5 Thin section of 24 h dental biofilm on plastic film. The microorganisms are more or less arranged in colonies of identical morphology. (Reproduced from Brecx *et al.*, 1983, with permission.) Bar = 1 μm.

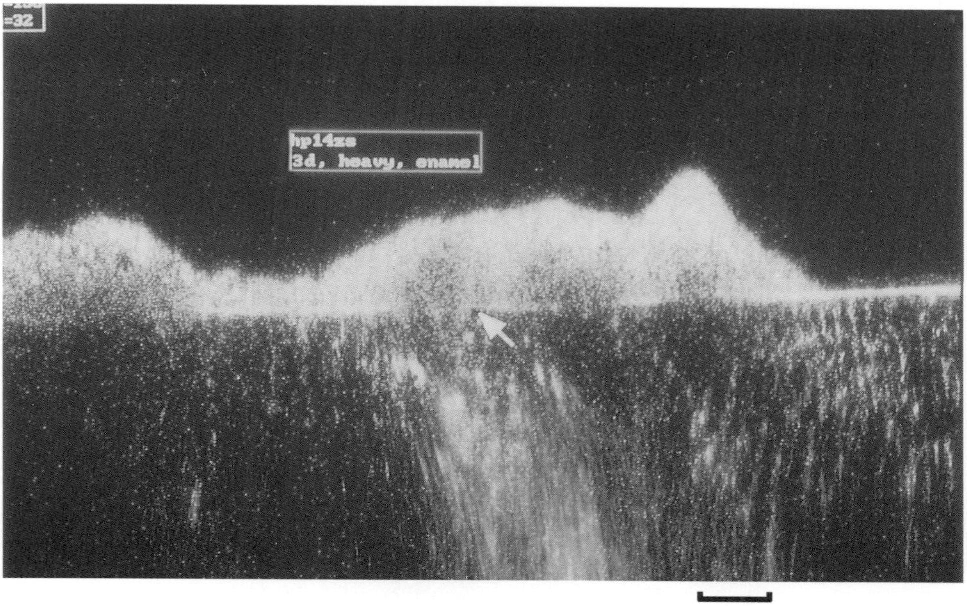

Figure 6 Confocal laser scanning microscopy photograph. Z-section of vital stained 3 d dental plaque on enamel from a heavy plaque former. Arrow = borderline between plaque and (self-fluorescent) enamel, the plaque consisting mainly of green living microflora ranging from 0 μm (right) to 32 μm (top of mountain-like structure). (Reproduced from Netuschil *et al.*, 1998, with permission.) Bar = 16 μm.

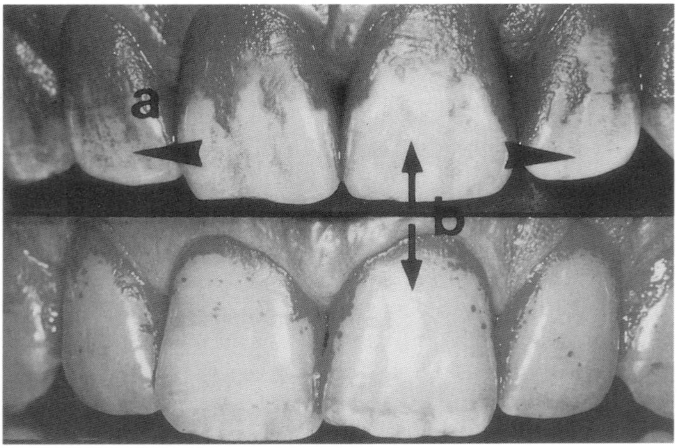

Figure 7 Clinical illustration of (a) (arrowheads) intra- and (b) (arrows) inter-subject variation in *de novo* plaque formation. Plaque was visualised with an erythrosin disclosing solution. Both photographs represent undisturbed plaque formation in the upper front region from two subjects with a healthy periodontium after 4 d. In one subject *(upper picture, heavy plaque former)*, nearly 50% of the tooth crown area was covered with plaque; in the other *(lower picture)* the amount of biofilm was relatively small. The formation of the biofilm preferentially occurred both along the gingival margin, from the interdental spaces, and from surface irregularities, where bacteria are sheltered against shear forces.

curve. The increase in microbial generation time (1 h for initial plaque, 12 h for 3 day-old plaque), on the other hand, might explain the levelling of the slope from day 4 onwards (Table 1).

During the night, plaque growth rate is reduced by some 50% (Quirynen *et al.*, 1989). This was a surprising finding since it would be hypothesised that reduced plaque removal and the decreased salivary flow (thus less antibacterial activity), at night would enhance plaque growth. The fact that the supragingival plaque obtains its nutrients mainly from the saliva (Carlsson, 1980) seems to be of more significance than the antibacterial activity of the latter.

Macrostructural Shielding and *de novo* Biofilm Growth

Early dental plaque formation on teeth follows a typical topographical pattern, with initial growth along the gingival margin and from interdental spaces, areas protected against shear forces, followed later by a further extension in the coronal direction (Mierau, 1984; Quirynen *et al.*, 1989). When the tooth surface demonstrates irregularities (Figure 7) biofilm formation does not necessarily start from the gingival margin but from any groove, crack or pit. Scanning electron microscopy studies (Lie, 1979; Lie and Gusberti, 1979; Nyvad and Fejerskov, 1987) clearly revealed that the early colonisation of the surface starts from irregularities (Figure 5), where bacteria escape shear forces allowing the time needed to change from reversible to irreversible binding (Marshall *et al.*, 1971; Quirynen and Bollen, 1995). By multiplication the bacteria subsequently spread out from these areas as a relatively even monolayer. Surface irregularities are also responsible for the so called "individualised" plaque growth pattern which is reproduced in the absence of sufficient oral hygiene (Mierau and Singer, 1978; Mierau, 1984). This phenomenon illustrates the importance of surface roughness in dental plaque growth.

Surface Micro-roughness

Besides the presence of pits or cracks, surface roughness (sr) as such affects the rate of supragingival dental plaque formation. Rough intra-oral surfaces (crowns, implant abutments and denture bases) accumulate and retain more plaque and calculus in terms of thickness, area, and colony forming units (for review see Quirynen and Bollen, 1995). This plaque also reveals an increased maturity of its bacterial component (characterised by an increased proportion of motile organisms and spirochetes) and or a denser packing (Figures 3, 4). Smoothing of an intra-oral surface decreases the rate of plaque formation. Below a certain surface smoothness ($R_a < 0.2$ µm) however, a further decrease in roughness does not result in an additional reduction in plaque formation (Bollen *et al.*, 1996; Quirynen *et al.*, 1996b). Thus there seems to be a threshold sr (R_a around 0.2 µm), above which bacterial adhesion will be facilitated (Bollen and Quirynen, 1998). The supragingival surface irregularities directly enhance initial bacterial adhesion but also favour plaque growth indirectly by sheltering the attached microorganisms from oral cleaning.

In a beagle dog study, Leknes *et al.* (1994a) studied the extent of subgingival colonisation in 6 mm pockets with smooth or rough dental root surfaces. They observed that smooth surfaces harboured significantly less plaque and concluded that subgingival irregularities shelter submerged microorganisms by impeding the cleaning action of the gingival crevicular fluid. Moreover, biopsies of the soft tissues showed a higher proportion of inflammatory cells in the junctional epithelium (and the underlying connective tissue) facing the rough surfaces (Leknes *et al.*, 1994b; 1996). Even subgingivally, sr plays an important role, at least on biofilm growth. The bacteria swimming/floating freely in the crevicular fluid of the pocket or adhering to the epithelial lining are evidently not influenced by this parameter. Recent studies (Quirynen *et al.*, 1993; 1996b; Bollen *et al.*, 1996) examined the influence of sr on subgingival plaque formation by comparing, within subjects, 3 or 12 month-old plaque on titanium abutments of oral implants with different sr values (R_a ranging form 0.1 to 0.8 µm). Smooth abutments ($R_a < 0.2$ um) were found to harbour 25 times fewer bacteria than rough ones, with a slightly higher density of coccoid (*i.e.* non pathogenic) cells.

Surface Free Energy

Glantz (1969) was the first to recognise *in vivo* the correlation between substratum surface free energy (sfe) and the retaining capacity for intra-oral bacteria. Undisturbed supragingival biofilm formation on test surfaces with different free energies, indicated a positive correlation between substratum sfe and the weight of accumulated plaque (measured at days 1, 3, and 7). Van Dijk *et al.* (1987) counted (in dogs) the number

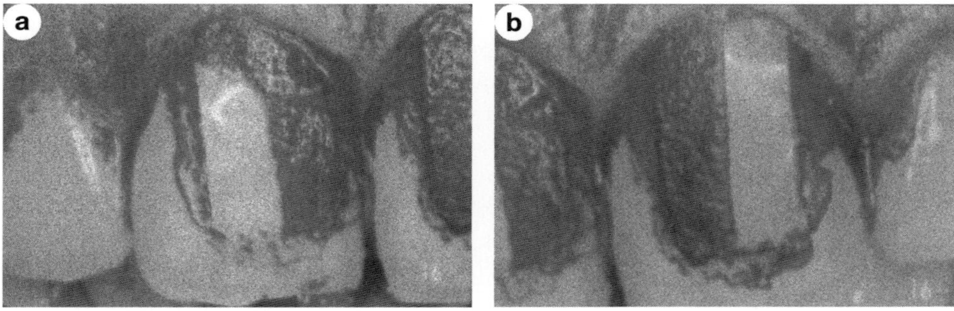

Figure 8 Photographs showing the relative clinical impact of sr and sfe on *de novo* plaque formation. Two small strips were glued to the central upper incisors of a patient who refrained from oral hygiene for 3 d. Each strip was divided in two halves, a rough region (R_a 2.0 μm) located mesially, and a smooth region (R_a 0.1 μm) distally located. The left strip (Figure 8a) was cellulose acetate (medium sfe: 58 erg cm^{-2}) and the right strip (Figure 8b) Teflon (low sfe: 20 erg cm^{-2}). Plaque was disclosed with 0.5% neutral red solution. The smooth regions show the decrease in biofilm formation due to the low sfe; the rough regions demonstrate the predominance of surface roughness, *i.e.* more plaque and no difference between the two materials.

of adhering microorganisms on solid surfaces with different sfes. After 2 h, low sfe surfaces (like Teflon and Parafilm) harboured fewer microorganisms than medium or high sfe surfaces (dentine, enamel, glass). Rölla *et al.* (1991) demonstrated that the application of a silicone oil to teeth to lower the sfe, resulted in a significant reduction in plaque formation.

Quirynen *et al.* (1989, 1990) studied the influence of sfe on undisturbed plaque growth in man over a 9-day period. Small polymer films with different sfes were glued on teeth (Figure 8) and allowed to accumulate bacteria. Hydrophobic surfaces (Teflon) harboured 10 times less plaque than hydrophilic surfaces (enamel). Moreover, hydrophobic surfaces were found to be less able to retain their plaque mass. A microbiological examination of 3-day old plaque samples from the above mentioned surfaces (Weerkamp *et al.*, 1989) indicated that Teflon was preferably colonized by bacteria with a low sfe whereas the opposite was observed for surfaces with a higher sfe (enamel). Moreover, strains of *S. sanguis I* isolated from Teflon were found to be significantly more hydrophobic than those isolated from higher energy surfaces (Weerkamp *et al.*, 1989). This suggests bacterial selection by, or adaptation to the surfaces, up to and even within the species level.

The effect of substratum sfe on plaque maturation was investigated by comparing 3 month-old plaque from implant abutments with either a high (titanium) or a low (Teflon coating) sfe. Low sfe substrata harboured a significantly less "mature" plaque both supra- and subgingivally, characterised by a higher proportion of cocci and a lower proportion of motile organisms and spirochetes (Quirynen *et al.*, 1994).

The above mentioned clinical observations confirmed *in vitro* adhesion experiments (in a flow chamber system) which showed that low sfe strains (*e.g. Streptococcus mitis*) adhered in higher numbers to hydrophobic substrata than to hydrophilic substrata, while the opposite was true for high sfe energy strains such as *S. mutans* (Uyen *et al.*, 1985; Sjollema *et al.*, 1988). It was also observed that the reversibility of bacterial adherence depended on substratum sfe (Van Pelt *et al.*, 1985; Busscher *et al.*, 1986).

In vivo, all intra-oral surfaces are covered with a surface coating (mainly consisting of salivary gycoproteins with, for example, sialic acid as a polar end). Substratum surface characteristics appear to be transferred from the substratum-protein interface to the protein-bacteria interface (Pratt-Terpstra *et al.*,1989; 1991) and thereby influence initial bacterial adhesion. Indeed, the physical and chemical natures of solid substrata were found to significantly affect the physico-chemical surface properties, the composition, packing, density, and/or the configuration of the surface coating (Lee *et al.*, 1974; Baier and Glantz, 1978; Fine *et al.*, 1984; Ruan *et al.*, 1986; Rykke *et al.*, 1991). Absolom *et al.*(1987) even observed a clear relationship between the type of proteins adsorbed and the substratum sfe. In an *in vitro* study Busscher *et al.* (1995) observed that the detachment of adhering bacteria might occur through a cohesive failure in the conditioning film (*e.g.* the pellicle) between bacteria and surface. This aspect should be kept in mind when adhesion or detachment of bacteria are considered in the oral biofilm.

Interaction Between Surface Micro-roughness and Sfe

The "relative" importance of the two parameters (sfe and roughness) on supragingival plaque formation has been examined *in vivo* (Quirynen *et al.*, 1990). Undisturbed plaque formation was followed on polymer strips with low and medium sfe where one half was smooth (R_a 0.1 μm) and the other half roughened (R_a > 2.0 μm). After 3 d of undisturbed plaque formation, as expected, significant inter-substratum differences were observed on the smooth regions, while the rough regions of the strips all demonstrated an abundant bacterial accumulation independent of the sfe (Figure 8). This indicates that, where bacterial adhesion is concerned, sr overrules the influence of sfe.

THE IMPACT OF VARIABLES ON INTRA-ORAL BIOFILM GROWTH

The rate of plaque formation on teeth differs significantly between subjects (Figure 7). There appear to be "heavy" (fast) and "light" (slow) plaque formers. Simonsson (1989) selected one group of fast and one group of slow plaque formers from a group of 133 individuals. Both groups were investigated for clinical, biochemical, biophysical and microbiological variables. In comparative analyses only minor differences appeared between the groups, and no single variable was considered as the only explanation to the great differences in the rate of dental plaque formation. A multiple regression analysis explained 90% of the variation in initial plaque formation by including the clinical wettability of the tooth surfaces, the saliva-induced aggregation of oral bacteria and the relative salivary flow conditions. The saliva from slow plaque formers reduced the colloidal stability of bacterial suspensions, such as for *Streptococcus sanguis* (Simonsson and Glantz, 1988). In a study by Zee *et al.* (1997) *de novo* plaque formation was followed on small enamel blocks that were bonded onto the teeth of slow and heavy plaque formers. After a 1-day fast, plaque formers showed more plaque with a more complex supragingival structure. However, from day 3 to 14 there were no discernible differences between the groups, except for a more prominent inter-microbial matrix in the rapid growth group. Moreover, rapid plaque formers showed higher proportions of Gram-negative rods (35% *vs* 17%) after 2 weeks (Zee *et al.*, 1996).

Variation within the Oral Cavity

Within the dental arch, large differences in plaque growth rate can be detected. In general, plaque grows more quickly in the lower jaw, the vestibular tooth surfaces and the interdental spaces (Lang *et al.*, 1973; Quirynen, 1986; Furuichi *et al.*, 1992).

Impact of Gingival Inflammation

Several studies clearly indicate that early *in vivo* plaque formation is more rapid on tooth surfaces facing inflamed gingival margins than on those adjacent to healthy gingiva (Saxton, 1973; Hillam and Hull, 1977; Brecx *et al.*,1980; Quirynen *et al.*, 1991; Ramberg *et al.*, 1994; 1995). These studies suggest that the increase in crevicular fluid production plays a keyrole in enhanced plaque formation, Probably, some substance(s) from this exudate (*e.g.* minerals, proteins or carbohydrates) favour both the initial adhesion and/or the growth of the early colonising bacteria (Hillam and Hull, 1977).

Impact of Patients' Age

A subject's age does not influence *de novo* plaque formation, either in terms of the amount produced or the composition of the plaque, except during the initial hours of biofilm accumulation where less bacteria are present on the teeth of elderly individuals (Brecx *et al.*, 1985). In a study no differences could be detected in *de novo* plaque formation between a group of young (20–25 years of age) and older (65–80 years of age) subjects who abolished mechanical tooth cleaning measures for 21 d (Fransson *et al.*, 1996). These observations largely confirm data by Holm-Pedersen *et al.* (1975) and Winkel *et al.* (1987). The developed plaque in the older patient group resulted, however, in a more severe gingival inflammation, which seems to indicate an increased susceptibility to gingivitis with ageing.

Inter-subject Variation

Besides differences in surface characteristics and/or in the degree of gingival inflammation (see above), inter-subject variation in plaque formation might also be explained by diet (Ainamo *et al.*, 1979), chewing fibrous food (Arnim, 1963), smoking (Sheiham, 1971), the presence of copper amalgam (Hyyppä and Paunio, 1977), tongue and palate brushing (Jacobson *et al.*, 1973), the colloid stability of bacteria in the saliva (Simonsson, 1989), anti-microbial factors present in the saliva (Adamson and Carlsson, 1982), the chemical composition of the pellicle (Simonsson *et al.*, 1987) and the retention depth of the dento-gingival area (Simonsson *et al.*, 1987). Rinsing with glucose or sucrose however, was found to have no detectable effect on biofilm growth during the first hours of plaque accumulation (Brecx *et al.*, 1981).

CONCLUSIONS

Intra-oral biofilm formation on a clean tooth/prosthetic surface follows an exponential growth curve. Large intra- and inter-individual differences appear. Surface characteristics (sr and to a lesser degree sfe) are responsible for the majority of dental plaque growth variability. The similarity with the observations made in other environments (*e.g.* larynx or cardiovascular prostheses, submarine surfaces, pipelines, etc.) is striking. Since the study of bacterial growth in the oral cavity does not involve elaborate or invasive techniques, it deserves more attention in the context of the study and control of biofilms.

REFERENCES

Absolom D.R., Zingg W., Neumann A.W. (1987). Protein adsorption to polymer particles: role of surface properties. *J Biomed Mater Res*, **21**, 161–171.

Adamson M., Carlsson J. (1982). Lactoperoxidase and tiocyanate protect bacteria from hydrogen peroxide. *Infect Immun*, **35**, 20–24.

Adriaens P.A., Edwards C.A., De Boever J.A., Loesche W.J. (1988). Ultrastructural observations on bacterial invasion in cementum and radicular dentin of periodontally diseased human teeth. *J Periodontol*, **59**, 493–503.

Ainamo J., Asikainen S., Ainamo A., Lathinen A., Sjöblom M. (1979). Plaque growth while chewing sorbitol and xylitol simultaneously with sucrose flavored gum. *J Clinical Periodontol*, **6**, 397–406.

Arnim S.S. (1963). The use of disclosing agents for measuring tooth cleanliness. *J Periodontol*, **4**, 217–245.

Axelsson P., Lindhe J. (1981). Effect of controlled oral hygiene procedures on caries and periodontal disease in adults. Results after 6 years. *J Clinical Petiodontol*, **8**, 239–248.

Axelsson P., Lindhe J., Nystrom B. (1991). On the prevention of caries and periodontal disease. Results of a 15-year longitudinal study in adults. *J Clinical Periodontol*, **18**, 182–189.

Baier R.E., Glantz P.-O. (1978). Characterization of oral *in vivo* films formed on different types of solid surfaces. *Acta Odontol Scand*, **36**, 289–301.

Bollen C.M.L., Quirynen M. (1998). The evolution of the surface roughness of different oral hard materials in comparison to the "threshold surface roughness". A review of the literature. *Dent Mater* (In press).

Bollen C.M.L., Papaioannou W., Van Eldere J., Schepers E., Quirynen M., van Steenberghe D. (1996). The influence of abutment surface roughness on plaque accumulation and peri-implant mucositis. *Clin Oral Implants Res*, **7**, 201–211.

Brecx M. (1997). Strategies and agents in supragingival chemical plaque control. *Periodontology 2000*, **15**, 100–108.

Brecx M., Theilade J., Attström R. (1980). Influence of optimal and excluded oral hygiene on early formation of dental plaque on plastic films. A quantitative and descriptive light and electron microscopic study. *J Clinical Periodontol*, **7**, 361–373.

Brecx M., Theilade J., Attström R. (1981). Ultrastructural estimation of the effect of sucrose and glucose rinses on early dental plaque formed on plastic films. *Scand J Dent Res*, **89**, 157–164.

Brecx M., Theilade J., Attström R. (1983). An ultrastructural quantitative study of the significance of microbial multiplication during early dental plaque growth. *J Periodontal Res*, **18**, 177–186.

Brecx M., Holm-Pedersen P., Theilade J. (1985). Early plaque formation in young and elderly individuals. *Gerodontics*, **1**, 8–13.

Brecx M., Winkler M., Netuschil L. (1994). Human dental plaque formation on plastic films. A quantitative S.E.M study. *J Western Soc Periodontol/Periodontal Abstracts*, **42**, 77–80.

Busscher H.J., Bos R., Van der Mei H.C. (1995). Initial microbial adhesion is a determinant for the strength of biofilm adhesion. *FEMS Microbiol Lett*, **128**, 229–234.

Busscher H.J., Uyen M.H.M.J.C., Weerkamp A.H., Postma W.J., Arends J. (1986). Reversibility of adhesion of oral streptococci to solids. *FEMS Microbiol Lett*, **35**, 303–306.

Carlsson J. (1980). Symbiosis between host and microorganisms in the oral cavity. *Scand J Infect Dis*, suppl **24** 74–79.

Carlsson J.S., Soderholm G., Almfeldt I. (1969). Prevalence of *Streptococcus sanguis* and *Streptococcus mutans* in the mouths of persons wearing full dentures. *Arch Oral Biol*, **14**, 43–249.

Carlsson J.S., Grahnen H., Jonsson G., Wikner S. (1970). Establishment of *Streptococcus sanguis* in the mouths of infants. *Arch Oral Biol*, **15**, 1143–1148.

Caufield P.W., Gibbons R.J. (1979). Suppression of *Streptococcus mutans* in the mouths of humans by a dental prophylaxis and topically-applied iodone. *J Dental Res*, **58**, 1317–1326.

Danser M.M. (1996). The prevalence of periodontal bacteria colonizing the oral mucous membranes. *ScD Thesis*, Universiteit van Amsterdam, The Netherlands.

Danser M.M., Timmerman M.F., van Winkelhoff A.J., van der Velden U. (1996). The effect of periodontal treatment on periodontal bacteria on the oral mucous membranes. ScD Thesis, *J Periodontol*, **67**, 478–485.

Evaldson G., Heimdahl A., Kager L., Nord C.E. (1982). The normal human anaerobic microflora. *Scandinavian J Infect Dis*, suppl **35**, 9–15.

Fine D.H., Wilton J.M.A., Caravana C. (1984). *In vitro* sorption of albumin, immunoglobulin G, and lysozyme to enamel and cementum from human teeth. *Infect Immun*, **44**, 332–338.

Fransson C., Berglundh T., Lindhe J. (1996). The effect of age on the development of gingivitis. Clinical, microbiological and histological findings. *J Clinical Periodontol*, **23**, 379–385.

Furuichi Y., Lindhe J., Ramberg P., Volpe A.R. (1992). Patterns of *de novo* plaque formation in the human dentition. *J Clinical Periodontol*, **191**, 423–433.

Gibbons R.J., Hay D.I., Cisar J.O., Clark W.B. (1988). Adsorbed salivary proline-rich protein 1 and statherin: receptors for type 1 fimbriae of *Actinomyces viscosus* T14V-J1 on apatitic surfaces. *Infect Immun*, **56**, 2990–2993.

Glantz P.-O. (1969). On wettability and adhesivenesss. *Odontol Revy*, **20**, suppl 17, 1–132.

Hillam D.G., Hull P.S. (1977). The influence of experimental gingivitis on plaque formation. *J Clin Periodontol*, **4**, 56–61.

Holm-Pedersen P., Agerbaek N., Theilade E. (1975). Experimental gingivitis in young and elderly individuals. *J Clin Periodontol*, **2**, 14–24.

Hsu S.D., Cisar J.O., Sandberg A.L., Kilian M. (1994). Adhesive properties of viridans streptococcal species. *Microb Ecol Health Dis*, **7**, 125–137.

Hyyppä T., Paunio K. (1977). The plaque-inhibiting effect of copper amalgam. *J Clinical Periodontol*, **4**, 1231–239.

Jacobson S.E., Crawford J.J., McFall W.R. (1973). Oral physiotherapy of the tongue and palate: relationship to plaque control. *J Am Dent Assoc*, **87**, 134–139.

Kolenbrander P.E., London J. (1992). Ecological significance of coaggregation among oral bacteria. *Adv Microb Ecol*, **12**, 183–217.

Kolenbrander P.E., Andersen R.N., Moore L.V.H. (1990). Intrageneric coaggregation among strains of human oral bacteria: potential role in primary colonization of the tooth surface. *Appl Environ Microbiol*, **56**, 3890–3894.

Kolenbrander P.E., Parrish K.D., Andersen R.N., Greenberg E.P. (1995). Intergeneric coaggregation of oral *Treponema* spp. with *Fusobacterium* spp. and intrageneric coaggregation among *Fusobacterium* spp. *Infect Immun*, **63**, 4584–4588.

Könönen E., Jousimies-Somer H., Asikainen S. (1992). Relationship between oral Gram-negative anaerobic bacteria in saliva of the mother and the colonization of her edentulous infant. *Oral Microbiol Immunol*, **71**, 273–276.

Lang N.P., Cumming B.R., Löe H. (1973). Toothbrushing frequency as it relates to plaque development and gingival health. *J Periodontol*, **44**, 396–405.

Lee R.G., Adamson C., Kim S.W. (1974). Competitive adsorption of plasma proteins onto polymer surfaces. *Thromb Res*, **4**, 485–490.

Leknes K.N., Lie T., Selvig K.A. (1994b). Root grooves: a risk factor in periodontal attachment loss. *J Periodontol*, **65**, 859–863.

Leknes K.N., Lie T., Wikesjb U.M.E., Bogle G.C., Selvig K.A. (1994a). Influence of tooth instrumentation roughness on subgingival microbial colonization. *J Periodontol*, **65**, 303–308.

Leknes K.N., Lie T., Wikesjö U.M.E., Böe O.E., Selvig K.A. (1996). Influence of tooth instrumentation roughness on gingival tissue reactions. *J Periodontol*, **67**, 197–204.

Liakoni H., Barber P., Newman H.N. (1987). Bacterial penetration of pocket soft tissues in chronic adult and juvenile periodontitis. An ultrastructural study. *J Clinical Periodontol*, **14**, 22–28.

Lie T. (1979). Morphologic studies on dental plaque formation. *Acta Odontol Scand*, **37**, 73–85.

Lie T., Gusberti F. (1979). Replica study of plaque formation on human tooth surfaces. *Acta Odontol Scand*, **37**, 65–72.

Liljemark W.F., Bloomquist C.G. (1996). Human oral microbial ecology and dental caries and periodontal diseases. *Crit Rev Oral Biol Med*, **7**, 180–198.

Liljemark W.F., Bloomquist C.G., Reilly B.E., Bernards C.J., Towsend D.W., Pennock A.T., LeMoine J.L. (1997). Growth dynamics in a natural biofilm and its impact on oral disease management. *Adv Dent Res*, **11**, 14–32.

Listgarten M.A., Helldén L. (1978). Relative distribution of bacteria at clinically healthy and periodontally diseased sites in humans. *J Clinical Periodontol*, **5**, 115–132.

Loesche W.J. (1976) Chemotheraphy of dental plaque formation. *Oral Sci Rev*, **9**, 65–107.

Loesche W.J. (1988). Ecology of the oral flora. In: Nisengard R.I., Newman M.G. (eds) *Oral Microbiology and Immunology*. W.B. Saunders Company, Philadelphia, pp. 307–319.

Marshall K.C., Stout R., Mitchell R. (1971). Mechanism of the initial events in the sorption of marine bacteria to surfaces. *J Gen Microbiol*, **68**, 337–348.

Mierau H.-D. (1984). Beziehungen zwischen Plaquebildung, Rauhigkeit der Zahnoberflache und Selbstreinigung. *Deutsche Zahnartzliche Zeitschrift*, **39**, 691–698.

Mierau H.-D., Singer D. (1978). Reproduzierbarkeit der Plaquebildung im dentogingivalen Bereich (3. Mitteilung). *Deutsche Zahnartzliche Zeitschrift*, **133**, 566–573.

Mikkelsen L. (1993). Influence of sucrose intake on saliva and number of microorganisms and acidogenic potential in early dental plaque. *Microb Ecol Health Dis*, **61**, 253–264.

Moore W.E.C., Moore L.V.H. (1994). The bacteria of periodontal diseases. *Periodontology 2000*, **5**, 66–77.

Naito Y., Gibbons R.J. (1988). Attachment of *Bacteroides gingivalis* to collagenous substrata. *J Dent Res*, **67**, 1075–1080.

Netushil L., Reich R., Unteregger G., Sculean A., Brecx M. (1998). A pilot study of confocal laser scanning microscopy for the assessment of undisturbed dental plaque vitality and topography. *Arch Oral Biol*, **43**, 277–285.

Nyvad B., Fejerskov O. (1987). Scanning electron microscopy of early microbial colonization of human enamel and root surfaces *in vivo*. *Scandinavian J Dent Res*, **95**, 287–296.

Nyvad B., Kilian M. (1987). Microbiology of the early colonization of human enamel and root surfaces *in vivo*. *Scandinavian J Dent Res*, **95**, 369–380.

Nyvad B., Kilian M. (1990). Comparison of the initial streptococcal microflora on dental enamel in caries-active and in caries-inactive individuals. *Caries Res*, **24**, 267–272.

Papaioannou W., Quirynen M., van Steenberghe D. (1996). The influence of periodontitis on the subgingival flora around implants in partially edentulous patients. *Clin Oral Implant Res*, **7**, 405–409.

Pratt-Terpstra I.H., Weerkamp A.H., Busscher H.J. (1989). The effect of pellicle formation on streptococcal adhesion to human enamel and artificial substrata with various surface free-energies. *J Dent Res*, **68**, 463–467.

Pratt-Terpstra I.H., Mulder J., Weerkamp A.H., Feijen J., Busscher H.J. (1991). Secretory IgA adsorption and oral streptococcal adhesion to human enamel and artificial solid substrata with various surface free energies. *J Biomater Sci Polymer Edition*, **2**, 239–253.

Quirynen M. (1986). Anatomical and inflammatory factors influence bacterial plaque growth and retention in man. ScD Thesis, Department of Periodontology, Catholic University, Leuven, Belgium.

Quirynen M., Bollen C.M.L. (1995). The influence of surface roughness and surface free energy on supra and subgingival plaque formation in man. A review of the literature. *J Clin Periodontol*, **22**, 1–14.

Quirynen M., Dekeyser C., van Steenberghe D. (1991). The influence of gingival inflammation, tooth type, and timing on the rate of plaque formation. *J Periodontol*, **62**, 219–222.

Quirynen M., Papaioannou W., van Steenberghe D. (1996a). Intra-oral transmission and the colonization of oral hard surfaces. *J Periodontol*, **67**, 986–993.

Quirynen M., Bollen C.M.L., Papaioannou W., Van Eldere J., van Steenberghe D. (1996b). The influence of titanium abutments surface roughness on plaque accumulation and gingivitis. Short term observations. *Int J Oral Maxillofac lmplants*, **11**, 169–178.

Quirynen M., Marechal M., Busscher H.J., Weerkamp A.H., Darius P.L., van Steenberghe D. (1990). The influence of surface free energy and surface roughness on early plaque formation. *J Clin Periodontol*, **17**, 138–144.

Quirynen M., Marechal M., Busscher H.J., Weerkamp A.H., Arends J., Darius P.L., van Steenberghe D. (1989). The influence of surface free energy on planimetric plaque growth in man. *J Dent Res*, **68**, 796–799.

Quirynen M., Van Der Mei H.C., Bollen C.M.L., Van den Bossehe L.H., Doornbusch G.I., van Steenberghe D., Busscher H.J. (1994). The influence of surface-free energy on supra- and subgingival plaque microbiology. An *in vivo* study on implants. *J Periodontol*, **65**, 162–167.

Quirynen M., Van der Mei H.C., Bollen C.M.L., Schotte A., Marechal M., Doornbusch G.I., Naert I., Busscher H.J., van Steenberghe D. (1993). An *in vivo* study of the influence of surface roughness of implants on the microbiology of supra- and subgingival plaque. *J Dent Res*, **72**, 1304–1309.

Ramberg P., Axelsson P., Lindhe J. (1995). Plaque formation at healthy and inflamed gingival sites in young individuals. *J Clinical Periodontol*, **2**, 85–88.

Ramberg P., Lindhe J., Dahlén G., Volpe A.R. (1994). The influence of gingival inflammation on *de novo* plaque formation. *J Clinical Periodontol*, **21**, 51–56.

Rölla G., Ellingsen J.E., Herlofson B. (1991). Enhancement and inhibition of dental plaque formation – some old and new concepts. *Biofouling*, **1**, 175–181.

Rönström A., Edwardsson S., Attström R. (1977). *Streptococcus sanguis* and *Streptococcus salivarius* in early plaque formation on plastic films. *J Periodontol Res*, **12**, 331–339.

Rotimi V.O., Duerden B.I. (1981). The development of the bacterial flora in normal neanates. *J Med Microbiol*, **14**, 51–62.

Ruan M.-S., Di Paola C., Mandel I.D. (1986). Quantitative immunochemistry of saliva proteins adsorbed *in vitro* to enamel and cementum from caries-resistant and caries-susceptible human adults. *Arch Oral Biol*, **31**, 597–601.

Rykke M., Ellingsen J.E., Sönju T. (1991). Chemical analysis and scanning electron microscopy of acquired pellicle formed *in vivo* on stannous fluoride treated enamel. *Scand J Dent Res*, **99**, 205–211.

Saxton C.A. (1973). Scanning electron microscope study of the formation of dental plaque. *Caries Res*, **7**, 102–109.

Scannapieco F.A., Torres G.I., Levine M.J. (1995). Salivary amylase promotes adhesion of oral streptococci to hydroxyapatite. *J Dent Res*, **74**, 1360–1366.

Schroeder H.E., De Boever J. (1970). The structure of microbial dental plaque. In: McHugh W.D. (ed) *Dental Plaque*. Livingstone, Edinburgh, pp. 49–74.

Sheiham A. (1971). Periodontal disease and oral cleanliness in tobacco smokers. *J Periodontol*, **42**, 259–263.

Simonsson T., Rönström A., Rundegren J., Birkhed D. (1987). Rate of plaque formation. Some clinical and biochemical characteristics of "heavy" and "light" plaque formers. *Scand J Dent Res*, **95**, 97–103.

Simonsson T., Glantz P.O. (1989). Influence of salivary from 'heavy' and 'light' plaque formers on the colloidal stability of bacterial suspensions. *Acta Odontol Scand*, **46**, 195–197.

Simonsson T. (1989). Aspects of dental plaque formation with special reference to colloid-chemical phenomena. *Swed Dent J*, **58**, 1–67.

Sjollema J., Busscher H.J., Weerkamp A.H. (1988). Deposition of oral streptococci and polystyrene latices onto glass in a parallel plate flow cell. *Biofouling*, **1**, 101–112.

Skopek R.J., Liljemark W.F., Bloomquist C.G., Rudney J.D. (1993). Dental plaque development on defined streptococcal surfaces. *Oral Microbiol Immunol*, **8**, 16–23.

Socransky S.S., Manganiello S.D. (1971). The oral microbiota of man from birth to senility. *J Periodontol*, **42**, 485–496.

Socransky S.S., Manganiello A.D., Propas D., Oram V., van Houte J. (1977). Bacteriological studies of developing supragingival dental plaque. *J Periodontol Res*, **12**, 90–106.

Socransky S.S., Haffajee A.D. (1992). The bacterial etiology of destructive periodontal disease: current concepts. *J Periodontol*, **63**, 322–331.

Socransky S.S., Haffajee A. (1997). Microbiology of periodontal disease. In: Lindhe J., Karring T., Lang N.P. (eds) *Clinical Periodontology and Implant Dentistry*. Munksgaard, Copenhagen, pp. 138–188.

Socransky S.S., Haffajee A.D., Cugini M.A., Smith C., Kent Jr R.L. (1998). Microbial complexes in subgingival plaque. *J Clin Periodontol*, **25**, 134–144.

Syed S.A., Loesche W.J. (1978). Bacteriology of human experimental gingivitis: effect of plaque age. *Infect Immun*, **21**, 821–829.

Tanzer J.M. (1992). Microbiology of dental caries. In: Slots J., Taubaum M.A. (eds) *Contemporary Oral Microbiology and Immunology*. St. Louis: Mosby-Year Book, pp. 377–424.

Tappuni A.R., Challacombe S.J. (1993). Distribution and isolation frequency of eight streptococcal species in saliva from predentate and dentate children and adults. *J Dent Res*, **72**, 31–36.

Theilade E., Wright W.H., Jensen S.B., Löe H. (1966). Experimental gingivitis in man. II. A longitudinal clinical and bacteriological investigation. *J Periodontal Res*, **1**, 1–13.

Uyen H.M., Busscher H.J., Weerkamp A.H., Arends J. (1985). Surface free energies of oral streptococci and their adhesion to solids. *FEMS Microbiol Lett*, **30**, 103–106.

Van Dijk J., Herkströter F., Busscher H., Weerkamp A., Jansen H., Arends J. (1987). Surface-free energy and bacterial adhesion, an *in vivo* study in beagle dogs. *J Clin Periodontol*, **14**, 300–304.

Van Pelt A.W.J., Weerkamp A.H., Uyen H.M.W.J.C., Busscher H.J., De Jong H.P., Arends J. (1985). Adhesion of *Streptococcus sanguis* CH3 to polymers with different surface free energies. *Appl Environ Microbiol*, **49**, 1270–1275.

Weerkamp A.H., Quirynen M., Marechal M., Van der Mei H.C., van Steenberghe D., Busscher H.J. (1989). The role of surface free energy in the early *in vivo* formation of dental plaque on human enamel and polymeric substrata. *Microb Ecol Health Dis*, 2, 11–18.

Weiger R., Netuschil L., von Ohle C., Schlagenhauf U., Brecx M. (1995). Microbial generation time during the early phases of supragingival dental plaque formation. *Oral Microbiol Immunol*, **10**, 93–97.

Winkel E.R., Abbas F., Van der Velden U., Vroom T.M., Scholte G., Hart A.A. (1987). Experimental gingivitis in relation to age in individuals not susceptible to periodontal disease. *J Clin Periodontol*, 14, 499–507.

Whittaker C.J., Klier C.M., Kolenbrander P.E. (1996). Mechanisms of adhesion by oral bacteria. *Annu Rev Microbiol*, **50**, 513–552.

Wolff L., Dahlèn G., Aeppli D. (1994). Bacteria as risk markers for periodontitis. *J Periodontol*, **65**, 498–510.

Zambon J.J. (1996). Periodontal diseases: microbial factors. *Ann Periodontol*, **1**, 879–925.

Zee K., Samaranayake L.P., Attstrom R. (1997). Scanning electron microscopy of microbial colonization of ‘rapid’ and ‘slow’ dental plaque formers *in vivo*. *Arch Oral Biol*, 42, 735–742.

Zee K., Samaranayake L.P., Attstrom R. (1996). Predominant cultivable supragingival plaque in Chinese “rapid” and “slow” plaque formers. *J Clin Periodontol*, **23**, 1025–1031.

12 Algal Biofilms

Maureen E. Callow

Algal biofilms will colonise any surface provided moisture and light are available. They are often complex, highly productive ecosystems held together by a matrix of extracellular polymeric substances (EPS), which are produced both by the algae themselves, and microorganisms which are ubiquitously present. Algal EPS mediates adhesion and gliding motility in raphid diatoms and some cyanobacteria, and is responsible for biostabilisation of sediments in rivers and estuarine environments. The current state of knowledge of the bioadhesive polymers which mediate attachment of cells and spores is discussed. Algal biofilms have complex carbon budgets and often exhibit high specific rates of metabolism with marked vertical gradients of photosynthesis/respiration and physico-chemical properties. In addition to contributing a significant amount to primary production in aquatic ecosystems, algal biofilms provide a major source of energy for higher trophic levels and play a major role in nutrient recycling. Algal biofilms are responsible for the biodeterioration of a variety of man-made structures in both the aerial and aquatic environments. Severe economic penalties result from the increase in hydrodynamic drag caused by biofilms on ships' hulls. Strategies to control algal biofouling are discussed.

KEY WORDS: algae, algal biofilm, diatoms, EPS, adhesive, biostabilisation, biofouling, biodeterioration

INTRODUCTION

Biofilms dominated by algae and/or cyanobacteria (blue-green algae) will grow on any surface provided there is moisture and light. In natural ecosystems, algal biofilms are familiar on trees and fences, on stones in rivers, and on the seashore. They are also found in extremely hostile environments from the ice and snow of the polar ice caps to the tropics. Algal biofilms on paintwork and buildings are considered unsightly whilst fouling of underwater structures such as ships' hulls causes substantial economic losses and may compromise safety.

Algal biofilms are complex, frequently highly productive, ecosystems, held together by a matrix of extracellular polymeric substances (EPS). Bacterial consortia are invariably present, whilst in aquatic habitats, other lower organisms such as fungi, grazing protozoa, flagellates and micrometazoa are frequently associated with the photosynthetic community (Bott, 1996; Lamberti, 1996). Such an aquatic biofilm community is often referred to as periphyton (see Wetzel, 1983). In aquatic environments, algal biofilms grow on stones, vegetation and sediments and when colonising these habitats, are often referred to as epilithic, epiphytic and benthic respectively (Moss, 1988) although the term 'benthic' is frequently applied to organisms which adhere to all types of substrata. The latter definition will be adopted in this chapter. Species diversity within natural assemblages can be very great and diatom communities in particular are proving useful bioindicators of water quality (see Lowe and Pan, 1996).

Algal biofilms often exhibit high specific rates of metabolism with marked vertical gradients of photosynthesis/respiration and physico-chemical properties (*e.g.* Kuhl *et al.*, 1996). In general, it appears there is a close-coupling of photosynthesis and respiration in biofilms, with the dissolved inorganic carbon (DIC) required by the photosynthesising members of the population being provided largely from respiration of the heterotrophs, which in turn, consume a significant proportion of the photosynthetically produced oxygen and organic carbon (Glud *et al.*, 1999). The surfaces of biofilms often have uneven topography, and their metabolism is strongly influenced by the diffusive boundary layer (Liehr *et al.*, 1989; 1990) Hydrodynamic conditions influence nutrient and gas exchange which in turn affect the structure of biofilms. Algal biofilms have complex carbon budgets due to the balance between photosynthesis, respiration and photorespiration of the algal cells combined with respiration of the heterotrophic community. EPS, which is produced by both algal and microbial cells and which cements the community together, has important ecological implications as it creates a diffusion barrier within the biofilm, which may limit the uptake of nutrients and CO_2 (Lock, 1981). However, there are advantages of diffusion resistance, *e.g.* the trapping of dissolved and colloidal organic matter. Bacteria also benefit from the O_2 and any fixed carbon released by the algal cells (Cooksey, 1992). EPS produced by diatoms, and to a lesser extent by cyanobacteria, is important in the biostabilisation of sediments, especially in estuarine environments (Paterson, 1994). Other attributes of EPS which benefit biofilm survival include resistance to desiccation and complexation of heavy metals (Brown *et al.*, 1988; Robinson *et al.*, 1992; Genter, 1996).

This chapter will describe the major types of biofilm-forming algae, their physiology and the importance of EPS in terms of their success in colonising new substrata. The role of algal biofilms, both in terms of natural ecological assemblages and in terms of the nuisance caused in the biofouling of man-made structures, will be discussed.

COLONISATION AND STRUCTURE OF BIOFILM ALGAE

Biofilm algae are usually unicellular, colonial or simple filamentous organisms. Biofilms dominated by diatoms are most common in aquatic environments, whilst green unicellular algae and cyanobacteria dominate in terrestrial environments. The species composition of individual biofilms depends primarily on qualitative and quantitative aspects of the inoculum, but light, the substratum, nutrient supply, competition and grazing are all important in determining whether the colonised surface will become a dynamic biofilm ecosystem, or whether higher organisms will become established to produce a climax community (see Lock, 1981; Biggs, 1996). Biofilms on exposed rocky shores are the major food source for grazing gastropods and their grazing activities prevent algal growth, thereby halting succession at the biofilm stage (Hawkins

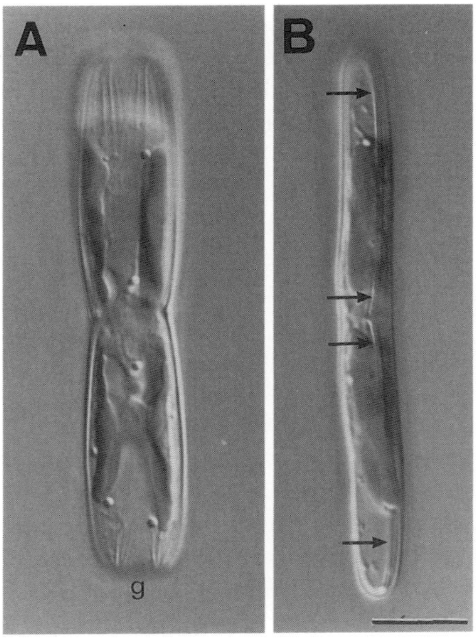

Figure 1 Light micrographs (Nomarski optics) of *Stauroneis decipiens* cells. The cell settling through medium lands with the girdle against the substratum (A), then flips over so that a valve face is in contact with the substratum (B). The silica frustule is composed of two valves and overlapping girdle bands (g). The raphes (uppermost raphe is arrowed in B), extend longitudinally on the face of both valves. Scale bar = 10 μm. (Reproduced from Lind *et al.*, 1997, with permission.)

and Hartnoll, 1983). Although the effect of herbivory is less marked in freshwater environments, it still plays a major role in regulating biofilm development (Steinman, 1996). Even within a dynamic biofilm ecosystem, a microsuccession and/or zones which may vary spatially and temporally on a micro- or macroscale can often be recognised.

The sequence of events following immersion of a clean surface in water starts with the adsorption of organic molecules (Baier, 1980) and adhesion of primary colonisers including bacteria and algal cells or spores. Diatoms are usually considered to be the most common and abundant of the early colonisers (Jackson and Jones, 1988). Diatoms are unicells or colonial algae, characterised by the presence of an elaborately ornamented silica frustule and chloroplasts containing the pigment fucoxanthin which masks the chlorophyll, hence their brown colour (Round *et al.*, 1990; van den Hoek *et al.*, 1995; Cox, 1996). Cells range in size from a few to several hundred micrometres. The diatom frustule is composed of overlapping halves (valves) and several rings of silica (girdle bands) which, together with a range of secreted EPS, encase the protoplast (Pickett-Heaps *et al.*, 1990; Round *et al.*, 1990). Pennate diatoms have bilateral symmetry and are the abundant types in benthic habitats, particularly genera which possess an elongate slit, the raphe, in one or both valves of the frustule (Figure 1). Raphid diatoms may be sessile or motile but in both cases, EPS is secreted *via* the raphe(s) and pores which in the former provides the means for adhesion whilst in the latter, it provides the mechanism for both adhesion and motility by gliding (Edgar and Pickett-Heaps, 1984). Although the term gliding is used to describe diatom locomotion, many move in a jerky, irregular manner, which is consistent with movement in highly viscous conditions (see Edgar and Pickett-Heaps, 1984). Average speeds of 1–25 μm s^{-1} have been recorded (Edgar and Pickett-Heaps, 1984) but speeds of 4–10 μm s^{-1} appear to be most typical (Cohn and Weitzell, 1996; Lind *et al.*, 1997). Raphid diatoms are probably the most important biofilm algae, being instrumental in the primary colonisation of submerged substrata. Initial contact with the substratum, "the first kiss" (Wetherbee *et al.*, 1998), is thought to represent an active commitment by raphid diatoms to adhere, and is thought to activate adhesion mechanisms specifically designed for subsequent binding to the substratum. Initial attachment is reversible as the cells adjust their position by gliding locomotion (Edgar and Pickett-Heaps, 1984). Secondary or permanent adhesion appears to be the result of the secretion of further EPS components *via* the raphe or pores, along with reorganisation and crosslinking reactions (Wetherbee *et al.*, 1998; Wustman *et al.*, 1998).

The EPS of some benthic diatoms may be secreted from points in the frustule other than the raphe, *e.g.* an apical pore field, becoming elaborated into structures such as pads and stalks (Daniel *et al.*, 1987; Hoagland *et al.*, 1993). Stalks and other extracellular structures that elevate the cell body above the substratum may confer advantages in competition for nutrients and light. Mucilage sheaths containing large numbers of cells may reach several centimetres in length, thereby assuming a filamentous appearance. In common with bacterial biofilms, the initial colonization may include only a few cells, as cell division rapidly gives rise to colonies which eventually become confluent to form a compact biofilm. Doubling times for diatom populations growing on glass slides in the sea at Miami, Florida were between 11.7 and 27.4 h depending on the time of year, and similar to those found in laboratory cultures (Cooksey *et al.*, 1984). Diatom biofilms are typically around 500 μm in thickness (see Figure 7) with approximately 2.7×10^5 cells cm^{-2} (Hendey, 1951). The gliding movement of raphid diatoms such as *Stauroneis* and *Amphora* also facilitates rapid colonisation of the surface, whilst the mucilage trails of EPS also provide a substratum for attachment of other colonisers (Stevenson and Peterson, 1989) and a food source for heterotrophs.

Green biofilm algae are predominantly unicellular, colonial or filamentous (van den Hoek *et al.*, 1995). Some common macroalgae *e.g.* the green alga *Enteromorpha* and the brown alga *Ectocarpus*, can adopt a diminutive form when growing in hostile conditions, for example, on antifouling paint on a ship's hull (see later). Many aquatic green and brown algae reproduce through the production of large numbers of motile spores, which leads to rapid colonisation of the substratum.

Cyanobacteria are simple prokaryotic organisms (Fogg *et al.*, 1973; Carr and Whitton, 1982) which are blue-green, sometimes red, in colour due to the presence of phycobilin pigments which mask chlorophyll. Many species are capable of fixing atmospheric nitrogen which facilitates growth in nutrient deficient environments (*e.g.* Toledo *et al.*, 1995). One of the most important roles that cynobacterial biofilms fulfil is N_2-fixation in paddy fields (Fogg *et al.*, 1973). In common with eukaryotic algae, cyanobacteria are able to perform oxygenic (oxygen-evolving) photosynthesis. Cyanobacteria exhibit a variety of growth forms from unicellular to variously "branched" filaments. Adhesion to the substratum is through the secretion of EPS, and many,

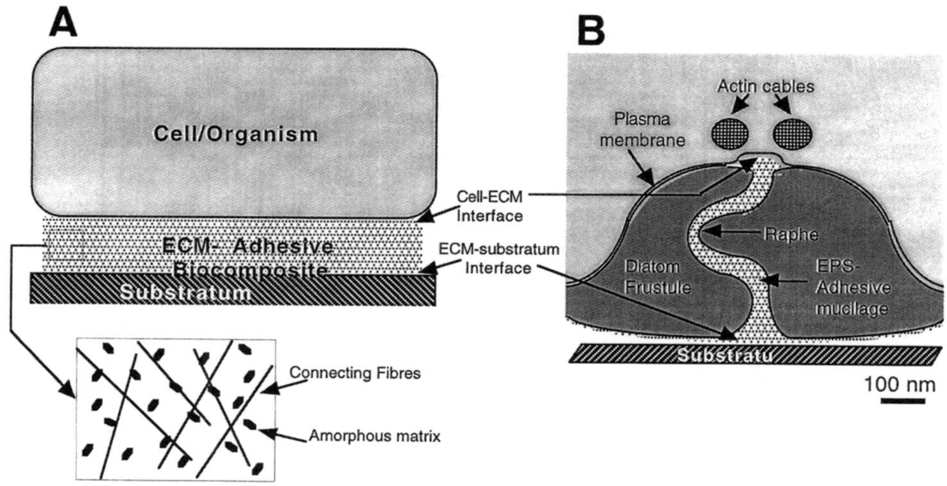

Figure 2 A = a diagrammatic representation of adhesion of a diatom cell to the substratum by means of the extracellular matrix (ECM) (= EPS) and showing the cell-ECM and ECM-substratum interfaces. B = a diagrammatic cross-section through the raphe of the diatom *Stauroneis decipiens*, adhered to the substratum by means of the ECM and showing the cell-ECM and ECM-substratum interfaces. Actin cables, part of the actin-based cytoskeletal system and located just beneath the plasma membrane, are thought to be directly involved in adhesion and motility. (Reproduced from Wetherbee *et al.*, (1998), by permission of the *Journal of Phycology*.)

whether single cells or filaments, are capable of gliding movements. Single cells tend to move jerkily but filaments *e.g. Osclillatoria*, glide in a more controlled way, moving backwards and forwards at speed of 2–11 µm s^{-1}. In common with some benthic diatoms, motile cyanobacteria leave a trail of EPS (Scott *et al.*, 1996). Gliding motility depends on the continuous secretion of EPS *via* specific pores (Hoiczyk and Baumeister, 1998). Cyanobacteria are frequently found as minor components in biofilms but in certain environments, they dominate *e.g.* in the spray zone, shallow water and sediments of freshwater environments, and in the upper intertidal and muddy or shallow sediments of salt marshes and estuaries. Cyanobacteria-dominated biofilms are also associated with hostile environments (*e.g.* Vincent *et al.*, 1993).

ALGAL EPS AND BIOADHESIVES

All benthic algae produce EPS which facilitates adhesion to the substratum, and locomotion in some raphid diatoms and cyanobacteria. Diatom EPS, referred to as the extracellular matrix (ECM) by Wetherbee *et al.* (1998) is a multicomponent, mucilaginous, organic bioadhesive complex found exterior to the plasma membrane (Figure 2). The major matrix components are acidic polysaccharides which are frequently carboxylated or sulphated (Daniel *et al.*, 1987; Hoagland *et al.*, 1993). Proteoglycans

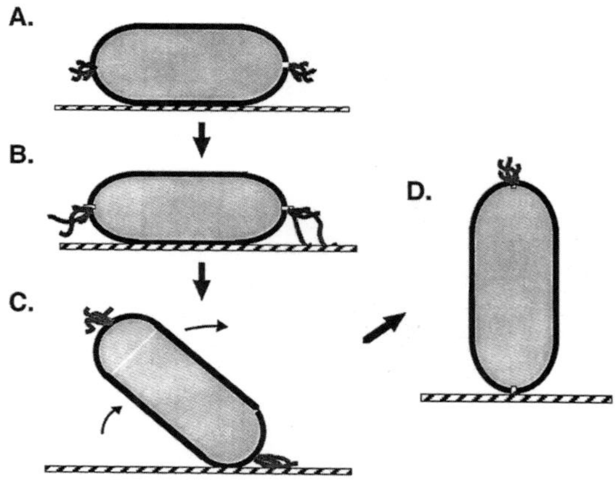

Figure 3 Diagrammatic representing the steps involved in initial adhesion of the raphid diatom *Stauroneis decipiens*. The cell, shown in cross-section, lands on its side (girdle). It is hypothesised that the adhesive mucilage containing components of the the adhesion complex is secreted and protrudes from the two raphes (A), apparently 'searching' for a substratum. Once contact is made (B), the motility mechanism interacts with the adhesion complexes to pull the cell up onto the raphe (C and D). The diatom is firmly attached (initial adhesion) once the adhesive mucilage has firmly adhered to the substratum and the adhesion complex is activated. (Reproduced from Wetherbee *et al.*, 1998, by permission of the *Journal of Phycology*.)

have been implicated in both adhesion and gliding motility of *Stauroneis decipiens*, a raphid diatom (Figure 1) (Lind *et al.*, 1997). Both processes are dependent on Ca^{2+} and metabolic energy (Cooksey and Wigglesworth-Cooksey, 1995). When a diatom cell initially adheres to a substratum, the cell-EPS interface and the EPS-substratum interface become united through an EPS-continuum that generates adhesion (Figure 2). Over time, the EPS is likely to be modified by secondary adhesion processes such as cross-linking and/or the addition of other molecules (Wetherbee *et al.*, 1998). The actin-based cytoskeleton is probably involved in the vesicular transport and secretion of EPS (Edgar and Pickett-Heaps, 1984; Wetherbee *et al.*, 1998). Diatoms may release themselves from the substratum, thereby discarding some of the secreted EPS as a diatom trail. Video microscopy of movement of *S. decipiens* in the presence of silica beads (Lind *et al.*, 1997) revealed that EPS originated at the raphe, moved along at the same speed as the cells and was secreted into trails. The EPS trails were initially sticky and elastic, and had the strength to impede or stop diatom movement unless the cell broke free. Lind *et al.* (1997) speculated that the surface EPS was fortified through the plasma membrane to the actin cytoskeleton though a series of molecular linkers which they called the adhesion complex (AC), the latter being capable of regulating both adhesion to and release from the substratum. The requirement for an AC is illustrated in Figure 3, which represents the sequence of events revealed by video microscopy which follow when a cell of *S. decipiens* lands on a surface on its side (girdle), rather than on its raphe region. Within 30–90 s, the cell goes through a series of manoeuvres that appears to activate the AC in firm, initial adhesion.

The cell appears to make contact with the substratum through extensions of the raphe mucilage and then use its motility apparatus to pull itself up so that one of its raphes is in direct contact with the substratum, at which point, the cell is firmly attached. If the cell lands on the raphe, adhesion is instantaneous. Once adhered, the diatom may glide parallel to the longitudinal axis of the cell whilst remaining attached (Wetherbee *et al.*, 1998).

Proteoglycans mediating adhesion and motility have been identified through the use of monoclonal antibody probes raised to EPS (Lind *et al.*, 1997; Wustman *et al.*, 1998). Lind *et al.* (1997) showed that the antibody StF.H4 bound to the cell surface, in the raphe and to adhesive trails. Functional studies showed that it inhibited adhesion to the substratum, the ability of cells to pull themselves up onto a raphe and gliding.

In some diatoms, *e.g. Achnanthes*, the secreted polymers which initially mediate motility are then organised into stalks which consist of three distinct regions, viz. a surface-adhered pad, a collar associated with the frustule at a terminal nodule or apical pore field, and an intervening shaft that separates the cell from the surface (Wang *et al.*, 1997). The three regions differ in both ultrastructure and composition (Wang *et al.*, 1997; Wustman *et al.*, 1997). The initial adhesive found in pads contains substantial amounts of G1cA and fucosyl residues, outer layers of the shaft contain G1cA, t-Fuc and non-sulphated D-Gal residues, while the inner core contains primarily sulphated galactosyl residues (Wustman *et al.*, 1997). Based on monoclanal antibody and lectin labelling, Wustman *et al.* (1998) concluded that the central core region of the shaft is derived from the raphe, and is related to material exuded from the raphe during cell motility.

The sticky domain of diatom EPS which confers adhesion is unknown, but it is likely to include components in addition to or other than the matrix acidic polysaccharides which constitute the bulk of the extracellular mucilages (Daniel *et al.* 1987; Hoagland *et al.*, 1993; Wustman *et al.*, 1997). The protein component of extracellular proteoglycans present in the EPS of *Stauroneis decipiens* may represent the sticky domain responsible for 'first-kiss' adhesion (Lind *et al.*, 1997; Wetherbee *et al.*, 1998). However, Wustman *et al.* (1997) provide evidence for the involvement of polysaccharides in motility and adhesion of *Achnanthes longipes* from a study employing inhibitors and consider one role of the protein component to be in crosslinking of the fucoglucuronogalactans (Wustman *et al.*, 1998).

A completely different type of bioadhesion mechanism is encountered in the majority of green and brown macroalgal members of biofilms. Here, the attaching stage is a motile spore which swims by means of 2 or 4 hair-like flagella, so that selection of a suitable substratum for settlement and thence permanent adhesion is possible. Many algal spores have the ability to respond to environmental conditions and the best known responses are to surface topography (thigmotactic response), light (phototactic response) and the presence of chemicals (chemotactic response) (Fletcher and Callow, 1992; Callow and Callow, 1998a). The change from a motile to a permanently adhered spore is fundamental to the colonisation of a new substratum. The swimming zoospore of the green alga, *Enteromorpha*, appears to 'sense' the surface via an apical papilla on which it rotates and it may, at this stage, become temporarily attached to the surface (Callow *et al.*, 1997). Permanent adhesion is characterised by discharge of Golgi-derived, adhesive-containing cytoplasmic vesicles, as the cell contracts against the surface (Evans and Christie, 1970). This phase of commitment is followed by exploitation of the surface by amoeboid-like, space-filling movements against the substratum and adjacent cells (Callow *et al.*, 1997). Monoclonal antibodies raised against settled zoospores which display the discharged adhesive indicate that the primary adhesive is a high molecular weight, polydisperse, *N*-linked glycoprotein (Stanley *et al.*, 1999).

AQUATIC ALGAL BIOFILMS

Biofilm Physiology

Algal-dominated biofilms are invariably heterogeneous ecosystems, the form being strongly dominated by the environment. Exposure to light can be extreme, from near zero in deep or turbid water to full sunlight in shallow clear water or terrestrial environments (see Hill, 1996). Within the biofilm, steep chemical gradients develop as illustrated by a number of elegant studies using microclectrodes (*e.g.* Revsbech *et al.*, 1983; Glud *et al.*, 1992; Kuhl and Jorgensen, 1992a; Kuhl *et al.*, 1996; Glud *et al.*, 1999). The boundary layer controls diffusion into and out of the biofilm, therefore CO_2 may not be supplied as quickly as it is consumed and it is often limiting, even when CO_2 in the bulk water is not limiting (Liehr *et al.*, 1990). Some CO_2 is provided by respiring heterotrophs in the lower layers but the carbon budget is also related to the thickness of the biofilm since penetration of light also controls the rate of photosynthesis. Photorespiration, defined as the oxidation of ribulose1,5-bisphosphate by the oxygenase activity of RUBISCO with the eventual production of CO_2 (Glud *et al.*, 1992), often occurs in biofilms with high rates of photosynthesis. The dense packing of cells producing O_2 and the restricted diffusion of CO_2 into, and O_2 out of the biofilm, produce high O_2: CO_2 ratios that favour photorespiration (Glud *et al.*, 1992). Photorespiration accounted for 17% of gross photosynthesis in a diatom biofilm (Glud *et al.*, 1992).

A study of a cyanobacterial biofilm using O_2 microelectrodes and fibre-optic microprobes showed that O_2 penetrated < 0.5 mm in darkness, but on illumination, photosynthesis caused the O_2 concentration to increase. At an irradiance of *ca.* 17.5 $\mu mol\ m^{-2}\ s^{-1}$, the O_2 compensation point was reached at the biofilm surface whilst at higher irradiance (200 $\mu mol\ m^{-2}\ s^{-1}$) the upper part of the biofilm became supersaturated with respect to O_2 and the oxygen penetration depth increased to 2 mm (Figure 4) (Kuhl *et al.*, 1996). Flux calculations used to determine respiration rates in light- and dark-incubated biofilms, showed that significantly higher areal O_2 consumption was associated with illuminated biofilms. This increase was only partly accounted for by photorespiration, the rest being ascribed to a deeper penetration of oxygen in the light, as well as an enhanced volumetric O_2 respiration in and below the photic zone (for a more detailed explanation, see Kuhl *et al.*, 1996).

Light is rapidly attenuated exponentially as it penetrates the water column (Hill, 1996), and further, within a biofilm (Liehr *et al.*, 1990; Kuhl *et al.*, 1992b). Thus, either because of low irradiance (Boston and Hill, 1991), or as a consequence of self-shading (Guasch and Sabater, 1995), photosynthesis is restricted to the upper cell layers. The euphotic zone of a cyanobacterial biofilm was less than 0.7 mm thick due

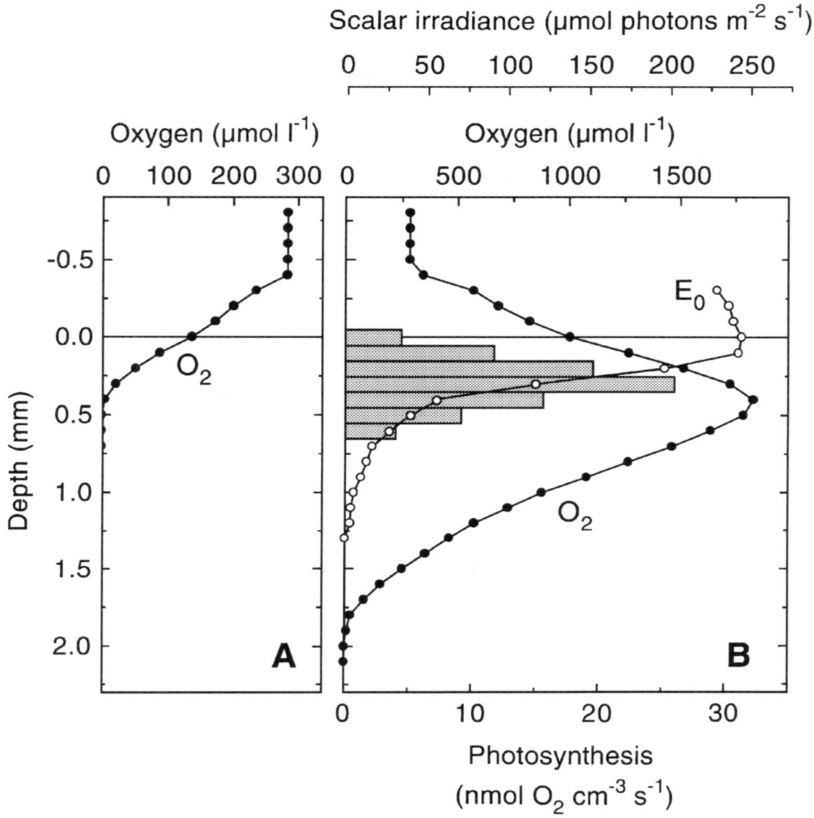

Figure 4 O_2 photosynthesis and light distribution in a cyanobacterial biofilm. (A) = steady-state O_2 profile in a dark-incubated biofilm; (B) = steady-state depth profiles of O_2, gross photosynthesis (shaded bars) and photon scalar irradiance E_0, in an illuminated biofilm. The incident photon irradiance at the biofilm surface was 200 µmol photons m^{-2} s^{-1}. (Reproduced from Kuhl *et al.* (1996), by permission of the *Journal of Phycology*.)

to the strong attenuation of photon scalar irradiance by the surface layers (Kuhl *et al.*, 1996). At high irradiances, the rapid attenuation of light is advantageous, as only cells at the surface experience photoinhibition (Hill, 1996). Cells exposed to high irradiance may develop increased levels of carotenoids to protect against photoinhibition (Guasch and Sabater, 1995). The spectral composition of light changes as it passes through a biofilm which may give rise to a vertical stratification of photosynthetic organisms with differing pigments (Kuhl *et al.*, 1994), although other factors such as the type of substratum and turbulence may also be of primary importance (Hoagland and Peterson, 1990). Light scattering also occurs within biofilms, increasing the total available light by up to 200% of the incident light received within the upper cell layers (Kuhl and Jorgensen, 1992b).

The migratory movement of biofilm algae in response to light has implications in terms of surface texture and hence the thickness of the diffusive boundary layer

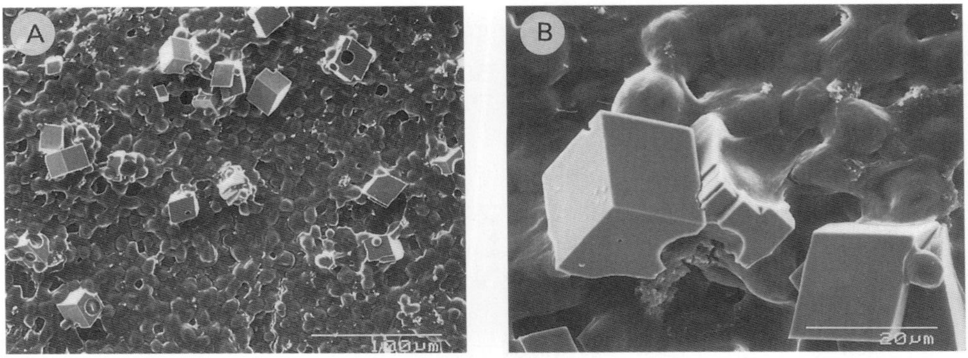

Figure 5 Scanning electron micrographs of the surface of a biofilm of the green alga *Chlorococcum* with calcite crystals seen at low (A) and high (B) magnification. The biofilm was grown for 8 weeks in Bold's basal medium (BBM) before transferring to artificial hardwater (0.25 mM calcium bicarbonate). After 1h illumination, crystals of calcite were visible on the surface of the photosynthesising biofilm. No crystals were formed on illuminated biofilms in BBM or biofilms in Ca bicarbonate in the dark. (Reproduced from Hartley *et al.*, 1996, with permission.)

(DBL). Cyanobacterial biofilms which are smooth at low-moderate light (< 100 µmol m^{-2} s^{-1}), migrate laterally and form dense tufts at higher irradiances (150–250 µmol m^{-2} s^{-1}) thereby causing an increase in the DBL from 250–400 µm (Kuhl *et al.*, 1996).

Hard Water Biofilms

Algal biofilms growing in hard waters are associated with the precipitation of calcite (calcium carbonate) which may form a substantial crust (Callow, 1993). The $CaCO_3$ content was shown to be strongly correlated with the amount of organic matter present (Heath *et al.*, 1993). Precipitation occurred at the surface, crystals becoming incorporated into the biofilm as the algae grew. Precipitation follows a diurnal cycle (House *et al.*, 1989) and is closely associated with photosynthesis (Hartley *et al.*, 1995). Photosynthesis utilises CO_2, thus raising the pH which leads to an increase in the proportion of CO_3^{-2}. Respiration produces CO_2 which lowers the pH leading to an increase in the proportion of carbonic acid and dissolved CO_2. In hard waters which are supersaturated with respect to calcite, the uptake of CO_2 during photosynthesis causes precipitation of calcite, whilst during darkness, respiration causes the pH to be lowered sufficiently to prevent precipitation and possibly to allow dissolution of some of the precipitated calcite (House *et al.*, 1989). Thus, algal biofilms in hard waters exert a large influence on the bulk water chemistry. The influence of algal photosynthesis on the rate of calcite precipitation was shown in a study employing pH and calcium microelectrodes which allowed chemical activity profiles to be measured above the biofilm surface (Hartley *et al.*, 1996). Pronounced gradients of pH and calcium concentration were demonstrated within a distance of 500 µm above the biofilm surface. The high degree of supersaturation caused by the low concentrations of CO_2 near the biofilm surface lead to the precipitation of calcite on the biofilm

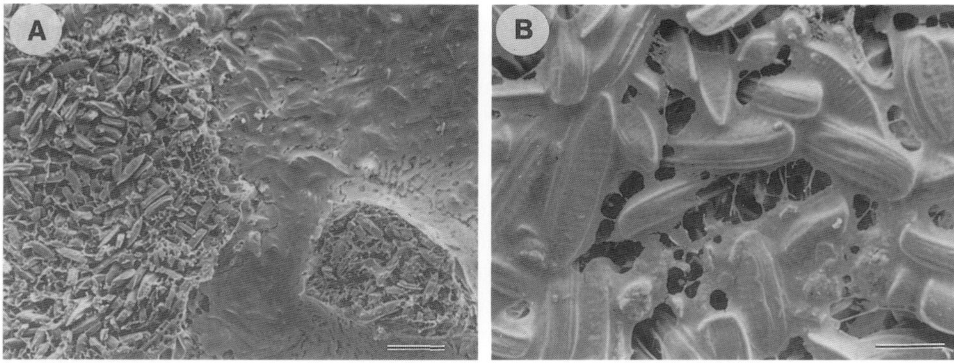

Figure 6 Low temperature scanning electron micrographs of the surface of diatom-dominated intertidal sediments. The EPS forms a smooth covering over large areas of cells (A). At higher magnification (B), details of the architecture of the diatom frustules (predominantly *Nitzchia epithemiodes*) can be seen beneath the covering of EPS. Scale bars: A = 100 μm; B = 20 μm. (Reproduced from Underwood *et al.*, 1995, with permission.)

(Figure 5). The photosynthetic activity of biofilms also has beneficial effects on water quality, since inorganic phosphate will co-precipitate with calcite, thus providing a self-cleansing mechanism through the removal of biologically available phosphorus (Murphy *et al.*, 1983; Koschel *et al.*, 1983; 1987; Kleiner, 1988; Kuchler-Krischun and Kleiner, 1990; Proft and Stutter, 1993; Hartley *et al.*, 1997). The presence of an algal biofilm on the surface of river sediments affects nutrient fluxes at the sediment-water interface as well as within the sediment and overlying water (Woodruff *et al.*, 1999a; 199b). Nutrients are removed effectively from polluted streams by algal biofilm communities (Vymazal, 1988). In addition, it is possible that particle-bound metals and organic contaminants *e.g.* pesticides, may be mobilised as a consequence of changes of pH and redox, mediated by the algal biofilm. The potential to remove metals by precipitation at the high pH values found inside photosynthesising algal biofilms has been demonstrated in a laboratory study (Liehr *et al.*, 1994).

Marine and Estuarine Biofilms

Mud flats are often visibly brown in colour, due to the luxuriant growth of diatoms (Whitton and Potts, 1982). The productivity is not great, amounting to approximately 100 g carbon fixed m^{-2} y^{-1} for Wadden Sea sediments (Cadee and Hegeman, 1974), being due in part, to light scattering by sediment particles (see Kuhl *et al.*, 1996). However, such diatom biofilms are extremely important in biostabilization of the sediments (Paterson, 1994). EPS produced by the diatoms binds the sediment particles together and forms a smooth surface layer (Figure 6) (Paterson, 1989; Paterson *et al.*, 1990; Underwood *et al.*, 1995) thereby increasing the critical shear strength of the sediments (Paterson, 1994). Diatoms in many intertidal habitats possess endogenous migratory rhythms which brings the cells to the surface at low tide during hours of daylight. For littoral species, avoidance of desiccation and photobleaching may also be important (Harper, 1977). Although such gliding locomotion is mediated by EPS (Edgar and Pickett-Heaps, 1984; Wetherbee *et al.*, 1998), its production appears not to be directly linked to primary production, being higher during darkness and prior to tidal cover than during periods of maximum photosynthesis (Underwood and Smith, 1998). Smith and Underwood (1998) concluded from ^{14}C-labelling studies that EPS production by intertidal epipelic diatoms is linked to migratory rhythms. Some benthic diatoms living in sediments are facultative heterotrophs and can use various organic substrates for growth at low light intensities or in darkness (Hellebust and Lewin, 1977) and facultative heterotrophy is considered to be a crucial survival mechanism for algae in light-limited conditions (Tuchman, 1996). Thus, the carbon budgets of diatom biofilms may vary according to the prevailing environmental conditions and they are also related to the presence of heterotrophic organisms. The diatom *Amphora* exhibits positive chemotaxis to some organic compounds (Cooksey and Cooksey, 1988; Wigglesworth-Cooksey and Cooksey, 1992), and chemotaxis may also play a role in the migratory movements of diatoms in sediments, allowing them to move towards high concentrations of nutrients (Cooksey and Wigglesworth-Cooksey, 1995).

Some intertidal mud flats can be green in colour where cyanobacteria flourish (van den Hoek *et al.*, 1995). Filamentous cyanobacteria may form a network in which sediment particles become trapped, thereby contributing to the stabilisation of the sediments (Stal, 1994). As with diatoms, cyanobacteria are also capable of altering their position in relation to the tidal cycle (Castenholz, 1982). At high tide migration into the sediments occurs, thereby avoiding being washed away. The gliding movement of cyanobacteria is also strongly influenced by light, positive and negative phototaxis being induced by dim and bright light respectively (Garcia-Pichel *et al.*, 1994; Kuhl *et al.*, 1994). A photophobic response enables a rapid reversal of movement if there is a sudden increase or decrease in light intensity. These responses ensure that cyanobacteria within biofilms are able to congregate in locations with near optimal conditions for photosynthesis and growth.

ALGAL BIOFILMS AND BIOFOULING

Aerial Environment

Terrestrial algae are predominantly unicellular or colonial green algae or cyanobacteria and most common genera can form lichens. Of the many genera commonly found in terrestrial habitats, biofilms dominated by the green alga *Pleurococcus* are ubiquitous in damp conditions in temperate climates. In humid tropical and subtropical localities, biofilms dominated by cyanobacteria and the green alga *Trentepohlia* flourish, the latter being characterised by their orange-yellow colour due to masking of chlorophyll by haematochrome pigments (Grant, 1982).

Algae colonising buildings and monuments give the surface a dirty, untidy appearance and are the forerunners of lichens which are capable of extensive corrosive activity (Bock and Sand, 1993; Morton and Surman, 1994). Cyanobacteria are considered to be the most significant group of pioneer colonisers (Grant, 1982) and have been reported to be the primary cause of disfigurement of surfaces throughout the world (Gaylarde and Morton, 1999), although the green alga *Trentepohlia* is reported to be the main cause of disfiguring paintwork and exterior surfaces in tropical regions (Wee and Lee, 1980). Preventative measures are primarily through the use of biocidal washes and biocides incorporated into surface coatings and paints (Lindner, 1997; Gaylarde and Morton, 1999).

Aquatic Environment

Algal biofilms are responsible for fouling a variety of submerged structures including pipes, ships' hulls, buoys and platforms. Algal biofilms in storage tanks and pipes used for potable water are undesirable because released DOC imparts tastes and odours (Hutson *et al.*, 1987). Algal biofilms reduce the thermal efficiency of industrial cooling towers and in extreme cases, result in the collapse of the splash-pack support systems (Callow, 1993). However, the major consequences of algal fouling are due to hydrodynamic drag and corrosion of metals (Callow and Edyvean, 1990). Most biofilms are heterogeneous in both species composition and thickness, which when colonising metal, leads to the formation of cathodic and anodic sites within the underlying metal (Edyvean and Videla, 1991; Hamilton, 2000; Lewandowski, 2000). The cathodic oxygen reduction reaction becomes dominant at more highly oxygenated sites and as a result less oxygenated sites become anodic. Thus an illuminated biofilm consisting of colonies of algal cells and bacteria will form cathodic and anodic sites respectively. Similarly, differences in the thickness of the algal biofilm can cause the setting up of differential cells, with thinner portions of the biofilm allowing the passage of oxygen and hence becoming cathodic and the thicker portions becoming anodic. Corrosion is particularly pronounced in sea water where the concentrated salt solution acts as an electrolyte that completes the circuit between anode and cathode. The flow of electrons between the two electrodes produces a measurable current and localised corrosion in the form of crevices and pits occurs in the metal at the anodic site. Mixed algal/bacterial films can also alter the concentrations of ions beneath the biofilm thereby producing chemical and pH concentration cells that act in a similar manner to oxygen concentration cells (Edyvean and Terry, 1983).

Algal fouling is less of a problem in freshwaters than in marine environments. The exception is in hardwaters where calcium salts, especially $CaCO_3$, are often deposited within algal biofilms (see Hard Water Biofilms). These calcareous deposits can form crusts up to several millimetres thick on the hulls of boats (Callow, 1993; Heath *et al.*, 1993). They increase drag and are difficult to remove, their removal often causing damage to the underlying substrate, especially on glass fibre hulls. The incorporation of 1, hydroxyethylene 1,1 diphosphonic acid (HEDP) into paints may reduce the formation of calcareous deposits (Heath *et al.*, 1995; 1996) but the benefits would need to be weighed against the release of phosphonate into rivers and static bodies of water.

In the marine environment, the major economic penalty imposed by algal biofilms, is the fouling of ships' hulls. The biofilm causes an increase in hull roughness, which leads in turn to an increase in hydrodynamic drag (Lewthwaite *et al.*, 1985; Schultz and Swain, 1998; Swain, 1998) as the hull moves through the water, thereby significantly increasing operational and maintenance costs (Bohlander, 1991; Alberte *et al.*, 1992). Diatoms, the green algae *Enteromorpha* and the brown alga *Ectocarpus* are the

Figure 7 Scanning electon micrographs of a diatom biofilm growing on a self-polishing copolymer antifouling paint containing cuprous oxide. (A) = the edge (between the vertical arrows) and surface (s) of the biofilm. (B) = the biofilm surface. The dominant diatoms are species of *Achnanthes* (arrows) and *Amphora* (arrowheads). Scale bars: (A) = 500 µm (B) = 50 µm.

most important ship fouling algae. The increase in skin friction in water flowing over algal biofilms depends not only on the thickness of the biofilm but also on the species composition and morphology (Schultz and Swain, 1998). Certain genera of diatoms are particularly important in terms of fouling of ships' hulls (Robinson *et al.*, 1985; Callow 1986); *Amphora* spp. are most common on copper paints while biofilms dominated by *Achnanthes* and *Amphora* (Figure 7) are commonly found on tributyltin self polishing copolymers (SPC) (Callow, 1996). Mixed species diatom biofilms also adhere to non-toxic, foul-release silicone elastomers (Callow *et al.*, 1986). The new generation of non-tin polishing antifouling paints (Callow, 1996) frequently become fouled with biofilms dominated by diatoms and *Enteromorpha*. The ability of certain species to form biofilms on toxic coatings is due to the resistance of particular species to biocides (see Callow, 1996; 1999a).

Increased environmental awareness and the regulation of biocidal antifouling paints are driving research and development of environmentally benign technologies to control biofouling. So far, only a few of the various methods proposed have been commercialised, and of those that have, none have matched the performance of antifouling coatings (see Milne, 1991; Swain, 1998). In the short- to medium-term, antifouling coatings employing copper and/or non-persistent organic biocides will be used for the majority of applications to control algal fouling (Callow and Willingham, 1996; Callow, 1999). Silicone foul-release coatings are proving effective for certain applications (Callow and Fletcher, 1994; Callow, 1996). However, whilst macrofouling organisms may detach readily, diatom slimes are more persistent and silicone elastomeric coatings are not sufficiently robust for many deep-sea vessels. A number of alternative technologies, such as the use of settlement inhibitors are being researched actively (Clare, 1996; Swain, 1998) including zosteric acid, a sulphoxy-phenolic acid, which at non-toxic concentrations, inhibits attachment of *Enteromorpha* spores (Callow and Callow, 1998b) as well as other fouling organisms (Todd *et al.*, 1993).

Long-term strategies for the development of non-toxic control of algal fouling, require an understanding of the physico-chemical processes at the interface where cells and spores attach. Information gained from these studies will assist in the development of molecular design criteria for surfaces that will resist biofouling.

CONCLUSIONS

Algal biofilms are the 'poor-relations' of the world of biofilms, in spite of their ubiquitous presence in damp or aquatic surroundings, and their importance in ecological assemblages and in biodeterioration. Biofilm algae ubiquitously coexist with heterotrophs, although the importance of these interactions is frequently ignored. Their role in modulating the transformation of many inorganics, notably C, N and P, into organic compounds is crucial to the well-being of the aquatic ecosystem. The high degree of genetic, and ultimately metabolic control, operating in biofilm algae must be considered in terms of the community as a whole, although so far, little research has been done in this area (see Amann and Kuhl, 1998 for a review of *in situ* methods). The importance of quorum sensing by N-acyl homoserine lactones (AHLs) in microbial biofilms has been established (Heys *et al.*, 1997) and the possibility there may be interactions between AHLs and metabolic products of algal cells is an intriguing question. Interactions between AHLs and products of eukaryotic organisms including algae have been reported recently (Gram *et al.*, 1996; Givskov *et al.*, 1996; Maximilien *et al.*, 1998). Future research may highlight the importance of intercellular signalling in the settlement and colonisation of substrata by cyanobacteria and eukaryotic algae.

ACKNOWLEDGEMENTS

Financial support for algal biofilm research has been received over a number of years from various organisations, notably the Natural Environment Research Council (NERC), International Coatings Ltd (Akzo Nobel) and the USA Office for Naval Research.

REFERENCES

Alberte R.S., Snyder S., Zahuranec B.J., Whetstone M. (1992). Biofouling research needs for the United States Navy; program history and goals. *Biofouling*, **6**, 91–95.

Amann R., Kuhl M. (1998). *In-situ* methods for assessment of microorganisms and their activities. *Curr Op Microbiol*, **1**, 352–358.

Baier R.E. (1980). Substrata influences on the adhesion of microorganisms and their resultant new surface properties. In: Bitton G., Marshall K. (eds) *Absorption of Microorganisms to Surfaces*. Wiley, New York, pp. 59–104.

Biggs B.J.F. (1996). Patterns of benthic algae of streams. In: Stevenson R.J., Bothwell M.L., Lowe R.L. (eds) *Algal Ecology: Freshwater Benthic Ecosystems*. Academic Press, London, pp. 30–55.

Bock E., Sand W. (1993). The microbiology of masonry biodeterioration. *J Appl Bact*, **74**, 503–514.

Bohlander G.S. (1991). Biofilm effects on drag: measurements on ships. In: *Polymers in a Marine Environment*. Institute of Marine Engineers, London, pp. 135–138.

Boston H.L., Hill W.R. (1991). Photosynthesis – light relations of stream periphyton communities. *Limnol Oceanogr*, **36**, 644–656.

Bott T.L. (1996). Algae in microscopic food webs. In: Stevenson R.J., Bothwell M.L., Lowe R.L. (eds) *Algal Ecology: Freshwater Benthic Ecosystems*. Academic Press, London, pp. 573–608.

Brown L.N., Robinson M.G., Hall B.D. (1988). Mechanisms for copper tolerance in *Amphora coffeaeformis* – internal and external binding. *Mar Biol*, **97**, 581–586.

Cadee G.C., Hegeman J. (1974). Primary production of the benthic misroflora living on tidal flats in the Dutch Wadden Sea. *Neth J Sea Res*, **8**, 260–291.

Callow M.E. (1986). Fouling algae from "in-service" ships. *Bot mar*, **24**, 351–357.

Callow M.E. (1993). A review of fouling in freshwaters. *Biofouling*, **7**, 313–327.

Callow M.E. (1996). Ship fouling: the problem and methods of control. *Biodeterior Abstr*, **10**, 411–421.

Callow M.E. (1999). The status and future of biocides in marine biofouling prevention. In: Fingerman M., Nagabhushanan R., Thompson M-F. (eds) *Recent Advances in Marine Biotechnology*, Volume III. Science Publishers Incorporated, USA, pp. 109–126.

Callow M.E., Edyvean R.G.J. (1990). Algal fouling and corrosion. In: Akatsuka I. (ed) *Introduction to Applied Phycology*. SPB Academic Publishing bv, The Hague, The Netherlands, pp. 367–387.

Callow M.E., Fletcher R.L. (1994). The influence of low surface energy materials on bioadhesion – a review. *Int Biodeterior & Biodegr*, **34**, 333–348.

Callow M.E., Willingham G.L. (1996). Degradation of antifouling biocides. *Biofouling*, **10**, 239–249.

Callow, M E., Callow J.A. (1998a). Enhanced adhesion and chemoattraction of zoospores of the fouling alga *Enteromorpha* to some foul-release silicone elastomers. *Biofouling*, **13**, 157–172.

Callow M.E., Callow J.A. (1998b). Attachment of zoospores of the fouling algae, *Enteromorpha* in the presence of zosteric acid. *Biofouling*, **13**, 87–95.

Callow M.E., Pitchers R.A., Milne A., (1986). The centrol of foulling by non-biocidal antifouling coatings. In: Evans L.V., Hoagland K.D. (eds) *Algal Biofouling*. Elsevier Applied Science, pp. 43–48.

Callow M.E., Callow J.A., Pickett-Heaps J.D., Wetherbee R. (1997). Primary adhesion of *Enteromorpha* (Chlorophyta, Ulvales) propagules: quantitative settlement studies and video microscopy. *J Phycol*, **33**, 938–947.

Carr N.G., Whitton B.A. (eds) (1982). *The Biology of Cyanobacteria*. Blackwell Scientific Publications, Oxford.

Castenholz R.W. (1982). Motility and taxes. In: Carr N.G., Whitton B.A. (eds) *The Biology of Cyanobacteria*. Blackwell Scientific Publications, Oxford.

Clare A.S. (1996). Marine natural product antifoulants: status and potential. *Biofouling*, **9**, 211–229.

Cohn S.A., Weitzell R.E. (1996). Ecological considerations of diatom cell motility. I. Characterization of motility and adhesion in four diatom species. *J Phycol*, **32**, 928–939.

Cooksey K.E. (1992). Extracellular polymers in biofilms. In: Melo L.F., Bott T.R., Fletcher M., Capseville B. (eds) *Biofilms, Science and Technology*. Nato ASI Series E: Applied Sciences, Volume 223. Kluwer Academic Publishers, The Netherlands, pp. 137–147.

Cooksey B., Cooksey K.E. (1988). Chemical signal response in diatoms of the genus *Amphora*. *J Cell Sci*, **92**, 523–529.

Cooksey K.E., Wigglesworth-Cooksey B. (1995). Adhesion of bacteria and diatoms to surfaces in the sea: a review. *Aquat Microb Ecol*, **9**, 87–96.

Cooksey B., Cooksey K.E., Miller C.A., Paul J.H., Rubin R.W., Webster D. (1984). The attachment of microfouling diatoms. In: Costlow J.D., Tipper R.C. (eds) *Marine Biodeterioration: an Interdisciplinary Study*. US Naval Institute, Annapolis, Maryland, pp. 167–171.

Cox E.J. (1996). *Identification of Freshwater Diatoms from Live Material*. Chapman & Hall, London.

Daniel G.F., Chamberlain A.H.L., Jones E.B.G. (1987). Cytochemical and electron microscopical observations on the adhesive materials of marine fouling diatoms. *Br Phycol J*, **22**, 101–118.

Edgar L.A., Pickett-Heaps J.D. (1984). Diatom locomotion. *Prog Phycol Res*, **3**, 47–88.

Edyvean R.G.J., Terry L.A. (1983). Polarization studies of 50D steel in cultures of marine algae. *Int Biodeterior Bull*, **19**, 1–11.

Edyvean R.G.J., Videla H.A. (1991). Biological corrosion. *Interdis Sci Rev*, **16**, 267–282.

Evans L.V., Christie A.O. (1970). Studies on the ship-fouling alga *Enteromorpha*. I. Aspects of the fine-structure and biochemistry of swimming and newly settled zoospores. *Ann Bot (Lond)*, **34**, 451–466.

Fletcher R.L., Callow M.E. (1992). The settlement, attachment and establishment of marine algal spores. *Br Phycol J*, **27**, 303–329.

Fogg G.E., Stewart W.D.P., Fay P., Walsby A.E. (1973). *The Blue-green Algae*. Academic Press, London.

Garcia-Pinchel F., Mechling M., Castenholz R.W. (1994). Diel migrations of microorganisms in a hypersaline microbial mat. *Appl Environ Microbiol*, **60**, 1500–1511.

Gaylarde C.C., Morton L.G.H. (1999). Deteriogenic biofilms on buildings and their control: a review. *Biofouling*, **14**, 59–74.

Genter R.B. (1996). Ecotoxicology of inorganic chemical stress to algae. In: Stevenson R.J., Bothwell M.L., Lowe R.L. (eds) *Algal Ecology: Freshwater Benthic Ecosystems*. Academic Press, London, pp. 403–468.

Givskov M., de Nys R., Manefield M., Gram L., Maximilien R., Eberl L., Molin S., Steinberg P.D., Kjelleberg S. (1996). Eukaryotic interference with homoserine lactone-mediated prokaryotic signalling. *J Bacteriol*, **178**, 6618–6622.

Glud R.N., Ramsing N.B., Revsbech N.P. (1992). Photosynthesis and photosynthesis-coupled respiration in natural biofilms quantified with oxygen microsensors. *J Phycol*, **28**, 51–60.

Glud R.N., Kühl M., Kohls O., Ramsing N.B. (1999). Heterogeneity of oxygen production and consumption in a photosynthetic microbial mat as studied by planar optodes. *J Phycol*, **35**, 270–279.

Grant C. (1982). Fouling of terrestrial substrates by algae and implications for control – a review. *Int Biodeterior Bull*, **18**, 57–64.

Gram L., de Nys R., Maximilien R., Grivskov M., Steinberg P. (1996). Inhibitory effects of secondary metabolites from the red alga *Dilsea pulchra* on swarming motility of *Proteus mirabilis*. *Appl Environ Microbiol*, **62**, 4284–4287.

Guasch H., Sabater S. (1995). Seasonal variations of photosynthesis-irradiance responses by biofilms in Mediterranean streams. *J Phycol*, **31**, 727–735.

Hamilton A. (2000) Microbially influenced corrosion in the context of metal microbe interactions. In: Evans L.V. (ed) *Biofilms: Recent Advances in their Study and Control*. Harwood Academic Publishers, pp. 419–434.

Harper M.A. (1977). Movements. In: Werner D. (ed) *The Biology of Diatoms*. Blackwell Scientific Publications, Oxford, pp. 224–249.

Hartley A.M., House W.A., Callow M.E., Leadbeater B.S.C. (1995). The role of a green alga in the precipitation of calcite and coprecipitation of phosphate in freshwater. *Int Rev Gesmaten Hydrobiol*, **80**, 385–401.

Hartley A.M., House W.A., Leadbeater B.S.C., Callow M.E. (1996). The use of microelectrodes to study the precipitation of calcite upon algal biofilms. *J Colloid Interface Sci*, **183**, 498–505.

Hartley A.M., House, W A., Callow M.E., Leadbeater B.S.C. (1997). Coprecipitation of phosphate with calcite in the presence of photosynthesising green algae. *Water Res*, **31**, 2261–2268.

Hawkins S.J., Hartnoll R.G. (1983). Grazing of intertidal algae by marine invertebrates. *Oceanogr Mar Biol Annu Rev*, **21**, 195–283.

Heath C.R. Leadbeater B.S.C., Callow, M E. (1993). Formation and calcification of biofilms on antifouling paints in hard waters. *Biofouling*, **7**, 29–55.

Heath C.R., Leadbeater B.S.C., Callow M.E. (1995). Effect of inhibitors on calcium carbonate deposition mediated by freshwater algae. *J Appl Phycol*, **7**, 367–380.

Heath C.R., Leadbeater B.S.C., Callow M.E. (1996). The control of calcification of antifouling paints in hard waters using a phosphonate inhibitor. *Biofouling*, **9**, 317–325.

Hellebust J.A., Lewin J. (1977). Heterotrophic nutrition In: Werner D. (ed) *The Biology of diatoms*. Blackwell, Oxford, pp. 169–197.

Hendey N.I. (1951). Littoral diatoms of Chichester Harbour with special reference to fouling. *J R Micros Soc*, **71**, 1–86.

Heys S.J.D., Gilbert P., Eberhard A., Allinson D.G. (1997). Homoserine lactones and bacterial biofilms. In: Wimpenny J., Handley P., Gilbert P., Lappin-Scott H., Jones M. (eds) *Biofilms: Community Interactions and Control*. BBC 3, Bioline, Cardiff, pp. 103–112.

Hill W. (1996). Effects of light. In: Stevenson R.J., Bothwell M.L., Lowe R.L. (eds) *Algal Ecology: Freshwater Benthic Ecosystems*. Academic Press, London, pp. 121–148.

Hoagland K.D., Peterson C.G. (1990). Effects of light and wave disturbance on vertical zonation of attached microalgae in a large reservoir. *J Phycol*, **26**, 450–457.

Hoagland K.D., Rosowski J.R., Gretz M.R., Roemer S.C. (1993). Diatom extracellular polymeric substances: function, fine structure, chemistry and physiology. *J Phycol*, **29**, 537–566.

Hoiczyk E., Baumeister W. (1998). The junctional pore complex, a prokaryotic secretion organelle, is the molecular motor underlying gliding motility in cyanobacteria. *Curr Biol*, **8**, 1161–1168.

House W.A., Shelley N., Fox A.M. (1989). Chemical modelling applications to experimental circulating streams. *Hydrobiologia*, **178**, 93–112.

Hutson R.A., Leadbeater B.S.C., Sedgwick R.W. (1987). Algal interference with water treatment processes. *Prog Phycol Res*, **5**, 265–299.

Jackson S.M., Jones E.B.G. (1988). Fouling film development on antifouling paints with special reference to film thickness. *Int Biodeterior*, **24**, 277–287.

Kleiner J. (1988). Coprecipitation of calcite with phosphate in lake water: a laboratory experiment modelling phosphate removal with calcite in Lake Constance. *Water Res*, **22**, 1259–1265.

Koschel R., Benndorf J., Proft G., Recknagel F. (1983). Calcite precipitation as a natural control mechanism of eutrophication. *Arch Hydrobiol*, **98**, 380–408.

Koschel R., Benndorf J., Proft G., Recknagel F. (1987). Model assisted evaluation of alternative hypotheses to explain the self-protection mechanism of lakes due to calcite precipitation. *Ecol Modell*, **39**, 59–65.

Kuchler-Krischun J., Kleiner J. (1990). Heterogeneously nucleated calcite precipitation in Lake Constance, a short resolution study. *Aquat Sci*, **52**, 176–197.

Kuhl M., Jorgensen B.B. (1992a). Microsensor masurements of sulphate reductin and sulphie oxidation in compact microbial communities of aerobic biofilms. *Appl Env Microbiol*, **58**, 1164–1174.

Kuhl M., Jorgensen B.B. (1992b). Spectral light measurements in microbenthic phototrophic communities with a fibre-optic microprobe coupled to a sensitive diode array detector. *Limnol Oceanogr*, **37**, 1813–1823.

Kuhl M., Lassen C., Jorgensen B.B. (1994). Optical properties of microbial mats: light measurements with fibre-optic microprobes. In: Stal L.J., Caumette P. (eds) *Microbial Mats*. NATO ASI Series Vol G35. Springer-Verlag Berlin, pp. 149–165.

Kuhl M., Glud R.N., Ploug H., Ramsing N.B. (1996). Micro environniental control of photosynthesis and photosynthesis-coupled respiration in an epilithic cyanobacterial biofilm. *J Phycol*, **32**, 799–812.

Lamberti G.A. (1996). The role of periphyton in benthic food webs. In: Stevenson R.J., Bothwell M.L., Lowe R.L. (eds) *Algal Ecology: Freshwater Benthic Ecosystems*. Academic Press, London, pp. 533–572.

Lewandowski Z. (2000). Structure and function of biofilms. In: Evans L.V. (ed) *Biofilms: Recent Advances in their Study and Control*. Harwood Academic Publishers, pp. 1–17.

Lewthwaite J.C., Molland A.F., Thomas K.W. (1985). An investigation into the variation of ship skin resistance with fouling. *Trans RINA*, **127**, 269–284.

Liehr S.K., Suidan M.T., Eheart J.W. (1989). Effect of concentration boundary layer on carbon limited algal biofilms. *J Environ Eng*, **115**, 320–335.

Liehr S.K., Suidan M.T., Eheart J.W. (1990). A modelling study of carbon and light limitation in algal biofilms. *Biotechnol Bioeng*, **35**, 233–243.

Liehr S.K., Chen H.J., Lin S.H. (1994). Metals removal by algal biofilms. *Water Sci Technol*, **30**, 59–68.

Lind J.L., Heimann K., Miller E.A., van Vliet C., Hoogrenradd N.J., Wetherbee R. (1997). Substratum adhesion and gliding in a diatom are mediated by estracellular proteoglycans. *Planta*, **203**, 213–221.

Lindner W. (1997). Studies on film preservatives: retention of DCMU in outdoor paints. *Biofouling*, **11**, 179–189.

Lock M.A. (1981). River epilithon – a light and organic energy transducer. In: Lock M.A., Williams D.D. (eds) *Perspectives in Running Water Ecology*. Plenum Press, pp. 3–39.
Lowe R.L., Pan Y. (1996). Benthic communities as biological monitors. In: Stevenson R.J., Bothwell M.L., Lowe R.L. (eds) *Algal Ecology: Freshwater Benthic Ecosystems*. Academic Press, London, pp. 705–739.
Maximilien R., de Nys R., Holmstrom C., Gram L., Givskov M., Crass K., Kjelleberg S., Steinberg P.D. (1998). Chemical mediation of bacterial surface colonisation by secondary metabolites from the red alga *Delisea pulchra. Aquat Microb Ecol*, **15**, 233–236.
Milne A. (1991). Ablation and after: the law and the profits. In: Smith R. (ed) *Polymers in a Marine Environment*. Institute of Marine Engineers, London, pp. 229–233.
Morton L.H.G., Surman S.B. (1994). Biofilms in biodeterioration. *Int Biodeterior Biodegr*, **34**, 203–221.
Moss B. (1988). *Ecology of Freshwater, Man and Medium*. Blackwell Scientific Publications.
Murphy T.P., Hall K.J., Yesaki I. (1983). Coprecipitation of phosphate with calcite in a naturally eitrophic lake. *Limnol Oceanogr*, **28**, 58–69.
Paterson D.M. (1989). Short-term changes in the erodibility of intertidal cohesive sediments related to the migratory nehavious of epipelic diatoms. *Limnol Oceanogr*, **34**, 223–234.
Paterson D.M. (1994). Microbial mediation of sediment structure and behaviour. In: Stal L.J., Caumette P. (eds) *Microbial Mats*. NATO ASI Series Vol G35. Springer Verlag, Berlin, pp. 97–109.
Paterson D.M., Crawford R.M., Little C. (1990). Subaerial exposure and changes in stability of intertidal estuarine sediments. *Estuarine Coastal Shelf Sci*, **30**, 541–556.
Pickett-Heaps J.D., Schmid, A-M., Edgar L.A. (1990). The cell biology of diatom valve formation. *Prog Phycol Res*, **7**, 1–168.
Proft G., Stutter E. (1993). Calcite precipitation in hard water lakes in calculation and experiment. *Int Rev Gesmaten Hydrobiol*, **78**, 177–199.
Revsbech N.P., Jorgensen B.B., Blackburn T.H., Cohen Y. (1983). Microelectrode studies of the photosynthesis and O_2, H_2S and pH profiles of a microbial mat. *Limnol Oceanogr*, **28**, 1062–1074.
Robinson M.G., Hall B.D., Voltolina D. (1985), SIime films on antifouling paints: short-term indicators of long-term effectiveness. *J. Coatings Technol*, **57**, 35–41.
Robinson M.G., Brown L.N., Quenneville M.L., Hall B.D. (1992). Aspects of copper tolerance and toxicity in *Amphora coffeaeformis. Biofouling*, **5**, 261–276.
Round F.E., Crawford R.M., Mann D.G. (1990). *The Diatoms*. CUP Cambridge.
Schultz M.P., Swain G. (1998). The effect of biofilms on turbulent boundary layer structure. *Proc Int Symp on Seawater Drag Reduction*. Newport, RI, July 22–23, pp. 175–181.
Scott C., Fletcher R.L., Bremer G.B. (1996). Observations on the mechanism of attachment of some marine fouling blue-green algae. *Biofouling*, **10**, 161–173.
Smith J.D., Underwood G.J.C. (1998). Exopolymer production by intertidal epipelic diatoms. *Limnol Oceanogr*, **43**, 1578–1591.
Stal L.J. (1994). Microbial mats in coastal environments. In: Stal L.J., Caumette P. (eds) *Microbial Mats*. Nato ASI Series Vol G35. Springer Verlag, Berlin, pp. 21–32.
Stanley M.S., Callow M.E., Callow J.A. (1999). Monoclonal antibodies to adhesive cell coat glycoprotein secreted by zoospores of the green alga *Enteromorpha*. *Planta*, **210**, 61–71.
Steinman A.D. (1996). Effects of grazers on freshwater benthic algae. In: Stevenson, RJ., Bothwell M.L., Lowe R.L. (eds) *Algal Ecology: Freshwater Benthic Ecosystems*. Academic Press, London, pp. 341–373.
Stevenson R.J., Peterson C.G. (1989). Variation in benthic diatom (Bacillariophyceae) immigration with habitat characteristics and cell morphology. *J Phycol*, **25**, 120–129.
Swain G. (1998). Biofouling control: a critical component of drag reduction. *Proc Int Symp Seawater Drag Reduction*. Newport, RI, 22–23 July, pp. 155–161.

Todd J.S., Zimmerman R.C., Crews P., Alberte R.S. (1993). The antifouling activity of natural and synthetic phenolic acid sulphate esters. *Phytochemistry*, **34**, 401–404.

Toledo G., Bashan Y., Soeldner A. (1995). In vitro colonization and increase in nitrogen-fixation of seedlings of black mangrove inoculated by a filamentous cyanobacterium. *Can J Microbiol*, **41**, 1012–1020.

Tuchman N.C. (1996). Role of heterotrophy in algae. In: Stevenson R.J., Bothwell M.L., Lowe R.L. (eds) *Algal Ecology: Freshwater Benthic Ecosystems*. Academic Press, London, pp. 299–319.

Underwood G.J.C., Smith J.D. (1998). *In situ* measurements of exopolymer production by intertidal epipelic diatom-dominated biofilms in the Humber estuary. In: Black K.S., Paterson D.M., Cramp A. (eds) *Sedimentary Processes in the Intertidal Zone*. Geological Society, London, Special Publications, *139*, pp. 125–134.

Underwood G.J.C., Paterson D.M., Parkes R.J. (1995). The measurement of microbial carbohydrate exopolymers from intertidal sediments. *Limnol Oceanogr*, **40**, 1243–1253.

van den Hoek C., Mann D.G., Jahns H.M. (1995). *Algae: an Introduction to Phycology*. CUP, Cambridge.

Vincent W.F., Castenholz R.W., Downes M.T., Howard-Williams C. (1993). Antarctic cyanobacteria, light, nutrients and photosynthesis in the microbial mat environment. *J Phycol*, **29**, 745–755.

Vymazal J. (1988). The use of periphyton communities for nutrient removal from polluted streams. *Hydrobiologia*, **166**, 225–237.

Wang Y., Lu, J-C., Mollet, J-C., Gretz M.R., Hoagland K.D. (1997). Extracellular matrix assembly in diatoms (Bacillariophyceae) II. 2,6-dichlorobenzonitrile inhibition of motility and stalk production in the marine diatom *Achnanthes longipes*. *Plant Physiol*, **113**, 1071–1080.

Wee Y.C., Lee K.B. (1980). Proliferation of algae on surfaces of buildings in Singapore. *Int Biodeterior Bull*, **16**, 113–117.

Wetherbee R., Lind J.L., Burke J. (1998). The first kiss: establishment and control of initial adhesion by raphid diatoms. *J Phycol*, **34**, 9–15.

Wetzel R.G. (1983). *Limnology*. 2nd Edn. Saunders College Publishing, New York.

Whitton B.A., Potts M. (1982). Marine littoral. In: Carr N.G., Whitton B.A. (eds) *The Biology of Cyanobacteria*. Blackwell Scientific Publications, Oxford.

Woodruff S.L., House W.A., Callow M.E., Leadbeater B.S.C. (1999a). The effects of biofilms on chemical processes in surfacial sediments. *Freshwater Biol*, **41**, 73–89.

Woodruff S.L., House W.A., Callow M.E., Leadbeater B.S.C. (1999b). The effects of a developing biofilm on chemical changes across the sediment-water interface in a freshwater environment. *Int Rev Hydrobiol*, **84**, 509–532.

Wustman B.A., Gretz M.R., Hoagland K.D. (1997). Extracellular matrix assembly in diatoms (Bacillariophyceae) I. A model of adhesives based on chemical characterization and localization of polysaccharides from the marine diatom *Achnanthes longipes* and other diatoms. *Plant Physiol*, **113**, 1059–1069.

Wustman B.A., Lind J., Wetherbee R., Gretz M.R. (1998). Extracellular matrix assembly indiatoms (Bacillariophyceae) III. Organization of fucoglucuronogalactans within the adhesive stalks of *Achnanthes longipes*. *Plant Physiol*, **116**, 1431–1441.

13 Food Industry Biofilms

John Holah and Hazel Gibson

The occurrence, nature and significance of biofilms on surfaces in food processing environments are discussed. A review of epifluorescence studies of factory environments show that surface populations may range from attached or dried on cells to microcolonies and biofilms. Extensive microcolony development and biofilm formation are usually limited to environmental surfaces. Factors affecting biofilm formation are discussed including the conditioning layer, the nature and roughness of the substratum, temperature, pH, nutrient availability and time available. This has led to the suggestion of a definition of a food factory biofilm. The composition of the surface population is usually heterogeneous, with pseudomonads and staphylococci being particularly common. Biofilms may be a source of spoilage and/or pathogenic organisms that may contaminate the product directly and indirectly. The significance of such contamination is dependent upon the risk category of the food product. The complexity of biofilms means that their control requires an integrated hazard analysis approach involving surface, equipment, and environment hygienic design and construction with appropriate and effective cleaning and disinfection schemes. The last section of the chapter identifies aspects for future consideration including the more extensive use of tracking methods using genetic fingerprinting techniques to determine the significance of particular strains. Other areas highlighted for the future are the investigation of the movement of bacteria across surfaces, targeting of cleaning and disinfection based on knowledge of the attachment/detachment process and use of novel surfaces to limit attachment and inactivate organisms.

KEY WORDS: food, biofilm, cleaning, disinfection, attachment, detachment

DO THEY EXIST?

Biofilm formation is a dynamic process and different mechanisms are involved in attachment and growth (Kumar and Anand, 1998). Initially cells are deposited and then attach and grow into microcolonies and this involves the formation of extracellular polymers (EPS) which are thought to be essential in the process (Allison and Sutherland, 1987). There have been several reviews of biofilm formation in the food industry (Pontefract, 1991; Holah and Kearney, 1992; Mattila-Sandholm and Wirtanen, 1992; Carpentier and Cerf, 1993; Zottola and Sasahara, 1994; Kumar and Anand, 1998). Studies of attached cells and biofilms involving swabbing and traditional microbiology give an indication of the viable microorganisms present, but do not provide any information on biofilm structure or detect physiologically stressed cells such as viable non-culturable (VNC) cells. Microscopy allows *in situ* analysis and

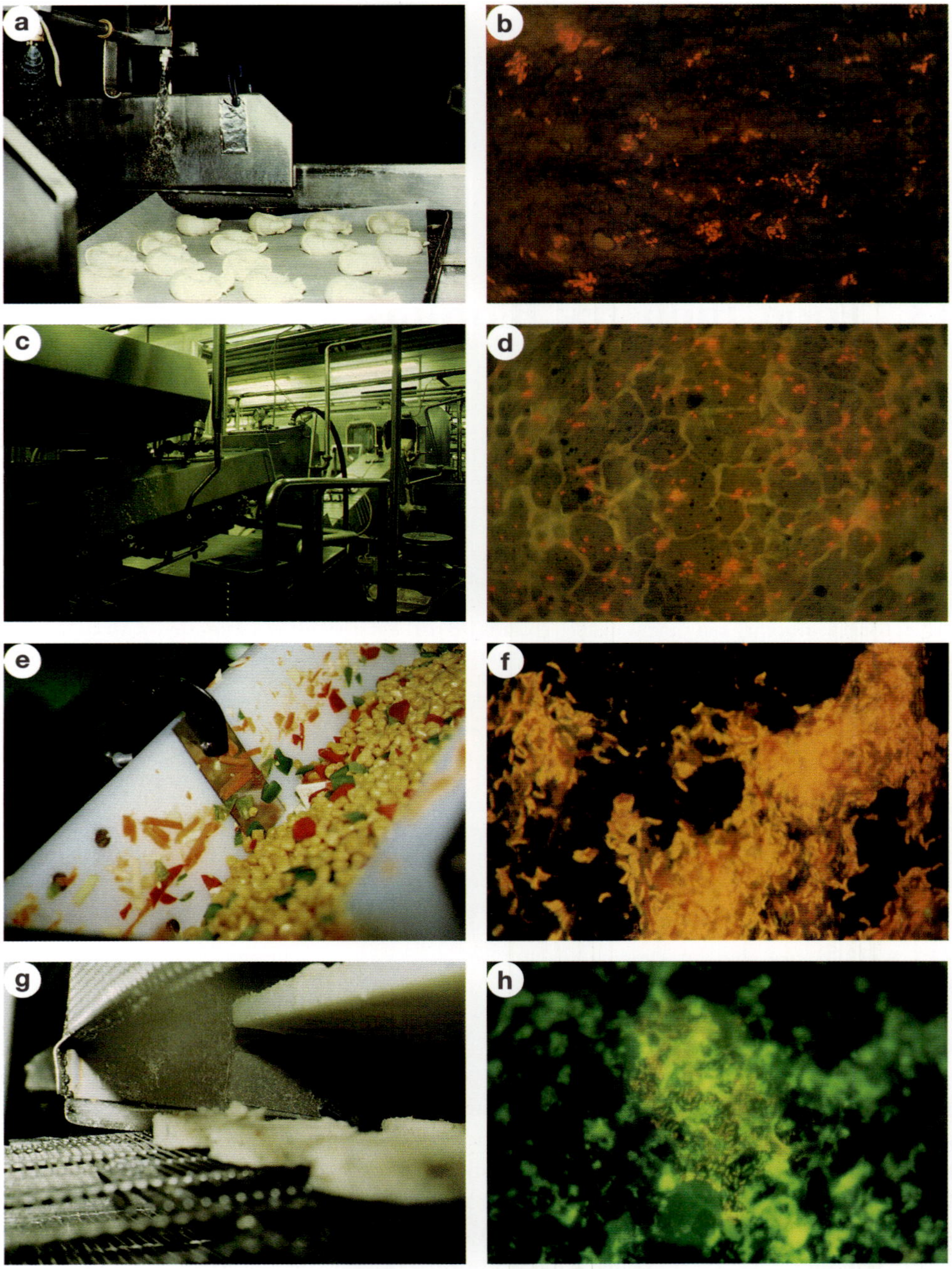

Figure 1 Individual cell and microcolony development on a number of food processing surfaces. a, b = individual cells developing on an egg glazie bath after 8 h; c, d = individual cells of *S. aureus* developing within a buttermilk line after 8 h; e, f = microcolony development after 6 h on a prepared salad line; g, h = development of microcolonies within 90 min on a fishcake forming machine.

numerous early studies used scanning electron microscopy (SEM) to study attachment (Zoltai *et al.*, 1981; Schwach and Zottola, 1984; Speers *et al.*, 1984; Stone and Zottola, 1985; Herald and Zottola, 1988; Mafu *et al.*, 1990). However, SEM requires sample preparation that may introduce artefacts such as shrinkage of EPS and more recent studies have used alternatives such as low temperature SEM, environmental SEM, confocal microscopy and epifluorescence microscopy to assess attachment and biofilm formation (Holah *et al.*, 1989; Little *et al.*, 1991; Sutton *et al.*, 1994; Hodgson *et al.*, 1995).

Holah *et al.* (1989) and later Gibson *et al.* (1995) were the first to use stainless steel test coupons. These were attached near to, or as part of, food contact surfaces and environmental locations in a variety of food processing environments. This was the first attempt to get an idea of biofilm formation under real conditions. The coupons were removed at intervals, stained and fixed immediately with acridine orange and returned to the laboratory to allow assessment of the rate and extent of attachment and biofilm formation. The sites selected were where biofilms were thought likely to develop, such as areas with visible moisture or condensation.

Figure 1a and b show attached microorganisms building up over an 8 h period on the surface of an egg glazing bath prior to baking. The level of single organisms was approximately 10^5 cells cm^{-2} and single organisms only were detected on 71.6% of the coupons examined. Figure 1c and d shows another single organism film after 8 h on a butter milk line and is significant because the attached organism was identified as the pathogen *Staphylococcus aureus*.

Figure 1e and f depicts microcolony development on a prepared salad production line whilst Figure 2a, b and c shows microcolonies on a pea process line. The level of microbial attachment shown is $>10^6$ cells cm^{-2}. Microcolonies were detected on 20.8% of coupons examined which, given more time, could have developed into biofilms. This is illustrated in Figure 2b where a thick visible biofilm can be seen next to the coupons attached to the conveyor side-guide. Figure 1g and h shows the development of microcolonies on a fishcake processing line and are notable because the degree of microbial attachment shown (together with the fishcake pieces) occurred in 90 min.

Extensive biofilms were only found on 6.6% of the coupons. Multilayer biofilms were found in the locations shown in Table 1; these were environmental surfaces rather than food contact surfaces (where there was sufficient time for biofilm formation) and they were associated with the presence of moisture or condensation. Typical of these biofilms is the development of a ceiling based biofilm in the condensation above a hot, potato mixer (Figure 2 d–f). The level of microbial attachment shown is $>10^7$ cells cm^{-2}. These results showed that generally biofilm formation was only found on environmental surfaces, and progression of attached cells to extensive biofilm on food contact surfaces was limited by regular cleaning and disinfection. The most extensive biofilm was found in a vegetable blancher extractor system, and reached almost complete coverage after 120 h (Table 1). This extractor system was cleaned once per week and therefore there was considerable time for biofilm formation. The conditions were such that rapid microbial growth would be expected, *i.e.* moisture in the form of steam and condensation, nutrients carried in the air flow from the blancher and slightly elevated temperature.

In a similar study, Hood and Zottola (1997b) also attached stainless steel chips near food contact surfaces and cast iron chips in the floor drains of four meat processing

Figure 2 Microcolony and biofilm development on two food processing surfaces. a, b = coupons attached to a pea processing line exiting a blancher; c = microcolony development in 12 h (note the thick visible biofilm to the left of the attached coupon); d, e = coupons attached to the ceiling above a hot, potato mixing operation; f = biofilm development on these after 24 h.

Table 1 Food processing environmental surfaces on which biofilms were detected in coupon studies (from Gibson *et al.*, 1995).

Factory type	*Area*	*Exposure time (hr)*	*Swab count (cfu cm^{-2})*	*% Area coverage*
Canned products	Waste can area	24	$>3.5 \times 10^7$	15.0
Canned products	Blancher extractor	24	$>4.7 \times 10^7$	26.5
			$>4.7 \times 10^7$	23.8
		24		20.2
		24		66.3
		48		85.2
		72		98.4
		120		
Meat substitute	Mixer – undersurface of ledge	20	$>5.9 \times 10^7$	56.1
Potato	Ceiling	24	5.5×10^5	3.6
Cod cakes	Conveyor	1.5	4.6×10^6	11.3
Biscuits	Steam clean room	16	$>1.4 \times 10^7$	20.2
Poultry products	Wall in rack washing area	6		
Peas	Inspection belt guard	48		

plants and examined the surfaces using epifluorescence microscopy. These authors found that stainless steel surfaces near food contact surfaces had attached cells at a concentration of approximately 10^4 cm^{-2} and therefore did not exhibit extensive biofilm formation; however, given time or inadequate cleaning and disinfection, these organisms could pose a biotransfer potential. Hood and Zottola (1997b) detected significant biofilms on the floor surfaces they examined. Other workers have also found that environmental surfaces are particularly associated with biofilm formation and could act as reservoirs for microbial contamination of foods, for example floor drains (Cox *et al.*, 1989; Nelson, 1990) and floors (Notermans *et al.* 1991). In addition, some food contact surfaces have been implicated as sites for biofilm formation; these include waste water pipes, bends in pipes, rubber seals, conveyor belts, stainless steel surfaces, Buna N and Teflon seals (Fletcher, 1985; Mafu *et al.*, 1990; Blackman and Frank, 1996).

Figure 3a–f illustrates one of the few biofilms on a food contact surface identified by the authors from a previous study (Holah *et al.*, 1989). These images indicate that under suitable conditions of nutrient supply and temperature, a multilayer biofilm can be generated in 16 h or less.

Studies of surfaces in food processing environments have shown that the nature of the surface population may vary from attached or dried-on cells to isolated microcolonies, a monolayer or multilayer biofilms. Attached cells and microcolonies, given sufficient time, can develop into extensive multilayer structures. However, the studies summarised above show that extensive biofilms are particularly associated with

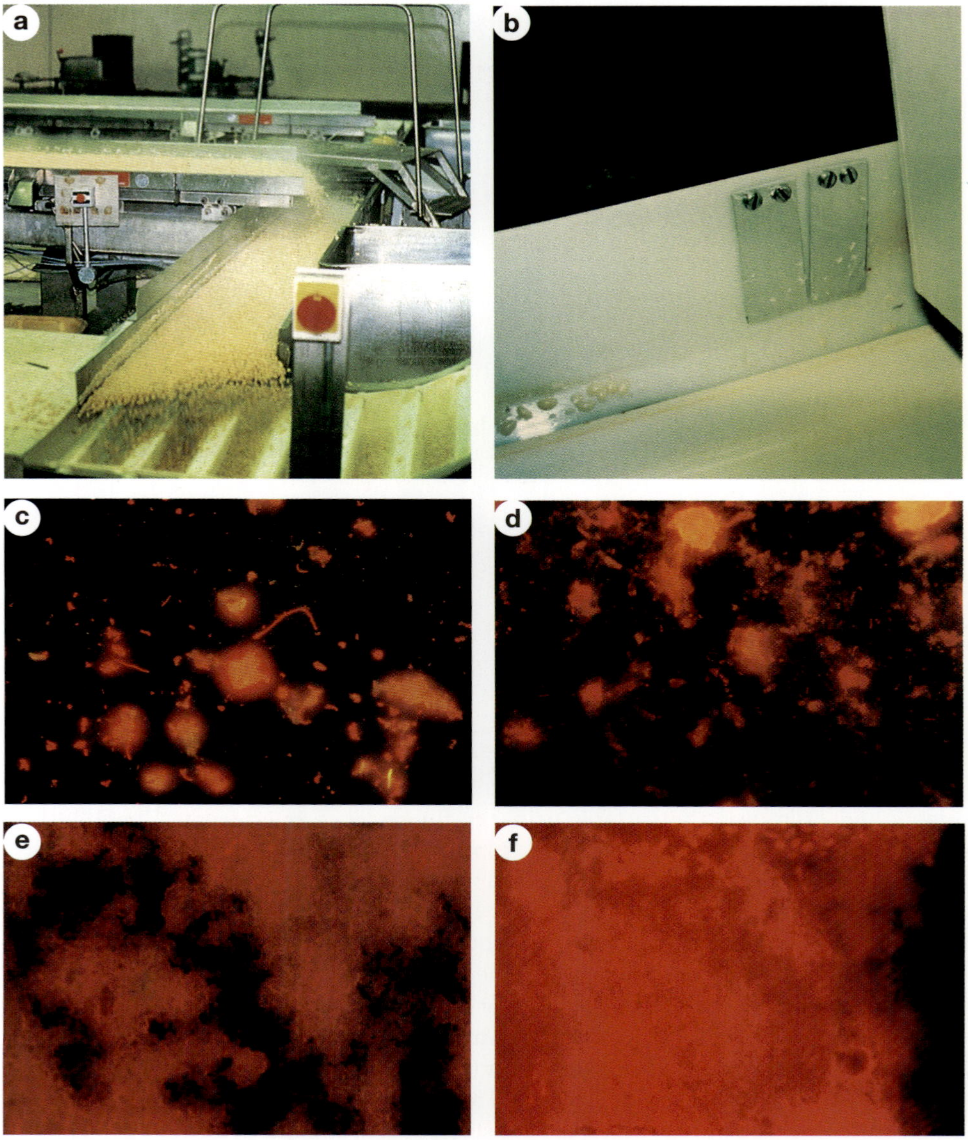

Figure 3 Biofilm development with time. a, b = coupons were attached to a baked bean inspection line and removed at various time periods; c, d, e, f = biofilm development over 4, 8, 12 and 16 h respectively.

non-food contact environmental surfaces and these may be a reservoir for spoilage and pathogenic organisms.

The question is, 'what is a food processing surface biofilm?' Clearly large biofilms several hundred microns to several millimetres in thickness, common in other indus-

Table 2 Summary of the frequency of genera among isolates identifed in 16 factory sites.

Genus	*Percentage (n=78)*
Pseudomonas	23
Staphylococcus	8.6
Enterobacter	8.6
Flavobacterium	7.7
Acinetobacter	7.7
Bacillus	6.5
Serratia	5.1
Klebsiella	5.1
Aeromonas	3.8
Vibrio	2.4
Citrobacter	2.4
Kluyvera	2.4
Agrobacterium	2.4
Hafnia	2.4
Providencia	1.2
Escherichia	1.2
Pasteurella	1.2
Proteus	1.2
Yersinia	1.2
Trichosporan	1.2

tries, rarely occur in food processing environments but because of the non-sterile nature of food processing, microorganisms are almost always present (at a low level) adhered to surfaces. Perhaps a food industry biofilm could be defined as a consortium of microorganisms developing within a defined time period, dependent on the cycle of cleaning and disinfection programmes, or, possibly, the core consortium surviving at low population densities (*e.g.* $<10^3$–10^4) remaining after such cleaning cycles (see section on Future Studies).

Although the composition of such biofilms is usually heterogeneous and dependent on the particular environment, there are certain trends. Analysis of the organisms present on surfaces in food processing environments has shown the presence of a variety of organisms. Table 2 shows genera isolated by Gibson *et al.* (1995) in studies on attached microorganisms in 17 different processing environments. Seventy nine percent of isolates were Gram-negative rods, 8.6% Gram-positive cocci, 6.5% Gram-positive rods and 1.2% yeast strains. The most common organisms were *Pseudomonas*, *Staphylococcus* and *Enterobacter* spp. Other studies on food processing environments have revealed similar trends. Speers *et al.* (1984) found *Pseudomonas* spp. and *Micrococcus* spp. and Zoltai *et al.* (1981) detected *Staphylococcus aureus* and *Streptococcus cremoris.* Hood and Zottola (1997b) isolated a variety of Gram-negative organisms associated with test surfaces in four meat processing plants with *Pseudomonas* and *Klebsiella* species being the most common and *Aeromonas* spp., *Citrobacter freundii* and *Hafnia alvei* also detected. These authors noted that the most common organisms were mucoid, indicating prolific EPS production.

Mettler and Carpentier (1998) studied the microflora associated with the surfaces in milk, meat and pastry sites and concluded that it was specific to the processing environment. *Pseudomonas* spp. predominated in the low temperature meat site and yeasts and *Leuconostoc* spp. in the pastry site. *Pseudomonas* spp. were found at all three sites and have been found in almost all food factory environments where biofilms have been studied (Mattila-Sandholm and Wirtanen, 1992; Gibson *et al.*, 1995). Pseudomonads are environmental psychrotrophic organisms that readily attach to surfaces and are common spoilage organisms in chilled foods. Other common Gram-negative bacteria that have been associated with surfaces are coliform organisms which are widely distributed in the environment and may be indicators of inadequate processing or post-process contamination.

Staphylococcus spp. were also found at all three sites in the study by Mettler and Carpentier (1998), which agrees with the findings of Gibson *et al.* (1995). In addition, other studies have found *Staphylococcus* sp. associated with surfaces (Mead and Scott, 1994). Staphylococci are associated with human skin and therefore their presence on surfaces may be as a result of transfer from food handlers.

These studies primarily rely on swabbing and traditional microbiology and therefore only represent a proportion of the culturable organisms that can be recovered from accessible sample areas.

Andrade *et al.* (1998) found that the thermoduric psychrotrophic lactic acid bacterium involved in milk spoilage readily attached to surfaces. Farrell *et al.* (1998) demonstrated the transfer of *Escherichia coli* O157:H7 from spiked meat samples to stainless steel surfaces in a meat grinder, thus demonstrating that the food product can be a source of pathogenic organisms that attach to surfaces and remain at low levels after cleaning treatments (50% of surfaces).

The organisms present on food processing surfaces can, therefore, be inoculated from the environment, from people and from the product. It is not clear under what circumstances the survival and development of microorganisms from each source are favoured, but the results to date suggest that pseudomonads and staphylococci are most frequently found and thus the environment is the most common source rather than the raw ingredients.

What Conditions are Required?

There are a number of factors that have been shown to affect bacterial attachment and biofilm formation, such as the conditioning layer, the nature and degree of roughness of the substratum, temperature, pH, nutrient availability and the time available. In their review of biofilm formation, Kumar and Anand (1998) conclude that adsorption of organic and inorganic molecules to the surface is the first stage, followed by transportation of the microorganisms to the surface by diffusion, sedimentation or turbulent flow. In the food processing environment, residues remaining on the surface after cleaning and disinfection act as a conditioning layer. Microorganisms attach to the conditioning layer and can develop into a complex biofilm community. Mettler and Carpentier (1998) showed that the conditioning layer found in milk premises could be simulated using milk; in meat and pastry premises the cleaning agents were found to recreate the conditioning layers, showing that cleaning chemicals are also important in the conditioning process. The conditioning layer alters the physico-chemical properties of the surface, therefore affects subsequent microbial attachment. For example, Johal (1988) found that conditioning of stainless steel surfaces with meat juices reduced the negative charge of the surface and therefore enhanced the potential accumulation of bacteria on the surfaces. Boulange-Petermann *et al.* (1997) showed how repeated attachment and cleaning cycles influenced attachment of *Streptococcus thermophilus*, so that the cells attached in clumps due to changes in the conditioning layer. The exact nature of the conditioning layer depends on the surface, the food matrix and the cleaning chemicals used.

The nature and microtopography of the surface is equally important in the attachment of bacteria. Stainless steel is the most common food contact material as it is easy to fabricate, durable, chemically and physiologically inert at a variety of food processing temperatures, corrosion resistant and generally easy to clean. However, the microtopography of stainless steel examined under a SEM reveals cracks and crevices which provide protection for attached bacteria from cleaning forces (Holah and Thorpe, 1990; Wirtenen *et al.*, 1995). SEM has shown that food borne pathogens and spoilage microorganisms accumulate as biofilms on a variety of surfaces including stainless steel, aluminium, glass, BunaN, Teflon seals and nylon materials found in food processing environments (Kumar and Anand, 1998). Sonak and Bhosle (1995) compared attachment of *Vibrio* spp. to stainless steel, polystyrene, copper and aluminium and found maximum attachment to aluminium. Blackman and Frank (1996) studied the ability of *Listeria monocytogenes* to attach to a variety of food processing surfaces and found that stainless steel, Teflon, nylon and polyester floor sealant all supported biofilm development.

Nutrient availability affects biofilm formation. Dewanti and Wong (1995) found that growth in a minimal salts medium resulted in a biofilm consisting of shorter cells of *E. coli* O157:H7 with a thicker layer of EPS compared to growth in a complex medium. In addition, reduced EPS production in the complex medium was associated with reduced biofilm stability, resulting in easier detachment of the biofilm. Similarly Hood and Zottola (1997a) found that growth medium significantly affected the adhesion of *Pseudomonas fragi*, *L. monocytogenes* and *Salmonella typhimurium*.

The temperature of the environment has been found to influence attachment. Chumkhunthod *et al.* (1998) found that *Pseudomonas putida* biofilm formation was more extensive at 22°C than at 30°C and Herald and Zottola (1988) reported that *Yersinia enterocolitica* adhered in higher numbers to stainless steel at 21°C than at 10°C or 35°C. Similarly, Blackman and Frank (1996) found that the attachment of *L. monocytogenes* to a variety of surfaces was reduced at 10°C compared to 21°C. In contrast, Smoot and Person (1998) found that attachment of *L. monocytogenes* to stainless steel and Buna-N rubber was maximal at 30°C. However, Wirtanen *et al.* (1996a) concluded that the effect of temperature was dependent on the organisms, with *P. fragi* and *Pediococcus inopinnatus* showing increased attachment at 6°C compared to 25°C, whereas *L. monocytogenes* showed increased attachment at 25°C and *Bacillus subtilis* attachment was unaffected by temperature.

Adhesion is also affected by pH, with maximum attachment generally at the optimum pH for metabolism. Stanley (1983) found maximum adhesion of *P. fragi* to stainless steel at pH 7–8, and Herald and Zottola (1988) found maximal adhesion of *Y. enterocolitica* at pH 8.

ARE THEY A PROBLEM?

As a generality the food industry can be divided into two groups in terms of potential problems with biofilms. The first group includes food product categories such as 1) shell eggs; 2) raw meat, poultry, and fish; fruit; 3) vegetables; 4) salads; 5) herbs; 6) nuts; 7) canned products; 8) aseptic products; 9) beverages including alcoholic drinks, soft drinks, fruit juices, coffee and tea; 10) baked goods; 11) dried goods; 12) confectionery; 13) snacks and breakfast cereals; and 14) oils and fats and food ingredients. In this category the products are either raw and thus already potentially contaminated with microorganisms and/or will be further processed (*e.g.* by heat treatment). Alternatively, the nature of their production, for example dry environments, means that biofilms do not form. In this category, therefore, biofilms are of little or no concern.

The second group includes 1) prepared salads; 2) dairy products including milk, cream, ice cream, cheese and dairy based desserts; 3) cooked meats and meat products; 4) smoked fish; 5) ready meals; 6) sandwiches; and 7) quiches and ready to eat products including soups, sauces, pasta, pizza, ethnic foods and salad dressings. In this group the food products may not be further processed and any subsequent microbiological contamination from biofilms during production could be unacceptable. However, hygiene standards in this category are much higher and microbiological cross-contamination from the processing environment is a recognised hazard which must be controlled.

Further to this, the food industry in Europe (and other parts of the world) are required by legislation (Council Directive on the Hygiene of Foodstuffs 93/43/EEC) to undertake a risk analysis of their process to ensure the safety of the food product and the most common such analysis is referred to as Hazard Analysis Critical Control Point (HACCP). The seven principles of HACCP, as suggested by the Codex Alimentarius Commission (1996) are 1) identify potential hazards and measures for their control; 2) determine critical control points (CCP); 3) establish critical limits which must be met to ensure that each CCP is under control; 4) establish a monitoring system; 5) establish the corrective action to be taken when monitoring indicates that a CCP is not under control; 6) establish verification procedures to confirm that the HACCP system is working effectively; and 7) establish documentation for procedures and records.

Following these principles, if the food processor believes that biofilms are a risk to the safety of the food product, appropriate control steps must be taken. These include providing an environment in which biofilm formation would be limited, undertaking cleaning and disinfection programmes as required, monitoring these programmes to ensure their success during their operation and verifying their performance by a suitable (usually microbiological) assessment.

Biofilms exist in food manufacturing environments and may have a number of detrimental physical effects such as causing a reduction in the efficiency of heat exchangers, an increase in fluid frictional resistance at the surface and an increase in the corrosion rate.

The attachment of bacteria to the food product or product contact surfaces can be an important source of potential contamination, leading to serious hygienic problems and economic losses due to food spoilage (Holah and Kearney, 1992; Mattila-

Sandholm and Wirtanen, 1992; Carpentier and Cerf, 1993; Boulange-Petermann *et al.*, 1997; Kumar and Anand, 1998). As previously discussed, the organisms attached to surfaces in food processing enviroments may be spoilage organisms. For example, pseudomonads and many other Gram-negative organisms detected on surfaces are the spoilage organisms of concern in chilled foods. The survival of organisms in biofilms may be a source of post process contamination, resulting in reduced shelf life of the product.

There have been numerous laboratory studies that have demonstrated the attachment of food borne pathogens to the types of surfaces found in food processing environments, *e.g. L. monocytogenes* (Herald and Zottola, 1988; Mafu *et al.*, 1990; Faber and Peterkin, 1991; Helke *et al.*, 1993; Mosteller and Bishop, 1993), *Y. enterocolitica* (Kumar and Singh, 1994), *Campylobacter jejuni* (Stern and Kazmi, 1989), *S. typhimurium* (Ronner and Wong, 1993), and *E. coli* O157 (Doyle 1991; Dewanti and Wong, 1995; Farrell *et al.*, 1998). Many of these studies have demonstrated the ability of these pathogens to form biofilms. However, these are generally laboratory based studies and there is scant published data on the presence of pathogens in biofilms in the food processing environment. Walker *et al.* (1991) found *L. monocytogenes* on approximately 5% of the surfaces examined and Lawrence and Gilmore (1995) and Destro *et al.* (1996) isolated *L. monocytogenes* from a range of food environments. Blackman and Frank (1996) concluded that the laboratory evidence for biofilm growth of *L. monocytogenes* was sufficient to suggest a substantial risk of this pathogen contaminating food contact surfaces if wet surfaces were not maintained in a sanitary condition. In addition, Troller (1993) estimated that 25% of food borne disease is caused by contaminated raw foods and contaminated material, equipment and utensils.

Both authors of the present article have been involved in factory based research and client confidential troubleshooting exercises which have demonstrated that pathogens, particularly at low temperatures, can survive for long periods on food processing surfaces and can resist the action of cleaning and disinfection programmes.

Resistance

A number of authors have shown that bacteria attached to various surfaces are generally more resistant to biocides than are organisms in suspension (Ridgeway and Olsen, 1982; Hugo *et al.*, 1985; Le Chevalier *et al.*, 1988; Frank and Koffi, 1990; Holah *et al.*, 1990; Lee and Frank, 1991; Wright *et al.*, 1991; Dhaliwal *et al.*, 1992; Andrade *et al.*, 1998, Das *et al.*, 1998). In addition, cells growing as a biofilm have been shown to be more resistant (Frank and Koffi, 1990; Lee and Frank, 1991; Ronner and Wong, 1993). Carpentier and Cerf (1993) found that peracetic acid, mercuric chloride, and formaldehyde had no effect on biofilms. The most effective biocides for suspended cells are not necessarily suitable for attached biofilm cells (Holah *et al.*, 1990).

Biocide resistance has been reported to increase with biofilm age (Anwar *et al.*, 1990; Frank and Koffi, 1990; Lee and Frank, 1991; Wirtanen and Mattila-Sandholm, 1992; Das *et al.*, 1998). However, Wirtanen *et al.* (1996b) found 3-day biofilms of *Bacillus cereus* were more sensitive than 1-day biofilms to the heat treatment and chemicals tested. In contrast, Chumkhunthod *et al.*, (1998) found there was no difference in susceptibility of *P. putida* biofilms over 1, 2 and 3 d.

The mechanism of resistance in attached and biofilm cells is unclear but may be due to physiological differences such as growth rate, membrane orientation changes due to attachment and EPS production. Equally, physical properties may have an effect *e.g.* protection of the cells within food debris or the material surface structure or causing changes in biocide diffusion to the cell/material surface The problem of resistance of biofilms is compounded by the possible presence of VNC cells. Rowe *et al.* (1998) found that the culturable form of *C.jejuni* was more resistant than the VNC form to quaternary ammonium compounds (QACs), QAC plus glutaraldehyde, and amphoterics, but VNC cells were more resistant to chlorine. In addition, in the food processing environment, attached cells are subject to a variety of stresses such as starvation, desiccation, chemicals and temperature extremes which may influence biocide susceptibility.

Routes of Transmission to Product Contact Surfaces

Biofilms may be a source of spoilage or pathogenic organisms and may transfer these organisms to food products both directly and indirectly. Direct mechanisms involve the detachment of cells from surfaces either *via* erosion (single cells) or sloughing (clusters of cells), both of which can involve physiological and mechanical processes. These processes have been reviewed by Willcock *et al.* (1997). Physiological erosion is based on cell division, where the daughter cells acquire a different hydrophobicity from the parent biofilm cells and are forced out of the biofilm to establish new communities. Physiological sloughing processes are based on the production of local weakness or 'fault lines' within the biofilm, caused either by hydrolytic cleavage *via* the production of extracellular polysaccharide lyase enzymes or amphipathic surfactant molecules. Mechanical processes induce both erosion and sloughing phenomena and are dependent on the particular characteristics of the local overlying product flow.

The nature of the detachment could be critical in terms of product safety and spoilage rates and a better understanding of this process would be of major benefit to food producers and to the necessary risk analyses. If the product contamination route was *via* erosion, this would have little implication to, for example, fresh produce as the additional few organisms entering each product lot may not change the overall product risk. However, it would have major implications to for example, ambient shelf stable products, as even one microorganism entering the product has sufficient time to grow to levels which could pose a spoilage or food safety hazard. Erosion, *via* which many lots become infected, is thus a major problem. Conversely, contamination by sloughing may be of little concern in ambient shelf stable foods as a larger number of cells entering only one lot may account for the current industry acceptable spoilage rate of >1 in 10,000 lots. Contamination by sloughing of fresh produce, however, may give rise to a single lot that has a high enough microorganism count to give concern.

Cells may be transferred from environmental surfaces to product and product contact surfaces indirectly by people, equipment, pests, cleaning and the air. Aerosols from floor drains have been cited as a potential source of contamination in food processing facilities (Heldman *et al.*, 1965). Holah *et al.* (1993) showed that high pressure water spray systems generated aerosols that could potentially disperse viable microorganisms over an extensive area and that these aerosols could carry spoilage or pathogenic organisms. For example, Spurlock and Zottola (1991) and Holah *et al.*

(1993) showed that *L. monocytogenes* could survive in aerosols. Factory studies by the authors have also indicated that pathogens can be transferred from the floor to product in close proximity to the floor, *e.g.* on racks or in bins, by the action of walking close to the product or by the splashing of trolley wheels.

HOW ARE THEY CONTROLLED?

The complexity of biofilms means that their control requires an integrated hazard analysis approach involving surface, equipment and environment design, construction and operation and appropriate effective cleaning and disinfection schemes.

With respect to microorganisms, hygienic design of food processing equipment is employed for three major reasons, *viz.* to ensure that all materials of construction are suitable, to eliminate areas in the design which could harbour food soil (product residues, microorganisms, dust, debris *etc.*) out of reach of cleaning agents, and to allow all liquids (product and cleaning chemicals) to be freely drained. The principles of hygienic design have been reviewed in general terms by the European Hygienic Equipment Design Group (EHEDG, 1993, 1995).

Boulange-Petermann *et al.* (1997) defines three types of surface with respect to surface topography and roughness, *viz.* polished surfaces where bacteria can be entrapped in grooves, pickled surfaces where bacteria are trapped/attach in the grain boundaries, and smooth surfaces. In their studies using an organic soil and spore mixture, Leclercq-Perlat and Lalande (1994) found that when surface topography was smooth on a microscopic scale, the cleanability differences observed were due to differences in surface chemical composition.

Stainless steel is the most commonly used construction material. Vanhaecke *et al.* (1990) showed that attachment of a hydrophilic strain of *P. aeruginosa* increased with the roughness average of stainless steel AISI 316L, but that there was no such relationship for the hydrophobic strain. Holah and Thorpe (1990) concluded that there was little difference in the cleanability of unused stainless steel, enamelled steel, mineral resin and polycarbonate surfaces in terms of removal of microorganisms. However, once the surfaces had been abraded, the mineral resin and polycarbonate were more difficult to clean than the more abrasion resistant stainless steel. Surface roughness and topography influence cleanability, with crevices and pits in the surface resulting in reduced cleanability (Holah and Thorpe, 1990). Wirtanen *et al.* (1995) concluded that roughness was more important in the removal of bacteria from surfaces than were organism type, surface material, cleaning procedure, time of cleaning or use of detergent.

The elimination of areas in which product and microorganisms can become trapped is ensured by addressing general equipment design and construction features for joints, fasteners, drainage, internal angles and corners, dead spaces, bearings and shaft seals, instrumentation, doors, covers and panels and controls (Holah, 1998). In addition, all surfaces should be designed such that fluids (product and cleaning fluids) can freely drain. This is to prevent subsequent contamination of food products by cleaning solutions and, especially for equipment to be used in high risk food manufacturing, to reduce moisture residuals which would aid microbial growth.

Providing that the processing environment and the production equipment are hygienically designed, then an appropriate effective cleaning and disinfection programme is the major control of surface accumulation. The cleaning phase is thought to be the most important stage for minimising microbial colonisation and removing attached microorganisms (Dunsmore *et al.*, 1981; Carpentier and Cerf, 1993). Factory trials on effectiveness have shown that cleaning may only be responsible for the removal of one log order of microorganisms (Gibson *et al.*, 1999). However, it is necessary for the removal of product soils and possibly EPS and therefore is critical for an effective disinfection stage.

Of the four factors/energies involved in cleaning, *i.e.* chemical, mechanical, temperature and time, the mechanical input has been shown by several workers to be important for biofilm removal (Blenkinsopp and Costerton, 1991; Mattila-Sandholm and Wirtanen, 1992; Wirtanen and Mattila-Sandholm, 1993; 1994; Gibson *et al.*, 1999). Techniques such as physical scrubbing of surfaces and the use of high pressure water sprays are particularly effective in terms of the removal of bacterial biofilms (Gibson *et al.*, 1999); however, high pressure systems generate aerosols that could potentially disperse viable microorganisms over an extensive area (Holah *et al.*, 1993).

Detergents are generally formulated to remove particular types of food soils rather than attached microorganisms. Wirtanen *et al.* (1995) and Gibson *et al.* (1999) found that detergents did not enhance the removal of bacterial biofilms in the absence of product debris. The role of detergents in the removal of bacteria from surfaces may have more significance in the presence of food debris, where in addition to direct attachment to conditioned stainless steel, bacteria may be attached to food particles. Wirtanen *et al.* (1996b) found that EDTA improved the removal of *Bacillus* biofilms from soiled and unsoiled surfaces. Detergents containing chelating agents such as EDTA help in biofilm removal, probably through the chelation of calcium and magnesium ions that destabilize the matrix. Gibson *et al.* (1999) found that an acidic and an alkaline detergent affected the viability of *S. aureus* and *P. aeruginosa* respectively, and therefore minimized the generation of aerosols of viable microorganisms and the spread of contamination.

The age of the biofilm has been found to influence the ease of removal. Wirtanen *et al.* (1996b) found that for *B. subtilis* and *P. fragi* the age of the biofilm increased the difficulty of detaching it from the surface. Other studies have shown that attachment for longer than 16 h increased the strength of attachment and resistance to removal and disinfection (Eginton *et al.*, 1998). In addition, temperature affected the ease of detachment of *L. monocytogenes*, with easier detachment at of 6°C than at 25°C. The effect of temperature on ease of detachment is organism dependent. Busscher *et al.* (1995) concluded that detachment and sloughing of a biofilm from a surface is dependent on the strength of interaction between the biofilm and the conditioning film, with the presence of a conditioning layer facilitating detachment and removal in some cases. The design of new non-adhesive materials should perhaps focus on materials from which it is easy to detach the attached cells (Busscher *et al.*, 1995).

Farrell *et al.* (1998) showed that cleaning using a combination of chemical and mechanical action reduced the level of *E. coli* O157:H7 attached to stainless steel chips. However, the organism was detected on 50% of surfaces after cleaning, showing a requirement for disinfection. After an effective appropriate cleaning treatment, bacteria remain on surfaces and there is a requirement, therefore, for a disinfection stage to reduce the surface population further and reduce the likelihood of rapid recolonisation of the surface. As discussed earlier, although attached bacteria have been shown to be more resistant to certain biocides, optimization of cleaning in terms of frequency and in enhancing susceptibility to disinfectants with informed choice of biocide can result in an effective disinfection stage.

Disinfection is temperature related, *i.e.* the higher the temperature, the greater the disinfection. This is not a problem for most food manufacturing sites that are operating outside ambient environmental temperature or higher, but it is critical in production areas running at chill temperatures. Taylor *et al.* (1999) examined the efficacy of 18 disinfectants at both 10°C and 20°C and demonstrated that for some chemicals, particularly quaternary ammonium based products, disinfection was much reduced at 10°C and they recommended that in chilled production environments, only products specifically formulated for low temperature activity should be used.

The practical undertaking of the stages (preparation, gross solids removal, rinsing, cleaning, rinsing, disinfection) and the sequence of the cleaning and disinfection procedure is most important (Holah, 1995). The correct sequence is essential to minimize contamination of the food product and is of particular importance for the category of foods requiring higher hygiene standards, defined earlier. The sequence of the cleaning and disinfection programme should be: 1) removal of gross debris from equipment surfaces; 2) removal of gross debris from environmental surfaces; 3) rinsing of environmental surfaces; 4) rinsing of equipment surfaces; 5) cleaning of environmental surfaces including drains; 6) rinsing of environmental surfaces; 7) cleaning of equipment surfaces; 8) rinsing of equipment surfaces; 9) disinfection of equipment surfaces; and 10) application of a disinfectant fog if required.

This sequence, combined with the selection of cleaning methods that limit aerosol generation, minimizes the contamination of the most critical areas, the product contact surfaces.

Assessment of efficacy of disinfection has been traditionally carried out using suspension based tests. However, these tests are unlikely to provide a realistic prediction of performance against attached bacteria. Gibson *et al.* (1995) and Holah *et al.* (1998) reviewed the methods that have been used to determine the efficacy of disinfectants, including suspension tests and tests using dried-on, attached and biofilm organisms. It was concluded that suspension tests allow screening and comparison of disinfectant products, whilst surface tests, although closer to the 'real' situation do not take into account the environmental stresses organisms may encounter in the processing environment (action of detergents, variations in temperature and pH and mechanical stresses) which may affect susceptibility. A novel protocol involving the generation of cells exhibiting a biofilm phenotype within a gel matrix has been described (Gilbert *et al.*, 1998; Wirtanen *et al.*, 1998). This was shown to be a simple reproducible method for the evaluation of the effectiveness of biofilms against organisms growing as a biofilm and the protocol could be adapted to include exposure to some of the stresses experienced on the food processing surface. These tests can only indicate likely anti-microbial performance in use, the ultimate test is a field trial.

WHAT OF THE FUTURE?

The field of biofilms in the food industry is still a very young science and from the previous three sections it can be surmised that the current state of knowledge is that, given suitable environmental conditions (temperature, nutrients, available water, time and adventitious microorganisms), microbial growth on surfaces will occur. Such growth is usually not an issue in low risk food production areas but may be a potential hazard in high risk processing areas. In addition, the removal and disinfection control of such growth may be problematical. There are many areas in this field, however, in which the food industry is seeking guidance and some of these areas are highlighted below to encourage further study.

Do Surface Adhered Microorganisms Matter?

Traditionally it has been believed that microorganisms adhering to food contact surfaces are able to enter the food and subsequently grow, posing a potential spoilage or safety hazard. Is this true and is there any evidence for it, other than the fact that periodically foodstuffs spoil and contain pathogens? In other words, has the spoilage/ pathogenic flora arisen from the processing surfaces or from another source?

An opposite view would be that it is also widely accepted that microorganisms have evolved to colonise certain environmental niches. Why then should a microorganism that has adapted to grow on a food processing surface be able to grow in a completely different niche, *e.g.* a food, and *vice versa*. Evidence that microorganisms may differ in their ability to colonise various sites was suggested by Michiels *et al.* (1997) who analysed 80 fluorescent pseudomonad strains obtained from minced turkey meat and a meat mincer. Random amplification of polymorphic DNA (RAPD) analyses followed by cluster analysis demonstrated that 17 mincer surface strains were significantly different to 63 meat strains. They suggested that only a small fraction of the *Pseudomonas fluorescens* strains comprising the raw meat population were able to form biofilms, but perhaps a different interpretation is that the biofilm strains were unable to grow in the product.

In a similar study in a shrimp processing plant, Destro *et al.* (1996) used RAPD and pulsed-field gel electrophoresis (PFGE) to identify 115 isolated *L. monocytogenes* strains. They also noted that none of the strains recovered from the environment, the process water or the utensils was recovered from the shrimps. There was, however, correlation between some strains on the shrimps and food workers' hands, though transfer could potentially have been either way.

A study by Lawrence and Gilmore (1995) isolated 289 *L. monocytogenes* strains from both the raw and cooked processing areas of a poultry plant and these were subsequently typed by RAPD. Analysis showed 18 profiles (A-R) of which the most common were A (64%) and B (14%). Both profiles A and B were found in the incoming raw poultry, in cooked poultry and on environmental samples from both raw and cooked processing areas. This suggests that it may be possible for some strains to grow both on the product and on environmental surfaces, though it is not clear whether the strains were isolated after cleaning or during processing where the strains identified from environmental surfaces may be product related. The presence of both strains in pre- and post-cooked product and on environmental surfaces may suggest either cross-contamination from the low risk environment to the high risk environment and then to the heat treated (pasteurised) product, or a heat process which is not entirely effective, so that raw poultry strains survive the process and then are found in the cooked product environment. What is worthy to note, however, is that the 6 profiles present in the cooked product environment only were not subsequently found in the cooked product. Similar findings of presence on surfaces and in the product, together with surface exclusive strains, was shown by Dodd *et al.* (1988) for *Staphylococcus aureus* strains differentiated using plasmid profiles on raw poultry at slaughter.

The use of microbiological fingerprinting techniques could ascertain whether, for the given species found, strain clusters are present which favour surface or food growth. Subsequent tests could then examine whether surface strains can readily grow in the corresponding food product and *vice versa*.

How Long will Microorganisms Survive in a Plant?

Through involvement with food companies during the development phases of some of the new genetic fingerprinting techniques, it has been possible to characterise the strains identified over several years. Thus the same strains of some pathogens, particularly *L. monocytogenes*, can present recurring problems in terms of product contamination within the same factory after an absence of several years. This has led to the suggestion that these strains may be present in raw materials supplied to the plant, that they are common in the environment surrounding the factory or that they may be 'hiding' within the factory and re-emerging at various times to cause problems. If the latter is true, it has major repercussions concerning both factory and processing equipment design and cleaning and disinfection practices.

Another finding was the number of similar strains of pseudomonads present on a dairy equipment surface at one sampling time. A pseudomonad in a batch of cream was thought to have come from the cooling water which was jacketing the cream in an old tank with suspect welds. Samples were taken from the tank surfaces, previous pipe work, the cooling water and the product and subjected to ribotyping. Whilst a correlation was found between the cooling water and the product in terms of the suspect microorganism, in excess of 20 similar pseudomonad strains were present on the processing equipment.

It has traditionally been believed that cleaning and disinfection, whilst not perfect, generally remove soil and microorganisms present at the end of the production period, leaving the surfaces clean for subsequent production but also available for subsequent colonisation (during production) from the product, the operatives or the environment. It would thus be expected that a point sample of the surface would reveal perhaps a few pseudomonads that had adhered during that production period rather than the large number found. This suggests that biofilms at the level of <100 cells cm^{-2} after cleaning and disinfection may be quite complex and resistant, and may hold a variety of microorganism types that have become specifically adapted to that environment. They thus form a focus of infection for the next production period and whilst their numbers cycle up and down through cleaning phases, they could remain present on the surface for substantial time periods.

Results of such further characterisation of the surface adhered flora may help identify whether surface microbial growth deserves concern, or whether nothing should be done that interferes with the formation of stable 'harmless' surface populations.

Can a Natural Plant Flora become Resistant to the Disinfectants Used, and How Often Should Disinfectant Types be Alternated?

This is perhaps the most often asked question regarding disinfection and has become topical in the food industry with the much publicised evidence of antibiotic resistance in public health (*e.g.* methicillin resistant *S. aureaus*, MRSA). Factory experience has shown that surface populations of microorganisms (particularly pseudomonads) can build up at various points of a number of processes to levels that cause concern. This was particularly prevalent when non-oxidative biocides, especially QACs (and to a lesser extent amphoterics), were used and it was assumed that the resulting population was resistant to the disinfectant. A further assumption was that disinfection resistance had evolved from a single cell *via* mutation, such that over a time period, a significant proportion of that species became resistant. The best advice was that QUATs should be changed periodically to an alternative disinfectant for a few days and then the QUAT reverted to.

Much work has been undertaken looking at the mechanisms of disinfectant resistance in pseudomonads and *Staphylococcus* spp., particularly by Sundheim (*e.g.* Sundheim *et al.*, 1998) and Russell (*e.g.* Russell, 1992). In addition, Lemaître *et al.* (1998) have demonstrated plasmid mediated resistance to antimicrobial agents, including common food industry disinfectants, in *Listeria* spp. It has not generally been possible to demonstrate resistance in the laboratory, however, at concentrations of disinfectant commonly used in the food industry. Coupled with this, non-oxidative biocides are now available which are much better formulated so that, as described earlier, they are more active, especially at lower temperatures. The best advice today would be to buy good quality disinfectants which should not need to be alternated or, if using low concentrations of cheaper products, to alternate more frequently as the now suggested mechanism of resistance adoption, plasmid transfer, is likely to happen more quickly and/or more frequently than mutation. The subject of plasmid transfer between bacteria is addressed by Angles and Goodman (2000).

Can Microorganisms 'Move' Across a Plant and Packaging Surfaces?

In food factories microorganisms of concern tend to be found in areas which are difficult to access, for example between components of equipment that have been jointed, *e.g.* by screws or bolts, but not been effectively sealed, *e.g.* metal to metal joints that are not welded, or plastic to plastic joints that are not glued. For thorough cleaning and disinfection, these joints have to be dismantled which is not often simple, and this is costly in terms of production down-time and may reduce the life of the equipment. Whilst more attention to hygienic equipment design would alleviate the problem in the short term, what is unknown is whether the penetration of microorganisms into such crevices is entirely physical, *e.g.* by pressure or capillary action, or whether microorganisms attached to surfaces are able to undertake lateral movement on the surface to a more advantageous niche.

Similarly, the ability of microorganisms to pass through micro-holes or channels in packaging materials into ambient shelf stable products, *e.g.* canned foods (referred to as leaker spoilage) is of concern. The movement of microorganisms through microholes and channels has recently been reviewed by Song and Hargraves (1998), who suggested that the micro-hole diameter, channel length, differential pressure, the presence of fluid and the external microorganism type (species, size, motility) and concentration were important factors controlling penetration. Again it is unknown whether there is a biological factor affecting movement through such holes and channels relating to surface-mediated microbial movement.

Will Future Cleaning Agents be Targeted to Bacterial Flora Rather than Food Soils?

Current two-stage cleaning and disinfection protocols require a thorough cleaning phase to remove food debris and some microorganisms, followed by a disinfection stage to further reduce microbial numbers. In certain food processing sectors, *e.g.* high risk short shelf-life foods, this may be insufficient and food manufacturers will seek additional methods to ensure that post-sanitation surfaces have even fewer numbers of viable microorganisms (especially pathogens).

Currently there is no detergent on the market, either a formulated surfactant or an enzyme based product, which is specifically targeted at microorganism removal. New cleaning products could be devised to reduce attachment strength, to enzymatically digest the microbial extracellular polymer matrix or to damage the cell so that subsequent disinfection application would be synergistic. If such a product became available, a three-stage protocol could be envisaged consisting of a cleaning phase to remove food debris, a cleaning phase to remove/damage bacteria and a final disinfection phase. If the new product also had some food soil removal ability it may be possible to combine the first two phases for cleaning of low soil environments.

Can Processing Surfaces be Modified to Reduce Biofilm Formation or to Ease its Removal?

Surfaces could be modified to reduce microbial attachment and/or enhance microbial removal or to incorporate a biocide to kill adhered cells.

The use of stainless steel as the material of choice for food processing surfaces has been established for a long time, though there are still many arguments about what the final finish should be. This includes the method of surface finishing, *e.g.* grit polishing or electropolishing, and surface roughness, though there is some agreement on the requirement for a roughness value of 0.8 μm Ra. It is not possible to use stainless steel for all applications, however, and polymers (*e.g.* plastics) and elastomers (*e.g.* rubbers) are also used in which the manufacturing process generally determines the surface finish.

What is still not known is the relationship between surface roughness and microbial adhesion. This is primarily a problem with the determination of the surface roughness value as the international standard approach for equipment manufacturers is based upon measurement by a diamond stylus, which is drawn across the surface. The stylus has a diameter of approximately 25 μm which is fine enough to measure surface finish for engineering purposes but is too coarse for microbiological purposes where imperfections of 2–3 m are large enough to harbour microorganisms. Lessons could be learned from the dental industry that uses laser stylus based instruments and also atomic force microscopy (AFM) to measure roughness.

Similarly, the relationship between other surface properties, for example charge or hydrophobicity, and attachment of microbial populations, as discussed in other chapters of this book, is not clear. Present understanding of the science, based on the alteration of surface charge, hydrophobicity or roughness, is thus insufficient to ensure the success of proposed surface modifications in reducing attachment or aiding cleanability.

With respect to the surface incorporation of biocides, there are a number of control issues that are well established in the food industry for current disinfection technology that will affect their success. These concern the choice of biocide, determination of biocide presence, measurement of biocide concentration, assessment of total biocide surface coverage, development of resistance and assessment of effect.

The choice of biocide is critical because it is known that there is no disinfectant that is effective against all microorganisms under all conditions (*e.g.* temperature, time, pH and presence of organic matter). The target microorganisms and the operating conditions within the food processing environment have to be considered. Similarly, the adoption of a resistant flora on the material surface could present major problems, including the incorporation of pathogens that may be sensitive to the biocide in isolation. The presence of a biocide initially in a material may be difficult to establish, as would the assurance that the biocide had been incorporated evenly over the whole of the material surface. Similarly, on leaching from the surface, the concentration of the incorporated biocide would have to be minimal.

Perhaps the two largest problems facing the longterm acceptance of biocide incorporated materials lie within psychology and testing. Firstly, the value of a biocide incorporating surface is that it has an additional benefit over and above that of a properly cleaned and disinfected surface in that additional disinfection may occur, though a properly cleaned surface would be required for this benefit to be realised. The food industry fear is that, if a product is marketed as being inherently biocidal, cleaning operatives may be less thorough in the application of their tasks. Secondly, there are no currently agreed test methods to determine the efficacy of biocides incorporated into surfaces. This could lead to performance claims that are not rigorously established.

The subject of biocidal surfaces, in the form of surface-catalysed hygiene is addressed by Gilbert and Allison (2000).

REFERENCES

Allison D.G., Sutherland I.W. (1987). The role of exopolysaccharides in adhesion of freshwater bacteria. *J Gen Microbiol*, 133, 1319–1327.

Andrade N.J., Bridgeman T.A., Zottola E.A. (1998). Bacteriocidal activity of sanitizers against *Enterococcus faecium* attached to stainless steel as determined by plate count and impedance methods. *J Food Prot*, 661, 833–838.

Angles M.L., Goodman A.E. (2000). Plasmid transfer between bacteria in biofilms. In: Evans L.V. (ed) *Biofilms: Recent Advances in their Study and Control*. Harwood Academic Publishers, pp. 81–99.

Anwar H., Dasgupta M.K., Costerton J.W. (1990). Testing the susceptibility of bacteria in biofilms to antibacterial agents. *Antimicrob Agents Chemother*, 34, 2043–2046.

Blackman I.C., Frank J.F. (1996). Growth of *Listeria monocytogenes* as a biofilm on various food processing surfaces. *J Food Prot*, **59**, 827–831.

Blenkinsopp S.A., Costerton J.W. (1991). Understanding bacterial biofilms. *Trends Biotechnol*, **9**, 138–143.

Boulange-Petermann L., Rault J., Bellon-Fontaine, M-N. (1997). Adhesion of *Streptococcus thermophilus* to stainless steel with different surface topography and roughness. *Biofouling*, **11**, 201–216.

Busscher H.J., Bos R., van der Mei H.C. (1995). Initial microbial adhesion is a determinant for the strength of biofilm adhesion. *FEMS Microbiol Letts*, **128**, 229–234.

Carpentier B., Cerf O. (1993). Biofilms and their consequences, with particular reference to the food industry. *J Appl Bacteriol*, **75**, 499–511.

Chumkhunthod P., Schraft H., Griffiths M.W. (1998). Rapid monitoring method to assess efficacy of sanitizers against *Pseudomonas putida* biofilms. *J Food Prot*, **61**, 1043–1046.

Codex Alimentarius Commission (1996). Hazard Analysis and Critical Control Point (HACCP) System and guidelines for its application. Draft report of the twenty-ninth session of the Codex Commission on Food Hygiene, Washington D.C., October 1996. Alinorm 97/13A, Appendix II, pp. 23–33.

Cox L.J., Kleiss T., Cordier J.L., Cordellano C., Konkel P., Pedrazzini C., Beumer R., Seibenga A. (1989). *Listeria* spp. in food processing, non-food and domestic environments. *Food Microbiol (Lond)*, **6**, 49–61.

Das J.R., Bhakoo M., Jones M.V., Gilbert P. (1998). Changes in biocide susceptibility of *Staphylococcus epidermidis* and *Escherichia coli* cells associated with rapid attachment to plastic surfaces. *J Appl Microbiol*, **84**, 852–858.

Destro M.T., Leitao M.F.F., Farber J.M. (1996). Use of molecular typing methods to trace the dissemination of *Listeria monocytogenes* in a shrimp processing plant. *Appl Environ Microbiol*, **62**, 705–711.

Dewanti R., Wong A.C.L. (1995). Influence of culture conditions on biofilm formation by *Escherichia coli* O157:H7. *Int J Food Microbiol*, **26**, 147–164.

Dhaliwal D.S., Cordier J.L., Cox L.J. (1992). Impedimetric evaluation of the efficacy of disinfectants against biofilms. *Lett Appl Microbiol*, **15**, 217–221.

Dodd C.E.R., Chaffey B.J., Waites W.M. (1988). Plasmid profiles as indicators of the source of contamination of *Staphylococcus aureus* endemic within poultry processing plants. *Appl Environ Microbiol*, **54**, 1541–1549.

Doyle M.P. (1991). *Escherichia coli* O157:H7 and its significance in foods. *Int J Food Microbiol*, **12**, 289–302.

Dunsmore D.G., Twomy A., Whittlestone W.G., Morgan H.W. (1981). Design and performance of systems for cleaning product contact surfaces of food equipment: a review. *J Food Prot*, **44**, 220–240.

Eginton P.J., Holah J., Allison P.S., Handley P. (1998). Changes in the strength of attachment following treatment with disinfectants and cleansing agents. *Lett Appl Microbiol*, **27**, 101–105.

EHEDG (1993). Hygienic design of closed equipment for the processing of liquid food processing. *Trends Food Sci Technol*, **4**, 375–379.

EHEDG (1995). Hygienic design of equipment for open processing of foods. *Trends Food Sci Technol*, **6**, 305–310.

Faber J.M., Peterkin P.I. (1991). *Listeria monocytogenes*: a food borne pathogen. *Microbiol Rev*, **55**, 476–571.

Farrell B.L., Ronner A.B., Wong C.L. (1998). Attachment of *Escherichia coli* O157:H7 in ground beef to meat grinders and survival after sanitation with chlorine and peroxyacetic acid. *J Food Prot*, **61**, 817–822.

Fletcher M. (1985). Effect of solid surfaces on the activity of attached bacteria. In: Savage D.C., Fletcher M. (eds) *Bacterial Adhesion*. Plenum Press, New York and London, pp. 339–361.
Frank J.F., Koffi R.A. (1990). Surface-adherent growth of *Listeria monocytogenes* is associated with increased resistance to surfactant sanitizers and heat. *J Food Prot*, **53**, 550–554.
Gibson H., Taylor J.H., Hall K.E., Holah J.T. (1995). Biofilms and their detection in the food industry. CCFRA R&D Report No. 1. CCFRA, Chipping Campden, UK.
Gibson H., Taylor J.H., Hall K.E., Holah J.T. (1999). Effectiveness of cleaning techniques used in the food industry in terms of the removal of bacterial biofilms. *J Appl Microbiol*, (In press).
Gilbert T., Allison D.G. (2000). Surface catalysed hygiene and biofilm control. In: Evans L.V. (ed) *Biofilms: Recent Advances in their Study and Control*. Harwood Academic Publishers, pp. 279–290.
Gilbert P., Jones M.V., Allison D.G., Heyes S., Maira T., Wood P. (1998). The use of poloxamer hydrogels for the assessment of biofilm susceptibility towards biocide treatments. *J Appl Microbiol*, **85**, 985–990.
Heldman D.R., Hendrick T.I., Hall C.W. (1965). Sources of airborne microorganisms in food processing area drains. *J Milk Food Technol*, **28**, 41–45.
Helke D.M., Somers E.B.and Wong A.C.L. (1993). Attachment of *Listeria monocytogenes* and *Salmonella typhimurium* to stainless steel and Buna-N in the presence of milk and milk components. *J Food Prot*, **56**, 479–484.
Herald P.J., Zottola E.A. (1988). Attachment of *Listeria monocytogenes* to stainless steel surfaces at various temperature and pH values. *J Food Prot*, **53**, 1549–1552.
Hodgson A.E., Nelson S.M., Brown M.R.W., Gilbert P. (1995). A simple *in vitro* model for growth control of bacterial biofilms. *J Appl Bacteriol*, **79**, 87–93.
Holah J.T. (1995). Food production areas. In *Disinfectants: actions and applications*. Office International Des Epizooties, Science and Technical Revue, **14**, 343–363.
Holah J.T. (1998). Hygienic design: international issues. *Dairy Food Environ San*, **18**, 212–220.
Holah J.T., Thorpe R.H. (1990). Cleanability in relation to bacterial retention on unused and abraded domestic sink materials. *J Appl Bacteriol*, **69**, 599–608.
Holah J.T., Kearney L.R. (1992). Introduction to biofilms in the food industry. In: Melo L.F., Bott T.R., Fletcher M., Capdeville B. (eds) *Biofiilms – Science and Technology*. Kluwer, Dordrecht, pp. 35–45.
Holah J.T., Betts R.P., Thorpe R.H. (1989). The use of epifluorescent microscopy to determine surface hygiene. *Int Biodeterior*, **25**, 147–153.
Holah J.T., Taylor J.H., Holder J.S. (1993). The spread of *Listeria* by cleaning systems. Technical Memorandum No. 673. CCFRA, Chipping Campden, UK.
Holah J.T., Lavaud A., Peters W., Dye K.A. (1998). Future techniques for disinfectant efficacy testing. *Int Biodeterior Biodegrad*, **41**, 273–279.
Holah J.T., Higgs C., Robinson S., Worthington D., Spenceley H. (1990). A Malthus based surface disinfection test for food hygiene. *Lett Appl Microbiol*, **11**, 255–259.
Hood S.K., Zottola E.A. (1997a). Growth media and surface conditioning influence the adherence of *Pseudomonas fragi*, *Salmonella typhimurium* and *Listeria monocytogenes* cells to stainless steel. *J Food Prot*, **60**, 1034–1037.
Hood S.K., Zottola E.A. (1997b). Isolation and identification of adherent Gram-negative microorganisms from four meat processing plants. *J Food Prot*, **60**, 1135–1138.
Hugo W.B., Pallent L.J., Grant D.J.W., Denyer S.P., Davies A. (1985). Factors contributing to the survival of *Pseudomonas cepacia* in chlorhexidine. *Lett Appl Microbiol*, **2**, 37–42.
Johal S. (1988). Bacterial adhesion to processing surfaces in the meat industry. PhD thesis, University of Surrey, UK.
Kumar C.G., Anand S.K. (1998). Significance of microbial biofilms in the food industry: a review. *Int J Food Microbiol*, **42**, 9–27.

Kumar C.G., Singh R.S. (1994). *Yersinia enterocolitica*, as an emerging foodborne pathogen — a review. *Indian J Dairy Sci*, **47**, 537–544.

Lawrence L.M., Gilmore A. (1995). Characterisation of *Listeria monocytogenes* isolated from poultry products and poultry-processing environment by random amplification of polymorphic DNA and multilocus enzyme electrophoresis. *Appl Environ Microbiol*, **61**, 2139–2144.

Leclercq-Perlat M.-N., Lalande M. (1994). Cleanability in relation to surface chemical composition and surface finishing of some materials commonly used in food industries. *J Food Eng*, **23**, 501–517.

LeChevalier M.W., Cawthorn C.D., Lee R.G. (1988). Inactivation of biofilm bacteria. *Appl Environ Microbiol*, **44**, 972–987.

Lee, S-L., Frank J.F. (1991). Effect of growth temperature and media on inactivation of *Listeria monocytogenes* by chlorine. *J Food Saf*, **11**, 65–71.

Lemaître, J-P., Echannaoui H., Michaut G., Davies C., Rousset A. (1998). Plasmid-mediated resistance to antimicrobial agents among Listeriae. *J Food Prot*, **61**, 1459–1464.

Little B., Wagner P., Ray R., Pope R., Scheetz R. (1991). Biofilms: an ESEM evaluation of artefacts introduced during SEM preparation. *J Ind Microbiol*, **8**, 213–222.

Mafu A.A., Roy D., Goulet J., Hagney P. (1990). Attachment of *Listeria monocytogenes* to stainless steel, glass, polypropylene and rubber surfaces after short contact times. *J Food Prot*, **53**, 742–746.

Mattila-Sandholm T., Wirtanen G. (1992). Biofilm formation in the food industry: a review. *Food Rev Int*, **8**, 573–603.

Mead G.C., Scott M.J. (1994). Coagulase negative staphylococci and coliform bacteria associated with mechanical defeathering of poultry carcasses. *Lett Appl Microbiol*, **18**, 62–64.

Mettler E., Carpentier B. (1998). Variations over time of microbial load and physicochemical properties of floor materials after cleaning in food industry premises. *J Food Prot*, **61**, 57–65.

Michiels C.W., Schellekens M., Soontjens C.F., Hauben K.J.A. (1997). Molecular and metabolic typing of resident and transient flora from a meat mincer. *J Food Prot*, **60**, 1515–1519.

Mosteller T.M., Bishop J.R. (1993). Sanitizer efficacy against bacteria in a milk biofilm. *J Food Prot*, **56**, 34–41.

Nelson J.H. (1990). Where are *Listeria* likely to be found in dairy plants? *Dairy Food Environ San*, **10**, 344–345.

Notermans S., Dormans J.A.M.A., Mead G.C. (1991). Contributions of surface attachment to the establishment of microorganisms in food processing plants: a review. *Biofouling*, **5**, 1–16.

Pontefract R.D. (1991). Bacterial adherence: its consequesnces in food processing. *Can Inst Food Sci Technol J*, **24**, 113–117.

Ridgeway H.F., Olsen B.H. (1982). Chlorine resistance patterns of bacteria from two drinking water distribution systems. *Appl Environ Microbiol*, **44**, 972–987.

Ronner A.B., Wong A.C.L. (1993). Biofilm development and sanitizer inactivation of *Listeria monocytogenes* and *Salmonella typhimurium* on stainless steel and Buna-N rubber. *J Food Prot*, **56**, 750–758.

Rowe M.T., Dunstall G., Kirk R., Loughney C.F., Cooke J.L., Brown S.R. (1998). Development of an image analysis system for the study of viable but non-culturable forms of *Campylobacter jejuni* and its use to determine their resistance to disinfectants. *Food Microbiol*, **15**, 491–498.

Russell A.D. (1992). Bacterial sensitivity and resistance. In: Russell A.D., Hugo W.B., Ayliffe G.A.J. (eds) *Disinfection, Preservation and Sterilization*. Blackwell Scientific Publications, pp. 211–229.

Schwach T.S., Zottola E.A. (1984). Scanning electron microscopic study on some effects of sodium hypochlorite on attachment of bacteria to stainless steel. *J Food Prot*, **47**, 756–759.

Smoot L.M., Person M.D. (1998). Effect of environmental stress on the ability of *Listeria monocytogenes* Scott A to attach to food contact surfaces. *J Food Prot*, **61**, 1293–1298.

Song Y.S., Hargraves W.A. (1998). Postprocess contamination of flexible pouches challenged by *in situ* immersion biotest. *J Food Prot*, **61**, 1644–1648.

Sonak S., Bhosle N.B. (1995). A simple method to assess bacterial attachment to surfaces. *Biofouling*, **9**, 31–38.

Speers J.G.S., Gilmour A., Fraser T.W., McCall R.D. (1984). Scanning electron microscopy of dairy equipment surfaces contaminated by two milk-borne organisms. *J Appl Bacteriol*, **57**, 139–145.

Spurlock A.T., Zottola E.A. (1991). The survival of *Listeria monocytogenes* in aerosols. *J Food Prot*, **54**, 910–912.

Stanley P.M. (1983). Factors affecting the irreversible attachment of *Pseudomonas aeruginosa* to stainless steel. *Can J Microbiol*, **29**, 1493–1499.

Stern N.J., Kazmi S.U. (1989). *Campylobacter jejuni*. In: Doyle M.P. (ed) *Foodborne Bacterial Pathogens*. Marcel Dekker, New York, pp. 71–110.

Stone L.S, Zottola E.A. (1985). Effect of cleaning and sanitizers on the attachment of *Pseudomonas fragi* to stainless steel. *J Food Sci*, **50**, 951–956.

Sundheim G., Langsrud S., Heir E., Holck A.L. (1998). Bacterial resistance to disinfectants containing quaternary ammonium compounds. *Int Biodeterior Biodegrad*, **41**, 235–239.

Sutton N.A., Hughes N., Handley P.S. (1994). A comparison of conventional SEM techniques, low temperature SEM and the Electroscan wet scanning microscope to study the structure of a biofilm of *Streptococcus crista* CR3. *J Appl Bacteriol*, **74**, 448–454.

Taylor J.H., Rogers S.J., Holah J.T. (1999). A comparison of the bactericidal efficacy of 18 disinfectants used in the food industry against *Escherichia coli* 0157:H7 and *Pseudomonas aeruginosa* at 10°C and 20°C. *J Appl Microbiol*, **87**, 718–725.

Troller J.A. (1993). *Sanitiation in Food Processing*. Academic Press, New York, pp. 79–110.

Vanhaeke E., Remon, J-P., Moors M., Raes F., De Rudder D., Van Peteghem A. (1990). Kinetics of *Pseudomonas aeruginosa* adhesion to 304 and 316 stainless steel; role of cell surface hydrophobicity. *Appl Environ Microbiol*, **56**, 788–795.

Walker R.L., Jensen L.H. Kinde H., Alexander A.V., Owens L.S. (1991). Environmental survey for *Listeria* species in frozen milk product plants in California. *J Food Prot*, **54**, 178–182.

Willcock L., Holah J., Allison D.G., Gilbert P. (1997). Population dynamics in steady-state biofilms: effects of growth environment upon dispersal. In: Wimpenny J., Handley P., Gilbert P., Lappin-Scott H., Jones M. (eds) *Biofilms: Community Interactions and Control*. Bioline, Cardiff, pp. 23–31.

Wirtanen G., Mattila-Sandholm T. (1992). Effect of the growth phase of foodborne biofilms on their resistance to a chlorine sanitizer. *Part II. Lebensm-Wiss Technol*, **25**, 50–54.

Wirtanen G., Mattila-Sandholm T. (1993). Epifluorescence image analysis and cultivation of foodborne bacteria grown on stainless steel surfaces. *J Food Prot*, **56**, 678–683.

Wirtanen G., Mattila-Sandholm T. (1994). Measurement of biofilm of *Pediococcus pentosacceus* and *Pseudomonas fragi* on stainless steel surfaces. *Colloids Surf B: Biointerfaces*, **2**, 33–39.

Wirtanen G., Ahola H., Mattila-Sandholm T. (1995). Evaluation of cleaning procedures in elimination of biofilm from stainless steel surfaces in open process equipment. *Inst Chem Eng, Part C: Food and Bioproducts Processing*, **73**, 9–16.

Wirtanen, G, Alanko T., Mattila-Sandholm T. (1996a). Evaluation of epifluorescence image analysis of biofilm growth on stainless steel surfaces. *Colloids Surf B: Biointerfaces*, **5**, 319–326.

Wirtanen G., Husmark U., Mattila-Sandholm T. (1996b). Microbial evaluation of the biotransfer potential from surfaces with *Bacillus* biofilms after rinsing and cleaning procedures in closed food-processing systems. *J Food Prot*, **59**, 727–733.

Wirtanen G., Salo S., Allison D.G., Mattila-Sandholm T., Gilbert P. (1998). Performance-evaluation of disinfectant formulation using poloxamer-hydrogel biofilm constructs. *J Appl Microbiol*, **85**, 965–971.

Wong A.C.L., Cerf O. (1995). Biofilms: implications for hygiene monitoring of dairy surfaces. *Bull Int Dairy Fed*, **302**, 40–44.

Wright J.B., Ruseska I., Costerton J.W. (1991). Decreased biocide susceptibility of adherent *Legionella pneumophila*. *J Appl Microbiol*, **71**, 531–538.

Zoltai P.T., Zottola E.A., McKay L.L. (1981). Scanning electron microscopy of microbial attachment to milk contact surfaces. *J Food Prot*, **44**, 204–208.

Zottola E.A., Sasahara K.C. (1994). Microbial biofilms in the food processing environment — should they be a concern? *Int J Food Microbiol*, **23**, 125–148.

14 The Role of Biosurfactants in Affecting Initial Microbial Adhesion Mechanisms

C.G. van Hoogmoed, H.C. van der Mei and Henk J. Busscher

Biosurfactants are amphiphilic compounds released by several microbial strains and species. Whereas some strains release biosurfactants in readily detectable amounts, it is hypothesized that many strains and species may release biosurfactants in minute amounts, only detectable by axisymmetric drop shape analysis by profile (ADSA-P). Biosurfactants have a tendency to adsorb to interfaces and adsorbed biosurfactants will affect the physico-chemical properties of the interface and in turn microbial adhesion to the interface through influencing the Lifshitz-Van der Waals-, electrostatic and acid-base interaction energy. Biosurfactant release by microorganisms adhering in a biofilm will interfere with the adhesion of other organisms. Examples are described of biosurfactant release by *Streptococcusmitis*, *Streptococcus thermophilus* and *Lactobacillus acidophilus* negatively interfering with pathogen adhesion in the oral cavity, the oropharynx and the urogenital tract, respectively.

KEY WORDS: biosurfactants, biofilm, microbial adhesion, adhesion mechanism

INTRODUCTION

Biosurfactants are released by living cells, in the majority of cases microorganisms, and characteristically have a hydrophobic and a hydrophilic moiety within the same molecule. This amphiphatic character imparts special properties including adsorption at interfaces, most notably the air-water interface. Surface activity is most often associated with the ability of a biosurfactant to reduce the surface tension of water.

Several hypotheses have been put forward for a physiological role for microbial biosurfactants, a well-known example being emulsan produced by *Acinetobacter calcoaceticus* RAG1 for which a role as an anti-adhesive has been described. *A.calcoaceticus* adheres to oil-droplets and utilizes the long chain *n*-alkanes for its growth. Once all long chain *n*-alkanes in an oil-droplet are utilized, the organism starts producing emulsan. Adsorption of the released emulsan to the oil droplets decreases the hydrophobicity of the oil surface, thereby stimulating desorption of the adhering organisms. At the same time, the reduced hydrophobicity of the depleted oil droplet is a marker for other organisms not to attach (Rosenberg, 1986).

Table 1 Biosurfactants of microbial origin.

Microorganism	*Biosurfactant*	*Reference*
	Glycolipids	
Pseudomonas aeruginosa	Rhamnose lipid	Hisatsuka *et al.*, 1971
Torulopsis bombicola	Sophorose lipid	Cooper and Paddock, 1984
Rhodococcus erythropolis	Trehalose lipid	Rapp *et al.*, 1979
Candida antarctica	Mannosylerythritol	Kitamoto *et al.*, 1990
	Peptide lipids	
Bacillus licheniformis JF2	Lichenysin	Javaheri *et al.*, 1985
Bacillus subtilis	Surfactin	Arima *et al.*, 1968
Bacillus polymyxa	Polymyxins	Rosado and Selding, 1993
Pseudomonas fluorescens	Viscosin	Laycock *et al.*, 1991
Serratia marcescens	Serrawettin	Matsuyama and Nakagawa, 1996
	Amino lipids	
Agrobacterium tumefaciens	Lysin lipid	Tahara *et al.*, 1976
Pseudomonas rubescens	Ornithine lipid	Wilkinson, 1972
	Glycerol lipids	
Thiobacillus thiooxydans	Phospholipide	Beeba and Umbreit, 1971
	Polysaccharide-protein-lipid complex	
Acinetobacter calcoaceticus RAG-1	Emulsan	Rosenberg, 1986
	Polysaccharide-protein complex	
Acinetobacter radioresistens	Alasan	Navon-Venezia *et al.*, 1995
Candida lipolytica	Liposan	Cirigliano and Carmen, 1984

Important biosurfactant types and the microbial species releasing them are listed in Table 1. Biochemically, biosurfactants can be divided into: (i) glycolipids (ii) peptide lipids (iii) amino lipids (iv) phospholipids and fatty acids and (v) polymeric biosurfactants, mostly complexes of carbohydrates, proteins and sometimes lipids. The best documented biosurfactant types are the glycolipids, including rhamnolipids released by *Pseudomonas aeruginosa* (Hisatsuka *et al.*, 1971). One of the most efficient biosurfactants known is surfactin released by *Bacillus subtilis*. Surfactin is a cyclic peptide lipid and at a concentration of 0.005% is capable of lowering the surface tension of water to 27 mJ m^{-2} (Arima *et al.*, 1968).

The release of biosurfactants by microbial strains can be monitored by a variety of methods. The simplest method involves examination of potential foaming of the culture medium after growth. Also, more sophisticated methods for exact measurement of the surface tension of biosurfactant solutions exist, such as the Wilhelmy plate or du Nouy ring method, although it has been argued that these methods are less appropriate for surfactant solutions (Lunkenheimer and Wantke, 1981). Since the development of axisymmetric drop shape analysis by profile (ADSA-P) for the measurement of surface tensions of surfactant solutions, this technique has been applied to 100 µl droplets of microbial suspensions of various bacterial strains, previously unknown to release biosurfactants. Thermophilic dairy streptococci (Busscher *et al*,

1994b), *Streptococcus mitis* strains (Van der Vegt *et al.*, 1991) and lactobacilli (Velraeds *et al.*, 1996a), have been described as biosurfactant releasing strains, owing to the minute amounts of biosurfactants that can be detected with ADSA-P.

Biosurfactants adsorbed to interfaces affect the properties of the interfaces that govern adhesion phenomena, including microbial adhesion, as a step in biofilm formation. In a biofilm, the release of minute amounts of biosurfactants by an organism is sufficient to cover a surface area 1000× its own geometrical area. It therefore provides a powerful mechanism to control the adhesion of other strains and species, or even organisms of its own strain, adhering in its vicinity. In this chapter, following a short description of the physico-chemical mechanisms applied by microorganisms to adhere to substratum surfaces, some hypotheses on how adsorption of microbially-released biosurfactants may influence the adhesive properties of substratum surfaces will be put forward. Finally, three examples will be given of an established role for biosurfactants in initial microbial adhesion as a step in biofilm formation in the oral cavity (*S. mitis* strains *versus Streptococcus mutans* adhesion), the oropharynx (*Streptococcus thermophilus versus* yeast adhesion) and the urogenital tract (lactobacilli *versus* uropathogen adhesion).

BIOFILM FORMATION

Microorganisms in aqueous environments have a tendency to adhere to surfaces, such as a rock in a stream, biomaterial implants or tooth surfaces in the human body, oil- and water pipelines and industrial surfaces such as heat exchangers. A biofilm is a complex microbial community, embedded in a matrix of polymers, adhering to a surface. The development of a biofilm is described as a sequence of events (Escher and Characklis, 1990; Van Loosdrecht *et al.*, 1990):

1 Conditioning of the substratum surface
When microorganisms and substratum surfaces are in an aqueous environment with dissolved organic molecules and/or microbial products, the microorganisms and organic matter are transported towards the substratum surface. However, before the first microorganisms arrive at the surface and adhere, the substratum surface is covered with a layer of organic molecules, known as "the conditioning film" (Gristina, 1987; Escher and Characklis, 1990).

2 Transport of microorganisms to a substratum
Depending on the system under consideration, microorganisms can reach the substratum surface by sedimentation, Brownian motion, diffusion, convection or chemotaxis. Transport can involve individual microorganisms, or coaggregates formed in suspension.

3 Initial adhesion
Once a microorganism is within the range of the interaction forces, reversible nonspecific initial adhesion occurs, followed by more irreversible, specific adhesion. Also co-adhesion, strong attractive interactions between already adhering, sessile and still planktonic microorganisms, may contribute to the formation of a biofilm at this stage (Lamont and Rosan, 1990). Reversible, nonspecific adhesion

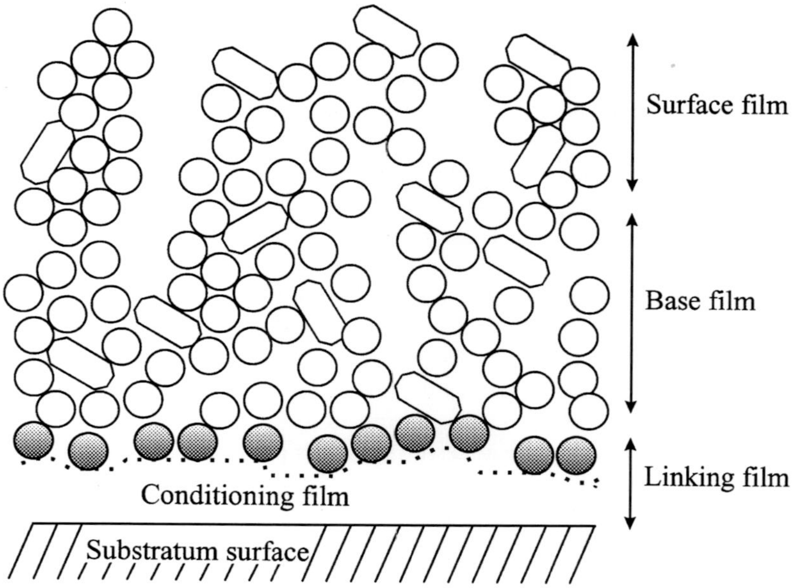

Figure 1 Biofilm model in which a surface –, base – and linking film are distinguished. The microorganisms that adhere initially (shaded circles) create a link between the substratum surface and the other organisms, constituting the biofilm (circles and rods).

occurs at large and intermediate separation distances (between 10 and 20 nm) and is mediated by Lifshitz-Van der Waals, acid-base and electrostatic interactions between the substratum surface and the microorganisms. Irreversible specific adhesion occurs at small separation distances (< 1.5 nm) and depends on adhesin-receptor interactions as well as on microbial carbohydrate polymer synthesis, as an expression of localized, mutually attractive physico-chemical interactions. The conditioning layer in particular, if present, may be involved in this type of interaction (Costerton *et al.*, 1987).

4 Growth
Finally, adhering microorganisms start to grow and colonize the substratum surface. Growth is the major factor contributing to the development of a mature biofilm, but continued adhesion to the biofilm surface needs to occur at all times, as detachment and parts of a biofilm are regularly sloughed off.

The initially adhering microorganisms have a special role in biofilm formation, although their relative number is low. During fluctuating shear and detachment forces, these organisms provide the contact with the substratum surface and must have sufficiently strong interaction in order to keep the entire biofilm in place. This pivotal role of the initial colonizers is emphasized in a new biofilm model (Busscher *et al.*, 1995), in which a linking film, base film and surface film are distinguished (see Figure 1).

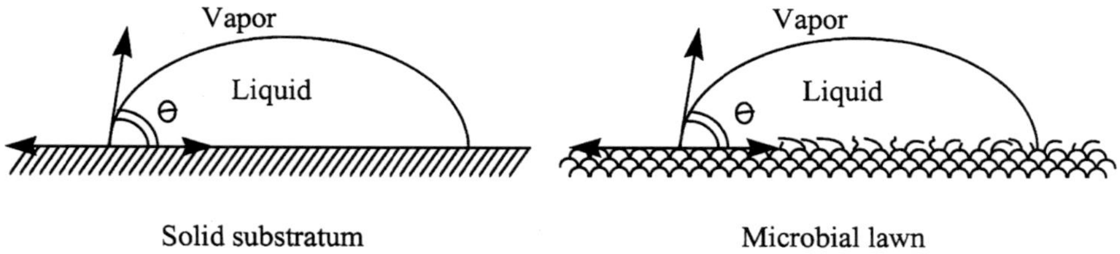

Figure 2 Contact angle measurements for a smooth and homogeneous solid surface (left) and for more heterogeneous bacterial cell surfaces (right), the latter presenting a "fuzzy coat" of collapsed structures.

INITIAL ADHESION: A PHYSICO-CHEMICAL APPROACH

Initial adhesion of microorganisms to surfaces can be described physico-chemically by a thermodynamic approach based solely on contact angles, and by the so-called extended DLVO-approach, including Lifshitz-Van der Waals, electrostatic and acid-base interactions. The two approaches will be briefly described.

The Thermodynamic Approach

In the thermodynamic approach, adhesion is described as the formation of one new interface between the substratum surface and adhering bacteria at the expense of the interfaces between the bacteria and the suspending liquid and the substratum-liquid interface. Each of the above interfaces contains an amount of interfacial free energy and provided adhesion is performed at constant pressure and temperature and if the molecular composition of the interface does not change, *e.g.* by adsorption of biosurfactants released by the microorganisms involved, the interfacial free energy of adhesion can be expressed as (Absolom *et al.*, 1983; Busscher *et al*, 1984)

$$\Delta G_{adh} = \gamma_{ms} - \gamma_{sl} - \gamma_{ml} \quad (1)$$

in which γ_{ms}, γ_{sl}, and γ_{ml} are the microorganism-solid, solid-liquid and microorganism-liquid interfacial free energies, respectively. The interfacial free energies occurring in Eqn. (1) can be obtained from contact angle measurements using Young's equation

$$\gamma_{sv} = \gamma_{sl} + \gamma_{lv} \cos\theta \quad (2a)$$

This equation relates the experimental assessable liquid-vapour interfacial free energy (γ_{lv}) and contact angle (θ) in the contact angle equilibrium on solid substrata (see Figure 2) with the solid-vapour (γ_{sv}) and the solid-liquid (γ_{sl}) interfacial free energies. Contact angles with liquids can not only be measured on solid surfaces but also on microbial lawns after they have been dried, and the contact angle reflects the "fuzzy coat" of collapsed microbial surface structures (see also Figure 2). In this case, Eqn. (2a) has to be replaced by

$$\gamma_{mv} = \gamma_{ml} + \gamma_{lv} \cos\theta \tag{2b}$$

Ultimately the interfacial free energies γ_{ij} can be evaluated from the Young equation by a variety of methods that either consider surface free energy as a single component thermodynamic quantity (the Neumann equation of state, Neumann *et al.*, 1974), or divide it into a dispersion and polar component (Owens and Wendt, 1969). Nowadays separating surface free energies in Lifshitz-Van der Waals and acid-base components, that consists in turn of an electron-accepting and electron-donating parameter (Van Oss *et al.*, 1988a) is gaining popularity, but it requires contact angle measurements with at least three liquids, while the equation of state (Neumann *et al.*, 1974) can be solved by measuring the contact angle with only one liquid.

In a thermodynamic approach to microbial adhesion it is assumed that the measured contact angles on bacterial lawns and solid substrata reflect the contact angles of the microbial and substratum area that are in contact. Considering the structural and chemical heterogeneity of microorganisms and (solid) surfaces, this may not necessarily be the case. In particular, the interpretation of contact angle data for more heterogeneous microbial surfaces is difficult (see Figure 2).

A justified application of interfacial thermodynamics requires reversible microbial adhesion. Often this aspect is neglected while in other studies reversibility is demonstrated by the passage of an air-liquid interface over adhering organisms (Busscher *et al.*, 1992; Pitt *et al.*, 1993), changing the microbial concentration in suspension (Busscher *et al.*, 1986) or altering the ionic strength of the suspending liquid (Rijnaarts *et al.*, 1995). Meinders *et al.* (1995) showed reversibility by measuring a small desorption of adhering microorganisms during deposition *in situ*.

The Extended DLVO-Approach

The DLVO-approach, named after Derjaguin, Landau, Verwey and Overbeek, describes adhesion as a result of attractive Lifshitz-Van der Waals, mostly repulsive electrostatic (Rutter and Vincent, 1980) and acid-base interactions (Van Oss *et al.*, 1988b) according to

$$\Delta G_T = \Delta G_{LW} + \Delta G_E + \Delta G_{AB} \tag{4}$$

in which ΔG_T, ΔG_{LW}, ΔG_E and ΔG_{AB} denote the total, the Lifshitz-Van der Waals, the electrostatic and the acid-base interaction energies, respectively. ΔG_{LW}, ΔG_E and ΔG_{AB} can be calculated from:

$$\Delta G_{LW} = \frac{-A}{6}\left\{\frac{a}{d} + \frac{a}{d+2a} + \ln\left[\frac{d}{d+2a}\right]\right\} \tag{5}$$

$$\Delta G_E = \frac{\epsilon a(\zeta_1^2+\zeta_2^2)}{4}\left\{\frac{2\zeta_1\zeta_2}{\zeta_1^2+\zeta_2^2}\ln\frac{1+\exp(-\kappa d)}{1-\exp(-\kappa d)}+\ln\{1-\exp(-2\kappa d)\}\right\} \quad (6)$$

$$\Delta G_{AB} = 2\pi a\lambda\Delta G_{mls}^{AB}\exp\left(\frac{l_0-d}{\lambda}\right) \quad (7)$$

in which A is the Hamaker constant, a the particle radius, d the interaction distance, ϵ the dielectric constant of the medium, ζ_1 and ζ_2 the zeta potentials of the substratum and the particle, respectively, κ the reciprocal Debye-Hückel length (Hiemenz, 1977), l_0 is the minium equilibrium distance ($\approx$0.157 nm) and λ the decay length ($\approx$0.6 nm) (Van Oss, 1995). ΔG_{mls}^{AB} is the acid-base interaction energy between a microorganism (m) and a solid surface (s) immersed in water (l) as follows from interfacial thermodynamics (see Eqn.1). The Hamaker constant can also be calculated from interfacial thermodynamics using

$$\Delta G_{LW,mls} = \frac{-A}{12\pi l_0^2} \quad (8)$$

Hamaker constants, however, can also be approximated from *ab initio* principles (Nir, 1976).

Microorganisms as well as most biological surfaces to which they adhere usually have a net negative charge under physiological conditions, leading to repulsive electrostatic interactions. Interaction energy calculated for the interaction of moderately hydrophilic bacteria (θ_{water} = 60°) with glass at low and high ionic strengths is shown in Figure 3. Under the influence of repulsive electrostatic interactions, and neglecting the acid-base interactions, the total interaction energy demonstrates a maximum at close approach that constitutes a barrier for adhesion in the primary interaction minimum (separation distances 1–1.5 nm). This barrier is suppressed by increasing the ionic strength of the medium, due to a decrease in the range and magnitude of the electrostatic interactions and at high ionic strengths the interaction barrier is absent and bacteria will adhere directly in the primary interaction minimum. The total interaction energy can also show a more shallow secondary minimum at larger separation distances (10–20 nm) at low and intermediate ionic strengths. It is thought that microorganisms adhere reversibly in the secondary minimum as its depth is only several kT-units and that irreversibly binding in the primary minimum will occur if the microorganisms have passed the energy barrier. In the case of oppositely charged microbial and substratum surfaces, ΔG_E and thus ΔG_T are negative at all separation distances. This results in direct primary minimum adhesion.

The acid-base interactions, when taken into account, are repulsive between hydrophilic glass and hydrophilic bacteria (see also Figure 3) and decay rapidly as a function of the separation distance. Consequently, the extended DLVO-approach can only account for secondary minimum interactions.

In the DLVO-approach it is assumed that microorganisms are smooth and homogeneous particles. Specific interactions between stereochemically complementary surface

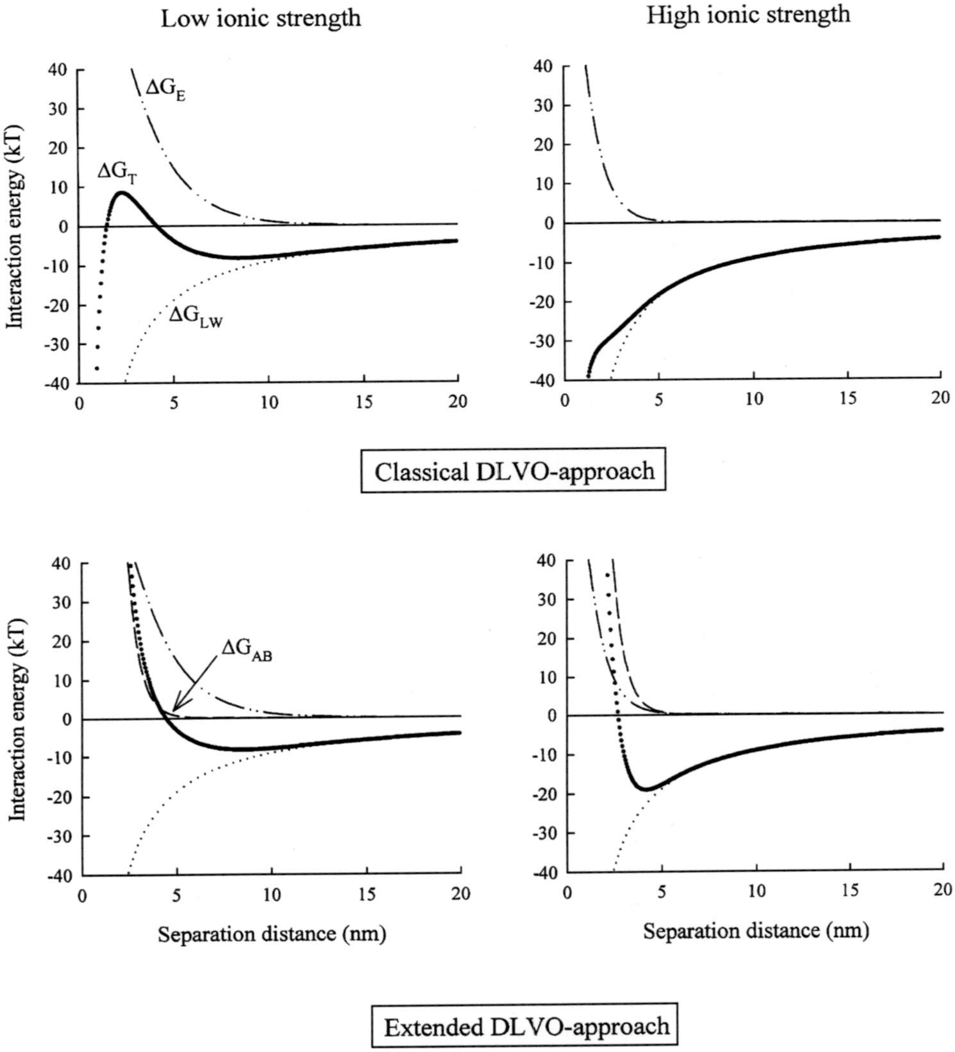

Figure 3 Interaction energies for glass and a hydrophilic microorganism (water contact angle 60°), according to the classical DLVO-approach (top) and the extended DLVO-approach (bottom). The interaction energies have been calculated for zeta potentials with equal sign at low (left) and high (right) ionic strengths.

components at short separation distances (< 1 nm), which are also active in determining the strength of adhesion in the primary minimum, are not incorporated in the DLVO-approach and the influence of microscopic, structural details of the cell surfaces is neglected.

THE INFLUENCE OF BIOSURFACTANTS ON ADHESION MECHANISMS

The interactions of biosurfactants with interfaces depend on the molecular structure of the hydrophobic and hydrophilic part and on the charge of the hydrophilic moiety, which may be anionic, cationic, nonionic or amphoteric. The charge of ionic biosurfactants is in turn influenced by the ionic conditions and pH of the surrounding liquid. Once an interface is conditioned with an adsorbed biosurfactant layer, its properties will be affected, thereby influencing adhesion of microorganisms, according to the extended DLVO-approach, in three distinguishable ways through (i) Lifshitz-Van der Waals interaction energy; the more hydrophobic a surface becomes the larger the Hamaker constant will be, resulting in a more negative total interaction energy and *vice versa*; (ii) electrostatic interaction; if the charge of the biosurfactant differs from the one of the bare interface, the electrostatic interaction will change; and (iii) acid-base interaction; a change in the hydrophobicity of an interface is accompanied by a change in wettability and the acid-base interactions will vary correspondingly.

The influence of a layer of adsorbed biosurfactant on the adhesion process becomes even more complicated when the biosurfactant is a polymer and polymer bridging or steric repulsion occurs (Rijnaarts *et al.*, 1995). Figure 4 schematically presents some of the modes of biosurfactant adsorption that may influence microbial adhesion. Biosurfactant adsorption can inverse the hydrophobicity of a substratum surface from hydrophobic to hydrophilic and *vice versa* and consequently change favourable conditions for microbial adhesion into unfavourable ones. Biosurfactant adsorption does not necessarily have to occur from solution, but the releasing organism itself can be adhering to the interface and release its biosurfactant from that position to alter the interface (see also Figure 4).

BIOSURFACTANT INTERFERENCE WITH INITIAL 'PATHOGEN' ADHESION

Three examples of biosurfactant-releasing microorganisms and their influence on pathogen adhesion in the oral cavity, the oropharynx and the urogenital tract are discussed. In the three examples, microbial biosurfactant release was established by ADSA-P.

The Oral Cavity

Pratt-Terpstra *et al.* (1989) pointed out that the oral bacterium *S. mitis* BMS releases substances which decrease the adhesion of *S. mutans* NS to glass (see Figure 5). Since *S. mutans* is an important etiological agent of coronal- and root surface caries (Syed *et al.*, 1975) this observation was of considerable interest in the context of the prevention of dental caries. Later these substances were recognised as biosurfactants because *S. mitis* BMS suspensions are able to decrease the surface tension by 26 mJ m^{-2} (Van der Vegt *et al.*, 1991). Later experiments showed that a non-biosurfactant

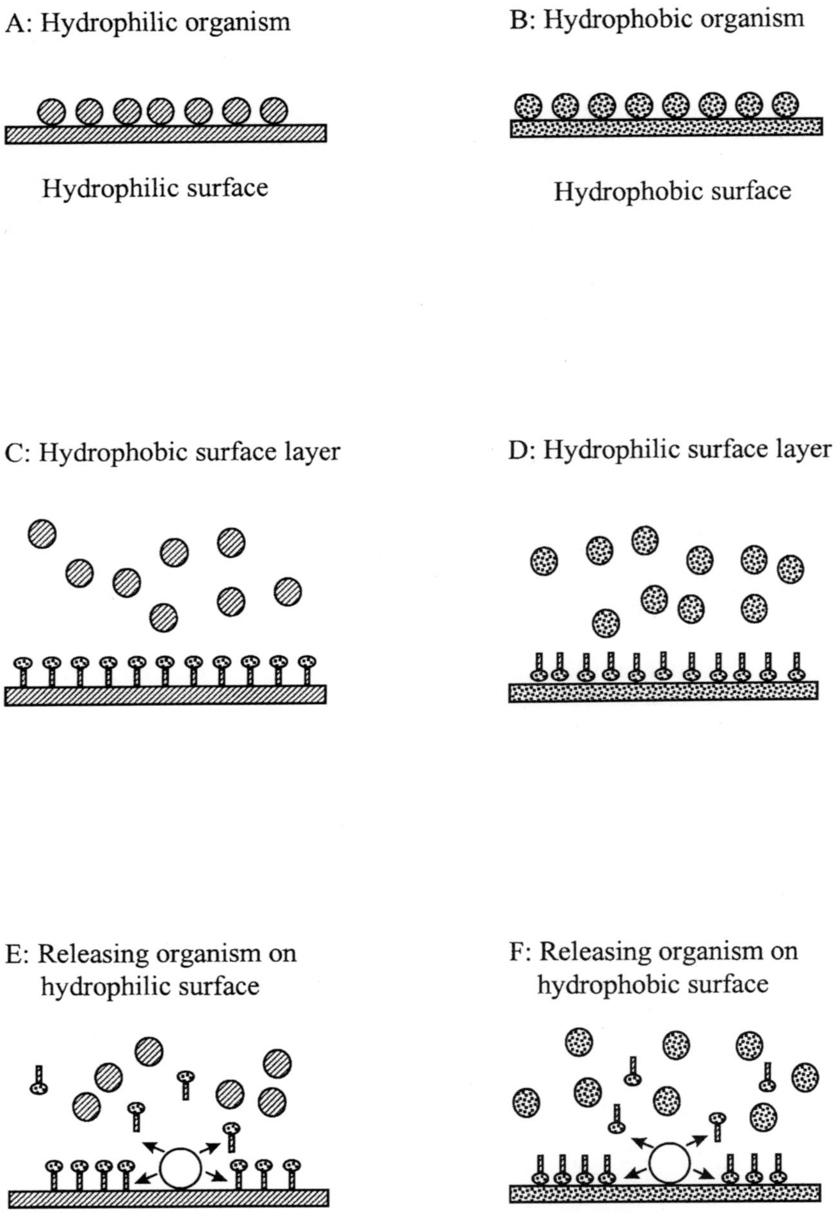

Figure 4 Adhesion of hydrophilic (hatched circles) and hydrophobic (dotted circles) microorganisms to a hydrophilic (hatched) and hydrophobic (dotted) interface (A and B) respectively. C and D = the formation of a biosurfactant conditioning film on the interface (in the absence of releasing organisms) inversing the hydrophobicity/hydrophilicity of the interface with an influence on microbial adhesion. E and F = the releasing microorganisms adhering to the interface. The hydrophobic and hydrophilic part of the biosurfactant are indicated as dotted and hatched areas respectively.

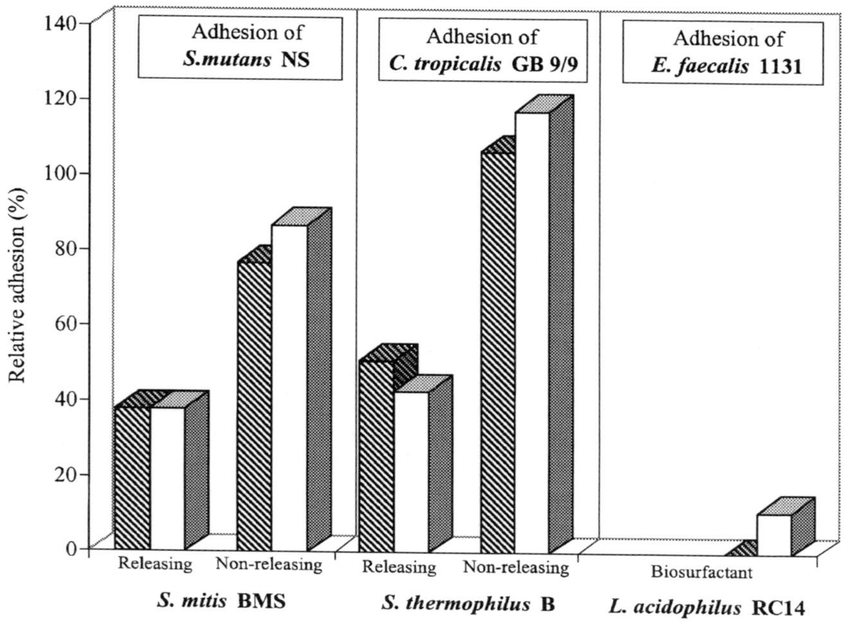

Figure 5 The relative initial deposition rate (hatched area) and the number adhering after 4 h (white area) of various pathogens in a parallel plate flow chamber to glass (*S. mutans* NS) and to silicone rubber (*C. tropicalis* GB 9/9 and *E. faecalis* 1131) in the presence of adhering biosurfactant-releasing bacteria and non-releasing variants at a surface coverage of 4% (Busscher *et al.*, 1997; Velraeds *et al.*, 1997). Pathogen adhesion is expressed relative to their adhesion to the substratum in the absence of other bacteria. Note that for *E. faecalis* 1131 adhesion only adsorbed biosurfactants (0.1 mg ml^{-1}) released by *L. acidophilus* RC14 are applied as a coating.

releasing *S. mitis* BMS variant lacked the ability to decrease the *S. mutans* NS adhesion (see also Figure 5).

The Oropharynx

Silicone rubber voice prostheses are often used for voice rehabilitation in laryngectomized patients after an oesophagal cancer. However, the silicone rubber becomes rapidly colonized with yeasts and bacterial strains resulting in a thick biofilm (Mahieu *et al.*, 1986) that impedes proper functioning of the prosthesis. The yeast adhesion is particularly troublesome, because attachment is accompanied by growth into the silicone rubber (Busscher *et al.*, 1994a). Patients reported that consumption of certain dairy products such as Turkish yoghurt prolonged the life time of voice prostheses. It was suggested that this effect was due to biosurfactant releasing *S. thermophilus* in yoghurt (Busscher *et al.*, 1997). At a low surface coverage (4%) on silicone rubber, biosurfactant-releasing *S. thermophilus* B resulted in both the initial deposition rate and the number of adhering *Candida tropicalis* after 4 h being significantly reduced (see Figure 5), regardless of the presence of a salivary conditioning film. Non-releasing *S. thermophilus* B variants were unable to decrease yeast adhesion (Busscher *et al.*, 1997).

The Urogenital Tract

Lactobacilli are the dominating microorganisms in the urogenital tract of healthy females (Pfau and Sacks, 1981). Lactobacilli are thought to play an important role in the defence against urinary tract infections by interference mechanisms such as competitive exclusion/displacement, coaggregation, production of lactic acid, hydrogen peroxide, bacteriocins and bacteriocin like substances (McGroarty and Reid, 1988; Reid and Tieszer, 1993) and the release of biosurfactants.

Velraeds *et al.* (1996a) showed that *Lactobacillus* strains release biosurfactants which are able to decrease the surface tension of water by 20–30 mJ m^{-2}. Some strains, including *Lactobacillus acidophilus* RC 14, release biosurfactants with a predominantly proteinaceous character (Velraeds *et al.*, 1996b). Adhesion of the uropathogen *Enterococcus faecalis* 1131, suspended in urine, to silicone rubber with an adsorbed *L.acidophilus* RC14 biosurfactant layer was reduced (see Figure 5), with a clear dose dependence (Velraeds *et al.*, 1997). Also adhesion of other uropathogens suspended in urine, such as other *E. faecalis* isolates, *Escherichia coli* and *Staphylococcus epidermis* species to silicone rubber, coated with biosurfactants released by lactobacilli, was inhibited (Velraeds *et al.*, 1998).

CONCLUDING REMARKS

In this chapter, a role for microbially released biosurfactants in inhibiting pathogen adhesion to surfaces is presented, based on experimental evidence. Adsorption of biosurfactants to surfaces can greatly alter the physico-chemical properties of an interface due to the amphiphilic nature of surfactants in general, and thereby the potential adhesion of microoorganisms. The inhibitory effect on pathogen adhesion by microbial biosurfactants has been described here for three different niches in the human body, and it is suggested that biosurfactant release is one of the means by which probiotic bacteria (Sanders, 1995) may exert their beneficial effects.

REFERENCES

Absolom D.R., Lamberti F.L., Policova Z., Zingg W., Van Oss C.J., Neumann A.W. (1983). Surface thermodynamics of bacterial adhesion. *Appl Environ Microbiol*, **46**, 90–97.

Arima K., Kakinuma A., Tamura G. (1968). Surfactin, a crystalline peptidelipid surfactant produced by *Bacillus subtilis*: isolation, characterization and its inhibition of fibrin clot formation. *Biochem Biophys Res Com*, **31**, 488–494.

Beeba J.L., Umbreit W.W. (1971). Extracellular lipid of *Thiobacillus thiooxidans*. *J Bacteriol*, **108**, 612–614.

Busscher H.J., Doornbusch G.I., Van der Mei H.C. (1992). Adhesion of mutans streptococci to glass with and without a salivary coating as studied in a parallel plate flow chamber. *J Dent Res*, **71**, 491–500.

Busscher H.J., Neu T.R., Van der Mei H.C. (1994b). Biosurfactant production by thermophilic dairy streptococci. *Appl Microbiol Biotechnol*, **41**, 4–7.
Busscher H.J., Bos R., Van der Mei H.C. (1995). Initial microbial adhesion is a determinant for the strength of biofilm adhesion. *FEMS Microbiol Lett*, **128**, 229–234.
Busscher H.J., Uyen M.H.W.J.C., Weerkamp A.H., Postma W.J., Arends J. (1986). Reversibility of adhesion of oral streptococci to solids. *FEMS Microbiol Lett*, **35**, 303–306.
Busscher H.J., Van Hoogmoed C.G., Geertsema-Doornbusch G.I., Van der Kuijl-Booij M., Van der Mei H.C. (1997). *Streptococcus thermophilus* and its biosurfactants inhibit adhesion by *Candida* spp. on silicone rubber. *Appl Environ Microbiol*, **63**, 3810–3817.
Busscher H.J., Weerkamp A.H., Van der Mei H.C., Van Pelt A.W.J., de Jong H.P., Arends J. (1984). Measurement of the surface free energy of bacterial cell surfaces and its relevance for adhesion. *Appl Environ Microbiol*, **48**, 980–983.
Busscher H.J., De Boer C.E., Verkerke G.J., Kalicharan R., Schutte H.K., Van der Mei H.C. (1994a). *In vitro* ingrowth of yeasts into medical grade silicone rubber. *Int Biodeter Biodegrad*, **33**, 383–390.
Cirigliano M.C., Carmen G.B. (1984). Isolation of a bioemulsifier from *Candida lipolytica. Appl Environ Microbiol*, **48**, 747–750.
Cooper D.G., Paddock D.A. (1984). Production of a biosurfactant from *Torulopsis bombicola. Appl Environ Microbiol*, **47**, 173–176.
Costerton J.W., Cheng K.J., Geesey G.G., Ladd T., Nickel J.C., Dasgupta M., Marrie T.J. (1987). Bacterial biofilms in nature and disease. *Annu Rev Microbiol*, **41**, 435–464.
Escher A., Characklis W.G. (1990). Modelling the initial events in biofilm accumulation. In: Characklis W.G., Marshall K.C. (eds) *Biofilms*. John Wiley & Sons Incorporated, New York, pp. 445–487.
Gristina A.G. (1987). Biomaterials-centered infection: microbial *vs* tissue integration. *Science*, **237**, 1588–1597.
Hiemenz P.C. (1977). Electrophoresis and other electrokinetic phenomena. In: Lagowski J.J. (ed) *Principles of Colloid and Surface Chemistry*. Marcel Dekker Incorporated, New York, pp. 452–487.
Hisatsuka K., Nakahara T., Sano N., Yamada K. (1971). Formation of rhamnolipid by *Pseudomonas aeruginosa* and its function in hydrocarbon fermentation. *Agric Biol Chem*, **33**, 686–692.
Javaheri M., Jenneman G.E., McInerney M.J., Knapp R.M. (1985). Anaerobic production of a biosurfactant by *Bacillus licheniformis* JF-2. *Appl Environ Microbiol*, **50**, 698–700.
Kitamoto D., Akiba S., Hioki C., Tabuchi T. (1990). Extracellular accumulation of mannosylerythritol lipids by a strain of *Candida antarctica. Agric Biol Chem*, **54**, 31–36.
Lamont R.J., Rosan B. (1990). Adherence of mutans streptococci to other oral bacteria. *Infect Immun*, **58**, 1738–1743.
Laycock M.V., Hildebrand P.D., Thibault P., Walter J.A., Wright J.L.C. (1991). Viscosin, a potent peptidolipid biosurfactant and phytopathogenic mediator produced by a pectolytic strain of *Pseudomonas fluorescens. J Agric Food Chem*, **39**, 483–489.
Lunkenheimer K., Wantke K.-D. (1981). Determination of the surface tension of surfactant solutions applying the method of Lecomte du Nouy (ring tensiometry). *Colloid Polym Sci*, **259**, 354–366.
Mahieu H.F., Van Saene J.J.M., Den Besten J., Van Saene H.K.F. (1986). Oropharynx decontamination preventing *Candida* vegetation on voice prostheses. *Arch Otolaryngol Head Neck Surg*, **112**, 321–325.
Matsuyama T., Nakagawa Y. (1996). Bacterial wetting agents working in colonization of bacteria on surface environments. *Colloids Surf B: Biointerfaces*, **7**, 207–214.
McGroarty J.A., Reid G. (1988). Detection of a *Lactobacillus* substance that inhibits *Escherichia coli. Can J Microbiol*, **34**, 974–978.

Meinders J.M., Van der Mei H.C., Busscher H.J. (1995). Deposition efficiency and reversibility of bacterial adhesion under flow. *J Colloid Interface Sci*, **176**, 329–341.

Navon-Venezia S., Zosim Z., Gottlieb A., Legmann R., Carmeli S., Ron E.Z., Rosenberg E. (1995). Alasan, a new bioemulsifier from *Acinetobacter radioresistens. Appl Environ Microbiol*, **61**, 3240–3244.

Neumann A.W., Good R.J., Hope C.J., Sejpal M. (1974). An equation-of-state approach to determine surface tensions of low-energy solids from contact angles. *J Colloid Interface Sci*, **49**, 291–304.

Nir S. (1976). Van der Waals interactions between surfaces of biological interest. *Prof Surf Sci*, **8**, 1–58.

Owens D.K., Wendt R.C. (1969). Estimation of the surface free energy of solids. *J Appl Polymer Sci*, **13**, 1741–1747.

Pfau A., Sacks T. (1981). The bacterial flora of the vaginal vestibule, urethra, and vagina in premenopausal women with recurrent urinary tract infections. *J Clin Microbiol*, **126**, 630–634.

Pitt W.G., Mcbride M.O., Barton A.J., Sagers R.D. (1993). Air-water interface displaced adsorbed bacteria. *Biomaterials*, **14**, 605–608.

Pratt-Terpstra I.H., Weerkamp A.H., Busscher H.J. (1989). Microbial factors in a thermodynamic approach of oral streptococcal adhesion. *J Colloid Interface Sci*, **129**, 568–574.

Rapp P., Bock H., Wray V., Wagner F. (1979) Formation, isolation and characterization of trehalose dimycolates from *Rhodococcus erythropolis* grown on n-alkane. *J Gen Microbiol*, **115**, 491–503.

Reid G., Tieszer C. (1993). Preferential adhesion of urethral bacteria from a mixed population to a urinary catheter. *Cells Mater*, **3**, 171–176.

Rijnaarts H.H.M., Norde W., Bouwer E.J., Lyklema J., Zehnder A.J.B. (1995). Reversibility and mechanism of bacterial adhesion. *Colloids Surf B: Biointerfaces*, **4**, 5–22.

Rosado A.S., Seldin L. (1993). Production of a potentially novel antimicrobial substance by *Bacillus polymyxa. World J Microbiol Biotechnol*, **9**, 521–528.

Rosenberg E. (1986). Microbial surfactants. *CRC Crit Rev Biotechnol*, **3**, 109–131.

Rutter P.R., Vincent B. (1980). The adhesion of microorganisms to surfaces: physico-chemical aspects. In: Berkeley R.C.W., Lynch J.W., Melling J., Rutter P.R., Vincent B. (eds) *Microbial Adhesion to Surfaces*. Ellis Horwood, Chichester, pp. 79–93.

Sanders M.E. (1995). Lactic acid bacteria and human health. In: Fuller R., Heidt P.J., Rusch V., Van der Waay D. (eds) *Probiotics: Prospects of use in Opportunistic Infections*. Institute for Microbiology and Biochemistry, Herborn-Dill, Germany, pp. 126–140.

Syed S.A., Loesche W.J., Pape H.L. Jr., Grenier E. (1975). Predominant cultivable flora isolated from human root surface caries plaque. *Infect Immun*, **11**, 727–731.

Tahara Y., Yamada Y., Kondo K. (1976). A new lipid: the ornithine and taurine-containing "Cerilipin". *Agric Biol Chem*, **40**, 243–244.

Van der Vegt W., Van der Mei H.C., Noordmans J., Busscher H.J. (1991). Assessment of bacterial biosurfactant production through axisymmetric drop shape analysis by profile. *Appl Microbiol Biotechnol*, **35**, 766–770.

Van Loosdrecht M.C.M., Lyklema J., Norde W., Zehnder A.J.B. (1990). Influences of interfaces on microbial activity. *Microbiol Rev*, **54**, 75–87.

Van Oss C.J. (1995). Hydrophobicity of biosurfaces-origin, quantitative determination and interaction energies. *Colloids Surf B: Biointerfaces*, **5**, 91–110.

Van Oss C.J., Chaudhury M.K., Good R.J. (1988b) Interfacial Lifshitz-Van der Waals and polar interactions in macroscopic systems. *Chem Rev*, **88**, 927–941.

Van Oss C.J., Good R.J., Chaudhury M.K. (1988a). Additive and nonadditive surface tension components and the interpretation of contact angles. *Langmuir*, **4**, 884–891.

Velraeds M.M.C., Van der Mei H.C., Reid G., Busscher H.J. (1996a). Inhibition of initial adhesion of uropathogenic *Enterococcus faecalis* by biosurfactants from *Lactobacillus* isolates. *Appl Environ Microbiol*, 62, 1958–1963.

Velraeds M.M.C., Van der Mei H.C., Reid G., Busscher H.J. (1996b). Physico-chemical and biochemical characterization of biosurfactants released by *Lactobacillus* strains. *Colloids Surf B: Biointerfaces*, 8, 51–61.

Velraeds M.M.C., Van der Mei H.C., Reid G., Busscher H.J. (1997). Inhibition of initial adhesion of uropathogenic *Enterococcus faecalis* to solid substrata by an adsorbed biosurfactant layer from *Lactobacillus acidophilus*. *Urology*, 49, 790–794.

Velraeds M.M.C., Van de Belt-Gritter B., Van der Mei H.C., Reid G., Busscher H.J. (1998). Interference in initial adhesion of uropathogenic bacteria and yeasts to silicone rubber by *Lactobacillus acidophilus* biosurfactant. *J Med Microbiol*, 47, 1081–1085.

Wilkinson S. (1972). Composition and structure of the ornithine-containing lipid from *Pseudomonas rubescens*. *Biochim Biophys Acta*, 27, 1–7.

15 Monitoring Biofilms by Fourier Transform Infrared Spectroscopy

Gill G. Geesey and Peter A. Suci

The use of infrared spectroscopy (IR) for chemical characterization of microbial biofilms developed from the success of this spectroscopic technique in detection and identification of the chemical constituents of organic conditioning films that formed upon submersion of a clean solid surface in aqueous media. Infrared spectra, collected in the attenuated total reflectance (ATR) mode using a Fourier transform infrared spectrometer (FT-IR), contain absorption bands that are contributed by the chemical constituents of surface-associated microorganisms. The unique chemistry of different bacterial species has been resolved by FT-IR, to the extent that the approach is used to validate other methods of bacterial identification. Both diffuse reflectance and ATR modes have been used to follow biofilm development under a variety of aqueous conditions. ATR/FT-IR offers the opportunity to follow bacterial attachment to and biofilm development on a variety of solid surface materials non-destructively in real-time. This approach has been particularly fruitful in assessing the contribution and identification of mechanisms of biologically influenced corrosion of metal surfaces. By combining ATR/FT-IR with microscopic techniques, it is now possible to relate biofilm structure to chemical reactions in developing biofilms.

KEY WORDS: biofouling, corrosion, attenuated total reflectance, surface chemistry

INTRODUCTION

Microbial biofilms have become an important biological component, intentionally or unintentionally, in many industrial water systems. The accumulation of microbial biomass on equipment surfaces in contact with aqueous media poses a variety of problems for water system operators ranging from under-deposit corrosion, to loss in heat transfer efficiency or hydraulic valve malfunction. Historically, industries have had to take a biofouled system off-line for periods of time either to mechanically remove the fouling layer or to replace the equipment when damaged beyond repair. Frequently, chemical biocides are used to "control" biofilm accumulation on industrial surfaces, but more often than not, a biofouling layer eventually appears, and if the layer cannot be subsequently removed by application of alternative biocides, the system must be taken off-line and mechanically cleaned (Strauss, 1985).

Monitoring techniques for microbiological growth have been developed primarily for measurements in the bulk aqueous phase. Since biofouling is a surface-localized phenomenon, conventional monitoring techniques are typically insensitive to most biofilm processes. A few on-line monitoring devices have been developed to measure surface biofouling which are based on fluid flow resistance or heat flux (Johnson and Howells 1981). Bryers and Characklis (1981) developed mathematical models based on system engineering parameters to predict biofouling and biofilm accumulation. Considerable biofilm development is required, however, before these devices detect surface fouling. By the time these devices sense that biofouling has occurred, the system may have become sufficiently fouled that subsequent remedial treatments are ineffective (Atkinson, 1979). Implementation of a sensitive and accurate biofilm monitoring system, on the other hand, permits operators to adjust control parameters to achieve long-term surface protection against biofilm accumulation. Accurate and sensitive monitoring facilitates prediction of when the system will deteriorate to an unacceptable level, so that corrective procedures or a scheduled maintenance activity can be effectively implemented.

Many industries faced with biofouling of equipment surfaces have monitored biofilm accumulation using surrogate coupons that are exposed to the same conditions as the equipment surfaces. In practice, these are sacrificed at intervals and analyzed for biofilm biomass by a variety of destructive or non-destructive methods. Coupons have been used in conjunction with side-stream loops to monitor surface fouling in parts of a system that are inaccessible (Poje *et al.*, 1982; McCoy *et al.*, 1981). While an improvement over gross system operational parameters, periodic sampling of coupons to assess biofouling is often inadequate and loss of system control can occur between periods of scheduled coupon sampling and analysis.

Some water systems, such as those used in the microelectronics and pharmaceutical industries, have such low tolerance for biofouling that coupons offer little benefit to an effective monitoring program (Patterson *et al.*, 1991). In other industries coupon retrieval poses problems. Coupons deployed in normally-accessible, spent nuclear fuel wet storage facilities have become so heavily contaminated with radionuclides through scavenging reactions of associated biofilms that they can no longer be recovered and analyzed due to safety regulations (Roberto, INEEL, personal communication). Thus, there is a real need for biofilm monitoring equipment that permits remote, sensitive, non-destructive, real-time analysis of surface biofouling. This will enable system operators to have a good understanding of the condition of the system surfaces at all times, as well as sufficient warning of a deteriorating condition in the system to implement effective remedial procedures that minimize costly, unscheduled shut-downs.

The same features of a biofilm monitor described above for practical surveillance in industry are also useful for more fundamental studies of microbial surface colonization. Most investigations of microbial surface colonization in the laboratory have involved the exclusive use of some form of visual observation or culture technique. The earliest biofilms studies utilized direct light microscopic techniques to qualitatively examine stained cells attached to transparent surfaces (Henrici, 1933). Later, phase contrast and transmitted differential interference contrast (DIC) optics were employed to examine microscopically unstained preparations in a non-destructive manner (Surman *et al.*, 1996). High resolution scanning electron microscopy (SEM) and transmission electron microscopy (TEM) offer additional information on biofilm structure and adhesive structures and matrix materials produced by microorganisms associated with biofilms (Mack *et al.*, 1975; Beech *et al.*, 2000). These methods are destructive in nature, however, and introduce distortions in spatial relationships between structures during sample dehydration. They are typically not amenable to routine biofilm monitoring due to the labor-intensive sample preparation and the non-quantitative nature of the information provided.

Commercialization of confocal scanning laser microscopy (CSLM) in recent years has made it possible to qualitatively describe biofilm structure, and quantitatively describe local spatial biofilm phenomena such as thickness, cell density and distribution non-destructively, in real time under fully hydrated conditions (Lawrence *et al.*, 1991). Environmental scanning electron microscopy (ESEM) and tapping mode atomic force microscopy (AFM) offer high resolution information on the outermost features of biofilms maintained under hydrated conditions (Bremer *et al.*, 1992; Collins *et al.*, 1993; Beech *et al.*, 2000). Spatially-resolved elemental information can also be from hydrated biofilms using ESEM.

Spectroscopic techniques offer unique opportunities to monitor biofouling at the molecular level. In contrast to most of the cellular and structural information provided by microscopic techniques, spectroscopy provides quantitative chemical information that offers new insight into the physiological state of biofilm populations, as well as the molecular composition and molecular interactions between their metabolic products and other constituents of the system. Reference databases containing infrared (ir) spectra of many molecules of biological importance are available to facilitate band assignment and identification of molecular structure.

A technique such as ir spectroscopy, when used in the attenuated total reflectance (ATR) or multiple internal reflectance (MIR) mode, offers the opportunity to monitor real-time chemical changes during various stages of biofilm development on a variety of different substrata. Recent developments in sample cell design permit concatenation of chemical spectroscopy, light microscopy and computer-assisted spectral and image analysis to establish structure-function relationships in biofilms.

In this chapter, the use of ir spectroscopy in the field of biofilm microbiology is described. While a brief review of the application of ir spectroscopy to the study of biofilms has recently been published (Naumann *et al.*, 1996), the present chapter emphasizes the novel, on-line, ir spectroscopic methods that have been applied to biofouling assessment and biofilm characterization.

EVALUATION OF SURFACE CONDITIONING FILMS BY ATR IR SPECTROSCOPY

ATR or MIR ir spectrometry, using a high refractive index, infrared-transparent internal reflection element (IRE) such as germanium (Ge), provides a thin sampling region at the surface of the IRE where it contacts the surrounding medium. Ir spectra of material deposited on the IRE surface can be obtained by this sampling technique. In one of the earliest applications of ir spectroscopy to biofilm monitoring, Baier (1973) employed a grating spectrometer and a rectangular, Ge IRE to characterize the chemical nature and adsorption kinetics of material that deposited on a clean surface upon immersion in aqueous environments. After only 10 min exposure to the water of Biscayne Bay, FL, protein-like molecules had adsorbed to the surface, based on the appearance of amide bands in the mid-ir (Goupil *et al.*, 1980). In other applications, the rate of adsorption of blood proteins to surfaces bathed in serum was even faster, with a uniform coating of protein of average thickness 50 nm forming within 5 s and a coating of average thickness of 100–200 nm forming after 60 s (Baier, 1973). These macromolecules appeared to be irreversibly adsorbed, as vigorous rinsing with distilled water failed to eliminate the respective ir peaks.

Using MIR ir spectroscopy in combination with ellipsometry and critical surface tension determinations Baier (1973) showed that surfaces containing conditioning films formed after only short periods of immersion in aqueous medium exhibit critical surface tensions similar to surfaces with conditioning films formed over longer periods of time. The similarities of the ir spectra, suggested that substratum surface properties did not influence the nature of the conditioning film.

Immersion of IREs in natural water for longer periods of time allowed Baier (1973) to follow colonization of the IRE surface by bacteria and other microorganisms. Band broadening and the appearance of peaks centered at 1050 cm^{-1} were common spectral features observed during bacterial surface colonization. The appearance of absorption bands around 1050 cm^{-1} suggested the presence of carbohydrate material. As a result of this information, Baier (1973) suggested that a glycoprotein layer is the first acquired modifying film obtained by a clean surface after immersion in aqueous media. In spite of the appearance of whole cells on the surface at later time periods, the spectra remained similar to that obtained initially in the presence of organic molecules adsorbed from the bulk aqueous phase (Baier, 1980). This early application of ATR ir spectroscopy established the now widely accepted paradigm that microbial attachment and biofilm formation on an inert substratum, regardless of composition, is preceded by the deposition of a spontaneously adsorbed, glycoprotein-dominated conditioning film.

FOURIER TRANSFORM IR SPECTROSCOPY

Ir spectroscopy became a more useful analytical method for biologists when Fourier transform techniques were developed to collect and process spectra. The same high quality spectra obtained with grating spectrometers could be collected and processed in about one-thousandth of the time using a Fourier transform infrared (FT-IR) spectrometer. The increased number of replicate spectra collected by FT-IR over that collected with a grating instrument leads to an increase in signal-to-noise ratio, which in turn increases sensitivity by a factor of 10–100. A key component of a FT-IR spectrometer is a Michelson interferometer. The interferometer consists of a fixed and moving mirror and a beamsplitter that are used to produce a path difference or a phase shift between all wavelengths of energy scanned, giving an interferogram which is related to the spectrum through its Fourier transform. In practice, a single-beam background transmittance spectrum is first measured and then a sample is placed in the beam path and another single-beam transmittance spectrum is measured. The ratio of these 2 spectra yields the sample transmittance spectrum, and the negative logarithm of this is usually computed to produce the absorbance spectrum. Details of how a FT-IR spectrum is obtained from an interferogram are provided by Griffiths (1983) and Griffiths and de Haseth (1986).

Additional advances in technology that contributed significantly to the application of FT-IR to the study of biological phenomena included 1) the development of the fast Fourier transform algorithm, 2) the laboratory minicomputer for rapid scanning, 3) incorporation of small reliable He-Ne lasers for high spectral resolution, 4) new detectors that allowed faster scan speeds and increased sensitivity, and 5) and mini- and micro-computer-based data systems that rapidly processed the spectral information (Griffiths, 1983). When affordable, FT-IR spectrometers with these features became commercially available in the mid-1970s, they quickly established their niche in microbiology and biofilm research.

Taxonomy and Identification of Microorganisms by FT-IR

FT-IR has been used to characterize microorganisms (Naumann *et al.*, 1991b; 1994). The sum or combinations of 50–60 spectral bands, resolved through enhancement techniques, has been used to establish strain-specific fingerprints to differentiate and identify unknown microorganisms. Among the bands that have been proposed to be of value in this regard is one centered at 1088 cm^{-1}, which is often assigned the symmetric vibration of the phosphodiester bond. That this particular molecular vibration can be used to distinguish microorganisms is unexpected since DNA, RNA and phopholipids all absorb in this region. In addition, the region between 900 and 600 cm^{-1} exhibits extremely rich information that has led to its designation as the "bacterial fingerprint region". Rarely are valid vibrational assignments made in this region, however.

The success of FT-IR in differentiating bacteria resides in the treatment of spectral fingerprints by such methods as principle-component analysis and artificial neural networks for data reduction and objective assessment of complex and composite data structures (Helm *et al.*, 1991). Grouping of bacteria on the basis of spectral similarity involves cluster analysis and other multivariate techniques using spectral reference libraries. These techniques all depend on the ability to establish pure cultures of the unknown microorganism.

In one of the early applications of combined FT-IR spectroscopy and microscopy, Naumann *et al.* (1991a; 1991b) identified microcolonies of bacteria containing less than 10^4 bacteria, using an ir-transparent plate as a stamp to transfer locally-separated microcolonies from an agar culture plate to an ir-sample holder. This replica technique permitted detection, enumeration and differentiation of populations from mixed cultures. Unlike some of the newer molecular techniques used to distinguish different microorganisms, ir microscopic differentiation necessitates the culture and isolation of populations before analysis. This bacterial fingerprinting technique is most useful as a complement to other approaches for bacterial differentiation and classification.

FT-IR Band Assignments from Intact Bacterial Cells

Naumann *et al.* (1996) point out that, with few exceptions, ir spectra of intact bacteria do not provide information on a single or a few specific cellular compounds because of the difficulty in making specific band assignments to specific structures from a complex assemblage of molecules. Nevertheless, the exceptions are worth noting inasmuch as they offer physiological information difficult to obtain by other techniques.

Bacteria produce a number of compounds not found in other biological systems. These compounds have become "signatures" of the microbial world. Naumann (1984) recognized fingerprint-like ir spectral features of the peptidoglycan from cell wall preparations from different bacteria. In Gram-positive bacteria, 90% of the cell wall consists of peptidoglycan (Brock and Madigan, 1991). In Gram-negative bacteria, only 5–20% of the cell wall is peptidoglycan, with LPS, phospholipid and protein contributing 30%, 20–25% and 45–50%, respectively to an outer membrane wall component (Boyd, 1988). Peptidoglycan contributes only 2% of the total cell mass of the Gram-negative bacterium *Escherichia coli* in balanced growth with a mass doubling time of 40 min (Niedhardt *et al.*, 1990). The contribution of cell wall components to total cell mass varies considerably, however, depending on the physiological status of the cell.

Using specular reflectance or internal reflectance sampling mode, dehydrated preparations of peptidoglycan from a variety of bacteria yielded common absorption bands (Naumann *et al.*, 1982). Furthermore, bands near 1730 cm^{-1}, 1600 cm^{-1}, and 1400 cm^{-1} afforded the opportunity to differentiate peptidoglycan among different species. Band intensities also offered information on degree of cross-linking and other structural features. By this approach, changes in peptidoglycan induced by the culture medium and chemical treatment to intact cells, or changes associated with stages of cell growth, can be detected by FT-IR (Naumann, 1984).

In other studies, freeze-dried preparations of different strains of oral streptococci, pressed into KBr pellets, when analyzed by transmission ir absorption spectroscopy, all exhibited similar absorption bands, but each could be distinguished on the basis of differences in the relative intensities of the bands (van der Mei *et al.*, 1989). Among the absorption bands detected, those at 1236 and 1082 cm^{-1}, arising from the asymmetric stretching mode of phosphates and carbonyl stretch of carbohydrate, respectively, were assigned to teichoic acids in the cell wall. Such assignments should be viewed with caution since phosphate groups in nucleic acids and carbohydrates in other cell structures also absorb in these regions. In *Bacillus subtilis*, teichoic acids represent 50% of the cell wall dry weight (Boyd, 1988). Teichoic acids are not found in walls of Gram-negative bacteria.

By normalizing the absorbance of bands contributed by teichoic acids as well as the amide I and amide II bands contributed by protein against the CH stretching region centered at 2930 cm^{-1}, van der Mei *et al.* (1989) were able to resolve the different strains of oral streptococci tested. An important conclusion drawn from the investigation was that transmission ir spectroscopy of freeze-dried bacterial cells yields surface-sensitive information comparable to x-ray photoelectron spectroscopy. However, these conclusions can only be justified if the contribution of intracellular and cytoplasmic membrane material is negligible in comparison to that of cell wall components. The insignificance of intracellular and membrane components to the ir spectrum of whole cells of Gram-positive bacteria remains to be confirmed.

FT-IR spectroscopy has been used to monitor metabolic events in bacterial cells relevant to agricultural, environmental and health sciences. For example, reversible spectral changes attributed to quantitative and structural changes in cell wall peptidoglycan were observed as a result of transfer of *Bradyrhizobium japonicum* from liquid to solid medium (Zeroual *et al.*, 1994). Spectra were obtained from samples dried onto zinc selenide (ZnSe) IREs to avoid water interference. Fourier transform self-deconvolution was used to resolve peptidoglycan bands in the presence of other biomolecules, which contribute a large numbers of vibrations in samples of whole cells. Absorption bands centered at 1154 cm^{-1} and 1027 cm^{-1}, assigned to the C-O stretching vibrations of carbohydrates, were shown to increase on the solid medium. These spectral changes were suggested to represent changes in the peptidoglycan structure based on electron microscopy of thin section preparations. Capsular polysaccharides, synthesized during growth on solid medium, may also have contributed to the changes in band intensities observed in this region of the spectrum. The difficulty in eliminating contributions from other cellular molecules present in whole cell preparations from a particular absorption band, even after application of spectra processing software, has complicated spectral interpretation and limited the use of FT-IR as a means of following the fate of specific microbial metabolites. Furthermore, since samples were manually deposited on the IRE and dehydrated before acquisition of spectra, it was not possible to observe the molecular changes in living cells in real time.

In addition to cell wall components, FT-IR spectral assignments have been made from whole cell preparations for dipicolinic acid during endospore formation, poly-β-hydroxybutyrate and glycogen-like storage material, and capsular polysaccharides (Naumann *et al.*, 1996). As spectrum deconvolution algorithms become more sophisticated and spectral data are matched to chemical data obtained by other independent analytical approaches, resolution of other metabolic products in whole cell preparations may be possible in the future.

FT-IR Analysis of Protein Conditioning Films and Microbial Biofilms Deposited *in situ*

Early studies with grating spectrometers suggested that irreversible adsorption of proteins from aqueous solutions to submerged surfaces is a key process in the formation of the conditioning film with which early surface-colonizing microbes interact. Protein interactions with the substratum may influence subsequent adhesion of surface-colonizing microbes. Conditioning film proteins may also control biofilm processes established at much later stages of surface biofouling. FT-IR offers the opportunity to resolve protein-surface interactions, not previously achievable with dispersive instruments, that may be important to biofouling.

Using near grazing incidence reflection-absorption FT-IR, where a dehydrated surface is sampled by a single external reflection of the incident beam, Taylor *et al.* (1996) observed little effect of substratum wettability (critical surface tension) on the quantity of a model protein, ribulose-1-5,-bisphosphate carboxylase-oxygenase (Rubisco) that adsorbed from solution. These results supported earlier work using a grating spectrometer and ATR sampling geometry (Baier, 1973). However, natural conditioning films deposited on surfaces of Fe, Al and Ti in nearshore tropical seawater yielded distinctly different spectra, although a protein signature was common to the conditioning films formed on all three substrata (Taylor *et al.*, 1996).

Using the reflectance-absorbance FT-IR spectroscopic technique above in combination with ellipsometry, Taylor *et al.* (1993, 1994), showed that the secondary structure of Rubisco was dependent on the extent of protein surface coverage, protein film thickness and the properties of the substratum to which the protein had adsorbed. The coverage and thickness of the adsorbed Rubisco was shown to have a significant influence on the surface properties of the substratum (Taylor *et al.*, 1994). The critical surface tension and surface free energy of Ti oxide and copper surfaces were significantly influenced by the amount of protein adsorbed.

Diffuse reflectance infrared Fourier transform (DRIFT) spectroscopic sampling of freeze-dried biofilms was used to identify molecular vibrations contributed by biomolecules that served as tracers of microbial biofilm development on surfaces (White *et al.*, 1988). Although the ability to monitor living biofilms is sacrificed, resolution is enhanced by this sampling mode. DRIFT was able to detect two physiological changes associated with the nutritional state of a biofilm population on a Ge IRE. The appearance and increase in intensity of bands assigned to vibrational modes of poly-hydroxyalkanoic acid (PHA) and acidic exopolysaccharides, two polymers associated with nutritional imbalance in bacterial cells, exemplify the utility of FT-IR to monitor the physiological state of a bacterial population. DRIFT has also been used to demonstrate an increase in biofilm protein and carbohydrate concentrations with increased applied shear on 316 stainless steel coupon mounted in a cell adhesion measurement module (Mittelman *et al.*, 1990).

Although these applications of FT-IR yielded important information on the nature and formation rates of organic conditioning films and the physiological response of surface-associated bacteria to environmental conditions, the samples required dehydration before acquisition of an ir spectrum. Dehydration has recently been shown to have a profound effect on the structure and topography of protein conditioning films (Baty *et al.*, 1997). Dehydration has long been known to disrupt the native structure of biofilms (Mack *et al.*, 1975). Utilization of a sampling technique that permits acquisition of high quality ir spectra of fully hydrated biofilms should provide the possibility for on-line monitoring of surface biofouling.

Integrating Attenuated Total Reflectance Sampling Geometry with FT-IR Spectroscopy

Because water absorbs strongly in the region of the ir spectrum where useful information on biologically-important molecular vibrations reside, the water bands typically mask the more subtle absorbance bands of proteins and other biological molecules. In the past, this problem was solved by several different approaches. Mattson *et al.* (1975) were successful in characterizing proteins adsorbed to multiple internal reflection elements under fully hydrating conditions by processing spectra collected with a grating spectrometer using a computer data processing system interfaced to the spectrometer. Water interference has also been overcome by substitution of D_2O for H_2O in the aqueous phase when grating spectrometers were used to collect spectra (MacCarthy *et al.*, 1975).

It is easier to subtract water absorbance from spectra collected with an FT-IR spectrometer than from those collected with grating spectrometers. Even with their superior scan speed, signal-to noise ratio and energy throughput, modern FT-IR spectrometers still required short path lengths to successfully minimize water interference. Short path lengths can be achieved with a conventional transmission liquid cell. Although transmission cells create path lengths of < 15 μm, it is difficult to reproduce this when reassembling the cell for replicate or comparative experiments. Alternatively, ATR sampling geometry utilize IREs that have a well defined path for a given angle of incidence and refractive indices of the IRE and surrounding medium (see Eqn 1, p. 264). The ATR sampling mode does not give rise to interference fringes common to transmission liquid cells, avoiding the masking of spectral features.

By limiting ir sampling to a thin region immediately adjacent to an IRE surface immersed in an aqueous medium, as is the case for the ATR beam path, it is possible to avoid the total loss of energy in the spectral region where maximum water absorption occurs. The depth of penetration, d_p, is defined as the distance into the solution where the evanescent field amplitude decays to e^{-1} of its magnitude at the IRE-solution interface. For Ge-water at a 45° angle of incidence, this distance is 0.064 λ_o where λ_o is the *in vacuo* wavelength (Mattson *et al.*, 1975). The d_p is approximately 0.4 μm for the water band at 1640 cm^{-1}, providing an effective path length of 4 μm, assuming the number of reflections is 10 (Iwaoka *et al.*, 1986). This sampling geometry, thus, provides the opportunity to isolate surface chemical information from bulk aqueous phase chemistry without surface dehydration. Further information on the ATR sampling mode and its applications to surface chemical characterization is provided by Wragg and White (1991).

ATR is a particularly useful sampling technique for non-destructive chemical characterization of protein behavior at a hydrated surface and chemical characterization of biological processes localized to the bottom few cell layers of a fully-hydrated biofilm. By combining ATR with FT-IR, it is possible to collect useful spectra rapidly from surfaces that require no dehydration. Water can be efficiently subtracted from the spectrum of the thin aqueous layer adjacent to the substratum containing the surface-associated conditioning film and/or microbial biofilm. This offers new opportunities to chemically characterize biological systems in their native state, non-destructively, in real-time.

The early IREs developed for ATR sampling were optimized for the beam geometry of grating spectrometers. Consequently, the rectangular-shaped ir beam entered the IRE across a rectangular surface. In contrast, FT-IR instruments produce a circular beam, which is not efficiently captured by or transferred through IREs with a rectangular entrance geometry. In 1983, SpectraTech Incorporated (Stamford, CT) introduced the CIRCLE accessory which efficiently captured the circular beam of the FT-IR spectrometer and transmitted it to a circular target in the detector (Wilks, 1982). With proper water subtraction, this cylindrical geometry permitted collection of FT-IR spectra of an organic phosphonic acid at solution concentrations as low as 100 mg l^{-1}. By the mid-1980s, ATR/FT-IR had evolved such that it could also offer new insights to chemical phenomena at hydrated surfaces. The situation improved even more with the availability of computer software that simplified water subtraction from the spectrum. Powell *et al.* (1986) described an algorithm that could be used to subtract water from FT-IR spectra of proteins adsorbed to an ATR crystal. Dousseau *et al.* (1989) described an algorithm, which allowed quantitative subtraction of water from transmission FT-IR spectra of proteins in aqueous solutions. All the technology was in place at this time to exploit non-destructive ATR/FT-IR chemical spectroscopy to characterize conditioning films and biofilms in ways never previously possible.

ATR/FT-IR of Hydrated Surface Conditioning Films

Nichols *et al.* (1985) were among the first to demonstrate the use of ATR/FT-IR to follow, in real time, the deposition of organic material from filtered seawater on a rectangular IRE under flow conditions. Their results showed that carbohydrate rather than protein was the predominant material adsorbed to the surface during the first 5 h of seawater immersion. These results contrast with those reported previously in which protein was the major component of the conditioning film on surfaces immersed in seawater (Baier, 1973). A possible explanation is that the carbohydrate fraction of adsorbed material, due to its lower affinity for the substratum, was lost during the dehydration procedure used in earlier ATR ir studies. Dehydration has also been shown to cause an increase in the ir absorbance of biomolecules adsorbed to surfaces over that obtained under fully hydrated conditions, as well as change the relative absorbance intensities of different chemical species associated with the surface (Geesey and Bremer 1990). Thus, the chemical nature of conditioning films needs to be reassessed using techniques that accommodate the presence of water.

Interactions between proteins and polysaccharides on surfaces in the presence of water have been monitored in real-time using ATR/FT-IR (Ishida and Griffiths, 1990). The acidic polysaccharide, alginic acid, adsorbs rapidly to a clean Ge surface from a flowing 1% aqueous solution at pH 7.4, based on changes in intensity of the 1034 cm^{-1} absorption band assigned to the C-O stretching of the sugar subunits. Desorption of the alginic acid from the Ge surface occurred slowly during surface rinsing over a 4-h period. Only 60% of that initially adsorbed desorbed over that period. Establishment of a protein film of β-lactoglobulin accelerated the rate of alginic acid adsorption to the Ge substratum compared to that observed in the absence of the protein film. More polysaccharide adsorbed to the "protein conditioned" surface than to the bare Ge surface. Protein was not displaced from the surface as a result of polysaccharide adsorption. The polysaccharide but not the protein desorbed from the surface during subsequent surface rinsing. A similar polysaccharide residual remained with the protein-conditioned surface as with the bare Ge surface. The neutral polysaccharride, dextran, unlike alginic acid, was excluded from the protein-conditioned Ge surface (Ishida and Griffiths, 1993a). Dextran exhibited a greater affinity for bare Ge than a protein-conditioned Ge surface. These studies suggest that neither acidic nor neutral polysaccharides, which resemble the matrix biopolymers of microbial biofilms, establish strong interactions with the protein conditioning film formed upon immersion of surfaces in aqueous media.

Proteins take on a different secondary and possibly tertiary structures when adsorbed to surfaces as compared to their structure in solution. Ishida and Griffiths (1993b) described how amide I/amide II ratios of proteins changed when spectra collected by a liquid transmission cell and a CIRCLE ATR cell were compared. They also described how to calculate protein film thickness and showed how solution phase pH influenced protein film thickness on a Ge IRE. Thicknesses of 5.6 nm, 7.2 nm and 6.7 nm were calculated for bovine serum albumin adsorbed at pH 7.0, 4.1 and 9.0, using spectra collected under fully hydrated conditions.

The interactions of a hydrated Ge surface and bulk aqueous phase detergents, used to control biofilm accumulation on membrane material, was investigated by ATR/FT-IR. The aliphatic tail of the surfactant was more firmly bound to the Ge surface than the ethoxy substituents (Ishida *et al.*, 1998). Studies of interactions between surfaces and compounds in the bulk aqueous phase are now a routine application for ATR/FT-IR.

Characterization of Fully Hydrated, Intact Microorganisms by ATR/FT-IR and Assignment of Ir Absorption Bands to Specific Biomolecules

One of the earliest applications of ATR/FT-IR to evaluate whole cells of microorganisms in aqueous medium was performed by Hopkinson *et al.* (1987). Using a flat, six-reflection ZnSe IRE (Specac, Orpington, Kent, UK), they suggested that the spectrum of an aqueous suspension of the yeast *Candida pseudotropicalis* was dominated by cell wall components. The broad band centered at 1070 cm^{-1} was suggested to be contributed by the C-O stretching vibrations corresponding to a composite of chitin and the main cell wall polysaccharides mannan and glucan. Their conclusions were based on evidence that these bands were significantly reduced in protoplast preparations, which preserve membrane envelope components (*i.e.* protein and lipids) but not these polysaccharide cell wall components. While this off-line application of ATR/FT-IR permits chemical characterization of hydrated intact microorganisms and other living material, it does not demonstrate the true analytical potential of the spectroscopic method for on-line monitoring of surface-associated biological processes.

Characterization of Microbial Biofilm Development on Surfaces in Real Time by ATR/FT-IR

The use of ATR/FT-IR to follow microbial biofilm development nondestructively, in real-time under fully hydrated conditions was first demonstrated by Geesey and Bremer (1990). Unlike earlier applications of ATR/FT-IR to biological systems which are short-term in nature (Gendreau *et al.*, 1981; Winters *et al.*, 1982), biofilm processes often require days or weeks to develop. To overcome spectral artifacts contributed by long-term fluctuations in energy throughput and temperature fluctuations, a double beam instrument was employed which could ratio out these effects on bacterial biofilm spectra. Two CIRCLE cells containing similar IREs were placed parallel to each other in the optical bench of a Perkin Elmer Model 1800 FT-IR spectrometer containing a medium-band HgCdTe (MCT) detector (Bremer and Geesey, 1991a). Both flow cells were sterilized with ethylene oxide. One CIRCLE cell was inoculated with microorganisms while the other was maintained in a sterile but otherwise identical state as the inoculated cell. The spectrum of the inoculated cell was then ratioed against that of the sterile control to eliminate the spectral artifacts that appear over the long term. Using this approach, it was possible to detect and identify the major chemical components of a developing microbial biofilm, as well as monitor changes in chemical composition during biofilm maturation under static and continuous flow conditions over a 196 h time span (Geesey and Bremer, 1990; 1991).

Nivens *et al.* (1993a) described a multi-channel ATR/FT-IR spectrometer that utilized a mid-ir liquid-cooled source, a Transept III interferometer, an optical system containing collection and focusing mirrors for three channels, three Harrick rectangular flow cells (Harrick Scientific, Ossining, NY), and a narrow-band MCT detector. They showed that reproducible ir spectra could be obtained from the three channels.

Nivens *et al.* (1993a) described two methodologies using the multi-channel spectrometer to obtain ATR/FT-IR spectra of living *Caulobacter crescentus* cells attached to flat trapezoidal Ge IREs. Spectra were obtained for attached bacteria in high purity water, which provided details of the attachment process without spectral interference from components of the bulk aqueous medium. They also followed growth of attached bacteria by using a culture medium that did not contribute ir absorption in the region 2000–1200 cm^{-1}. Using the amide II band as a marker for biofilm biomass, they determined the detection limit to be approximately 5×10^5 cells cm^{-2}. Sterile controls gave rise to an ir absorption band at 1080 cm^{-1}, which was attributed to precipitation of inorganic salts onto the IRE. Adhesion of *C. crescentus* to the IRE produced ir absorption bands at 1648, 1550, 1306 cm^{-1} (amide I, amide II and amide III, respectively), and at 1454, 1397, 1246, 1080 cm^{-1} (C-H bend, C-O stretch, P = O stretch or amide III, C-O stretch of alcohols and carbohydrates, respectively). It should be noted, however, that there is a strong contribution from the phophodiester linkage in DNA and RNA cellular constituents in the region 1080–1088 cm^{-1}, and this is likely to dominate vibrations contributed by the C-O stretch. The intensity of all bands increased during the first 3 h, suggesting that bacterial cells had accumulated on the surface over that period of time. In other studies, White *et al.* (1991) used ATR/FT-IR to monitor colonization of a Ge IRE surface by *C. crescentus*. They related the intensities of ir bands assigned to amide I, amide II and the C-O stretching region of carbohydrates to bacterial cell densities on the surface determined by acridine orange direct count microscoppy over an 80 h period.

The sampling depth of the evanescent wave of radiation D_p can be calculated using the equation,

$$D_p = \lambda/2\pi n_1[\sin^2\theta - (n_2/n_1)]^{1/2} \qquad (1)$$

where λ is the wavelength of the radiation, n_1 is the refractive index of the IRE, n_2 is the refractive index of the medium in contact with the IRE, and θ is the angle of incidence. For water in contact with a Ge IRE cut at a 45° angle, D_p is calculated to be 0.5 μm at the wavelength where amide II absorbs. Thus, by using a Ge IRE as a substratum, the sampling depth is optimized to detect only the layer of bacterial cells in a biofilm that is in contact with the substratum.

By comparing intact biofilm spectra with those of isolated microbial products, it was possible to time-resolve the biosynthesis and deposition of the carbohydrate matrix polymer glue that anchored the bacterial cells to the substratum (Geesey and Bremer, 1990). According to the Beer-Lambert Law, the absorbance of a component in the spectrum is proportional to the concentration of the component at a given path length and an extinction coefficient. Quantitative as well as qualitative information can therefore be extracted from the spectra. Fink *et al.* (1987) demonstrated a good correlation between amide I or amide II band intensities and amount of adsorbed protein. Correspondingly, the amide II peak area from the ATR/FT-IR spectra provided a good measure of bacterial cell density on the surface of an IRE (Suci *et al.*, 1997).

Use of ATR/FT-IR to Characterize Interactions between Microbial Adhesins and Surfaces

Few studies have established a clear role for conditioning films in microbial adhesion to and biofilm formation on substrata immersed in aqueous environments. Using [^{3}H]-leucine labeled cells of *Pseudomonas fluorescens* and an *Acinetobacter sp.*, Pringle and Fletcher (1986) determined that conditioning films comprised of the protein bovine serum albumin or bovine glycoprotein or lipopolysaccharide from *Escherichia coli* inhibited attachment to polystyrene. Similarly, Frolund *et al.* (1996) found that a conditioning film comprised of a mussel adhesive protein (MAP) inhibited attachment of cells of *Hyphomonas* MHS-3 to a Ge substratum. These results were consistent with the affinity of purified polysaccharide adhesin from this bacterium for the Ge surface in the presence and absence of MAP conditioning films. The binding data were obtained by ATR/FT-IR. While these results suggest that some model monomolecular conditoning films interfere with the adhesion process in some bacteria, extrapolation to natural conditions is premature.

The differences in the ATR/FT-IR spectrum of *C. crescentus* cells attached in high purity water and growth medium enabled Nivens *et al.* (1993b) to detect different holdfast organelles under different adhesion conditions. Such information would be difficult to obtain using other analytical techniques. While these investigators did not pursue the identity of the holdfast molecules responsible for the spectrum differences, these and the studies described above demonstrate that ATR/FT-IR offers many new opportunities to better understand the molecular basis of microbial adhesion and biofilm development (Suci and Geesey, 1998).

Use of ATR/FT-IR to Characterize Interactions between Antimicrobial Agents and Biofilm Populations

Microbial cells within biofilms are more resistant or recalcitrant to antimicrobial agents (AA) than when suspended as individual cells in aqueous media (Nickel *et al.*, 1985; Evans and Holmes, 1987; Gilbert *et al.*, 1990). The word recalcitrant is preferred since resistance implies that gene mutations have been identified which cause an alteration in the AA target, or genes have been acquired that produce products which inactivate the AA. Biofilm recalcitrance to AA is not as well understood as classical resistance mechanisms utilized by suspended cell populations. Theories of biofilm recalcitrance fall into two categories that are non-exclusive. The first is based on the hypothesis that microorganisms in biofilms have physiological characteristics that make them less susceptible to the lethal dose. For example, biofilm microorganisms replicate and/or metabolize more slowly than cultures of individual cells in aqueous suspension (Gilbert *et al.*, 1990). Alternatively, transport of the AA may be hindered in biofilms such that the lethal dose does not reach certain portions of the biofilm (Nichols *et al.*, 1988; Hodges and Gordon, 1991; Hoyle *et al.*, 1992).

Different mechanisms of AA resistance by bacterial cells in biofilms have been evaluated using ATR/FT-IR. Jass (1990) was the first to consider this approach to monitor β-lactam antibiotic penetration in biofilms of *Pseudomonas aeruginosa* and was able to isolate the ir absorbance of the antibiotics from that contributed by the microbial cells. The results suggested that the biofilm matrix did not impede the transport of either pipericillin or ticarcillin. Suci *et al.* (1994) used ATR/FT-IR to follow the penetration of ciprofloxacin into *P. aeruginosa* biofilms. Vrany *et al.* (1997) used results obtained by ATR/FT-IR to mathematically model transport diffusion coefficients for the fluoroquinolone antibiotics levofloxacin and ciprofloxacin, binding site density, and adsorption and desorption rates for biofilms of *P. aeruginosa*. Like Jass (1990), these investigators concluded that transport of the antibiotics could not explain the recalcitrance of the biofilm population and suggested that physiological factors were responsible. In both studies, ATR/FT-IR was used to quantify non-destructively the accumulation of the antibiotics at the biofilm-substratum interface in the presence of a complex spectral contribution from living bacterial cells, in real time, under fully hydrated conditions.

IRE SURFACE MODIFICATION FOR ATR/FT-IR

Once ATR became the sampling mode of choice for real-time ir studies on surface-associated conditioning film and biofilm processes under hydrated conditions, a need for IREs with different substratum surface properties became apparent. ZnSe, Ge and KRS-5 (thallium bromoiodide) are the only materials used to fabricate commercially-available IREs that offer reasonable ir transparency in the biologically-relevant region of the ir spectrum. Silicon and sapphire IREs offer only limited opportunities in this regard.

The notion that an IRE surface could be modified by coating with an ultrathin film of a different material without compromising IRE performance or spectrum quality was evaluated by Baier and Loeb (1971). They silanized Ge IREs cleaned by radio frequency glow-discharge in air to produce different surface free energies. These surfaces were then exposed to natural and artificial freshwater and seawater for varying periods of time, dehydrated and analyzed both spectroscopically and microsopically to determine the influence of surface energy on conditioning film and biofilm accumulation. The results indicated that differences in substratum surface free energy exerted no detectable influence on the formation of the macromolecular organic conditioning film (Baier *et al.*, 1983). However, very tightly adsorbed and densely compressed interfacial films were preferentially attracted to Ge surfaces displaying an extremely high surface energy, whereas, more loosely attached and easily removed films were associated with highly-methylated, silanized surfaces (DePalma and Baier, 1978).

Thin films of hydroxyapatite (HA), a multicomponent compound which forms the major constituent of the outermost part of the tooth surface, were deposited on a Ge IRE by radiofrequency sputtering and used to characterize the adsorption of salivary components to the tooth surface (Ruckenstein and Gourisankar, 1983). The films were thin enough (< 20 nm) to detect ir absorption bands assigned to protein after dehydration and analysis by multiple internal reflection ir spectroscopy. No attempt was made to evaluate the stability of the HA film in aqueous environments, however.

IRE surfaces have also been modified to display the properties of membrane material. Ishida *et al.* (1998) dip-cast thin films of cellulose acetate (CA) onto Ge IREs in order to study membrane fouling by organic molecular films and microbial biofilms.

They found that the CA film became slowly hydrated when in contact with aqueous media. The unusually high degree of hydration as measured by the increase in intensity of the water absorption band at 1639 cm^{-1} was attributed to a factor other than normal polymer expansion, possibly separation of the film from the IRE. CA film hydration was accompanied by a shift to lower wavenumbers of the carbonyl band from 1749 to 1747 cm^{-1}, and a shift to higher wavenumbers for the C-O-C acetate ester band from 1230 to 1232 cm^{-1}. By exposing the CA film to solutions of alternating ionic strength, these investigators were able to demonstrate the passage of water across the film. The film appeared to act as an osmotic pressure cell. Thus, ATR/FT-IR offers a novel, sensitive way to evaluate membrane flux.

ATR/FT-IR Evaluation of Behavior of Molecules and Cells Adsorbed to Chemically-Modified IRE Surfaces

Silanized Ge IREs have been implanted in rabbits and the chemical nature of the tissue components that grew in contact with the implant surface subsequently evaluated by ATR ir spectroscopy and other surface analytical techniques (Baier *et al.*, 1984). These experiments indicated that protein-dominated films of greater thickness developed on implant surfaces with lower surface energy than the films that developed on intermediate and high surface energy surfaces. Infrared spectral analysis also revealed that material adsorbed to the lower energy surfaces was less substantially altered from its natural solution state configuration than that adsorbed to higher energy surfaces.

The behavior of proteins, polysaccharides and surfactants on IRE surfaces coated with thin films of CA was also evaluated by Ishida *et al.* (1998). Electrostatic interactions were suggested to dominate adsorption behavior of the various organic molecules tested. Feedwater to a reverse osmosis treatment plant consisting of secondary treated wastewater was also flowed across the CA thin film to evaluate the types of compounds that adsorb and contribute to membrane fouling. Adsorption bands contributed by proteins and carbohydrates dominated the ATR/FT-IR spectrum. During a rinse with deionized water, the carbohydrates desorbed rapidly whereas, proteins desorbed more slowly. These results corroborate earlier studies indicating the lower affinity for surfaces of carbohydrates than proteins.

The notion that ATR/FT-IR could be used to evaluate the stability of metal surfaces in the presence of adsorbed biomolecules was demonstrated by Iwaoka *et al.* (1986). Using chemical vapor deposition (PVD), they deposited ultrathin (10–20 nm), continuous films of metallic copper on Ge IREs that were sufficiently thin to permit penetration of the evanescent wave of ir radiation into the surrounding medium. Cu films (6.7 nm nominal thickness) deposited on Ge IREs by PVD were characterized by several surface analytical techniques (Bremer *et al.*, 1991). Films were shown to be continuous and to possess a Cu(I) oxide layer by x-ray photoelectron spectroscopic analysis. The surfaces were not significantly altered by ethylene oxide sterilization or exposure to water.

Ishida and Griffiths (1990) compared protein adsorption at different pHs on bare Ge IREs and Ge IREs coated with a thin film (3–4 nm) of either Cu or Ni deposited by PVD. They determined that the net charge on albumin appears to be more significant than the nature of the substratum in controlling how much protein accumulates of the surface. Correspondingly, surface charge effects were suggested to control protein adsorption rates on these substrata.

Ishida and Griffiths (unpublished results) conducted a thorough investigation of the influence of different Cu film thicknesses on the optical properties of the stratified media and the resulting ir spectra of water at the Cu film-water interface. Films deposited by PVD < 8 μm thick were found to be discontinuous. Even films > 10 μm in thickness were found to erode in a non-uniform manner in the presence of some aqueous media such that metal islands were produced. Water band shifts were attributed to Cu island effects, whereas, band shape changes were attributed to surface enhanced infrared absorption. These phenomena complicate ir spectrum band assignment and interpretation.

Advances in thin film technology have made it possible to deposit well-defined, stable metal films on IREs for subsequent studies in aqueous environments. Pedraza *et al.* (1989) described the use of XeCl laser treatment to enhance adhesion of 80 nm thick copper films on sapphire substrates relative to as-sputtered films. Laser treatment also improved the smoothness of the films.

Advanced thin film technologies have also been used to deposit alloys of stainless steel (ss) on IREs for biofilm studies. The phase structure of the ss thin film was influenced by the nature of the underlying substratum (Godbole *et al.*, 1993). The phase structure of 316L ss sputtered on Ge substrates was controlled by careful selection of annealing temperature (Godbole *et al.*, 1992). Stability in water of 316L ss thin films deposited on Ge IREs required the deposition of a 2 nm-thick chromium oxide bonding layer on the Ge prior to sputtering the ss (Pedraza *et al.*, 1993). Films deposited in this manner were found to be stable under non-aggressive aqueous conditions for at least 1000 h (Suci *et al.*, 1993). The conditions included exposure to and colonization by a consortium of bacteria.

Comparing ir spectra and microscopic images of biofilms collected on IRE-modified high and low energy surfaces, Baier (1980) found similar densities of surface-associated bacteria. Bacteria associated with the low energy surface appeared to be positioned on top of the glycoprotein matrix rather than embedded within it as was the case on the high energy surface.

Ishida *et al.* (1998) studied the fouling of CA thin films deposited on Ge IREs. Bacterial cell attachment was monitored in real time by the intensity of the amide II band at 1547 cm^{-1} as a solution containing a bacterial inoculum flowed over the CA film surface. Within 20 min, bacterial adhesion could be detected. A cellular adhesion rate of 7.38×10^{-4} h^{-1} was obtained using this approach. The cell accumulation rate on the CA film decreased after the fluid flowing across the film surface was replaced with liquid containing no bacterial inoculum. At the end of the experiment, the bacterial density on the film was determined by direct epifluorescent microscopic enumeration after staining cells with a fluorogenic compound that binds cellular DNA.

Advances in surface chemistry have thus expanded the utility of ATR/FT-IR in assessment of the influence of substratum properties on molecular and cellular interactions. Such investigations can now be carried out in the presence of water in real time, yielding useful thermodynamic and kinetic information.

MONITORING CORROSION BY FT-IR SPECTROSCOPY

Use of Diffuse Reflectance FT-IR to Monitor Microbiologically Influenced Corrosion

White *et al.* (1986) used the off-line DRIFT sampling method to correlate increased electrochemically-derived corrosion rates with the appearance of surface-associated bacterial products on metal coupons exposed to aqueous bacterial suspensions. They related the increased corrosion rates to an increase in accumulation of surface-associated material displaying an ir absorbance centered at 1440 cm^{-1} when a 304 stainless steel surface was exposed to a marine bacterial culture in artificial seawater. The material was subsequently suggested to be extracellular calcium hydroxide, possibly associated with an organic matrix, based on ir spectral comparison with an authentic standard (Nivens *et al.*, 1986).

ATR/FT-IR Using IREs Coated with Thin Metal Films to Monitor Microbiologically Influenced Corrosion

Iwaoka *et al.* (1986) and Jolley *et al.* (1989) showed that the water absorbance intensity measured by ATR/FT-IR was a very sensitive way to monitor the integrity (average film thickness) of thin Cu films deposited on cylindrical IREs by various methods. This approach was able to detect dissolution or ionization of only a few atomic layers of metallic Cu in contact with the aqueous phase. After formation of a Cu oxide layer, the underlying metallic Cu film appeared to be stable for extended periods of time in the absence of aggressive conditions. Such sensitivity allowed these and other investigators to evaluate the aggressive nature of biomolecules towards the hydrated metallic Cu film (Iwaoka *et al.*, 1986; Jolley *et al.*, 1989). The analytical approach proved useful in demonstrating a novel mechanism for microbially influenced corrosion of copper involving different matrix polysaccharides excreted by the biofilm bacterial populations (Geesey *et al.*, 1986; 1987).

Using ATR/FT-IR, Ishida and Griffiths (1990) evaluated the corrosive nature of the protein bovine serum albumin (BSA) when adsorbed to Cu and Ni thin (3–4 nm) films deposited on IREs by PVD. BSA did not exert a corrosive effect on any of the metal substrata to which it adsorbed. The albumin promoted adsorption of acidic polysaccharides to the metal films but their effect on film stability (corrosion) was not assessed.

Cylindrical IREs containing copper thin films were used to demonstrate the aggressive action of specific strains of bacteria growing as biofilms on the copper surface in real time using ATR/FT-IR. Corrosion of the copper film was detected by an increase in water band intensity at 1640 cm^{-1}, following exposure to a aqueous suspension of bacteria which formed a biofilm on the copper film (Geesey and Bremer, 1991; Bremer and Geesey, 1991b). Analysis the spectra revealed that an increase in the C-O stretching bands of carbohydrates at 1062 and 1229 cm^{-1} associated with biofilm development coincided with the increase in water absorption band at 1640 cm^{-1}, suggesting that the polysaccharide matrix was linked to corrosion of the copper. That corrosion occurred *via* localized attack was indicated by the appearance of discrete areas of discoloration on the copper film upon removal of the biofilm. It was also shown that not all microbial biofilms promote localized attack (Geesey and Bremer, 1991). Some strains of bacteria, if given the opportunity to form the initial biofilm, protect the copper surface from aggressive attack by other biofilm-forming bacterial strains that colonize the surface later (Bremer and Geesey 1991b). This on-line monitoring technique made it possible to characterize interfacial chemical reactions over the long-term without disturbing the living organisms growing at the interface.

Corrosion Product Identification by ATR/FT-IR

Inorganic corrosion products formed on metal coupons have been identified by FT-IR after pressing the dehydrated coupons against multiple internal reflection elements (Borgard *et al.*, 1988). Since this procedure does not protect against sample oxidation and dehydration, corrosion products may become chemically altered prior to analysis. Thus, this analytical approach offers little additional information that could not be obtained by x-ray diffraction or high vacuum spectroscopic techniques.

Electrochemically Modulated Infrared Spectrometry

A technique that enhances the sensitivity of internal or external reflection approaches is electrochemically modulated infrared spectroscopy (EMIRS). A "pseudo double beam" approach is used to normalize spectra and enhance surface sensitivity. Spectra are constructed by difference in reflectivity, ΔR, of a surface at two potentials. The modulation is between two potentials, V_1 and V_2, which causes a change in the adsorbed material. This, in turn, gives rise to changes in the optical constants of the interface affecting the reflectivity, resulting in a signal in phase with modulation. A spectrum is obtained by recording this change in DR as the wavelength of the probing radiation is slowly scanned.

When used in conjunction with FT-IR, this sampling technique is referred to as subtractively normalized interfacial FT-IR spectroscopy (SNIFTERS). For SNIFTERS, the sum of all interferograms taken at V_1 (the reference level, at a level of zero adsorbance for the species under consideration) are subtracted from those taken at V_2 and then divided by V_1 to normalize the spectra. The most useful ΔR corresponds to vibrations of species that cycle through surface adsorption and desorption or oxidized and reduced states at the two potentials. EMIRS, when used in conjunction with the ATR sampling mode, has the potential to identify which biofilm microbial metabolites or adsorbed organic species participate in a particular corrosion reaction at a surface. Although the technique can detect very subtle changes in molecular behavior at surfaces, it is usually necessary to have a good understanding of the reactions anticipated. SNIFTERS can be performed on-line when used in conjunction with ATR sampling mode, although this has not been widely documented. A more detailed discussion of EMIRS and SNIFTERS, and their applications to corrosion reaction characterization and monitoring, is provided by Wragg and White (1991). Although, EMIRS has not yet been used to characterize or monitor biocorrosion, it represents an extremely powerful alternative approach to current methods used in the field.

CONCATENATION OF IR SPECTROSCOPIC AND LIGHT MICROSCOPIC TECHNIQUES TO INTEGRATE BIOCHEMISTRY AND BIOLOGICAL STRUCTURE OF MICROBIAL BIOFILMS

To date, it has been difficult to obtain chemical information on intact biofilms. Light microscopic techniques typically provide bacterial cell density, diversity and distribution determinations, microcolony size, diversity and density determinations and even associations between populations or cells, but little chemical information in particular on fully hydrated samples. Ir microscopy provides the opportunity to obtain spatially-resolved chemical information on dehydrated samples, but because of the effects dehydration imposes on spatial relationships as well as ir spectra, it has contributed limited insight to biofilm chemistry. Ir microscopy has been used to collect DRIFT spectra of freeze-dried biofilms on a metal corrosion coupon. This approach permitted the mapping of ratios of band intensities contributed by cell protein, PHA and acidic exopolysaccharides at a resolution of spot diameters of 20 μm (White *et al.*, 1988).

Recent developments in Raman microscopy make this technique a potentially more useful tool for resolving spatially-dependent, chemical information on biofilm structures (Schaeberle *et al.*, 1995; Morris *et al.*, 1996). Since water does not absorb strongly in the Raman spectrum, useful, high-resolution biochemical information should be obtainable from fully hydrated specimens. Autofluorescence from biological samples can compromise Raman spectral quality, however.

Another approach to relate chemical and structural information on biofilms involves the combined ATR/FT-IR spectroscopy and light microscopy. The first attempt to combine these instruments was reported by Suci *et al.* (1997). They described a flat plate, flow-channel reactor containing a flat trapezoidal Ge IRE as a surface colonized by microorganisms introduced in an aqueous stream, which passes through the reactor. The reactor also contained a viewing window that accommodated reflected differential interference contrast (DIC) or epifluorescence microscopic observation of the Ge surface through a water immersion objective lens when mounted on the stage of a microscope. The reactor could be transferred to the optical bench of an FT-IR spectrometer housing a Harrick Horizon Cell mirror assembly (Harrick Scientific, Ossing, NY) for collection of ir spectra without disturbance to biological processes taking place on the Ge surface. They demonstrated how surface-associated bacterial cell density determinations by DIC microscopy correlated with surface-associated biomass determinations based on the intensity of the amide II protein band during different stages of biofilm development. A similar system has been developed by Ishida *et al.* (1998) to follow biofouling of reverse osmosis membrane material. While these concatenated techniques do not yet offer the opportunity to obtain spatially-resolved chemical information due to the averaging of molecular vibrations across the IRE surface, they do allow comparisons between biofilms formed on different IREs that are run in parallel.

SUMMARY AND CONCLUSIONS

FT-IR has proven to be a useful analytical approach for on-line monitoring of chemical and biological changes at surfaces in contact with aqueous media. ATR sampling geometries have isolated the acquisition of ir spectra to the solid-liquid interface where organic conditioning films form as a result of adsorption of protein and carbohydrate material from the bulk aqueous phase. The same sampling geometries have been successfully used to detect the attachment of bacteria from the bulk aqueous phase to the substratum surface and to characterize the adhesive holdfast material excreted by the attached microorganisms. Because this sampling approach does not disturb biological processes, it has been used to follow microbial biofilm development over long time periods. It has proved to be very useful in monitoring chemical and biochemical changes at the base of biofilms. Through this approach, new mechanisms of microbiologically influenced corrosion have been demonstrated, proposed mechanisms of biofilm recalcitrance to antimicrobial agents have been evaluated and changes in physiological status of cells observed. None of these discoveries could have been as easily made using other analytical approaches. Moreover, because ir absorbance follows the Beer-Lambert equation, quantitative information can be obtained in many cases for molecules present at an interface. Although, ir spectroscopy has long been used for chemical identification of relatively pure samples, it offers limited capabilities when sampling complex biological systems such as mixed populations of intact bacterial cells growing on surfaces. Nevertheless, FT-IR has proved useful in detecting differences between different types of bacteria and has provided taxonomic information on microorganisms.

FT-IR as an approach to on-line monitoring of microbiological phenomena is still in its infancy. As the need to define biological processes at surfaces expands, new applications of FT-IR will evolve in this area. Since most microbiological processes in nature occur in the presence of surfaces, exceptional opportunities exists for new discoveries to be made by evaluating microbiological reactions at surfaces. The ATR sampling mode, when combined with FT-IR, offers a unique window to the chemical and biochemical reactions underlying the microbiological processes that develop on surfaces.

REFERENCES

Atkinson D.L. (1979). Controlling and predicting cooling tower water quality. *Ind Water Engineer*, **16**, 37–40.

Baier R.E. (1973). Influence of the initial surface condition of materials on bioadhesion. In: Acker R.F. (ed) *Proc 3rd Int Congr Marine Corrosion and Fouling*. Northwestern University Press, Evanston, IL, pp. 633–639.

Baier R.E.(1980). Substrata influences on the adhesion of microorganisms and their resultant new surface properties. In: Bitton G., Marshall K.C. (eds) *Adsorption of Microorganisms to Surfaces*. Wiley and Sons, New York, pp. 59–104.

Baier R.E., Loeb G.I. (1971). Multiple parameters characterizing interfacial films of a protein analogue, polymethylglutamate. In: Craver C.D. (ed) *Polymer Characterization: Interdisciplinary Approaches*. Plenum Press, New York, pp. 79–96.

Baier R.E., Meyer A.E., DePalma V.A., King R.W., Fornalik M.S. (1983). Surface microfouling during the induction period. *J Heat Transfer*, **105**, 618–624.

Baier R.E., Meyer A.E., Natiella, J.R., Natiella R.R., Carter J.M. (1984). Surface properties determine bioadhesive outcomes: methods and results. *J Biomed Mater Res*, **18**, 337–355.

Baty A.M., Leavitt P.K., Siedlecki C.A., Tyler B.J., Suci PA., Marchant R.A., Geesey G.G. (1997). Adsorption of adhesive proteins from the marine mussel, *Mytilus edulis*, on polymer films in the hydrated state using angle dependent x-ray photoelectron spectroscopy and atomic force microscopy. *Langmuir*, **13**, 5702–5710.

Borgard B., Jones D., Heidersbach R. (1998). The use of FTIR and Raman spectroscopies for corrosion product analysis. Paper No. 154, *Corrosion 88*. National Association of Corrosion Engineers, Houston, TX.

Boyd R.F. (1988) *General Microbiology*, 2nd edn. Times Mirror/ Mosby College Publishers, St. Louis.

Bremer P.J., Geesey G.G. (1991a). An evaluation of biofilm development utilizing non-destructive attenuated total reflectance Fourier transform infrared spectroscopy. *Biofouling*, **3**, 89–100.

Bremer P.J., Geesey G.G. (1991b). Laboratory-based model of microbiologically induced corrosion of copper. *Appl Environ Microbiol*, **57**, 1956–1962.

Bremer P.J., Geesey G.G., Drake B. (1992). Atomic force microscopy examination of the topography of a hydrated bacterial biofilm on a copper surface. *Curr Microbiol*, **24**, 223–230.

Bremer P.J., Geesey G.G., Drake B., Jolley J.G., Hankins M.R. (1991). Characterization of a thin copper film to investigate microbial biofilm formation. *Surf Interface Anal*, **17**, 767–772.

Brock T.D., Madigan M.T. (1991) *Biology of Microorganisms*, 6th edn. Prentice Hall, Englewood Cliffs, NJ.

Bryers J., Characklis W. (1981). Early fouling biofilm formation in a turbulent flow system: overall kinetics. *Water Res*, **15**, 483–491.

Collins S.P., Pope R.K., Scheetz R.W., Ray R.I., Wagner P.A., Little B. (1993). Advantages of environmental scanning electron microscopy in studies of microorganisms. *Microbiol Res Tech*, **25**, 398–405.

Cormensana A., Tindall B.J. (eds) *Bacterial Diversity and Systematics*. Plenum, New York, pp. 67–85.

DePalma V.A., Baier R.E. (1978). Microfouling of metallic and coated metallic flow surfaces in model heat exchange cells. In: Gray R. (ed) *Proceedings of the Ocean Thermal Energy Conversion Biofouling and Corrosion Symposium*. Department of Energy, Washington, DC, pp. 89–106.

Dousseau F., Therrien M., Pezolet M. (1989). On the spectral subtraction of water from the FT-IR spectra of aqueous solutions of proteins. *Appl Spectrosc*, **43**, 538–542.

Evans R.C., Holmes C.J. (1987). Effect of vancomycin hydrochloride on *Staphylococcus epidermidis* biofilm associated with silicone elastomer. *Antimicrob Agents Chemother*, **31**, 889–894.

Fink D.J., Hutson T.B., Chittur K.K., Gendreau M.R. (1987). Quantitative surface studies of protein adsorption by infrared spectroscopy. *Anal Biochem*, **165**, 147–154.

Frolund B., Suci P.A., Langille S., Weiner R.M. Geesey G.G.(1996). Influence of protein conditioning films on binding of bacterial polysaccharide adhesin from *Hyphomonas* MHS-3. *Biofouling*, **10**, 17–30.

Geesey G.G., Bremer P.J. (1990). Application of Fourier transform infrared spectrometry to studies of copper corrosion under bacterial biofilms. *Mar Technol Soc J*, **24**, 36–43.

Geesey G.G., Bremer P.J. (1991). Evaluation of copper corrosion under biofilms. Paper No. 111, *Corrosion 91*. National Association of Corrosion Engineers, Houston, TX.

Geesey G.G., Iwaoka T., Griffiths P.R. (1987).Characterization of interfacial phenomena occurring during exposure ot a thin copper film to an aqueous suspension of an acidic polysaccharide. *J Colloid Interface Sci*, **120**, 370–376.

Geesey G.G., Mittelman M.W., Iwoaka T., Griffiths P.R. (1986). Role of bacterial exopolymers in the deterioration of metallic copper surfaces. *Mater Perform*, **25**, 37–40.

Gendreau R.M., Winters S., Leininger R.I., Fink D., Hassler C.R., Jakobsen R.J. (1981). Fourier transform infrared spectroscopy of protein adsorption from whole blood: *Ex vivo* dog studies. *Appl Spectrosc*, **35**, 353–357.

Gilbert P., Collier P.J., Brown M.R.W. (1990). Influence of growth rate on susceptibility to antimicrobial agents: biofilms, cell cycle, dormancy, and stringent response. *Antimicrob Agents Chemother*, **34**, 1865–1868.

Godbole M.J., Pedraza A.J., Allard L.F., Geesey G. (1992). Characterization of sputter-deposited 316L stainless steel films. *J Mater Sci*, **27**, 5585–5590.

Godbole M.J., Pedraza A.J., Park J.W., Geesey G.G. (1993). The crystal structure of stainless steel films sputter-deposited on austenitic stainless steel substrates. *Scr Metall Mater*, **28**, 1201–1206.

Goupil D.W., DePalma V.A., Baier R.E. (1980). Physical/chemical characteristics of the macromolecular conditioning film in biological fouling. In: *5th Int Congr Marine Corrosion and Fouling*. Barcelona, Spain, Department of Energy, Division of Solar Energy, pp. 401–410.

Griffiths P.R. (1983). Fourier transform infrared spectrometry. *Science*, **222**, 297–302.

Griffiths P.R., de Haseth J.A. (1986) *Fourier Transform Infrared Spectrometry*. John Wiley and Sons, New York.

Helm D., Labischinski H., Schallehn G., Naumann D. (1991). Classification and identification of bacteria by Fourier-transform infrared spectroscopy. *J Gen Microbiol*, **137**, 69–79.

Henrici A.T. (1933). Studies of freshwater bacteria I. A direct microscopic technique. *J Bacteriol*, **25**, 277–286.

Hodges N.A., Gordon C.A. (1991). Protection of *Pseudomonas aeruginosa* against ciprofloxacin and β-lactams by homologous alginate. *Antimicrob Agents Chemother*, **35**, 2450–2452.

Hopkinson J.H., Moustou C., Reyonlds N., Newbury J.E. (1987). Application of attenuated total reflection in the infrared analysis of carbohydrates and biological whole cell samples in aqueous solutions. *Analyst*, **112**, 501–505.

Hoyle B.D., Alcantara J., Costerton J.W. (1992). *Pseudomonas aeruginosa* biofilm as a diffusion barrier to piperacillin. *Antimicrob Agents Chemother*, **36**, 2054–2056.

Ishida K.P., Griffiths P.R. (1990). *In situ* investigation of the adsorption of proteins and polysaccharides at aqueous/solid interfaces by infrared internal reflection spectrometry. In: Schueing D.R. (ed) *Fourier Infrared Spectroscopy in Colloid and Interface Science*. Am Chem Soc Symp Ser No. 447, American Chemical Society, Washington, DC, pp. 208–224.

Ishida K.P., Griffiths P.R. (1993a). Comparison of the amide I/II intensity ratio of solution and solid-state proteins sampled by transmission, attenuated total reflectance, and diffuse reflectance spectrometry. *Appl Spectrosc*, **47**, 584–589.

Ishida K.P., Griffiths P.R. (1993b). Investigation of polysaccharide adsorption on protein conditioning films by attenuated total reflection infrared spectrometry. 1 Germanium surfaces. *J Colloid Interface Sci*, **160**, 190–200.

Ishida K.P., Bold R.M., Ridgway H.F. (1998). Influence of molecular conditioning films on microbial colonization of synthetic membranes determined by internal reflection spectrometry. Final report to National Water Research Institute Project No MRDP 699-508-95, Orange County Water District, Fountain Valley, CA, 99p.

Iwaoka T., Griffiths P.R., Kitasako J.T., Geesey G.G. (1986). Copper-coated cylindrical internal reflection elements for investigating interfacial phenomena. *Appl Spectrosc*, **40**, 1062–1065.

Jass J. (1990). β-lactam penetration through *Pseudomonas aeruginosa* biofilms monitored by ATR-FTIR spectroscopy. MS Thesis, University of Calgary, Canada.

Johnson C., Howells M. (1981) Biofouling: new insights, new weapon. *Power*, **125**, 90–91.

Jolley J.G., Geesey G.G., Hankins M.R., Wright R.B., Wichlacz P.L. (1989). *In situ*, real-time FT-IR/CIR/ATR study of the biocorrosion of copper by gum arabic, alginic acid, bacterial culture supernatant and *Pseudomonas atlantica* exopolymer. *Appl Spectrosc*, **43**, 1062–1067.

Lawrence J.R., Korber D.R., Hoyle B.D., Costerton J.W. (1991). Optical sectioning of microbial biofilms. *J Bacteriol*, **173**, 6558–6567.

MacCarthy P., Mark H.B., Griffiths P.R. (1975). Direct measurement of the infrared spectra of humic substances in water by Fourier transform infrared spectroscopy. *J Agric Food Chem*, **23**, 600–602.

Mack W.N., Mack J.P., Ackerson A.O. (1975). Microbial film development in a trickling filter. *Microb Ecol*, **2**, 215–226.

Mattson J.S., Smith C.A., Paulsen K.E. (1975). Infrared internal reflection spectrometry of aqueous protein films at the germanium-water interface. *Anal Chem*, **47**, 736–738.

McCoy W.F., Bryers J.D., Robbins J., Costerton J.W. (1981) Observations of fouling biofilm formation. *Can J Microbiol*, **27**, 910–917.

Mittelman M.W., Nivens D.E., White D.C. (1990). Differential adhesion, activity, and carbohydrate:protein ratios of *Pseudomonas atlantica* monocultures attaching to stainless steel in a linear shear gradient. *Microb Ecol*, **19**, 269–278.

Morris, H.R., Hoyt C.C., Miller P., Treado, P.J. (1996). Liquid crystal tunable filter Raman chemical imaging. *Appl Spectrosc*, **50**, 805–811.

Naumann D. (1984). Some ultrastructural information on intact, living bacterial cells and related cell-wall fragments as given by FTIR. *Infrared Phys*, **24**, 233–238.

Naumann D., Helm D., Labischinski H. (1991a). Microbiological characterizations by FT-IR spectroscopy. *Nature (Lond)*, **351**, 81–82.

Naumann D., Helm D., Schultz C. (1994). Characterization and identification of micro-organisms by FT-IR spectroscopy and FT-IR microscopy. In: Priest F.G., Ramos-Cormenzane A., Tindall B.J. (eds) *Bacterial Diversity and Systematics*. Plenum Press, New York, pp. 67–85.

Naumann D, Schultz CP, Helm D (1996). What can infrared spectroscopy tell us about the structure and composition of intact bacterial cells? In: Mantsch H.H., Chapman D. (eds) *Infrared Spectroscopy of Biomolecules*. Wiley-Liss, New York, pp. 279–310.

Naumann D., Helm D., Labischinski H., Giesbrecht P. (1991b). The characterization of microorganisms by Fourier-transform infrared spectroscopy. In: Nelson W.H. (ed) *Modern Techniques for Rapid Microbiological Analysis*. VCH, New York, pp. 43–96.

Naumann D., Barnickel G., Bradaczek H., Labischinski H., Giesbrecht P. (1982). Infrared spectroscopy, a tool for probing bacterial peptidoglycan. *Eur J Biochem*, **125**, 505–515.

Nickel J.C., Ruseska I., Wright J.B., Costerton J.W. (1985). Tobramycin resistance of *Pseudomonas aeruginosa* cells growing as a biofilm on urinary catheter material. *Antimicrob Agents Chemother*, **27**, 619–624.

Nichols P.D., Henson M.J., Guckert J.B., Nivens D.E., White D.C. (1985). Fourier transform-infrared spectroscopic methods for microbial ecology: analysis of bacteria, bacteria-polymer mixtures and biofilms. *J Microbiol Methods*, **4**, 79–94.

Nichols W.W., Dorrington S.M., Slack M.P.E., Walmsley H.L. (1988). Inhibition of tobramycin diffusion by binding to alginate. *Antimicrob Agents Chemother*, **32**, 518–523.

Niedhardt F.C., Ingraham J.L., Schaechter M. (1990). *Physiology of the Bacterial Cell*. Sinauer Associates Incorporated, Sunderland, MA.

Nivens D.E., Nichols P.D., Henson J.M., Geesey G.G., White D.C. (1986). Reversible acceleration of the corrosion of AISI 304 stainless steel exposed to seawater induced by growth and secretions of the marine bacterium *Vibrio natriegens*. *Corrosion*, **42**, 204–210.

Nivens D.E., Schmit J., Sniatecki J., Anderson T., Chambers J.Q., White D.C. (1993a). Multichannel ATR/FT-IR spectrometer for on-line examination of microbial biofilms. *Appl Spectrosc*, 47, 668–671.

Nivens D.E., Chambers J.Q., Anderson T.R., Tunlid A., Smit J., White D.C. (1993b). Monitoring microbial adhesion and biofilm formation by attenuated total reflection/ Fourier transform infrared spectroscopy. *J Microbiol Methods*, 17, 199–213.

Patterson M.K., Husted G.R., Rutkowski A. (1991). Bacteria: isolation, identification, and microscopic properties of biofilms in high-purity water distribution systems. *Ultrapure Water*, **8**, 18–24.

Pedraza A.J., Godbole M.J., Lowndes D.H., Thompson J.R. (1989). Enhanced metal-ceramic adhesion by sequential sputter deposition and pulsed laser melting of copper films on sapphire substrates. *J Mater Sci*, 24, 115–123.

Pedraza A.J., Godbole M.J., Bremer P.J., Avci R., Drake B., Geesey G.G. (1993). Stability in aqueous media of 316L stainless steel films deposited on internal reflection elements. *Appl Spectrosc*, 47, 161–166.

Poje G., O'Connor J.M., Ginn T.C. (1982). Physical simulation of power plant condenser tube passage. *Water Res*, 16, 921–928.

Powell J.R., Wascz F.M., Jakobsen R.J. (1986). An algorithm for the reproducible spectral subtraction of water from the FT-IR spectra of proteins in dilute solutions and adsorbed monolayers. *Appl Spectrosc*, 40, 339–344.

Pringle J.H., Fletcher M. (1986). Influence of substratum hydration and adsorbed macromolecules in bacterial attachment to surfaces. *Appl Environ Microbiol*, **51**, 1321–1325.

Ruckenstein E., Gourisankar S.(1983). A nondestructive approach to characterize deposits on various surfaces. *J Colloid Interface Sci*, **96**, 245–250.

Schaeberle M.D., Karakatsanis C.G., Lau, C.J. and Treado, P.J. (1995). Raman chemical imaging: noninvasive visualization of polymer blend architecture. *Anal Chem*, 67, 4316–4321.

Strauss S.D. (1985). Condenser-biofouling control looms large in light of toxic-discharge deadline. *Power*, **129**, 83–85.

Suci P.A., Geesey G.G. (1998). Molecular level approach in microbial adhesion to inert surfaces. *Recent Res Dev Microbiol*, 2, 107–113.

Suci P.A., Pedraza A.J., Godbole M.J., Geesey G.G. (1993).Use of sputter-deposited 316L-stainless steel ultrathin films for microbial influenced corrosion studies. In: Interrante C.G., Pabalan R.T. (eds) *Materials Research Society Symposium Proceedings, Vol. 294, "The Scientific Basis for Nuclear Waste Management XVI"*. Materials Research Society, Pittsburgh, PA, pp. 381–385.

Suci P.A., Mittelman M.W., Yu F.P., Geesey G.G. (1994). Investigation of ciprofloxacin penetration into *Pseudomonas aeruginosa* biofilms. *Antimicrob Agents Chemother*, **38**, 2125–2133.

Suci P.A., Siedlecki K.J., Palmer R.J., White D.C., Geesey G.G. (1997). Combined light microscopy and attenuated total reflection Fourier transform infrared spectroscopy for integration of biofilm structure, distribution and chemistry at solid-liquid interfaces. *Appl Environ Microbiol*, 63, 4600–4603.

Surman S.B., Walker J.T., Goddard D.T., Morton L.H.G., Keevil C.W., Weaver W., Skinner A., Hanson K., Caldwell D., Kurtz J. (1996). Comparison of microscopic techniques for the examination of biofilms. *J Microbiol Methods*, **25**, 57–70.

Taylor G.T., Troy P.J., Nullet M., Sharma S.K., Liebert B.E. (1994). Protein adsorption from seawater onto solid substrata: II. Behavior of bound protein and its influence on interfacial properties. *Mar Chem*, 47, 21–39.

Taylor G.T., Troy P.J., Nullet M., Sharma S.K., Liebert B.E., Mower H.F. (1993). Spectroscopic examination of protein adsorption from seawater onto titanium. *Appl Spectrosc*, 47, 1140–1151.

Taylor G.T., Zheng D., Lee M., Troy P.J., Gyananath G., Sharma S.K. (1996). Influence of surface properties on accumulation of conditioning films and marine bacteria on substrata exposed to oligotrophic waters. *Biofouling*, **11**, 31–57.

van der Mei H.C., Noordmans J., Busscher H.J. (1989). Molecular surface characterization of oral streptococci by Fourier transform infrared spectroscopy. *Biochem Biophys Acta*, **991**, 395–398.

Vrany J.D., Stewart P.S., Suci P.A. (1997). Comparison of recalcitrance of ciprofloxacin and levofloxacin exhibited by *Pseudomonas aeruginosa* biofilms displaying rapid transport characteristics. *Antimicrob Agents Chemother*, **41**, 1352–1358.

White D.C., Buchanan, R.A., Dowling N., Danko J.C., Zhang X. (1988). Microbiological corrosion of stainless steel. In: DeLucia D.E., Diller T.E., Prager M. (eds) *Bioprocess Engineering Symposium – 1988*. Am Soc Mech Engineer New York, NY, pp. 89–95.

White D.C., Nivens D.E., Mittelman M.W., Chambers J.Q., King J.M.H., Sayler G.S. (1991). Non-destructive on-line monitoring of MIC. Paper No. 114, *Corrosion 91*. National Association of Corrosion Engineers, Houston, TX.

White D.C., Nivens D.E., Nichols P.D., Kerger B.D., Henson M.J., Geesey G.G., Clarke C.K. (1986). Corrosion of steels induced by aerobic bacteria and their exocellular polymers. In: Dexter S.C. (ed) *Biologically Induced Corrosion*. National Association of Corrosion Engineers, Houston, TX, pp. 233–243.

Wilks P.A. (1982). Sampling method makes on-stream ir analysis work. *Ind Res Dev*, Sept, 132–135.

Winters S., Gendreau R.M., Leininger R.I., Jakobsen R.J. (1982). Fourier transform infrared spectroscopy of protein adsorption from whole blood: *ex vivo* sheep studies. *Appl Spectrosc*, **36**, 404–409.

Wragg J.L., White H.W. (1991). *In situ* infrared spectroscopy. Paper No. 77, *Corrosion 91*. National Association of Corrosion Engineers, Houston, TX.

Zeroual W., Choisy C., Doglia S.M., Bobichon H., Angiboust J.-F., Manfait M. (1994). Monitoring of bacterial growth and structural analysis as probed by FT-IR spectroscopy. *Biochim Biophys Acta*, 1222, 171–178.

16 Surface Catalysed Hygiene and Biofilm Control

Peter Gilbert and David G. Allison

Reaction-diffusion limitation across the extracellular polymeric matrices that comprise the bulk of microbial biofilms is probably the major cause of failure of oxidising biocides to sanitise contaminated surfaces. Strategies have been developed by the authors that overcome this problem through the catalytic generation of highly-active biocidal species, at the colonised surfaces. Transition metal catalysts such as cobalt phthalocyanine break down peroxides and persulphates to give active oxygen-species. The inclusion of these catalysts within materials, or as surface coatings markedly increases the effectiveness of such agents against biofilms formed upon them. Additionally, there are significant improvements in the ease with which the surfaces may be subsequently cleansed. Improved hygiene is generated not only through a concentration of active species at the biofilm-substratum interface but also through the creation of a catalysis-driven diffusion pump. The latter increases the flux of treatment agent across the biofilm and provides enhanced penetration and kill. This general approach is also applicable to situations where the catalysts are enzymes and generate active killing species from glucose and halogens in an analogous fashion to the bactericidal actions within phagosomal inclusions.

KEY WORDS: catalysis, hydrogen peroxide, potassium monopersulphate, *Pseudomonas aeruginosa*, poloxamer hydrogels, cobalt phthalocyanine, copper phthalocyanine

INTRODUCTION

Biofilms are notoriously difficult to control by the use of conventional chemical agencies or antibiotics and can often survive on a surface through relatively harsh hygienic cleaning protocols (Characklis, 1990; Little *et al.*, 1990; Holah *et al.*, 1994). Much of this recalcitrance has been attributed to a reaction diffusion limitation to the passage of chemically reactive biocides such as peroxides, isothiazolones, and halogens, across the glycocalyx (De-Beer *et al.*, 1994; Huang *et al.*, 1995). The action of such agents has also been noted to change the visco-elastic properties of the glycocalyx, in some instances firmly cementing the biofilms to the treated surface. This renders the cleansing of the surface more problematic and leaves residues that promote re-establishment of a new biofilm post-treatment. Other factors that contribute to the reduced susceptibility of the biofilm community relate to the much reduced growth rate displayed by deeply-seated cells brought about by nutrient-deprivation. This sub-population of cells may, in many instances, be regarded as metabolically dormant (Lewandowski *et al.*, 1991; Fletcher, 1992; Mozes and Rouxhet, 1992). Slow-growing and dormant cells are co-incidentally less susceptible to the actions of most chemical antimicrobials and antibiotics (Costerton *et al.*, 1987; Brown *et al.*, 1990; Gilbert *et al.*, 1990).

Many different strategies have been adopted in order to increase the delivery of hygiene (killing and removal) to surfaces that are contaminated with biofilms (Stickler, 1997). These include (i) modification of the physico-chemical properties of the surface (surface free-energy, charge, roughness) rendering it less 'sticky' with respect to microorganisms (McAllister *et al.*, 1993), (ii) incorporation of antimicrobial agents within the surface or within a coating material (Bayston *et al.*, 1989; McLean *et al.*, 1993; Bayston, 1994; Stanton and Bayston, 1999), (iii) application of water-erodible coatings from which the release of biocide into attached cells facilitates hydrolytic ablation of the surface and removal of accumulated soils (Eastwood, 1994; Suzangar *et al.*, 1999), (iv) combinations of ii and iii (Eastwood, 1994), and (v) screening for specific anti-biofilm/anti-adhesion agents. Such methods have met with varying degrees of failure. In this respect surface-modifications made to reduce microbial attachment have been shown only to delay the onset of colonisation, since reservoirs of antimicrobial and erodible surfaces become depleted with time, and the search for agents with specific anti-biofilm activity has been largely unsuccessful until the recent discovery of anti-quorum-sensing chemicals such as the furanones (de Nys *et al.*, 1995; Givskov *et al.*, 1996).

Alternative approaches have been developed by the authors where the biocide is generated by catalysis at the colonised surface from a relatively innocuous treatment agent (Gilbert and Jones, 1996; Wood *et al.*, 1996; 1998). Provided that catalytic activity is maintained and fresh treatment agent is available, such approaches will continually provide hygiene agent to the interactive surface between the biofilm and the substratum. The catalysts might be inorganic chemicals or enzymes or enzyme combinations and they may be incorporated within the materials which comprise the surface, or applied as a conditioning film.

METHODS

Growth of Biofilms

A stable, mucoid clinical isolate of *Pseudomonas aeruginosa* (PaWH) was used throughout. Biofilms were variously formed on catalyst-containing and control discs (see below) by growth in tryptone soya broth (Oxoid CM129) at 37°C in submerged liquid culture (Wood *et al.*, 1996), or in a Constant Depth Film Fermentor (CDFF, 100μm thickness) (Wimpenny *et al.*, 1993). Biofilms were grown for confocal microscopy in Perspex flow-cells (Wolfaardt *et al.*, 1994), constructed with catalyst-containing and control coverslips prepared from Trylon resin (see below).

Preparation of Test Materials

Trylon resin coupons

Trylon resins (Trylon Ltd, Wollaston, UK), with and without the catalysts cobalt sulphonated phthalocyanine, (CoPC 100 μg ml^{-1}) or copper sulphonated phthalocyanine, (CuPC, 100 μg ml^{-1}), were cast as cylinders (2.5 cm and 5 mm diameter). These were machined into 3 mm thick discs (Wood *et al.*, 1996) suitable for use in batch culture and the CDFF respectively. Resin sheets were also cast between glass plates separated by 0.15 mm spacers. Such sheets had optical properties that were suitable as microscope cover slips. Slips, with and without catalyst, were cut and mounted on flow chambers (see above).

Alginate sheet

Sodium alginate (Sigma, 3%w/v), with and without the catalysts (100 μg ml^{-1}), was poured onto levelled glass plates and allowed to dry in air at room temperature for 48h. Sheets, approximately 150 μm thick were cut into 2 × 2 cm squares and immersed separately into 0.5M calcium acetate (10 min, constant agitation). The cross-linked sheets were washed three times in distilled water, autoclaved at 121°C in water, and stored.

Poloxamer hydrogels

Poloxamer F127 (ICI, Wilton, UK) is a di-block co-polymer of polyoxyethylene and polyoxypropylene. Aqueous solutions show thermo-reversible gellation, being liquid at temperatures <15°C and robust gels at temperatures >15°C. Such gels may be used to generate biofilm constructs for testing purposes (Gilbert *et al.*, 1998; Wirtanen *et al.,* 1998). Poloxamer flakes were made up to 30w/v in tryptone soya broth (Oxoid), with and without catalysts (100 μg ml^{-1}) and refrigerated overnight in order for hydration to take place. Chilled poloxamer solutions were inoculated with 10^4 cfu ml^{-1} of a stationary phase culture of *P. aeruginosa.* Drops (100 μl) were placed onto borosilicate glass or alginate cover slips. These were held within sealed Petri dishes containing moistened cotton wool and incubated for 16 h at 37°C. During this time cell numbers increased exponentially up to 5×10^{10} ml^{-1}. Viable counts could be performed on the gels by transferring to chilled diluents.

Susceptibility Testing of Biofilm and Planktonic Bacteria

The susceptibility towards exposure to potassium monopersulphate (KMPS) and hydrogen peroxide (0–4 mg ml^{-1}), of variously grown *P. aeruginosa* cultures was assessed by serial dilution and surface plate count, following exposure to the agents for 30 min at room temperature and neutralisation in sodium thiosulphate (9 mg ml^{-1}).

Intact biofilms, cultured in batch culture or within the CDFF, together with their substrate/support-resins were placed directly into solutions of the treatment agents.

After exposure, viable counts were performed on the neutralised reaction mixtures, on suspensions of biofilm cells prepared in neutraliser (10 ml) by vigorous shaking in a Griffin flask shaker, and upon the treated surfaces by the plate blot succession method of Eginton *et al.* (1995).

Biofilms prepared within poloxamer resins were exposed by immersion in treatment agent at 37°C, removed after 10 min. and transferred to neutraliser solutions at 10°C. After *ca* 5 min the poloxamer gels had dissolved and further dilutions and viable counts of the cell suspension could be made.

Confocal Microscopy

Biofilms were cultured within flow cells and visualised on the underside of catalyst-containing and control Trylon-resin cover-slips using a Nikon FXA Microphot epifluorescence microscope mounted on a BioRad MRC-600 laser scanning confocal imaging system employing an argon laser source. Horizontal optical thin sections were made at 5 μm increments and/or saggital sections through the depth of the biofilm (*ca* 40 μm) at randomised fields.

After various periods of growth or treatment, the input lines to the flow chambers were clamped and 1 ml of either fluorescein (0.1%w/v, Molecular Probes, Oregon,

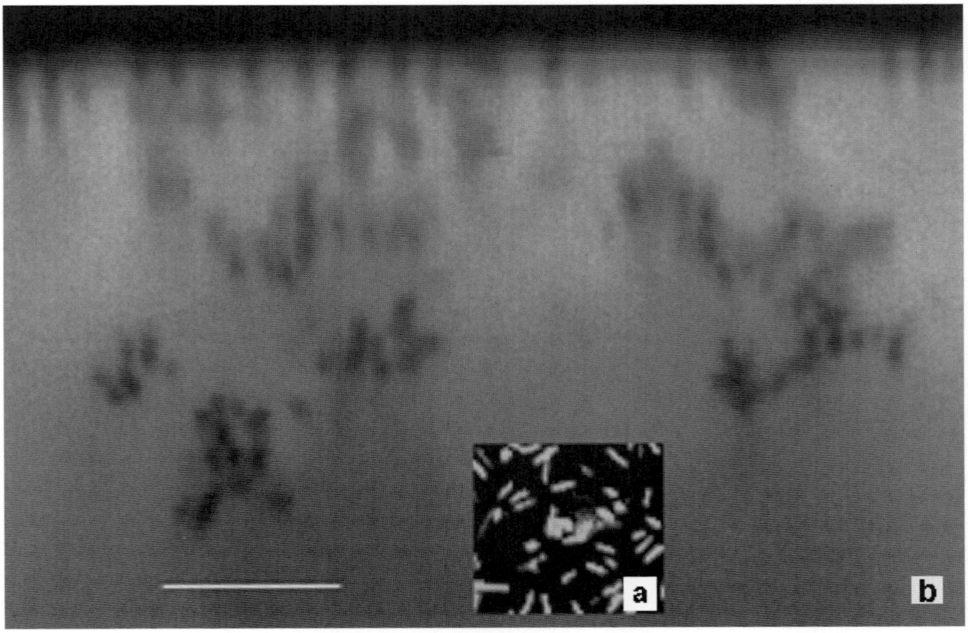

Figure 1 Laser confocal image of a *P. aeruginosa* biofilm formed on a cobalt phthallocyanin – containing Trylon cover slip forming part of a parallel-plate flow chamber. (a) transverse section through the biofilm, 2 μm from the surface of the cover slip, stained with BacLight viability stain and showing the live channel (b) in saggital section through the biofilm and cover slip, negative stained with fluorescein. Bars = 10 μm.

USA), BacLight viability stain (Molecular Probes, Oregon, USA), or 9-((acridinecarbonyl) amino)-2,2,6,6-tetramethylpiperidin-1-oxyl (Tempo-9-AC, Molecular Probes, Oregon, USA) was added to the chamber. BacLight can distinguish between metabolically active and inactive cells. In this instance chambers were flushed to remove residual treatment agent, re-clamped and incubated with the stain for 10 min before viewing. Negative staining with fluorescein and visualisation of free radicals (Temp-9-AC) was immediate and could be conducted in the presence of treatment agent.

RESULTS AND DISCUSSION

Biofilm formation upon the Trylon-resin discs was reduced by *ca* 90%, from 3×10^6 cfu cm^{-2} to 3×10^5 cfu cm^{-2}, by inclusion of CoPC or CuPC respectively. The presence of these catalysts, however, had no significant effects upon either biofilm formation within the CDFF or upon the growth of *P. aeruginosa* within incubated poloxamer hydrogels. In the flow chambers, confocal microscopy showed the biofilms to uniformly coat the surfaces (Figure 1a) and to project microcolonies, approximately 30–40 μm thick into the surrounding medium (Figure 1b). Biofilms grown on control Trylon discs were markedly less susceptible to KMPS and hydrogen peroxide than the planktonic suspensions (Figure 2). Susceptibility of the biofilms to KMPS and

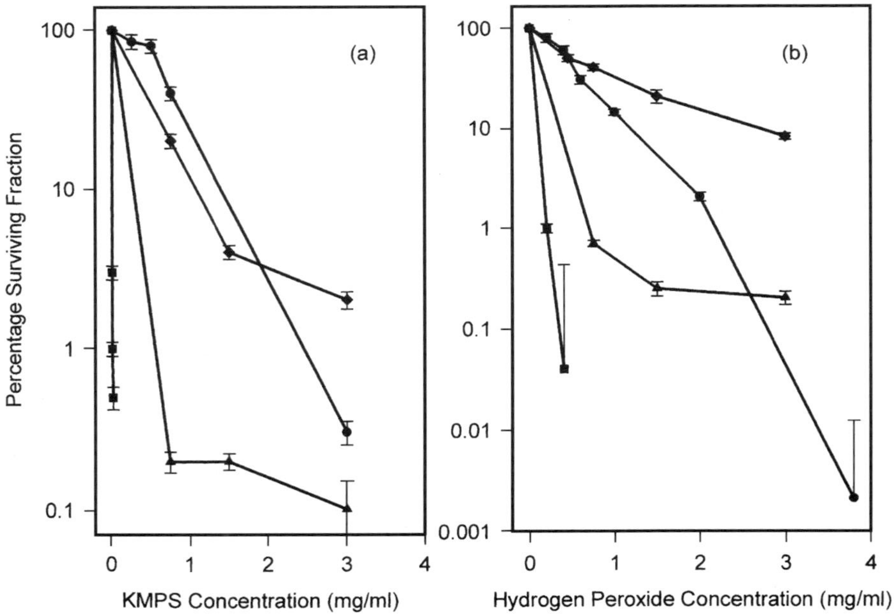

Figure 2 Effect of exposure for 30 min at room temperature to various concentrations of KMPS (a) and H_2O_2 (b) on the survival of *P. aeruginosa* grown planktonically (●) and as biofilms on Trylon discs without (◆) and with incorporation of a CuPC (▲) or CoPC (■) catalyst. Survivors of treated biofilms were estimated from the numbers of cfu removed from saline during rinsing and vigorous shaking. Bars = SEs.

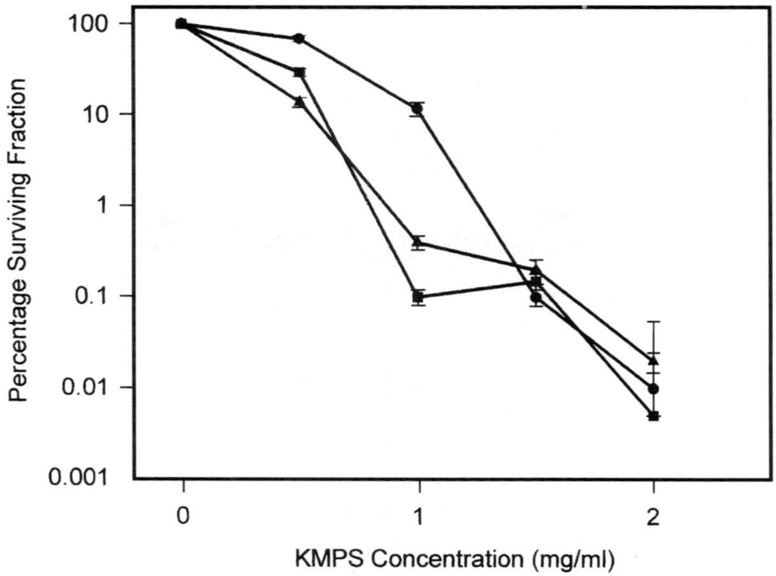

Figure 3 Effects of exposure for 30 min at ambient temperature to various concentrations of KMPS on the survival of *P. aeruginosa* grown in a constant depth film fermenter on Trylon-discs without (●) and with incorporation of CoPC (■) or CuPC (▲) catalysts. Survivors of treated biofilms were estimated as those recovered by scraping. Bars = SEs.

peroxide was however markedly enhanced whenever catalyst was incorporated in the substratum (Figure 2). In all instances catalysed killing of biofilm bacteria occurred at lower concentrations of treatment agent than were required for killing even the planktonic bacteria. In such respects CoPC was more effective than CuPC both as a catalyst for the decomposition of peroxides and persulphates, and as an enhancer of hygiene delivery. Significant improvements in delivered hygiene were apparent even with the relatively thick biofilms (100 μm) generated in the CDFF (Figure 3).

BacLight stains were employed to visualise the killing of biofilm bacteria formed on a catalyst coated cover slip exposed to KMPS (100 μg ml^{-1}) in the flow chamber (Figure 4). After a 10 min exposure cells within 5 μm of the catalyst containing surface, but not those further into the biofilm, had been killed. This indicated that the killing process had commenced at the substratum and then migrated outwards towards the fluid phase. After 30 min exposure all the cells within the 40–μm-thick microcolonies had been killed. Likewise, relatively low concentrations of KMPS (20 μg ml^{-1}) and peroxide (500 μg ml^{-1}) on CoPC-containing surfaces did not permit surviving cells to be recovered from the batch grown biofilms after 30 min treatment. Reactive species such as those generated in these systems would not be expected to survive long enough to travel more than a few micrometres from their point of generation. This was confirmed by confocal micrographs taken through biofilms, in the presence of KMPS, using the general free-radical probe Tempo-9-AC (Figure 5) which demonstrated a concentration of free radicals to a depth of approximately 5μm from the surface and

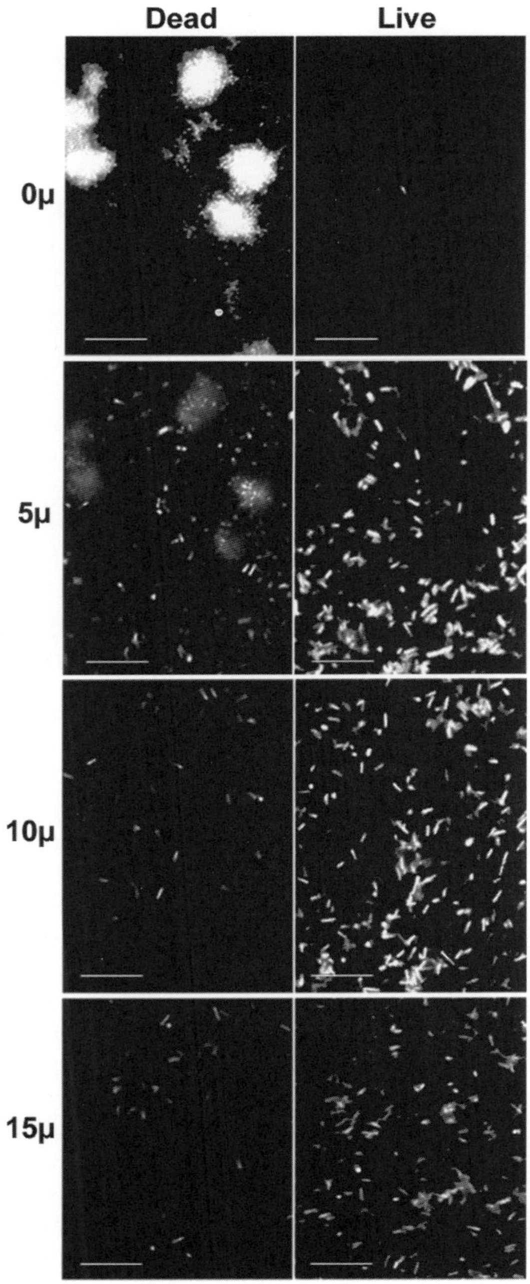

Figure 4 Laser confocal images of transverse sections taken at distances of 0–15 µm from a CoPC-containing surface colonised with *P. aeruginosa*. The biofilms had been exposed to potassium monopersulphate (100 µg ml^{-1}) for 10 min prior to staining with BacLight viability stain. Images in the right hand column represent the live channel; images in the left hand column represent the dead channel. Bars = 10 µm.

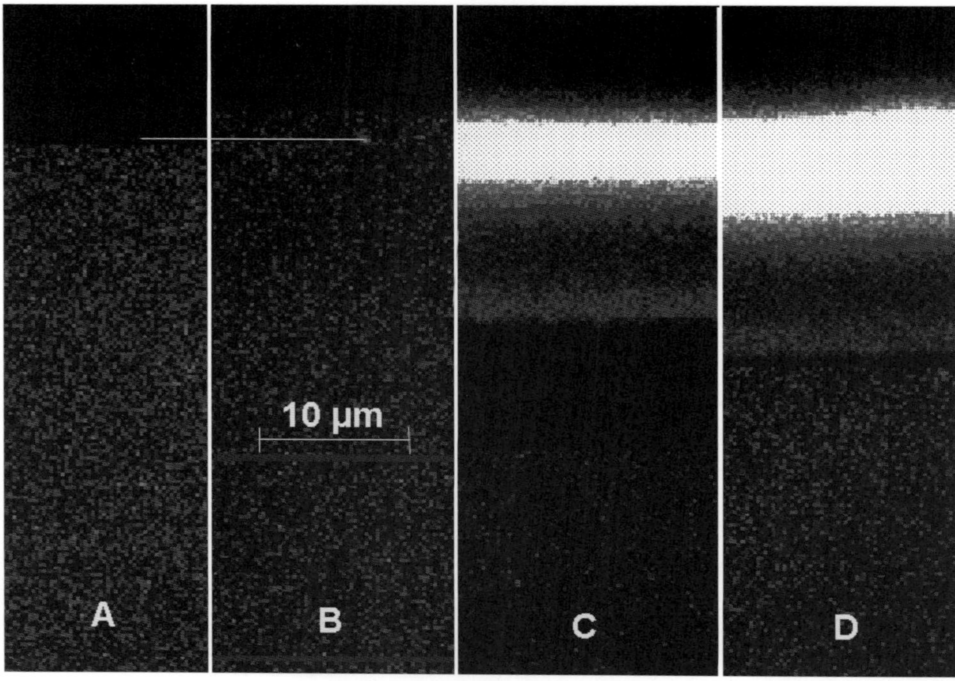

Figure 5 Laser confocal images of z-plane sections taken through a CoPC-containing (B, C and D) and control (A) surfaces colonised with *P. aeruginosa*. The biofilms had been exposed to potassium monopersulphate (100 µg ml^{-1}) in the presence of TEMPO-9-AC free radical probe for 30 min (A) and (D), 3 min (B), and 10 min (C). Bars = 10 µm.

within the resin body itself. If the enhancement of killing by the presence of catalyst was due solely to the generation of active oxygen species at the biofilm substratum interface it would likely only relate to a small fraction of cells immediately in contact with the surface. Vital staining however, indicated this not to be the case. An explanation of this apparent anomaly was that the catalytic degradation of the treatment agent at the substratum would establish a diffusion pump whereby further treatment agent would be drawn through the biofilm and delivered to the substratum. This would increase the 'flux' of treatment agent across the biofilm and, where the treatment agent is inherently active, lead to an increased delivery of hygiene throughout the film. This hypothesis was examined through the use of biofilm-constructs (Gilbert *et al.*, 1998; Wirtanen *et al.*, 1998), several millimetres in thickness, placed upon catalyst-containing alginate sheets. Constructs were prepared with poloxamer hydrogels. These exhibit thermo-reversible gellation and exist as gels at temperatures >15°C, but are liquid at temperatures <15°C. *P. aeruginosa* cells were incorporated into poloxamer hydrogels made with tryptone soya broth as diluent and incubated at 37°C in a humid environment. The enveloped cells showed a classic growth curve with lag, log and stationary phases. Cell densities achieved in the gels at stationary phase were *ca* 5×10^{10} cfu ml^{-1}. These populations possessed an outer membrane

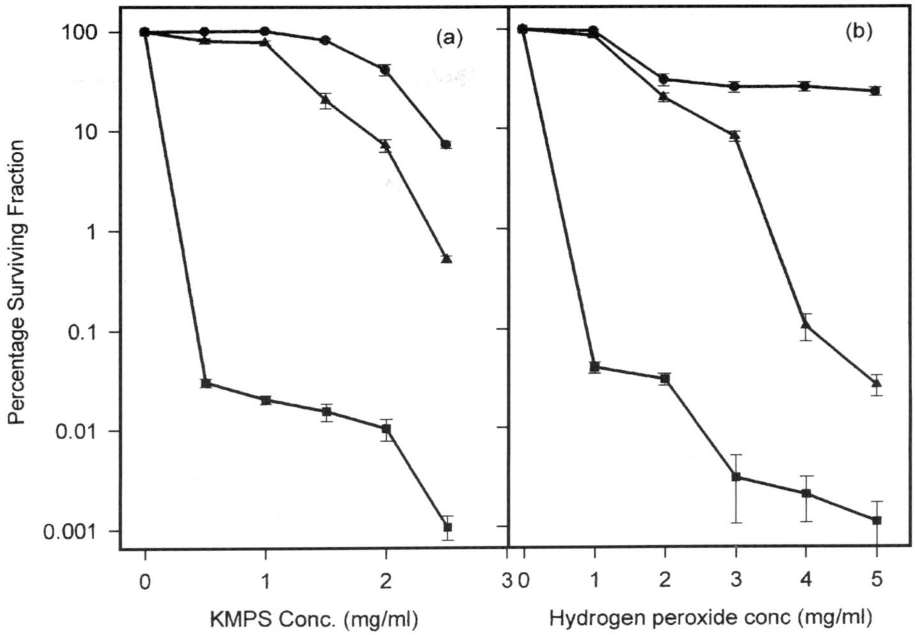

Figure 6 Survival of *P. aeruginosa* following exposure to various concentrations and (a) potassium monopersulphate and (b) H_2O_2 for 20 min. Cultures had been incorporated into a Poloxamer hydrogel containing growth medium and incubated as a gel for 16 h. Hydrogels were formed onto alginate sheets with (▲) or without (●) CoPC catalyst (100 µg ml^{-1}). Hydrogels where the CoPC catalyst was included within the hydrogel (■) were also tested. Bars = SEs.

protein composition characteristic of the biofilm phenotype and they were also resistant to a range of treatment agents, including quaternary ammonium compounds, halogens and peroxide (Gilbert *et al.*, 1998). The generated gels were dome shaped with diameters of *ca* 1.2 cm and a maximum thickness at the centre of 2–3 mm. Hydrogels, together with their support material (alginate with and without catalyst) were immersed in KMPS or peroxide solutions for up to 20 min, removed, and the cells recovered by dilution in cold diluent. The results are presented in Figure 6. Inclusion of the catalyst within the alginate substratum led to significant enhancements in killing (1–3 logs) over the controls. If the additional hygiene delivered in these systems was due to generation of active oxygen species *per se* then action would be confined to a 5 µm thick layer adjacent to the alginate sheet. In thick biofilms such as these, such active species could reach less than 0.005% of the hydrogel-enveloped cells and would not therefore be expected to generate a measurable enhancement of killing. The additional hygiene generated must therefore be attributed to an increased 'flux' of biocide through the polymer matrix. Significantly, when the catalyst was incorporated in the hydrogel-matrix sterilisation of the construct could be achieved within 20 min by relatively low concentrations of hydrogen peroxide (6 mg ml^{-1}).

This indicates that active oxygen species had been generated throughout the enveloped cell population, and in close proximity to the cell surfaces. One implication of these results is that if a catalyst could be impregnated into a mature biofilm before exposure to treatment agent then effective killing might be delivered to biofilms associated with almost any surface. Inorganic catalysts such as those described in this chapter would not suffer reaction-diffusion problems of access through the biofilm. Impregnation of a surface might be achieved by using the poloxamer hydrogels to deposit thin gels upon a biofilm coated pre-warmed, hard surface. Secondary treatments with dilute peroxide would inactivate the catalyst-encased biofilm, and the residual materials might then be dissolved off the surface by the application of cold water. This is a very similar approach to that of Wilson (1995) where oral biofilms are impregnated with a photosensitising agent prior to the application of laser light at an appropriate wavelength. Both approaches benefit from continuous delivery of an extrinsic, highly reactive species within the depths of the biofilm.

CONCLUSIONS

Generation of a biocidal species from a less active treatment agent at a colonised surface significantly enhances the effectiveness of sanitisation treatments. Whilst the principle of surface catalysed hygiene has been demonstrated with inorganic catalysts of the degradation of peroxides, similar systems where enzyme catalysts, such as glucose oxidase and chloroperoxidase, are used to generate biocidal species from innocuous treatment agents such as glucose, oxygen and chloride can be envisaged. It is possible that many types of surface could be manufactured containing a variety of catalysts and these might cover applications as diverse as extruded plastic pipework, chopping boards, laminated work-surfaces, ceramic glazes, grouting materials, and paints non-woven webs.

REFERENCES

Bayston R. (1994). A process for prevention of device-related infection: results of challenge *in vitro* and duration of antimicrobial activity. In: Wimpenny J.T., Handley P.S.H., Gilbert P., Lappin-Scott H.M. (eds) *The Life and Death of Biofilm*. Bioline Press, Cardiff, UK, pp. 149–152.

Bayston R., Grove N., Siegel J., Lawellin D., Barsham S. (1989). Prevention of hydrocephalus shunt catheter colonisation *in vitro* by impregnation with antimicrobials. *J Neurol Neurosurg Psychiatry*, 52, 605–609.

Brown M.R.W., Collier P.J., Gilbert P. (1990). Influence of growth rate on susceptibility to antimicrobial agents: modification of the cell envelope and batch and continuous culture studies. *Antimicrob Agents Chemother*, 34, 1623–1628.

Characklis W.G. (1990). Microbial fouling. In: Characklis W.G., Marshall K.C. (eds) *Biofilms*. John Wiley, New York, USA, pp. 523–584.

Costerton J.W., Cheng K.J., Geesey K.G., Ladd P.I., Nickel J.C., Dasgupta M., Marrie T.J. (1987). Bacterial biofilms in nature and disease. *Annu Rev Microbiol*, 41, 435–464.

De-Beer D., Srinivasan R., Stewart P.S. (1994). Direct measurement of chlorine penetration into biofilms during disinfection. *Appl Environ Microbiol*, 60, 4339–4344.

De Nys R., Steinberg P.D., Willemsen P., Dwarjanyn S.A., Gabelish C.L., King R.J. (1995). Broad spectrum effects of secondary metabolites from the red alga *Delisea pulchra* in antifouling assays. *Biofouling*, **8**, 259–271.

Eastwood I.M. (1994). Problems with biocides and biofilms. In: Wimpenny J.T., Nichols W.W., Stickler D., Lappin-Scott H.M. (eds) *Bacterial Biofilms and their Control in Medicine and Industry*. Bioline Press, Cardiff, UK, pp. 169–172.

Eginton P.J., Gibson H., Holah J., Handley P.S., Gilbert P. (1995). Quantification of the ease of removal of bacteria from surfaces. *J Ind Microbiol*, **15**, 305–310.

Fletcher M. (1992). Bacterial metabolism in biofilms. In: Melo L.F., Bott T.R., Fletcher M., Capdeville B. (eds) *Biofilms – Science and Technology*. Kluwer Academic Publishers, Dordrecht, pp. 113–124.

Gilbert P., Jones M. (1996). Articles, compositions and process for cleaning surfaces by using a catalyst at the surface. UK patent 8th February 1996 (C3597); International (PCT) Patent PCT/EP/96/00443 (31st January 1996); India Patent 85/BOM/96 (8th Feb 1996); South Africa Patent 96/0736 (31st January 1996); Thailand Patent 29948/96 (6th Feb 1996).

Gilbert P., Collier P.J., Brown M.R.W. (1990). Influence of growth rate on susceptibility to antimicrobial agents: biofilms, cell cycle, dormancy, and stringent response. *Antimicrob Agents Chemother*, **34**, 1865–1868.

Gilbert P., Jones M.V., Allison D.G., Heys S. Maira T., Wood P. (1998). The use of poloxamer hydrogels for the assessment of biofilm susceptibility towards biocide treatments. *J Appl Microbiol*, **85**, 985–991.

Givskov M., de Nys R., Manefield M., Gram L., Maximilien R., Eberl L., Molin S., Steinberg P., Kjelleberg, S. (1996). Eukaryotic interference with homoserine lactone-mediated prokaryotic signalling. *J Bacteriol*, **178**, 6618–6622.

Holah J.T., Bloomfield S.F., Walker A.J., Spenceley H. (1994). Control of biofilms in the food industry. In: Wimpenny J.T., Nichols W.W., Stickler D., Lappin-Scott H.M. (eds) *Bacterial Biofilms and their Control in Medicine and Industry*. Bioline Press, Cardiff, UK, pp. 163–168.

Huang C.T., Yu F.P., McFeters G.A., Stewart P.S. (1995). Non-uniform spatial patterns of respiratory activity within biofilms during disinfection. *Appl Environ Microbiol*, **61**, 2252–2256.

Lewandowski Z., Walser G., Characklis W.G. (1991). Reaction Kinetics in biofilms. *Biotechnol Bioeng*, **38**, 877–882.

Little B.J., Wagner P.A., Characklis W.G., Lee W. (1990). In: Characklis W.G., Marshall K.C. (eds) *Biofilms*. John Wiley, New York, USA, pp. 635–670.

McLean R.J., Hussain A.A., Sayer M., Vincent P.J., Hughes D.J., Smith T.J. (1993). Antibacterial activity of multi-layer silver-copper surface films on catheter material. *Can J Microbiol*, **39**, 895–99.

McAllister E.W., Carey L.C., Brady P.G., Heller R., Kovacs S.G. (1993). The role of polymeric surface smoothness of biliary stents in bacterial adherence, biofilm deposition, and stent occlusion. *Gastrointest Endosc*, **39**, 422–425.

Mozes N., Rouxhet P.G. (1992). Influence of surfaces on microbial activity. In: Melo L.F., Bott T.R., Fletcher M., Capdeville B. (eds) *Biofilms – Science and Technology*. Kluwer Academic Publishers, Dordrecht, pp. 125–36.

Stanton C., Bayston R. (1999). Use of antimicrobial polymers to prevent device related infections. In: Wimpenny J.T., Gilbert P., Walker J., Brading M., Bayston R. (eds) *Biofilms, the Good, the Bad and the Ugly*. Bioline Press, Cardiff, UK, pp. 65–71.

Sticker D. (1997). Chemical and physical methods of biofilm control. In: Wimpenny J.T., Handley P., Gilbert P., Lappin-Scott H.M., Jones M. (eds) *Biofilms: Community Interactions and Control*. Bioline Press, Cardiff, UK, pp. 215–226.

Suzangar S., Allison D.G., Eastwood I., Gilbert P. (1999). An evaluation of the colonisation resistance of erodable biocide-containing coatings. In: Wimpenny J.T., Gilbert P., Walker J., Brading M., Bayston R. (eds) *Biofilms, the Good, the Bad and the Ugly*. Bioline Press, Cardiff, UK, pp. 53–64.

Wilson M. (1995). Local sensitisation of *Streptococcus sanguis* biofilms. In: Wimpenny J.T., Handley P.S.H., Gilbert P., Lappin-Scott H.M. (eds) *The Life and Death of Biofilm*. Bioline Press, Cardiff, UK, pp. 143–147.

Wimpenny J.W.T., Kinniment S.L., Scourfield M.A. (1993). The physiology and biochemistry of biofilm. In: Denyer S.P., Gorman S.P., Sussman M. (eds) *Microbial Biofilms: Formation and Control*. Blackwell Scientific Publications, Oxford, UK, pp. 51–94.

Wirtanen G., Salu S., Matilla-Sandholm T., Allison D.G., Gilbert P. (1998). Performance evaluation of disinfectant formulations using poloxamer-hydrogel biofilm-constructs. *J Appl Microbiol*, **85**, 965–972.

Wolfaardt G.M., Lawrence J.R., Robarts R.D., Caldwell S.J., Caldwell D.E. (1994). Multicellular organization in a degradative biofilm community. *Appl Environ Microbiol*, **60**, 434–446.

Wood P., Jones M., Bhakoo M., Gilbert P. (1996). A novel strategy for control of microbial biofilms through generation of biocide at the biofilm-surface interface. *Appl Environ Microbiol*, 62, 2598–2602.

Wood P., Caldwell D.E., Evans E., Jones M., Korber D.R., Wolfhaardt G., Wilson M., Gilbert P. (1998). Surface catalysed hygiene of thick *Pseudomonas aeruginosa* biofilms. *J Appl Microbiol*, **84**, 1092–1099.

17 *Legionella* Biofilms: their Implications, Study and Control

J. Barry Wright

In the more than two decades since the Gram-negative bacterium *Legionella pneumophila* was first isolated and named, considerable progress in understanding this organism and its related species has occurred. Since initial isolation, these bacteria have been demonstrated to be nearly ubiquitous aquatic organisms that can live in a variety of harsh environments. They have been shown to possess an exopolysaccharide that allows them to live in biofilms and this, together with their ability to infect and survive within amoebae, has been postulated to confer upon them the ability to survive in harsh environments. Their mode of growth also seems to play a role in their ability to infect susceptible hosts. A variety of molecular techniques has been developed to assist in the study of these unique organisms and these have allowed a significant increase in understanding of their ability to grow in various environments as well as to infect humans. Since they often grow in environments associated with institutions housing large numbers of individuals, efforts have been made to eradicate them from these environments. A number of methodologies have been proposed, including a new method that has elicited cautious optimism concerning the control of these bacteria. This review will describe recent advances in understanding of the mode of growth of these organisms in biofilms, methods to study them, and methods currently in use to control their growth.

HISTORY

In the summer of 1976 following a convention of the American Legion in Philadelphia, USA, a number of attendees fell ill with a febrile disease that progressed into a fulminate pneumonia. The pneumonia progressed rapidly and did not respond to normal chemotherapy. Over the course of a relatively short time, a number of the individuals who attended the meeting succumbed to the infection. As a result of the number of individuals who became ill and their common attendance at the convention, investigators at the Centers for Disease Control in Atlanta, USA, began an investigation to identify the causative agent (Fraser *et al.*, 1977). Following a number of fortuitous experiments, the causative agent of the Philadelphia disease was identified as a Gram-negative bacillus that was exceptionally fastidious in its nutritional requirements in the laboratory (McDade *et al.*, 1977). The name *Legionella pneumophila* was adopted for this previously undescribed organism (Brenner *et al.*, 1979).

Subsequent investigations utilizing frozen tissue samples revealed that *Legionella* spp. were not "new" organisms, as had originally been supposed, since cases of legionellosis were retrospectively diagnosed as having occurred at least as early as the 1940s (McDade *et al.*, 1979; Hebert *et al.*, 1980). In addition, it was discovered during the retrospective serological study that the pneumonic form of legionellosis (Legionnaires' disease) was not the only manifestation of an infection with legionellae. In 1968 an outbreak of a relatively mild febrile illness afflicted a number of individuals at a county health office in Pontiac, USA (Glick *et al.*, 1978). This self-limiting disease, described as Pontiac fever, was not associated with any mortalities and remained of an unknown aetiology until following the identification of the causative agent of the Philadelphia disease. Once the organism was identified and isolated, serum samples stored from the time of the Pontiac outbreak revealed that the causative agent was *Legionella pneumophila*.

During the two decades following the identification of *L. pneumophila*, at least 42 species of *Legionella* have been isolated and classified. In spite of the fastidious nature of the members of this genus when cultivated in the laboratory, the organisms have been demonstrated to be nearly ubiquitous inhabitants of natural (Fliermans *et al.*, 1981; Spino *et al.*, 1984; Fliermans, 1985) and man-made (Dondero *et al.*, 1980; Marrie *et al.*, 1987; States *et al.*, 1987) fresh water systems. Thus *Legionella* spp. appear to thrive in most freshwater sources, and the variety of locations in which they thrive demonstrates that they have a broad spectrum of nutrient concentration tolerance. These environments also represent a wide variety of temperatures; legionellae may be isolated from deep wells, where the water temperature is approximately 5°C, to hot water distribution systems where the water temperature is typically above 50°C. A few descriptions of the isolation of members of this genus from brackish water habitats have also appeared in the literature (Palmer *et al.*, 1993). In addition, the bacteria have been linked to the potable water supplies of buildings in which cases of legionellosis occurred (Stout and Yu, 1997). These water supplies are typically well chlorinated and have a low bacterial load. Another common source of legionellae is cooling tower water and this has also been linked with institutional outbreaks of disease. These environments, although less oligotrophic than potable water supplies are treated with biocides to minimize microbial growth. In addition, the cooling towers are generally located away from areas to which the general population has ready access.

LEGIONELLAE AS BIOFILM ORGANISMS

In the mid-1980s the possibility that legionellae, like the majority of bacteria, grow associated with biofilms began to be explored in detail. This proposition made inherent sense based on the general understanding of biofilms at the time.

The environments in which legionellae are found vary significantly in their general hospitality towards microorganisms. In the case of many surface water environments, numerous amoebae and other predators are present. Potable water supply systems are generally well chlorinated, minimizing the ability of microorganisms to survive and proliferate. Similarly, cooling towers are typically dosed with biocides on a regular basis to minimize the proliferation of bacteria and other water-borne organisms. Resistance to predators and chemical biocides, growth in oligotrophic environments

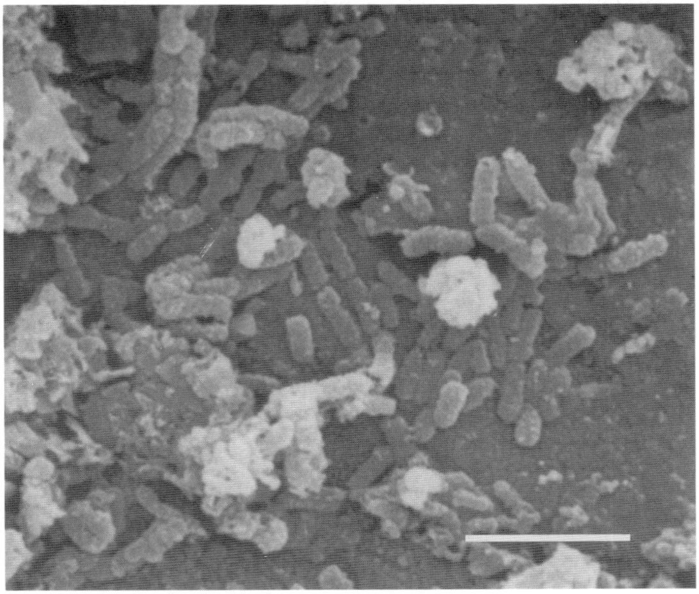

Figure 1 SEM photograph of *L. pneumophila* growing on a coupon placed in a influent water line supplying a hospital cooling tower. Bar = 5 μm. (Reproduced from Wright *et al.*, 1989, with permission.)

and survival under a wide range of temperatures are all characteristics of biofilm-associated bacteria.

Legionella spp. were eventually demonstrated to grow well in biofilms in the laboratory (Schofield and Locci, 1985). The bacteria were also demonstrated to grow in biofilms *in situ* in well water distribution pipes, on potable water distribution system components, and on cooling tower components (Wright *et al.*, 1989) (see Figure 1). Laboratory experiments eventually demonstrated that these organisms, like other biofilm-forming bacteria, elaborate extensive exopolysaccharide material (Wright and Costerton, 1986) that is postulated to be of significant importance in binding the cells to the substrate. The exopolymer has been demonstrated to be present on cells growing in pure culture in the laboratory, in pure and mixed culture on laboratory surfaces, as well as on engineered surfaces in a variety of industrial settings. Arguably, the exopolymer may play a role in allowing the bacteria to survive in such a wide diversity of environments. It is well documented that the growth of bacteria in exopolysaccharide-encapsulated microcolonies or in biofilms mediate a degree of protection to the organisms in that they demonstrate a significantly enhanced resistance towards disinfection by a variety of chemical agents (Ruseska *et al.*, 1982; Wright *et al.*, 1991).

Legionellae found in natural environments are also regularly associated with higher microorganisms, most notably amoebae (Rowbotham, 1984; Newsome *et al.*, 1998). In many instances, these bacteria have been demonstrated to be pathogenic for the amoebae. The infection and subsequent pathogenesis of amoebal infection by legionellae has been demonstrated to be very similar to the sequence of events during infection of mammalian macrophages (Horwitz, 1983; Newsome *et al.*, 1985). The association of legionellae with both biofilms and freshwater amoebae has a number of implications to both the survival and infectivitity of these potential pathogens.

INFECTION HYPOTHESES

High Density of Planktonic Cells

A number of scenarios exist that attempt to explain the ability of legionellae to survive in a large number of aquatic habitats and also be moderately successful as opportunistic pathogens. One of the first suggested that the bacteria, normally present in the water system associated with a building (either potable or cooling), reach a certain level in the system and are aerosolized and subsequently inhaled by an individual who may develop the disease. This explanation received early credence from animal experiments that demonstrated the infectivity of aerosolized *L. pneumophila* (Fitzgeorge *et al.*, 1983).

Amoebae

Another hypothesis for the mode of *Legionella* infection that has gained widespread acceptance is linked to the association of these bacteria with freshwater amoebae. The ability of various strains of *Legionella* spp. to infect different strains of amoebae in a manner analogous to the invasion of macrophages led to the suggestion that amoebae may harbour the bacteria and provide them with a vehicle for transmission. This is an especially important concept in view of the fact that *Legionella* spp. are not particularly resistant to desiccation (Berendt, 1980; Hambleton *et al.*, 1985). When it is considered that, in some instances of outbreaks of legionnaire's disease, the bacteria are transported over significant distances as an aerosol, it is apparent that they must be less susceptible to desiccation than some laboratory studies would indicate. Thus, encapsulation of virulent organisms within amoebal cysts would provide them with a means of traversing significant distances without losing viability as a result of desiccation within the comparatively dry air. Encapsulation in amoebal cysts would also provide the bacteria with a significant measure of resistance to disinfection, as will be discussed subsequently.

Biofilm

A third mechanism by which legionellae are able to enhance their survival and which may also account for their ability to survive and remain virulent following aerosolization in a dehydrating environment, is linked to their ability to grow adherent to a variety of surfaces as biofilms. It is possible that this mode of growth may also provide also a mechanism for the spread of infectious particles of bacteria. As the bacteria grow in a biofilm they, at some point, may become dislodged from the surface and aerosolized. If the bacteria are dislodged in the form of small pieces of biofilm (or microcolonies) the bacteria would then be in a form that demonstrates enhanced

Table 1 Effect of exposure to desiccating conditions on the viability of aggregates and single cells of *Legionella pneumophila*, n ≥ 5 (adapted from Wright, 1988).

Desiccation Time (H)	*Biofilm Cells* $Log_{10}(CFU) \pm SE$	*Microcolonies* $Log_{10}(CFU) \pm SE$	*Dispersed Cells* $Log_{10}(CFU) \pm SE$
0	7.30 ± 0.01	6.17 ± 0.03	7.43 ± 0.05
0.08	ND+	5.78 ± 0.21	5.95 ± 0.02
0.25	ND	6.11 ± 0.02	4.83 ± 0.09
0.5	ND	6.04 ± 0.05	BDL
1.0	ND	5.88 ± 0.07	BDL
2.0	6.73 ± 0.03	5.96 ± 0.04	
3.0	ND	BDL*	
4.0	6.48 ± 0.05	BDL	
6.0	6.2 ± 0.05		
8.0	5.89 ± 0.85		
10.0	5.66 ± 0.10		
12.0	5.56 ± 0.11		
14.0	5.56 ± 0.02		

*BDL = below detectable limits; +ND = not determined

survival in an aerosol. The microcolony, complete with surrounding exopolymer (comprised of 99% water [Sutherland, 1971]), may be able to withstand more significant exposure to desiccating conditions than their single cell counterparts. In addition, the rate of dehydration of the bacterial cells may also be controlled to some degree by the presence of the exopolymer. Fry and Greaves (1951) suggested that the rate of dehydration of bacteria cells may be a significant factor in maintaining their viability. In addition, assuming that the microcolony exists as a multi-cellular array of bacteria, it is possible that even though the bacteria in the outer portions of the microcolony become completely dehydrated, those on inside of the microcolony may survive much longer, allowing them to remain viable and infective until they reach a suitable host.

Evidence suggesting that bacteria in microcolonies (or biofilm chunks) are less susceptible to desiccation than planktonic cells was obtained in the laboratory (Wright, 1988). *L. pneumophila* cells were allowed to colonize Douglas fir (commonly used to make cooling tower slats) surfaces in an artificially hardened water medium. Washed, dispersed, planktonic bacteria, or a suspension of *L. pneumophila* microcolonies recovered by scraping colonized surfaces, were applied to water-soaked pieces of fir at a density of approximately 10^7 colony forming units (CFU) per disc. Similarly, colonized fir discs, with an approximate cell density of 10^7 CFU $disc^{-1}$ were also rinsed with water. The discs with the three types of bacterial inocula were exposed to the laboratory environment in loosely covered Petri plates for up to 14 h. During this time the bacteria were allowed to dehydrate under ambient laboratory conditions; the temperature was approximately 22°C with a relative humidity of approximately 65%. The Petri plates were covered to eliminate variance in the air currents passing over the samples during the course of the study. Periodically throughout the experimental period, multiple discs were sampled and the numbers of viable cells remaining on the discs were determined. As evidenced in Table 1, there was a substantial difference in the survival of the bacteria exposed to similar conditions of dehydration depending on their degree of aggregation on the discs. The cultivable planktonic cells were reduced to levels below detection in a matter of minutes. The cells that were present on the discs in the form of biofilm "chunks" or microcolonies were observed to remain viable and cultivable for approximately 2 h. The biofilm cells, conversely, remained viable for at least 14 h. These results suggest that microcolonies are significantly more resistant to desiccation than single cells of this organism.

Infectivity of *Legionella* Particles

Another line of evidence supporting microcolonies as a possible mode of transmission of infective particles is work done with artificial (agar-entrapped) microcolonies of *Legionella* (Wright, 1988). *L. pneumophila* serogroup 1 cells, either as dispersed single cells or artificial microcolonies, were instilled into the lungs of guinea pigs and the course of clinical disease was monitored. A mildly virulent strain of *L. pneumophila* serogroup 1 was chosen for these experiments. Infection with single cells required 2–3 orders of magnitude more bacteria to induce a similar level of pulmonary injury (or percentage of animals succumbing to the infection) than if the bacteria were instilled as artificial microcolonies. Animals infected with an inoculum of cells that was not lethal when instilled as single cells, demonstrated signficant mortality if the bacteria were instilled as articifial microcolonies. Additionally, animals inoculated with the same sublethal inoculum of single cells together with an equivalent aliquot of sterile agar beads also demonstrated no mortality.

Other authors have suggested that legionellae encapsulated in amoebal cysts are provided with protection from environmental stresses (Barbaree *et al.*, 1986) and that these cysts may provide a means for the bacteria to be transmitted from a source to a host (Rowbotham, 1980; Brieland *et al.*, 1997). It has also been reported that individuals, asymptomatic for any form of legionellosis, commonly demonstrate antibody titres to amoebal antigens (Powell *et al.*, 1993). As in the work of Wright (1988) with artificial microcolonies of *L. pneumophila*, Brieland *et al.*(1997) also demonstrated that *L. pneumophila*-infected amoebae were more pathogenic than an equivalent number of bacteria or a co-inoculum of the bacteria and amoebae. Given the seeming propensity for legionellae to infect and multiply within some species of amoebae, this may provide another mechanism for the bacteria to survive in relatively harsh environments and also be able to infect their host.

THE STUDY OF *LEGIONELLA* SPP.

Culture

Legionellae have proved to be difficult organisms to study. The bacteria were first demonstrated by McDade *et al.* (1977) who utilized techniques for working with rickettsia to prove that the causative agents of Legionnaires' disease were bacteria. Following demonstration of the nature of the causative agents, efforts were made to learn how to cultivate them in the laboratory. Initial efforts to recover the organisms from infected tissue relied heavily upon the use of animals and/or embryonated eggs

(McDade *et al.*, 1977). Although the method allowed for the recovery of the bacteria, it was not an ideal laboratory method. Over time, additional, more acceptable, methods for the cultivation of the bacteria were found. Following a series of iterations, a solid laboratory culture medium was developed that allowed cultivation of the bacteria. This was a buffered charcoal yeast extract agar (BCYE; Pasculle *et al.*, 1980; Edelstein, 1981). The medium could also be supplemented with a variety of antibiotics to inhibit overgrowth by non-target organisms (*e.g.* Wadowsky and Yee, 1981; Dennis *et al.*, 1984). In addition to the use of antibiotics, other steps, including elevated temperature (Dennis *et al.*, 1984) and/or acidic pH (Bopp *et al.*, 1981) preincubations have been employed to reduce the non-legionellae present in water (and tissue) samples. It remains an interesting facet of *Legionella* spp. that they require such optimized conditions for growth in the laboratory and yet are nearly ubiquitous colonizers of a variety of aquatic environments.

Once the bacteria have been isolated, it is possible to grow them in pure culture on laboratory agar. To grow the bacteria in large quantities, a yeast extract broth was developed (Ristroph *et al.*, 1980) that supports rapid and luxuriant growth of the bacteria. Many of the additives, such as iron and cysteine, required for the growth of the bacteria on agar surfaces are also required for the cultivation of the bacteria in liquid culture. However, a relatively poor supply of nutritional supplements is required to support the survival and proliferation of these organisms, even in pure culture, under laboratory conditions that more closely reflect the conditions under which they are found in industrial waters. This is fortuitous since, in studies conducted to examine the efficacy of various biocides against planktonic and sessile legionellae, it is important to maintain conditions similar to those found in the system where the biocide is intended to be used. Therefore, to simulate cooling tower water, for example, artificially hardened water has been employed (Wright *et al.*, 1989; 1991). This water has been found to support the growth of *L. pneumophila* with only the addition of an extract derived from the medium in which *Fischerella* sp., a cyanobacterium, has been grown (Pope *et al.*, 1982). This formulation demonstrates the incongruencies associated with the growth of members of the genus. In some laboratory settings very complex media appear to be required but in "natural" settings and under laboratory conditions designed to reflect these conditions, a very much less complicated medium is required.

Fluorescent Antibodies for Identification

As a result of the difficulties associated with the growth of legionellae in the laboratory, there is an associated difficulty with the positive identification of the bacteria that grow and are presumptively identified as *Legionella* spp. The use of pH or temperature pre-treatments, combined with antibiotic-supplemented media that inhibit the growth of non-legionellae aids identification and in addition, some species are relatively unique in their appearance when grown on BCYE, for example, the bluish-white autofluorescence of *L. gormanii* and *L. dumoffi* (Johnson and Rinehart, 1984). Nonetheless, positive identification remains difficult. One of the first methods employed to identify these bacteria was the use of fluorescently labelled antibodies. This technique rapidly gained acceptance for identifying species of legionellae cultivated in the laboratory, particularly as they pertained to potential clinical infections. This method could also be employed for positively identifying *Legionella* spp. growing, in pure or mixed culture, *in situ* on industrial surfaces as well as on laboratory surfaces (Wright *et al.*, 1989).

Although fluorescein-labelled antibodies proved to be an excellent method of positively identifying the legionellae against which the antibodies were directed, the method is less likely to be used now except to confirm the identity of the bacteria recovered from an experimental system inoculated with a known species of bacteria. The antibody technique may also be used as a first attempt to identify an unknown, presumptive *Legionella* sp. However, the use of fluorescent antibodies may be impaired by the non-specific binding of the antibody to other bacterial species or to phenotypic variation within the species for which the antibody was prepared (Wilkinson *et al.*, 1990). In addition, the visualization of *Legionella* spp. within thick biofilm, *in situ*, may also be hampered by the inability of the antibody to penetrate to the deeper layers of the biofilm (Swerinski *et al.*, 1985).

Molecular Identification Techniques

In addition to the inherent difficulties associated with the fluorescein-labelled antibody technique, two other coincident occurrences have also served to diminish the importance of fluorescent antibodies in the detection and identification of legionellae. First, since the initial identification of *L. pneumophila*, there has been an explosion in the numbers of recognized species within the family Legionellaceae. Currently, there are at least 42 recognized *Legionella* species (Benson *et al.*, 1996). In addition, the predominant species (*L. pneumophila*) is further subdivided into 13 separate serogroups (Kwaik *et al.*, 1998). These facts make the use of highly specific antibodies less appealing as many different antisera would potentially be required to identify a particular isolate. In conjunction with the numerical growth of recognized members of the genus, concurrent advances have been made in the utilization of DNA- and RNA-based identification methods. These are becoming increasingly important for identifying isolates as well as for demonstrating the rationale for placing a *species novum* within the genus *Legionella*.

Among the various molecular methods available for identifying and classifying bacteria, genotypic methods including DNA-DNA hybridization (Brenner, 1986) and ribotyping (Grimont *et al.*, 1989) are available. Recently, considerable attention has been given to polymerase chain reaction (PCR)-based methodologies. These have included protocols that target the 16S rRNA gene (Jonas *et al.*, 1995), and the 23S–5S spacer region for identification of legionellae (Robinson *et al.*, 1996). In one study, van Belkum *et al.* (1996) used the 16S–23S spacer region to identify *L. pneumophila*. However, the majority of these studies have employed the 5S RNA gene to identify the genus (Saint, 1998). To further examine an isolate and determine if it is *L. pneumophila*, most researchers rely on the *mip* (macrophage infectivity potentiator) gene as a species-specific marker (Maiwald *et al.*, 1994; Fricker and Fricker, 1995). Presti *et al.* (1998) have also reported work involving a large number of type strains as well as environmental and clinical isolates that allowed them to conclude that random amplified polymorphic DNA (RAPD) analysis is useful for rapidly identifying new isolates to the species level. The same report also concluded that the RAPD method was useful for the clustering of isolates that may represent new species. An earlier study (Bansal and McDonnell, 1997) suggested that this technique (RAPD) might be useful for further identifying isolates to the serogroup level. However, identification of isolates to this level by RAPD has been suggested to be potentially over-optimistic (Presti *et al.*, 1998) and may reflect an insufficient number of tested isolates. Nonetheless, the use of PCR-based techniques demonstrates that relatively rapid identification of legionella isolates is feasible.

Two disadvantages of PCR-based techniques is that they may require the cultivation of the bacteria in the laboratory prior to identification and they do not allow for *in situ* visualization of the bacteria, concurrent with their identification. As noted previously, one of the problems associated with the growth of legionellae is that they exhibit a very fastidious nature in the laboratory environment. They have also been demonstrated to form viable but non-cultivable cells which cannot be grown in the laboratory without prior passage through a suitable host (Hussong *et al.*, 1987; Paszko-Kolva *et al.*, 1992; Steinert *et al.*, 1997). These characteristics make methods that rely on culturing the bacteria prior to identification less than ideal in some instances.

As a solution to the problems associated with molecular techniques fluorescent, *in situ* hybridization (FISH) of whole cells with rRNA-targeted probes is currently used. This technique has become a very valuable tool for specific detection of individual cells without the need for first growing them on laboratory media (DeLong *et al.*, 1989; Amann *et al.*, 1995). In 1995, Manz *et al.* reported the use of a 16S RNA-directed oligonucleotide probe (LEG705) suitable for the *in situ* identification of members of the family *Legionellaceae*. Broadly specific probes of this kind are crucial for detecting members of this family, as they are found *in situ* in a variety of domestic and industrial applications. However, of the recognized species of *Legionella*, 80% of the isolates important in disease are *L. pneumophila*. Additionally, of the number of different serogroups within this species, 95% of infections are caused by *L. pneumophila* serogroup 1. Therefore, from a medical standpoint, this particular serogroup is by far the most important member of the family. As a result of this, together with a number of unique characteristics shared by members of this family (*e.g.* intracellular proliferation (Horwitz, 1987)) as well as characteristics unique to *L. pneumophila* (*e.g.* induction of coiling phagocytosis (Horowitz, 1984)), *L. pneumophila* serogroup 1 is the most well-studied organism within this family. Hence, efforts have been made to make probes specific for this serogroup. Grimm *et al.* (1998) have reported the construction of an oligonucleotide 16S rRNA-directed probe useful for detecting single cells of *L. pneumophila* serogroup 1. This unique probe (LEGPNE1) seems to discriminate between different species and serogroups of *Legionella* (See Table 2). In addition to being able to identify single cells of either *Legionella* spp. or *L. pneumophila* serogroup 1, the FISH technique that these probes were developed for have been successfully used to demonstrate the presence of *Legionella* (and to differentiate between species) infecting *Acanthamoeba* (Manz *et al.*, 1995; Grimm *et al.*, 1998). These results further demonstrate the power of this technique.

Confocal Microscopy

Another technique that has proven to be a great boon in the study of biofilms in a quasi-natural state is the use of confocal scanning laser microscopy (Caldwell *et al.*, 1992; Wolfaardt *et al.*, 1994a; 1994b). This technique provides the ability to docu-

Table 2 Target organisms and reference strains used for fluorescent *in situ* hybridization (reproduced from Grimm *et al.*, 1998).

Organism	*Hybridization signal with the following probes*	
	LEG705	*LEGPNE1*
Legionella pneumophila		
Philadelphia 1 (serogroup 1)	+	+
Corby	+	+
Serogroup 6 (environmental) isolate	+	+
Serogroup 4 (patient) isolate	+	+
Bloomington (serogroup 3)	+	+
Los Angeles (serogroup 4)	+	+
Legionella bozemanii	+	–
Legionella hackeliae	+	–
Legionella longbeachae	+	–
Legionella micdadei	+	–
Legionella anisa	+	–
Burkholderia cepacia	–	–
Escherichia coli	–	–
Pseudomonas aeruginosa	–	–

+ = strong hybridization signal; – = no hybridization signal

ment occurrences within a biofilm and map, in three-dimensions, the processes that are occurring. Combined with FISH techniques, as well as other specific probes, details regarding the interactions occurring within legionellae-containing biofilms may be elucidated.

Biofilm Development

The techniques available to study *Legionella* biofilms presuppose the existence of appropriate methods to develop biofilms for study. A variety of test systems exist that can be used to study the growth of bacteria within biofilms in simulated water distribution systems (Walker, 1997) and cooling towers (McCoy *et al.*, 1986), both commonly identified sources of legionellae. Model systems may include sacrificial pipe sections on which biofilm develops and which may be studied following the removal of the section from the model system. Alternatively, pipes with removable and replaceable surfaces may also be employed to monitor biofilms in model systems (McCoy and Costerton, 1982; Nickel *et al.*, 1985; Wright *et al.*, 1989; 1991). Regardless of the method chosen to form the biofilm, it is important to maintain conditions that allow for representative biofilms to form. Similarly, it is important to choose surfaces for colonization that reflect the surface being modelled.

Although it is desirable to maintain laboratory conditions as close to field conditions as possible, some latitude is required in order to study biofilm processes by various techniques and to isolate the effects of various manipulations. In addition, it is necessary to demonstrate that the conditions within the model reflect, in a reasonably accurate manner, the conditions in the original system. If these factors are kept in mind it is possible to create test rigs that closely mimic the system under study

(Wright, 1997) allowing the model system to predict occurrences within the system being modelled. The predictive capability of a model system is particularly important if the results are to be indicative of the results of a particular manipulation within a given system. Inappropriate models may lead to erroneous conclusions regarding the anticipated effect of similar manipulations in the actual system. Examples of this were evident in early work examining biocides to control *Legionella* in cooling towers where tested compounds proved to be less efficacious under field conditions than anticipated based on laboratory experiments (*e.g.* England *et al.*, 1982). These results underscore the need to mimic "reality" when developing test systems.

CONTROL OF *LEGIONELLA* SPP.

Cooling Tower Biocides

Various methods of controlling legionellae in their various reservoirs have been examined and practiced since 1976. Cooling towers, particularly those associated with large institutions that may house a large population of immunocompromised people, have been identified as one of these reservoirs. The method of choice for controlling bacterial populations within these devices remains the use of industrial biocides. These may be oxidizing or non-oxidizing biocides (Buecker and Post, 1998). New biocidal actives are slow to emerge due to regulatory and environmental concerns. However, novel methods of delivery and new synergistic combinations of existing biocides (*e.g.* Hsu, 1997; McCoy, 1998) are continually being investigated for use in controlling microorganisms in various process streams, including cooling towers. The World Health Organization has also published guidelines relating to the control of legionellae in various aquatic environments (WHO, 1990).

Potable Water

Although cooling towers were among the first documented reservoirs of legionellae to be associated with disease transmission (Garbe *et al.*, 1985), legionellae have also been found associated with potable water distribution systems in a variety of different institutions and private homes (Moreno *et al.*, 1997). Nineteen consecutive outbreaks of nosocomial Legionnaire's disease in the United Kingdom have been linked to bacterial dispersion from potable water distribution systems (Joseph *et al.*, 1994). Surveys by a number of investigators have demonstrated that hospital water distribution systems may be colonized with *Legionella* (*e.g.* Marrie *et al.*, 1994; Lin *et al.*, 1998b).

The hot-water distribution system has been most frequently linked to the survival and dissemination of these bacteria. The factor most commonly observed to promote proliferation of legionellae in this part of the water distribution system is the elevated temperature (Rogers *et al.*, 1994). This is, in part, due to the demonstrated thermotolerance of these organisms (Dennis *et al.*, 1984). Other growth promoting conditions include the age and type of the calorifier (Alary and Joly, 1992; Stout *et al.*, 1992), dead legs within the distribution system (WHO, 1985), accumulation of scale and sediment (Lin *et al.*, 1998b), elevated concentrations of calcium and magnesium (Vickers *et al.*, 1987), and colonization of pipes with bacteria and protozoa (Lin *et al.*, 1998b).

The need for effective means of control is obvious for hospital environments and other institutions where large numbers of aged or debilitated persons may reside (Patterson *et al.*, 1997). A variety of methods have been described for the treatment of water distribution systems in an effort to control the growth of legionellae. The most commonly described methods of controlling legionellae in water distribution systems will be described, along with some indication of their relative efficacy.

Hot water flush

The first method successfully employed for disinfecting potable water lines involved elevating the hot water temperature and flushing the hot water system. In utilizing this method, it is necessary for the hot water temperature at each outlet to reach and sustain a temperature of 60°C. Recommendations by the Centers for Disease Control and Prevention (Atlanta, USA) had indicated that a sustained 5 min flush was sufficient to decontaminate a system (Tablan *et al.*, 1994). However, the experience at two American hospitals has led some authors to suggest that a 5 min flush is probably inadequate and a 30 min flush is required (Lin *et al.*, 1998b). If, following the elevated temperature flush of the water system, the water temperature is maintained at 60°C, clinical experience suggests that it will remain culture-negative for *Legionella* spp. and there will be an on-going reduction in the incidence of hospital-acquired Legionnaires' disease (Muraca *et al.*, 1990). However, if water temperatures are returned to normal, recolonization of the distribution system and a recurrence of nosocomial Legionnaires' disease may occur within months (Ezzeddine *et al.*, 1989).

Oxidants

The most common method of controlling the bacterial load in potable water is through the use of chlorine. *Legionella* spp., like many other organisms commonly found in domestic water supplies, are relatively quickly inactivated by an appropriate concentration of chlorine (Yabuuchi *et al.*, 1995). However, as for many other bacterial species, the biofilm mode of growth exhibited by legionellae also helps these organisms survive in the presence of relatively high concentrations of chlorine (Lin *et al.*, 1998a). Hyperchlorination of water systems, by one of two methods, has been used to control colonization of water distribution systems by legionellae (Lin *et al.*, 1998b). The first method involves a periodic (shock) hyperchlorination where the residual chlorine level is elevated to 20–50 mg l^{-1} throughout the system. After a 60–120 min contact time, the system is drained and re-mixed with incoming water so that residual chlorine levels return to the normal 0.5–1 mg l^{-1} level. Alternatively, continuous hyperchlorination involves the continuous injection of chlorine into the system to maintain a chlorine level in excess of the normal level of approximately 1 mg l^{-1} level. Regular or sustained elevation of the concentration of chlorine in a distribution system will increase the corrosiveness of the water, resulting in an increase in the number of pipe leaks (Grosserode *et al.*, 1993).

In addition to increasing the rate of pipe corrosion, other factors make hyperchlorination less than ideal as a means of water treatment. Among these is the objectionable odour of highly chlorinated water. However, the possibility also exists that normally achieved levels of chlorination may only suppress the bacteria rather than kill them (Kuchta *et al.*, 1983; Lin *et al.*, 1998a). Another potential reason for the persistence of legionellae in chlorinated water may be that they are protected within sediment or biofilm. A third possible survival tactic that may account for their survival in chlorinated water as well as for their dissemination from their aquatic habitat is that they may exist within cysts derived from *Acanthamoeba* spp. (Berk *et al.*, 1998). These protozoan cysts have been documented to survive free chlorine levels of up to 50 mg l^{-1} (Kilvington and Price, 1990). Survival of legionallae within the potable water system, even when a good chlorination programme is in place, seems to account for their rapid re-growth once residual chlorine levels return to normal (Lin *et al.*, 1998a).

Ultraviolet light

Another attractive method for controlling members of the family *Legionellaceae* in potable water supplies is through the use of ultraviolet light (Muraca *et al.*, 1987). Intense UV light may cause irreversible (and lethal) DNA damage to the bacteria, preventing them from reproducing. Ultraviolet light, however, is not useful as a "systemic" disinfection or control system because it is only able to disinfect water that passes immediately through the region being treated; it is therefore useful solely as a continuous treatment for water near its point of use. In addition, a system needs to be in place to prevent the accumulation of scale on the lamps (Lin *et al.*, 1998a). The utilization of UV light, combined with some mechanism to prevent chemical fouling of the lamps, has been most successfully used in near-point-of-use applications (Liu *et al.*, 1995) such as in transplant units (Matulonis *et al.*, 1993) where the risks associated with potential nosocomial legionellosis are particularly high.

Copper-silver ionization

Recently, a new method has been reported in the literature as being a potentially effective means for controlling legionellae in water distribution systems. This system is based on the simultaneous release of copper and silver ions into water as it passes through a flow cell containing copper and silver electrodes. The release of ions from the electrodes is controlled by a microprocessor that regulates the direct current supplied across the electrodes (States *et al.*, 1998).

The method has been demonstrated to be effective in a hospital where thermal eradication had been unsuccessful (Miuetzner *et al.*, 1997). Repeated heat flushes decreased the rate of recovery of the bacteria from sentinel faucets throughout this institution to 0%, from an initial approximately 80% rate of positive culture. However, shortly after (29 days) this success, the rate of positive cultures obtained from these same indicator faucets returned to near base-line levels. The authors (Miuetzner *et al.*, 1997) reported that the silver-copper ionization method successfully reduced the rate of positive cultures within the first month of treatment and maintained a rate of positivity between 0% and 4% throughout the following 22 month follow-up period.

A similar study by Lin *et al.* (1998b) demonstrated a similar type of effect in another two hospital buildings. In these cases, the rate of culture positivity at sentinel sites decreased from a rate of 50–70% to 0% within 4–12 weeks. These authors also reported the inability to culture bacteria at the sentinel sites remained for an additional 6–8 weeks following cessation of the treatment protocol. These results suggest that the method may provide a more long-lasting effect than either thermal eradication or hyperchlorination. Part of the success of this system in maintaining an apparently *Legionella*-free water supply for some period following cessation of the copper-silver ionization treatment has been postulated to be related to the higher levels of copper found associated with the biofilm population (Lin *et al.*, 1998b). This may result in the inactivation of a significant proportion of the biofilm-associated legionellae and thereby retard the rate of recolonization of the distribution system.

This new method of controlling *Legionella* spp. in water distribution systems is cause for cautious optimism regarding its potential for significantly reducing the rate of institutionally acquired legionelloses (Goetz and Yu, 1997). However, there are some potential concerns with this methodology. The first is that some authors have demonstrated that there are transient fluctuations in the levels of copper and silver ions in solution (Miuetzner *et al.*, 1997).

These fluctuations occasionally elevate the silver ion concentration to levels in excess of the United States Environmental Protection Agency (USEPA) treatment action level (USEPA, 1993). Miuetzner *et al.* (1997) also reported that copper and silver concentrations in water recovered from hot water storage tanks could significantly exceed the EPA standards. Another observation was that this method appears to be relatively specific for legionellae. States *et al.* (1998) observed that the *Legionella*-decontamination efficacy did not always completely eradicate the legionellae, and did not seem to impact other heterotrophic bacteria in the system. These authors also reported a minor decrease in the percentage of sites that were positive for amoebae; there was also no correlation between sites positive for *Legionella* and those positive for amoebae. These results indicated that the copper-silver ionization method inhibits legionellae directly rather than by controlling the protozoa that may support proliferation of these bacteria. With only limited time studies on this subject, it also remains to be seen whether bacteria colonizing institutional water systems develop any resistance to these heavy metals. Certainly, bacteria exposed to a variety of heavy metals eventually develop a level of resistance to these bactericides (Li *et al.*, 1997).

CONCLUSION

More than two decades have passed since their identification, yet *Legionella* spp. remain and continue to cause occasional outbreaks of pneumonic illnesses, particularly in institutions housing immunocompromised individuals. This is in spite of considerable research (and resultant recommendations) regarding methods to control their growth in the man-made environments linked to their transmission to susceptible individuals. The importance of the biofilm mode of growth in the survival of these bacteria in various water supplies and possibly in their ability to cause disease has been established and will become clearer as new methods that aid in their study are more commonly employed. This should result in an increase in understanding, thereby reducing the likelihood of the organisms causing disease.

REFERENCES

Alary M., Joly J.R. (1992). Factors contributing to the contamination of hospital water distribution systems by legionellae. *J Infect Dis*, **165**, 565–569.

Amann R.I., Ludwig W., Schleifer, K-H. (1995). Phylogenetic identification and *in situ* detection of individual microbial cells without cultivation. *Microbiol Rev*, **59**, 143–169.

Bansal N.S., McDonnell F. (1997). Identification and DNA fingerprinting of *Legionella* strains by randomly amplified polymorphic DNA analysis. *J Clin Microbiol*, **35**, 2310–2314.

Barbaree J.M., Fields B.S., Feeley J.C., Gorman G.W., Martin W.T. (1986). Isolation of protozoa from water associated with a legionellosis outbreak and demonstration of intracellular multiplication of *Legionelia pneumophila*. *Appl Environ Microbiol*, **51**, 422–424.

Benson R.F., Thacker W.L., Daneshvar M.I., Brenner D.J. (1996). *Legionella waltersii* sp. nov. and an unnamed *Legionella* genomospecies isolated from water in Australia. *Int J Syst Bacteriol*, **46**, 631–634.

Berendt L.F. (1980). Influence of blue-green algae (Cyanobacteria) on survival of *Legionella pneumophila* in aerosols. *Infect Immun*, **32**, 690–692.

Berk S.G., Ting R.S., Turner G.W., Ashburn R.J. (1998). Production of respirable vesicles containing live *Legionella pneumophila* cells by two *Acanthamoeba* spp. *Appl Environ Microbiol*, **64**, 279–286.

Bopp C.A., Sumrer J.W., Morris G.K., Well J.G. (1981). Isolation of *Legionella* spp. from environmental water samples by low-pH treatment and use of a selective medium. *J Clin Microbiol*, **13**, 714–719.

Brenner D.J. (1986). Classification of Legionellaceae: current status and remaining questions. *Isr J Med Sci*, **22**, 620–632.

Brenner D.J., Steigerwalt A.G., McDade J.E. (1979). Classification of the Legionnaires' disease bacterium: *Legionella pneumophila, a genus novum, species nova* of the family Legionellaceae, *familia nova*. *Ann Intern Med*, **90**, 656–658.

Brieland J.K., Fantone J.C., Remick D.G., LeGendre M., McClain M., Engleberg N.C. (1997). The role of *Legionella pneumophila*-infected *Hartmannella vermiformis* as an infectious particle in a murine model of Legionnaires' disease. *Infect Immun*, **65**, 5330–5333.

Buecker B., Post R. (1998). Control of biofouling in evaporative cooling systems. *Chem Eng Prog*, **Sept**, 45–50.

Caldwell D.E., Korber D.R., Lawrence J.R. (1992). Confocal laser microscopy and digital image analysis in microbial ecology. *Adv Microb Ecol*, **12**, 1–67.

DeLong E.F., Wickham G.S., Pace N.R. (1989). Phylogenetic stains: ribosomal RNA-based probes for the identification of single microbial cells. *Science*, **243**, 1360–1363.

Dennis P.J., Bartlett C.L.R., Wright A.E. (1984). Comparison of isolation methods for *Legionella* spp. In: Thornsberry C., Balows A., Feeley J.C., Jakubowsky W. (eds) *Legionella: Proceedings of the Second International Symposium*. American Society for Microbiology, Washington, pp. 294–296.

Dondero T.J., Rendtroff R.C., Mallison G.F., Weeks R.M., Levy J.S., Wong E.W., Schaffner W. (1980). An outbreak of Legionnaires' disease associated with a contaminated air-conditioning cooling tower. *N Engl J Med*, **302**, 365–370.

Edelstein P.H. (1981). Improved semiselective medium for isolation of *Legionella pneumophila* from contaminated clinical and environmental specimens. *J Clin Microbiol*, **14**, 298–303.

England III A.C., Fraser D.W., Mallison G.F., Mackel D.C., Skaliy P., Gorman G.W. (1982). Failure of *Legionella pneumophila* sensitivities to predict culture results from disinfectant-treated air-conditioning cooling towers. *Appl Environ Microbiol*, **43**, 240–244.

Ezzeddine H., Van Ossel C., Delmée M., Wauters G. (1989). *Legionella* spp. in a hospital hot water system: effect of control measures. *J Hosp Infect*, **13**, 121–131.

Fliermans C.B. (1985). Ecological niche of *Legionella pneumophila*. In: Katz S.M. (ed) *Legionellosis*. CRC Press, Boca Raton, pp. 75–116.

Fliermans C.B., Harvey R.S. (1984). Effectiveness of 1-bromo-3-chloro-5,5-dimethylhydantoin against *Legionella pneumophila* in a cooling tower. *Appl Environ Microbiol*, **47**, 1307–1310.

Fliermans C.B., Cherry W.B., Orrison L.H., Smith S.J., Tison D.L., Pope D.H. (1981). Ecological distribution of *Legionella pneumophila*. *Appl Environ Microbiol*, **41**, 9–16.

Fitzgeorge R.B., Baskerville A., Broster M., Hambleton P., Dennis P.J. (1983). Aerosol infection of animals with strains of *Legionella pneumophila* of different virulence: comparison with intraperitoneal and intranasal routes of infection. *J Hyg*, **90**, 81–89.

Fraser D.W., Tsai T.R., Orenstein W., Parkin W.E., Beecham H.J., Sharrar R.G., Harris J., Mallison G.F., Martin S.M., McDade J.E., Shepard C.C., Brachman P.S. (1977). Legionnaires' disease. Description of an epidemic of pneumonia. *N Engl J Med*, **297**, 1189–1197.

Fricker E.J., Fricker C.R. (1995). *Detection of Legionella* sp. using a commercially available polymerase chain reaction test. *Water Sci Technol*, **31**, 407–408.

Fry R.M., Greaves R.I.N. (1951). The survival of bacteria during and after drying. *J Hyg*, **49**, 220–246.

Garbe P.L., Davis B.J., Weisfeld J.S., Markowitz L., Miner P., Garrity F., Barabee J.M., Reingold A.L. (1985). Nosocomial Legionnaires' disease: epidemiologic demonstration of cooling towers as a source. *J Am Med Assoc*, **254**, 521–524.

Glick T.H., Gregg M.B., Berman B., Mallison G., Rhodes W.W., Kassanoff L. (1978). Pontiac fever. An epidemic of unknown etiology in a health department. I. Clinical and epidemiologic aspects. *Am J Epidemiol*, **107**, 149–160.

Goetz A., Yu V.L. (1997). Copper-silver ionization: cautious optimism for Legionella disinfection and implications for environmental culturing. *Am Infect Control*, **25**, 449–451.

Grimm D., Merkert H., Ludwig W., Schleifer, K-H., Hacker J., Brand B.C. (1998). Specific detection of *Legionella pneumophila*: construction of a new 16S rRNA-targeted oligonucleotide probe. *Appl Environ Microbiol*, **64**, 2686–2690.

Grimont F., Lefevre M., Ageron E., Grimont P.A.D. (1989). rRNA gene restriction patterns of *Legionella* species: a molecular identification system. *Res Microbiol*, **140**, 615–626.

Grosserode M., Helms C., Pfaller M. (1993). Continuous hyperchlorination for control of nosocomial legionnaires' disease: a ten-year follow-up of efficacy, environmental effects, and cost. In: Barbaree J.M., Breiman R.F., Dufour A.P. (eds) *Legionella – Current Status and Emerging Perspectives*. American Society for Microbiology, Washington, pp. 226–229.

Hambleton P., Bailey N.E., Fitzgeorge R.B., Baskerville A. (1985). Clinical chemical responses to experimental airborne legionellosis in the guinea pig. *Br J Exp Pathol*, **66**, 173–183.

Hebert G.A., Moss C.W., McDougal L.K., Bozeman F.M., McKinney R.M., Brenner D.J. (1980). The rickettsia-like organisms Tatlock (1943) and Heba (1959) bacteria phenotypically similar but genetically distinct from *Legionella pneumophila* and the Wiga bacterium. *Ann Intern Med*, **92**, 45–52.

Horwitz M.A. (1983). Formation of a novel phagosome by the Legionnaires' disease bacterium (*Legionella pneumophila*) in human monocytes. *J Exp Med*, **158**, 1319–1331.

Horwitz M.A. (1984). Phagocytosis of the Legionnaires' disease bacterium (*Legionella pneumophila*) occurs by a novel mechanism: engulfment within a pseudopod coil. *Cell*, **36**, 27–33.

Horwitz M.A. (1987). Characterization of avirulent mutant *Legionella pneumophila* that survive but do not multiply within human monocytes. *J Exp Med*, **166**, 1310–1328.

Hsu J.C. (1997). Synergistic microbicidal combinations using 4,5-dichloro-2-octyl-3-isothiazolone and certain commercial biocides. United States Patent 5,591,760, issued January 7, 1997.

Hussong D., Colwell R.R., O'Brien M., Weiss E., Pearson A.D., Weiner R.M., Burge W.D. (1987). Viable *Legionella pneumophila* not detectable by culture on agar media. *BioTechnology*, **5**, 947–950.

Johnson W., Rinehart K. (1984). Isolation and partial chemical characterization of membrane-bound fluorescent compounds from *Legionella gormanii* and *Legionella dumoffi*. In: Thomsberry C., Balows A., Feeley J.C., Jakubowsky W. (eds) *Legionella: Proceedings of the Second International Symposium*. American Society for Microbiology, Washington, pp. 87–89.

Jonas D., Rosenbaum A., Weyrich S., Bhakdi S. (1995). Enzyme-linked immunoassay for detection of PCR-amplified DNA of legionellae in bronchoalveolar fluid. *J Clin Microbiol*, **33**, 1247–1252.

Joseph C.A., Watson J.M., Harrison H.G., Bartlett C.L. (1994). Nosocomial Legionnaires' disease in England and Wales, 1980–1992. *Epidemiol Infect*, **112**, 329–345.

Kilvington S., Price J. (1990). Survival of *Legionella pneumophila* within cysts of *Acanthamoeba polyphaga* following chlorine exposure. *J Appl Bacteriol*, **68**, 519–525.

Kuchta J.M., States S.J., McNamara A.M. (1983). Susceptibility of *Legionella pneumophila* to chlorine in tap water. *Appl Environ Microbiol*, **46**, 1134–1139.

Kwaik Y.A., Gao L.-Y., Stone B.J., Venkataraman C., Harb O.S. (1998). Invasion of protozoa by *Legionella pneumophila* and its role in bacterial ecology and pathogenesis. *Appl Environ Microbiol*, **64**, 3127–3133.

Li X.Z., Nikaido H., Williams K.E. (1997). Silver-resistant mutants of *Escherichia coli* display active efflux of Ag^+ and are deficient in porins. *J Bacteriol*, **179**, 6127–6132.

Lin Y.-S., Stout J.E., Yu V.L., Vidic R.D. (1998a). Disinfection of water distribution systems for *Legionella. Semin Respir Infect*, **13**, 147–159.

Lin Y.-S., Vidic R.D., Stout J.E., Yu V.L.(1998b). *Legionella* in water distribution systems. *J Am Water Works Assoc*, **90**, 112–121.

Liu Z., Stout J.E., Yu V.L. (1995). Efficacy of ultraviolet light in preventing *Legionella* colonization of a hospital water distribution system. *Water Res*, **29**, 2275–2280.

Maiwald M., Kissel K., Srimuang S., von Knabel Doeberitz M., Sonntag H.-G. (1994). Comparison of polymerase chain reaction and conventional culture for the detection of legionellas in hospital water samples. *J Appl Bacteriol*, **76**, 216–225.

Manz W., Amann R.I., Szewzyk R., Szewzyk U., Stenström T.-A., Hutzler P., Schleifer K.-H. (1995). *in situ* identification of *Legionellaceae* using 16S rRNA-targeted oligonucleotide probes and confocal laser scanning microscopy. *Microbiology (Reading)*, **141**, 29–39.

Marrie T.J., Gass R., Sumarah R., Yates L. (1987). *Legionella pneumophila* in a physiotherapy pool. *Eur J Clin Microbiol*, **6**, 212–213.

Marrie T.J., Green T., Burbridge S. (1994). *Legionellaceae* in the potable water of a Nova Scotia hospital and Halifax residences. *Epidemiol Infect*, **112**, 143–150.

Matulonis U., Rosenfeld C.S., Shadduck R.K. (1993). Prevention of *Legionella* infection in a bone marrow transplant unit: mulifaceted approach to decontamination of a water system. *Infect Control Hosp Epidemiol*, **14**, 571–575.

McCoy W.F. (1998). Imitating natural microbial fouling control. *Mater Perform* **Apr**, 45–48.

McCoy W.F., Costerton J.W. (1982). Growth of sessile *Sphaerotilus natans* in a tubular recycle system. *Appl Environ Microbiol*, **43**, 1490–1494.

McCoy W.F., Wireman J.W., Lashen E.S. (1986). Efficacy of methylchloro/methylisothiazolone biocide against *Legionella pneumophila* in cooling tower water. *J Indust Microbiol*, **1**, 49–56.

McDade J.E., Brenner D.J., Bozeman F.M. (1979). Legionnaires' disease bacterium isolated in 1947. *Ann Intern Med*, **90**, 659–661.

McDade J.E., Shepard C.C., Fraser D.W., Tsai T.R., Redus M.A., Dowdle W.R. (1977). Legionnaires' disease: isolation of a bacterium and demonstration of its role in other respiratory disease. *N Engl J Med*, **297**, 1197–1203.

Miuetzner S., Schwille R.C., Farley A., Wald E.R., Ge J.H., States S.J., Libert T., Wadowsky R.M. (1997). Efficacy of thermal treatment and copper-silver ionization for controlling *Legionella pneumophila* in high-volume hot water plumbing systems in hospitals. *Am J Infect Control*, **25**, 452–457.

Moreno C., de Blas I., Miralles F., Apraiz D., Catalan V. (1997). A simple method for the eradication of *Legionella pneumophila* from potable water systems. *Can J Microbiol*, **43**, 1189–1196.

Muraca P., Stout J.E., Yu V.L. (1987). Assessment of chlorine, heat, ozone, and UV light for killing *Legionella pneumophila* within a model plumbing system. *Appl Environ Microbiol*, **53**, 447–453.

Muraca P., Yu V.L., Goetz A. (1990). Disinfection of water distribution systems for *Legionella*: a review of application procedures and methodologies. *Infect Control Hosp Epidemiol*, **11**, 79–88.

Newsome A.L., Baker R.L., Miller R.D., Arnold R.R. (1985). Interactions between *Naegleria fowleri* and *Legionella pneumophila. Infect Immun*, **50**, 449–452.

Newsome A.L., Scott T.M., Benson R.F., Fields B.S. (1998). Isolation of an amoeba naturally harboring a distinctive *Legionella* species. *Appl Eviron Microbiol*, **64**, 1688–1693.

Nickel J.C., Wright J.B., Ruseskal I., Marrie T.J., Whitfield C., Costerton J.W. (1985). Antibiotic resistance of *Pseudomonas aeruginosa* colonizing a urinary catheter *in vitro. Eur J Clin Microbiol*, **4**, 213–218.

Palmer C.J., Tsai Y.-L., Paszko-Kolva C., Mayer C., Sangermano L.R. (1993). Detection of *Legionella* species in sewage and ocean water by polymerase chain reaction, direct fluorescent antibody and plate culture methods. *Appl Environ Microbiol*, **59**, 3618–3624.

Pasculle A.W., Feeley J.C., Gibson R.J., Cordes L.G., Meyerowitz R.L., Patton C.M., Gorman G.W., Carmack C.L., Ezzell J.W., Dowling J.N. (1980). Pittsburgh pneumonia agent: direct isolation from human lung tissue. *J Infect Dis*, **141**, 727–732.

Paszko-Kolva C., Shahamat M., Colwell R.R. (1992). Long term survival of *Legionella pneumophila* serogroup 1 under low-nutrient conditions and associated morphological changes. *FEMS Microbiol Ecol*, **102**, 45–55.

Patterson W.J., Hay J., Seal D.V., McLuckie J.C. (1997). Colonization of transplant unit water supplies with *Legionella* and protozoa: precautions required to reduce the risk of legionellosis. *J Hosp Infect*, **37**, 7–17.

Pope D.H., Soracco R.J., Gill H.K., Fliermans C.B. (1982). Growth of *Legionella pneumophila* in two-membered cultures with green algae and cyanobacteria. *Curr Microbiol*, **7**, 319–322.

Powell E.L., Newsome A.L., Allen S.D. (1993). Free-living amoeba antigens recognized by naturally occurring human antibodies. *Abstract, Annual Meeting, American Society of Microbiology*, p. 512.

Presti F.L., Riffard S., Vandenesch F., Etienne J. (1998). Identification of *Legionella* species by random amplified polymorphic DNA profiles. *J Clin Microbiol*, **36**, 3193–3197.

Ristroph J.D., Hedlund K.W., Allen R.G. (1980). Liquid medium for growth of *Legionella pneumophila. J Clin Microbiol*, **11**, 19–21.

Robinson P.N., Heidrich B., Tiecke F., Fehrenbach F.J., Rolfs A. (1996). Species-specific detection of *Legionella* using polymerase chain reaction and reverse dot-blotting. *FEMS Microbiol Lett*, **140**, 111–119.

Rogers J., Dowsett A.B., Dennis P.J., Lee J.V., Keevil C.W. (1994). Influence of temperature and plumbing material selection on biofilm formation and growth of *Legionella pneumophila* in a model potable water system containing complex microbial flora. *Appl Environ Microbiol*, **60**, 1585–1592.

Rowbotham T.J. (1980). Preliminary report on the pathogenicity of *Legionella pneumophila* for fresh water and soil amoebae. *J Clin Pathol*, **33**, 1179–1183.

Rowbotham T.J. (1984). Legionellae and amoebae. In: Thomsberry C., Balows A., Feeley J.C., Jakubowsky W. (eds) *Legionella: Proceedings of the Second International Symposium*. American Society for Microbiology, Washington, pp. 325–327.

Ruseska I., Robbins J., Lashen E.S., Costerton J.W. (1982). Biocide testing against corrosion-causing oilfield bacteria help control plugging. *Oil Gas J*, **8**, 253–264.

Saint C.P. (1998). A colony based confirmation assay for *Legionella* and *Legionella pneumophila* employing the EnviroAmp^{Tm} *Legionella* system and agglutination. *Lett Appl Microbiol*, **26**, 377–381.

Schofield G.M., Locci R. (1985). Colonization of components of a model hot water system by *Legionella pneumophila. J Appl Bacteriol*, **58**, 151–162.

Spino D.F., Rice E.W., Geldreich E.E. (1984). Occurrence of *Legionella* spp. and other aquatic bacteria in chemically contaminated ground water treated by aeration. In: Thornsberry C., Balows A., Feeley J.C., Jakubowsky W. (eds) *Legionella: Proceedings of the Second International Symposium*. American Society for Microbiology, Washington, pp. 318–320.

States S.J., Conley L.F., Kuchta J.M., Oleck B.M., Lipovich M.J., Wolford R.S., Wadowsky A.M. (1987). Survival and multiplication of *Legionella pneumophila* in municipal drinking water systems. *Appl Environ Microbiol*, **53**, 979–986.

States S., Kuchta J., Young W., Conley L., Ge J., Costello M., Dowling J., Wadowsky R. (1998). Controlling *Legionella* using copper-silver ionization. *J Am Water Works Assoc*, **90**, 122–129.

Steinert M., Emödy L., Amann R., Hacker J. (1997). Resuscitation of viable but non-culturable *Legionella pneumophila* Philadelphia JR32 by *Acanthamoeba castellanii. Appl Environ Microbiol*, **63**, 2047–2053.

Stout J.E., Yu V.L., Yee Y.C., Vaccarella S., Diven W., Lee T.C. (1992). *Legionella pneumophila* in residential water supplies: environmental surveillance with clinical assessment for Legionnaires' disease. *Epidemiol Infect*, **190**, 49–57.

Stout J.E., Yu V.L. (1997). Current concepts: legionellosis. *N Engl J Med*, **337**, 682–684.

Sutherland I.W. (1971). Bacterial exopolysaccharides: their nature and composition. In: Sutherland I.W. (ed) *Surface Carbohydrates of the Prokaryotic Cell*. Academic Press, London, pp. 27–96.

Swerinski H., Gaiser S., Bardtke D. (1985). Immunofluorescence for the quantitative determination of nitrifying bacteria: Interference of the test in biofilm reactors. *Appl Microbiol Biotechnol*, **21**, 125–128.

Tablan O.C., Anderson L.J., Ardern N.H., Breiman R.F., Butler J.C., McNeil M.M. (1994). Guidelines for prevention of nosocomial pneumonia. *Amer J Infect Control*, **22**, 247–292.

United States Environmental Protection Agency Office of Water Advisories. (1993). *Health Advisories for Drinking Water Contaminants*. Lewis Publishers, Boca Raton.

van Belkum A., Maas H., Verburgh H., Leeuwen N. (1996). Serotyping, ribotyping, PCR-mediated ribosomal 16S-23S spacer analysis and arbitrarily primed PCR for epidemiological studies on *Legionella pneumophila. Res Microbiol*, **147**, 405–413.

Vickers R.M., Yu V.L., Hanna S.S., Muraca P., Diven W., Carmen N., Taylor F.B. (1987). Determinants of *Legionella pneumophila* contamination of water distribution systems: 15-hospital prospective study. *Infect Control*, **8**, 357–363.

Wadowsky R., Yee R.B. (1981). Glycine-containing selective medium for isolation of *Legionellaceae* from environmental specimens. *Appl Environ Microbiol*, **42**, 768–772.

Walker J.T. (1997). Monitoring biofouling and *Legionella pneumophila. Waste Water Treat*, **Sept**, 16.

Wilkinson I.J., Sangster N., Ratcliff R.M., Mugg P.A., Davos D.E., Lanser J.A. (1990). Problems associated with identification of *Legionella* species from the environment and isolation of six possible new species. *Appl Environ Microbiol*, **56**, 796–802.

Wolfaardt G.M., Lawrence J.R., Robarts R.D., Caldwell D.E. (1994a). Multicellular organization in a degradative biofilm community. *Appl Environ Microbiol*, **60**, 434–436.

Wolfaardt G.M., Lawrence J.R., Robarts R.D., Caldwell D.E. (1994b). The role of interactions, sessile growth, and nutrient amendments on the degradative efficiency of a microbial consortium. *Can J Microbiol*, **40**, 331–340.

World Health Organization. (1985). *Environmental Aspects of the Control of Legionellosis*. World Health Organization, Copenhagen, p. 11.

World Health Organization. (1990). Epidemiology, prevention and control of legionellosis: memorandum from a WHO meeting. *Bull World Health Org*, **68**, 155–164.

Wright J.B. (1988). The ecological and pathological importance of the mode of growth of *Legionella pneumophila* serogroup 1. PhD Thesis, University of Calgary, Canada.

Wright J.B. (1997). Significantly reduced toxicity approach to paper machine deposit control. In: 1997 Engineering & Papermakers Conference Proceedings. TAPPI Press, Atlanta, pp. 1083–1088.

Wright J.B., Costerton J.W. (1986). Visualization of the glycocalyx surrounding cells of *Legionella pneumophila*. Abstract Annual Meeting, Alberta Heritage Foundation for Medical Research, Edmonton, Canada.

Wright J.B., Ruseska I., Costerton J.W. (1991). Decreased biocide susceptibility of adherent *Legionella pneumophila*. *J Appl Bacteriol*, **71**, 531–538.

Wright J.B., Ruseska I., Athar M.A., Corbett S., Costerton J.W. (1989). *Legionella pneumophila* grows adherent to surfaces *in vitro* and *in situ*. *Infect Control Hosp Epidemiol*, **10**, 408–415.

Yabuuchi E., Wang L., Yamayoshi T. (1995). Bactericidal effect of chlorine on strains of *Legionella species*. *J Jpn Assoc Infect Dis*, **69**, 151–157.

18 Biofilms in Drinking Water Treatment and Distribution

Anne K. Camper

The low nutrient environment present in drinking water treatment plants and distribution systems would not appear to be a hospitable environment for bacterial growth. However, biofilms are found on almost every submerged surface in treatment plants and distribution systems. In treatment, biofilms can be used to reduce the concentration of organic matter that forms disinfection by-products as well as produce biologically stable water that reduces biofilm growth in distribution systems. Biological filters do not appear to support the growth of pathogenic bacteria. Distribution system biofilms are deleterious, in that they can release indicator organisms, and heterotrophic bacteria, and may cause taste and odor problems. Control of these biofilms is difficult. Disinfection alone is usually ineffective. Reduction of organic matter, improved disinfection, and the implementation of corrosion control if unlined iron pipes are present, or a combination of these methods is helpful in controlling distribution system biofilms.

KEY WORDS: biofilms, biological treatment, drinking water, distribution systems

INTRODUCTION

Biofilms occur universally on submerged surfaces in drinking water treatment plants and in distribution systems (Figure 1). Their activity and presence is usually considered to be undesirable, leading to deleterious changes in water quality. Such biofilms are influenced by system operating parameters and environmental variables such as water quality, water demand and the types of surfaces present.

Because biofilms may release organisms of regulatory concern into the water, the industry has been very interested in understanding mechanisms for controlling biofilm processes in distribution systems. The belief has been that all organisms in drinking water are undesirable, and treatment methods have been optimized to reduce the presence of organisms to the lowest levels possible. However, increasing disinfectant concentrations or contact times to control microbial activity has become increasingly difficult due to regulations governing the concentrations of disinfection by-products in water. For this reason, the option of using biofilms in a beneficial manner in drinking water treatment is now gaining acceptance. Since biofilms will inevitably develop in water systems, it is better to design a treatment process where they can

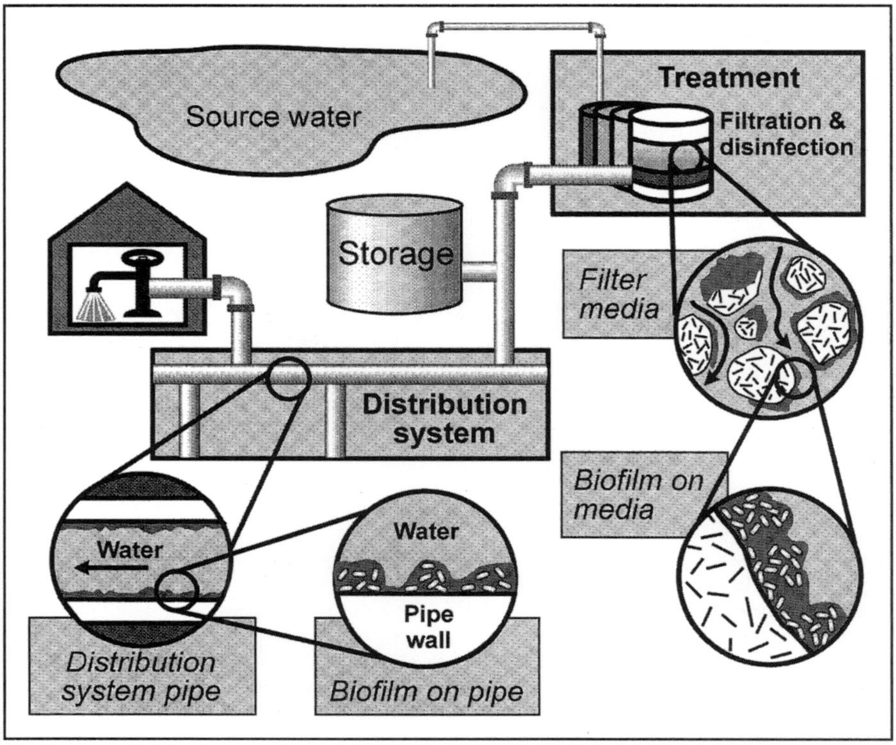

Figure 1 Schematic denoting relevant areas of biofilm activity in drinking water treatment and distribution. Biological treatment involves the encouragement of biofilm growth on filter media. Filtration may be preceded by ozonation to improve removal of organic carbon. The filter media may be anthracite, sand or granular activated carbon. Colonized filter media can be released from the filter and enter the distribution system. A disinfectant is added downstream of the filters before or as the water enters the distribution system to reduce the load of organisms and help mitigate contamination events. Regardless of the level of treatment or the type of disinfectant used, biofilms will grow on pipe surfaces. Note the relatively reduced level of biofilm thickness on the pipe surface compared to that on the filter media. The type of pipe material, the concentration and composition of the disinfectant, corrosion control practices as well as reduction of the organic matter all have the potential to influence the amount of biofilm present on pipe surfaces.

be positively manipulated rather than allow them to cause problems downstream where control is expensive or impossible.

Biological treatment utilizes natural bacterial populations to reduce the organic matter in water that would contribute to both disinfection by-product formation and downstream biofilm formation in distribution systems. The concept of biological treatment has been gaining favor, with a considerable amount of research being done to optimize performance and to address the potential for filters to harbor organisms of public health concern.

BIOFILMS IN WATER TREATMENT

Almost all filters in drinking water treatment have some level of biological activity. In the past, this activity has been viewed as undesirable because the organisms are released into the water and may cause increased microbiological counts. Therefore, typical water treatment plant operation has included practices intended to minimize biofilm growth on the filters, including carrying a disinfectant residual onto the filters and using chlorinated backwash water. With increasingly stringent regulations on the presence of disinfection by-products and the growth of biofilms in distribution systems, the potential to enhance biological activity in filters to improve water quality has gained favor.

Drinking water that has been subjected to microbial activity in a controlled manner in a treatment plant is more "biologically stable" or less likely to contribute to microbial proliferation in the distribution system (Cipparone *et al.*, 1997). Biologically treated water typically has lower disinfectant demand and disinfection by-product formation potential than conventionally treated water if the source water is high in organic carbon (Huck and Anderson, 1992; Cipparone *et al.*, 1997). As utilities move to using ozone as a primary disinfectant and for taste/odor/color control, biological filters may be necessary to reduce the concentrations of biodegradable organic carbon entering the distribution system. In some regards, slow sand filtration represents biological treatment, but the term typically refers to the optimization of biofilm activity on rapid rate filters.

Designed and engineered drinking water biological treatment was first implemented in France and other western European countries nearly 20 years ago (Sontheimer *et al.*, 1978; 1979a; 1979b). In the most traditional form, separate granular activated carbon (GAC) filters are located downstream from conventional treatment. In conventional treatment, particle removal is optimized through coagulation, flocculation, sedimentation, and filtration. The water is then ozonated and passed through filters that are optimized for microbial utilization of a portion of the natural organic matter remaining in the water. Biological filters are typically operated with exhausted carbon, that is, the chemisorptive capacity of the GAC has been exceeded. The surfaces of the filter media act as supports for microbial attachment and growth, resulting in a biofilm adapted to using the organic matter found in that particular water. Total organic carbon removals in these filters range from 5 to 75% (Bouwer and Crowe, 1988).

The discussion presented below focuses primarily on secondary biological treatment, *i.e.* biological filters after conventional filtration. However, as will be seen in the last section on this topic, there is a growing interest in using a single filter for both particle removal and biological filtration. Dual operation presents its own unique set of considerations for operation and performance, but many of the principals used to design and operate second stage filters are applicable to single filters also.

Ozonation

A common treatment step before biological treatment is ozonation. Ozone may be applied to reduce taste and odor compounds, remove color, provide primary disinfection for protozoan cysts, or to reduce disinfection demand/disinfection by-products by oxidizing some of the organic matter. Water that has been preozonated often has elevated levels of lower molecular weight organic compounds; these compounds have been associated with increased biofilm development downstream (van der Kooij *et al.*, 1989; Price, 1994; LeChevallier *et al.*, 1996b). Goel *et al.* (1995) reported that the fraction of recalcitrant natural organic matter in water made available for microbial growth was increased after ozonation, but the numerical value varied from site to site. This has also been substantiated by van der Kooij *et al.* (1982), Werner and Hambsch (1986), Servais *et al.* (1987) and Speitel *et al.* (1993). Because biofilms can form either in controlled treatment processes (biological filters) or in uncontrolled deleterious locations (distribution systems), the drinking water industry considers biological filtration after ozonation, regardless of the original intent of ozone application.

Importance of Filter Media

GAC vs other media

The selection of filter media (sand, anthracite or GAC) for use in biological filters has been of interest primarily due to capital cost and performance issues. Sand and anthracite are usually less expensive than GAC, but it is important to optimize biological activity and the resultant performance of the filter in terms of net removal of organic carbon from the water. For instance, although rapid sand filters do have the capacity to biologically remove carbon (Eberhardt *et al.*, 1977; Sontheimer *et al.*, 1978; Borbiogot *et al.*, 1982; van der Kooij and Hijnen, 1985) it has been found that GAC typically has superior performance (LeChevallier *et al.*, 1992; *c.f.* DeWaters and DiGiano, 1990 and Hozalski *et al.*, 1995). This is presumed to be the result of the higher amount of biomass that attaches to GAC *vs* anthracite (Niquette *et al.*, 1998). LeChevallier *et al.* (1992) demonstrated that there were more bacteria per unit surface area on GAC than on sand, and that the total organic carbon (TOC) removal rate was 51% *vs* 26%. Another advantage of GAC over other media is that the attached microbial population is less prone to shock from changes in water quality, down time, or accidental application of disinfectant (Bablon *et al.*, 1988; Krasner *et al.*, 1993).

Iron oxide coated sand

Even though the emphasis in drinking water has been on the use of GAC, there is evidence to suggest that iron oxide coated media may be a better choice for removal of natural organic matter (Jacangelo *et al.*, 1995; Owen *et al.*, 1995). Iron oxides have a large potential for the sorption of natural organic matter (McCarthy *et al.*, 1993; Parfitt *et al.*, 1977; Zhou *et al.*, 1994). Under abiotic conditions, humic material is irreversibly held on the surface of iron oxides (Gu *et al.*, 1994; 1996). This property has been used to develop a technique for the removal of natural organic matter (NOM) from water by coating sand particles used in slow sand filter beds with iron oxides (McMeen and Benjamin, 1997). Circumstantial evidence indicates that the bound organic matter is potentially available for biofilm bacteria, since these same investigators stated that the iron oxide-coated olivine used in their filtration studies continued to remove NOM for a 16 month time period; they suggested that the adsorption sites were being "bioregenerated."

Importance of Operational Conditions

There are at least three operational conditions that have been shown to be important in the performance of biological filters, *viz.* empty bed contact time (EBCT), temperature, and backwash conditions. The EBCT is the residence time of the fluid in the filter calculated as though the entire volume occupied by the filter medium is occupied by water. Both temperature and EBCT have an impact on the amount of organic carbon removal. In the first case, longer contact times allow for longer potential reaction times. Temperature influences the rate at which the reactions occur. If temperatures are higher, the contact time needed to achieve a specific removal can be reduced. Operationally, the contact time can be adjusted, but temperature is extremely difficult if not impossible to control. Backwash strategies will influence the amount of biofilm left on the filter media and therefore impact the organic carbon removal efficiencies after the filter is put back on line.

Empty bed contact time

In most cases, TOC removal in biological filters can be improved by increasing the EBCT. Because of the very large volumes treated by a drinking water plant, a small reduction in EBCT results in substantial savings in filter volume. Experimental EBCTs in biological filters have varied from 2 to 30 min. Sontheimer and Hubel (1987) demonstrated an increase in dissolved organic carbon removal from 27 to 41% when the EBCT increased from 5 to 20 min. LeChevallier *et al.* (1992) reported a 29% removal of TOC with a 5 min EBCT and a 51.5% reduction at 20 min. Prevost *et al.* (1990) suggest that a 20 min EBCT is required for 90% removal of biodegradable organic carbon. However, there are instances where increased EBCT is not beneficial, which is probably a result of the biodegradability of the organic matter present in the water (Hozalski *et al.*, 1995).

Temperature

As stated above, the EBCT required for biological removal of TOC will be temperature dependent. This has been demonstrated at a full-scale biological filtration plant, where 12 min of EBCT was required at 0.5°C for the same percentage removal obtained in 6 min at 10–12°C (Niquette *et al.*, 1998). The importance of temperature has been demonstrated for the removal of biodegradable organic matter as well as for the removal of specific ozonation by-products (Krasner *et al.*, 1993; Coffey *et al.*, 1995).

Backwashing

Optimizing backwash for biological filters may be different than for conventional particle removal filters. The backwash cycle must be designed so that biomass loss is balanced with the need for decreased headloss and the control of undesirable organisms (*e.g.* nematodes) in the filter.

DiGiano (1992), Miltner *et al.* (1992) and Ahmad *et al.* (1998) evaluated the impact of chlorinated *vs* non-chlorinated backwash water on the performance of filters. In all cases, filters backwashed with chlorinated water had lower biomass and in general effluent quality suffered.

The mode of backwash may have an effect on biomass removal and subsequent removal of organic carbon. In a model system, Hozalski and Bouwer (1998) demonstrated that water wash without a disinfectant removed only 20–40% of the biomass from filter media, and the retained biomass was capable of maintaining good biological removal of TOC. In pilot plant studies, Niquette *et al.* (1998) showed that filter performance of biologically active carbon filters was not decreased after nonchlorinated backwash, and biomass density increased. This was presumably due to the removal of inorganic material that acted as a diffusion barrier for microbial growth.

In the operation of biological filters, retention of biomass must be balanced with the ability to retain adequate filter run times and reasonable headloss in biological filters. Ahmad *et al.* (1998) determined that water wash alone was not sufficient to clean biological filters. The end result was a buildup of excessive headloss over time. In contrast, air plus subfluidization with water flow followed by a water wash maintained filter run times without seriously impacting biological performance for organic carbon removal. The retention of biological activity was partially explained by the observation that nonbiological particles are more easily removed from filters during backwash than bacteria (Ahmad and Amirtharajah, 1998).

Filters for Particle Removal and Biological Activity

Due to the costs associated with installation of a complete set of dedicated biological filters, treatment plants considering the implementation of biological treatment may wish to promote biofilm development on the existing rapid rate single or dual media filters. This option is being considered by facilities that presently have difficulty meeting disinfection by-product regulations, those that have implemented ozonation to attain primary disinfection requirements, or those that experience regrowth in their distribution systems. It is important to note that particle removal in these filters cannot be compromised, and biological treatment is therefore of secondary importance. This is an area of growing interest, as evidenced by the review article by Urfer *et al.* (1997), which concludes with a section on research needs in the area.

Downstream Effects of Biomass Released from Biological Filters

The end result of biological filtration is conversion of organic carbon in the water into bacterial biomass. Ideally, this biomass is immobilized on the filter media and removed during the backwash cycle. It is possible, however, for colonized filter fines or bacterial aggregates to be shed from the filter media and enter the distribution system. There is evidence to suggest that particle associated bacteria are less susceptible to disinfection (LeChevallier *et al.*, 1984) and that these organisms can then pass the disinfection barrier (Camper *et al.*, 1986; Stewart *et al.*, 1990). There has also been concern expressed by the drinking water industry that biological filters may support the proliferation and/or concentration of organisms of public health or regulatory importance.

Release of colonized filter media

The type of filter medium and filter operations influences the release of colonized filter medium particles. In a comparison between bacterial counts on released filter fines from biologically activated sand, anthracite, and biologically activated carbon (BAC), the BAC fines contained significantly higher numbers of heterotrophs (LeChevallier *et al.*, 1992). When the release of fines from full-scale anthracite and sand filters was evaluated (Camper *et al.*, 1987), particle counts were similar to those from GAC filters. In the same study, where particles released from GAC filters throughout a filter run were enumerated, higher filtration rates, deeper GAC filter beds, and higher applied water turbidity resulted in higher particle release. The age of the carbon did not influence particle release. These findings were confirmed by Stringfellow *et al.* (1993) who demonstrated that the number of particles released from sand or GAC contactors were similar. They also observed that these particles were frequently colonized by heterotrophs, although no coliforms were detected. Amirtharajah and Westein (1980) showed that an elevated number of particles were released both immediately prior to backwash and shortly after the filter was put back in operation if proper filter-to-waste procedures were not followed. It is probable that colonized filter fines or detached bacteria are also released during these two events, as illustrated by an elevated number of coliforms detected in the filtrate immediately after backwash (Bucklin *et al.*, 1991). Moran *et al.* (1993) showed that there can be breakthrough of turbidity due to detachment of particles from deep laboratory filters at the end of a filter run, and these particle sizes were comparable to the range associated with *Cryptosporidium* oocysts.

In field studies, Camper *et al.* (1987) collected released filter fines from GAC filters by passing water from the underdrain during an entire filter run through a gauze filter in a 47 mm Swinnex. Recognizing that small fines may not have been efficiently retained in the gauze, the average number of particles in a sample was 2,333 ranging in size from 1.0 to 3.5×10^3 μm in diameter. Seventeen percent of the filter runs released carbon fines that contained coliform bacteria, and 28% of these coliforms exhibited the fecal biotype. It should be noted, however, that none of these utilities experienced elevated coliform numbers in their distribution systems.

In a similar study, Stewart *et al.* (1990) used a modified Swinnex with a polycarbonate filter to trap carbon particles released from a pilot GAC filter. An average of 36 particles l^{-1} were detected with sizes ranging from 2 to >40 μm. It was determined that 200 to 7,000 viable bacteria could be recovered from 1000 particles. The numbers of coliforms were low, with one reported fecal coliform isolated from the released filter fines.

Proliferation of organisms of public health concern

The drinking water industry has expressed concern about the potential for pathogenic organisms to grow on biological filters and then be released into the water. For such organisms to be found in water emanating from a biological filter they must successfully colonize and compete with the indigenous organisms (or be retained in large numbers), be released from the filter, and penetrate the disinfection barrier.

To colonize and persist on filter media, the pathogens must be able to compete successfully with the heterotrophic bacterial populations. Competition with the existing microflora may be a key parameter in preventing proliferation of pathogenic organisms. In previous research using laboratory columns containing GAC, Camper *et al.* (1985) found that a suite of pathogenic bacteria could survive on GAC when fed a sterile source of surface water. However, if the pathogens were challenged with organisms present in unsterilized surface water, the numbers of pathogens declined. If pathogens were added to the carbon simultaneously with the autochthonous heterotrophs, they declined more rapidly than in the first instance, but the decline of pathogens in the filter and filter effluent was most rapid if pathogens were added to previously colonized GAC. In other studies using groundwater and laboratory columns with virgin GAC, GAC that had been in operation in a filter for a few months, and BAC from a full-scale plant, coliform elimination was most rapid from the BAC filter (LeChevallier *et al.*, 1998). Other laboratory studies (Rollinger and Dott, 1987) with several pathogens on GAC gave similar results. The pathogens persisted when introduced to sterile GAC and fed sterile water, but were eliminated from the medium when they were subsequently challenged by autochthonous bacteria from tap water.

The advent of molecular techniques has led to more detailed investigations of pathogen persistence on filters. The use of PCR can demonstrate pathogen presence even if the organisms cannot be cultured. A recently completed project utilized laboratory columns of filter media inoculated with enteric bacterial pathogens (*Salmonella*, *E. coli 0157:H7*). Culturable cells were not detected within 1 d. Bacteria were detected using PCR and fluorescent antibody techniques for 2 weeks, but the pathogen numbers decreased rapidly after inoculation. In companion pilot scale tests, the organisms in filter effluents were reduced three to four logs within 48 h. *Giardia* and *Cryptosporidium* showed similar behavior, suggesting that the bacterial pathogens were not attaching to the filter media in significant numbers. There was no evidence to support the concept that bacteria were growing in biofilms on the filters (Burr *et al.*, 2000).

Impact of released cells/colonized filter fines on the distribution system

At first consideration, it may seem that biological filters would release organisms and filter fines that may have an adverse affect on the distribution system. However, before bacteria from filters reach the distribution system, they must pass the disinfection barrier. They must then be transported in water containing disinfectant, attach to the pipe wall, and proliferate. Evidence from pilot studies has shown that chlorination after biological treatment produced water with lower bacterial numbers than that from conventional treatment (LeChevallier *et al.*, 1998). Presumably this is because the use of chlorine on the conventional filter (including backwash) selected for organisms less susceptible to disinfection. Even if these organisms reach the biofilm, there is ecological evidence showing that high levels of inoculation are required for the colonization and persistence of an allochthonous organism (Warren *et al.*, 1992). Work with filter fines colonized with a coliform has shown that the particles can attach to existing biofilms, but the organisms on their surfaces have no selective advantage (Morin *et al.*, 1996). In these experiments there was a selective release of the filter fines and most of the coliforms from the biofilm when a disinfectant was applied.

All of the evidence gathered to date from laboratory and pilot experiments as well as full-scale experience points to the ability of biological treatment to produce microbiologically safe drinking water, especially if post-disinfection is practiced.

BIOFILMS IN DISTRIBUTION SYSTEMS

In the last decade the traditionally held view that water quality is solely the result of treatment been refuted. Treatment processes have been optimized to produce water that meets regulatory and end-user requirements, often at great expense. However, this well-treated water may arrive at the consumer's tap modified in taste, odor and microbial content, as well as in other important aspects. Likewise, highly purified and treated industrial water may become compromised in distribution and unsuitable for downstream processes. Many of these changes in quality are the result of the growth of biofilms on the surfaces of distribution systems.

It is reasonably well documented that the increase of bacterial counts through distribution is the result of the detachment of biofilm cells rather than growth of organisms in the water. Published accounts by van der Wende *et al.* (1989) and LeChevallier *et al.* (1990b) demonstrated that elevated bacterial counts in water could not be attributed to replication of suspended cells, but rather was due to biofilm growth. A simple calculation based on mass loading of a distribution system demonstrates the potential for a small amount of organic carbon to produce a large number of organisms. A hypothetical treatment plant produces water with a concentration of total organic carbon of 2 mg l^{-1}, of which 10% can be used for bacterial growth. The treatment plant produces 20, 000 m^3 d^{-1}, and all this water must pass through the distribution system. If a yield of 0.1 g dry cell mass g^{-1} carbon is assumed, this system could produce approximately 400 grams or 10^{14} new cells each day if the biofilm is at steady state. Another interpretation is that the system would produce enough released biofilm to account for 10^8 total cells ml^{-1}. A distribution system should be viewed as a complex reactor system where many factors, including the growth of biofilms, contribute to water quality deterioration.

Problems Associated with Biofilms in Distribution Systems

Coliform and heterotroph regrowth

The initial interest in biofilm growth in distribution systems arose from the observation that total coliform bacteria were appearing in distributed water that met all the criteria for microbiological quality at the plant. Coliform bacteria are used world wide as indicator organisms, and their presence has been associated historically with fecal contamination of water. When these bacteria were found in systems with well-run treatment plants and properly maintained distribution systems, it was hypothesized that the bacteria were growing in biofilms on the surface and being released into the water. The fact that bacteria colonize pipe material was reported in the 1970's and early 80's (O'Connor *et al.*, 1975; Allen *et al.*, 1980; Olson, 1982) but it was not until 1984 in the state of Connecticut that the growth of colifoms on distribution system materials was linked to their presence in the water (Centers for Disease Control, 1985). This incident promoted the study of biofilms in distribution systems across the U.S. and Europe.

The diagnosis of a coliform regrowth event is typically done by eliminating all other potential sources of contamination. It is very difficult to prove conclusively that the organisms are growing in the distribution system. Even when coliform counts are high, coliforms may not be found on the surfaces of excavated distribution system materials (LeChevallier *et al.*, 1987; Characklis, 1988). Presumably this is because the organisms are in discrete locations rather than uniformly distributed throughout the biofilm (LeChevallier *et al.*, 1987; Camper *et al.*, 1996).

The regrowth of heterotrophs can be also be of concern, especially for European communities which are required to monitor their presence. Some U.S. utilities will routinely monitor heterotrophs as a general indicator of microbial quality, and may be required to assess their numbers if chlorine residuals are too low. The general heterotrophic population is usually not of public health concern, but with the growing immunocompromised population, many utilities are interested in minimizing the presence of these organisms in their water. Some heterotrophs may be opportunistic pathogens, and for this and other reasons their control is desirable.

Colonization by pathogens and opportunistic pathogens

There is limited information available on the presence of pathogens in distribution system biofilms, primarily because detection is difficult. In spite of this difficulty, there are instances where opportunistic pathogens have been detected, including *Aeromonas* spp., *Mycobacterium* spp. and *Helicobacter pylori*. These organisms are presently on the U.S. EPA's Contaminant Candidate List (USEPA, 1998), and as such, there is much interest in determining the levels at which they exist in distribution systems.

Aeromonas spp. have been found in distribution system biofilms (van der Kooij, 1988; Havelaar *et al.*, 1990). Although these organisms are known to survive under low nutrient conditions, little is known about the factors that govern their growth. There is a positive correlation of growth with residence time (Havelaar *et al.*, 1990), but a mixed correlation with temperature (Burke *et al.*, 1984; Havelaar *et al.*, 1990) in water systems with little or no disinfectant. In controlled laboratory experiments, low level disinfection lead to the elimination of the organism from mixed population biofilms (Camper *et al.*, 1998).

It is known that organisms of the genus *Mycobacterium* can grow in biofilms, and that there is the potential for selection for these organisms in water due to their resistance to chlorine (Collins *et al.*, 1984; Schulze-Robbecke and Fischeder, 1989; Briganti and Wacker, 1995). These organisms were found in biofilms on the surfaces of pilot systems that received conventionally and biologically treated water, with a lower frequency of detection when the systems received biologically treated water. These studies showed that no mycobacteria were found in pipes receiving biologically treated water followed by chlorination (LeChevallier *et al.*, 1998). Of particular concern are the MAC, or *Mycobacterium avium* complex, implicated in infections in the immunocompromised population, particularly those with acquired immunodeficiency syndrome (Horsburgh, 1991; Nightingale *et al.*, 1992).

There is presently a great deal of interest in determining the ability of *Helicobacter pylori* to survive in biofilms. Laboratory studies have shown that it can persist in water for extended time periods (West *et al.*, 1992; Shahamat *et al.*, 1993) and it has been isolated from drinking water (Klein *et al.*, 1991). Mackay *et al.* (1999) found that in a mixed population biofilm, this organism can be detected by PCR for up to 192 h. This suggests that it can at least persist if the inoculum is sufficiently high.

There is even less information about the survival of frank bacterial pathogens (*Salmonella* spp., *Escherichia coli* O157:H7, *Vibrio cholerae*) in distribution system biofilms than for opportunistic pathogens. Pathogens in distribution system biofilms would be subjected to the same ecological pressures as were described above for biological filters. It is therefore unlikely that pathogens would persist. The rationale presented above for the pathogen survival in biological filters would likely apply to distribution systems also.

There are no known instances where frank bacterial pathogens have been recovered in distribution system biofilms, even after that system has suffered a water-borne disease outbreak. There are laboratory studies where pathogens have been shown to persist, but these systems may not be representative of drinking water conditions. For example, *Salmonella enteritidis* has been shown to grow in pure culture biofilms when fed glucose at a concentration of 100 mg l^{-1} (Jones and Bradshaw, 1996). *Campylobacter* spp., as detected by flourescent antibody staining, could survive for over 42 d in mixed population biofilms fed filter sterilized tap water when inoculated at a concentration of *ca* 10^6 cells ml^{-1} (Buswell *et al.*, 1998). Studies done with *E. coli* O157:H7 showed that even after a high population inoculum, the organisms could not be detected by fluorescent antibody methods in mixed population biofilms after a few days. In contrast, *Salmonella typhimurium* persisted for over 50 d, but there was no evidence of proliferation (Camper *et al.*, 1998). All these studies suggest that the organisms may be present, but presence alone must be separated from viability and/or infectivity.

Control of Biofilms in Distribution Systems

Almost every water distribution system is prone to biofilm formation, regardless of the purity of the water, type of pipe material, or biocide treatment used. It is known that there can be substantial changes in metal concentrations, bacterial populations, disinfectant residuals and disinfectant by-products, and esthetic qualities (taste, odor, and color) in the water through distribution. There is an interaction between surface-mediated reactions (corrosion, biocide/disinfectant demand, immobilization of substrates for bacterial growth), mass transfer and mass transport processes, and bulk fluid properties (*e.g.* concentration and type of biocides, general water chemistry, and organic concentration) on the microbial ecology of the biofilm and subsequent water quality changes. These interactions can be exceedingly complex, which means that control measures are not obvious and are often system specific.

There are a variety of operational and environmental parameters that appear to encourage the growth of biofilms in distribution systems. Surveys of the industry have shown that temperature, organic carbon levels, the concentration and type of disinfectant, and the presence or absence of a corrosion control regime when corrodible materials are used in the distribution system are of importance (LeChevallier, 1990; Smith *et al.*, 1990). Some of these factors have been studied, and evidence to support their importance is given below.

Reduction of organic carbon concentrations

Regulations on disinfection by-product formation and the desire to produce biologically stable water has lead to the identification of options to reduce the concentration of organic matter in water. The three most likely options are enhanced coagulation, activated carbon adsorption, and biological filtration. For any of these options, identifying design criteria for the amount of organic carbon in water that is capable of supporting biofilm growth has been of key interest.

There have been several methods developed to measure the amount of organic carbon in water that is available for bacterial growth. The assimilable organic carbon (AOC) test was developed by van der Kooij *et al.* (1982) and measures the increase in numbers of a known bacterial culture when grown in unamended drinking water. Threshold levels of AOC to limit heterotrophic biofilm growth have been set at 10 μg C l^{-1} for heterotrophs (van der Kooij, 1992) and a recommended 50 μg C l^{-1} for coliform control (LeChevallier *et al.*, 1991). A second measurement is the biodegradable organic carbon (BDOC) test, where the water is inoculated with an undefined mixed bacterial population in suspension or on sand particles and the change in dissolved organic carbon is measured over time. A guideline BDOC value of 0.15 mg l^{-1} (Servais *et al.*, 1993) for biological stability has been suggested, with Joret (1994) recommending that the levels be adjusted for temperature.

It is important to note that these measurement techniques were developed by European investigators responsible for the monitoring and operation of European systems where many of the confounding variables influencing regrowth have been minimized. For example, European systems are operated without a substantial disinfectant residual. It is not uncommon for the upper limit for disinfectant leaving European sources to be set at 0.2–0.4 mg l^{-1}, while US facilities typically use much higher concentrations (1–4 mg l^{-1}). For this and other reasons, the reasonable success of these methods in European systems has not been as straightforward in the US. There are instances where associations between AOC/BDOC and biofilm are not clear cut. In a field survey, regrowth was seen in systems with average AOC levels both > and <100 μg l^{-1} (LeChevallier *et al.*, 1996a). In pilot experiments, there was a weak correlation between biofilm and influent AOC concentrations, but no correlation with the concentration of AOC in the reactors (Camper, 1996). Studies where known organic carbon sources were used also demonstrated similar results; increased organic carbon loading did not always result in higher biofilm cell numbers (Ellis *et al.*, 1999). There appear to be interactions in the distribution system that complicate the ability to use AOC/BDOC as an independent assessment of regrowth potential.

Regardless of how appropriate or inappropriate a specific test or the values from these tests are for predicting regrowth, the concept of reducing the amount of organic carbon entering a distribution system is still valid. It is only logical to assume that limiting the amount of substrate for heterotrophs will lead to less biofilm growth, fewer detached cells, and improved water quality. This has been demonstrated in pilot studies where, in the absence of a disinfectant, biological treatment has reduced the number of organisms on pipe surfaces (LeChevallier *et al.*, 1998; Volk *et al.*, 1998). Likewise, a detailed study of two full-scale systems demonstrated that biological treatment produced water that had the same microbiological quality as conventional treatment, with less residual disinfectant required (LeChevallier *et al.*, 1998).

Controlling biofilms with disinfectants

Biofilms are notoriously less susceptible to disinfectants than suspended cells, and the literature contains many references to this effect. It is generally believed that increasing the concentration of a disinfectant should control regrowth, but many instances exist where the opposite effect is seen (Martin *et al.*, 1982; Hudson *et al.*, 1983; Reilly and Kippen, 1984; CDC, 1985; Oliveri *et al.*, 1985; LeChevallier *et al.*, 1987). Hudson *et al.* (1983) found no correlation between free chlorine residuals and the number of heterotrophic plate count (HPC) organisms per unit surface area. An interesting observation has been that the presence of chlorine can enhance the numbers of coliforms in biofilms (Camper *et al.*, 1996; 1999).

There are several reasons why this lack of disinfection efficacy may occur. First, there may be mass transfer limitations that prevent the penetration of the disinfectant into the biofilm (van der Wende and Characklis, 1990; LeChevallier *et al.*, 1990a; 1993; deBeer *et al.*, 1994; Srinivasan *et al.*, 1995; Koudjonou *et al.*, 1997). Disinfectants may react with the total organic carbon in the water, making it more available for microbial growth (Bryant *et al.*, 1992). Low level chlorination (*ca* 0.2 mg l^{-1} residual) has also been observed to increase the growth rate of heterotrophs in biofilms (Ellis *et al.*, 1999). It has also been shown that biofilm bacteria are physiologically different from suspended cells, and this may account for their reduced susceptibility to disinfection (Gilbert and Brown, 1985). It is also possible that the disinfectants increase corrosion, which, as shown below, is an important factor in biofilm development on pipe surfaces.

With the increased emphasis on controlling disinfection by-products as well as controlling regrowth, utilities have become more interested in using monochloramine as a secondary disinfectant. Monochloramine is typically not used as a primary disinfectant because of its high concentration × time (CT) requirements, but it has been found to be a superior disinfectant over free chlorine for biofilms (Griebe *et al.*, 1994; Srinivasan *et al.*, 1995; Olson, 1996; Ollos *et al.*, 1997). When full scale distribution systems were studied, chloramines were more effective at reducing the number of biofilm total coliforms and HPC than chlorine (Neden *et al.*, 1992). Donlan and Pipes (1988) demonstrated that elevated monochloramine concentrations were associated with reduced attached bacterial populations. A field study of 31 utilities showed that systems that used chloramines had 0.51% coliform positives in 35, 159 water samples as compared to 0.97% of 33, 196 samples in chlorinated systems. The same study showed that the average density of coliforms in chlorinated systems was 35 times higher than that in the chloraminated systems (LeChevallier *et al.*, 1996b).

Monochloramine may also have an advantage over chlorine if corrodible surfaces are present. A study in a model pipe loop system composed of several materials showed that biofilms on galvanized, copper, or PVC surfaces were readily disinfected by free chlorine or monochloramine (1 mg l^{-1}) while iron pipe surface-associated bacteria were more susceptible to monochloramine than free chlorine (4 mg l^{-1}) (LeChevallier *et al.*, 1990a). Abernathy (1998) showed that monochloramine was more effective at reducing bacterial counts on ductile iron surfaces than free chlorine, and the effect was more pronounced if a corrosion inhibitor was also used.

The importance of materials

Distribution system pipe materials apparently have a marked effect on biofilm formation. In particular, unlined iron (mild steel, cast and ductile iron) pipes have a profound positive effect on the number of attached bacteria (Camper, 1996; Camper *et al.*, 1996; LeChevallier *et al.*, 1993; 1996a; 1996b). A utility survey has shown a positive relationship between the number of miles of unlined metal pipes and coliform occurrences (LeChevallier *et al.*, 1996b). The influence of iron pipes can be substantial, with one experiment demonstrating a >100 fold increase in biofilm cell numbers on iron compared to polyvinylchloride (PVC) surfaces (LeChevallier *et al.*, 1998) and another showing a two-fold increase on the same surfaces (van der Kooj and Oorhuizen, 1997). Other pilot system experiments have supported an interaction between corroding iron pipes and biofilms. Neden *et al.* (1992) found that bacterial populations on unlined cast iron were the highest while PVC was colonized with the lowest number, while Block (1992) determined that there was a progressive decrease in bacterial densities on surfaces from cast iron, tinned iron, cement lined cast, to stainless steel. Delanoue *et al.* (1997) also noted that unlined iron pipes had higher biofilm densities than non-ferrous materials, and that mild steel was more heavily colonized than cast and ductile iron. In another study where cast iron, medium density polyethylene and unplasticized polyvinyl chloride surfaces were tested, the cast iron surface had the highest doubling time for biofilm bacteria, an average of 97% more organisms than the other materials, and the highest species diversity (Kerr *et al.*, 1999).

The same influence of materials has been demonstrated for coliforms. Laboratory experiments have shown that mild steel surfaces supported ten-fold more heterotrophs and two to ten fold more coliforms than polycarbonate surfaces. The effect was also seen in effluent cell counts. The presence of a small amount of mild steel (10% on the basis of surface area) in an otherwise plastic system resulted in the same elevated biofilm counts on all surfaces (Camper *et al.*, 1996).

There are several reasons why iron surfaces may be more conducive to bacterial growth than inert materials. It is probable that the metal surfaces exert a disinfectant demand (Knocke, 1988; Knocke *et al.*, 1994; Vasconcelos *et al.*, 1996), which would limit disinfectant efficacy against biofilms. There is circumstantial evidence for this effect, because biofilms on ferrous metal surfaces have been found to be less susceptible to free chlorine than those on inert materials (LeChevallier *et al.*, 1987; 1988; 1990a; 1998; Chen *et al.*, 1993; Kerneis *et al.*, 1994). In addition, increased corrosion rates decrease the efficacy of free chlorine against biofilm organisms (LeChevallier *et al.*, 1993).

It is also possible that corrosion products change the availability of natural organic matter to biofilm bacteria. Aquatic humic substances readily adsorb to iron oxide surfaces (Tipping, 1981; Gu *et al.*, 1994; Parfitt *et al.*, 1997; Varadachari *et al.*, 1997), and this may change the chemical and biological properties of the molecules. When attached to surfaces, humic substances uncoil due to collapse and binding with the surface, which results in the exposure of sugars and peptides previously concealed in the hydrophobic molecule (Gu *et al.*, 1994). Experiments in the author's laboratory have demonstrated that humic substances can be used as a sole carbon and energy source for biofilms (Camper *et al.*, 1999). In addition, the presence of an iron oxide surface with bound humic substances enhances the attachment of bacteria and the subsequent growth rate when humics are provided in the bulk water (Qui, 1999).

Corrosion control

If corroding surfaces contribute to increased biofilm accumulation, it should follow that corrosion control may reduce biofilm cell numbers. There is evidence to demonstrate that corrosion control has mitigated coliform regrowth in full scale systems (Martin *et al.*, 1982; Hudson *et al.*, 1983; Lowther and Moser, 1984; Schreppel *et al.*, 1997). The mechanism by which corrosion control influences regrowth could be based on less reactivity of the disinfectant with the surface, fewer bacterial attachment sites, or other factors. For example, LeChevallier *et al.* (1993) showed that corrosion control reduced biofilm numbers but attributed the response to increased chlorine efficacy. It may be that control is dictated by water quality and the type of disinfectant used. For example, an increase in pH (a common corrosion control scheme) may decrease chlorine efficacy because of the shift in the equilibrium towards the less effective hypochlorite ion. This effect would not be seen if phosphate was used to control corrosion. It may also be that corrosion control supercedes disinfection; Martin *et al.* (1982) showed that increasing the pH to 9 in a chlorinated system reduced bacterial counts. Other studies have cited that corrosion control was more important than disinfection in reducing biofilm densities (LeChevallier *et al.*, 1998). As supporting evidence, the author and colleagues have noted that at near neutral pH, in the absence of corrosion control, the presence of low levels of disinfectant increases biofilm density on ductile or steel surfaces. This is presumably because corrosion was enhanced and the disinfectant consumed at the surface (Camper, 1996; Abernathy, 1998). Abernathy (1998) has shown that biofilm density is directly proportional to the mass of corrosion products present. The reduction of corrosion product concentration and a concomitant decrease in biofilm was accomplished by adding phosphates or increasing the pH or by changing from chlorine to monochloramine. The best results were obtained when phosphates were used in conjunction with monochloramine. In another pilot experiment, the results indicated that corrosion control was more important in reducing bacterial numbers in biofilms than decreasing the amount of useable organic matter in the water or the maintenance of a free chlorine residual (LeChevallier *et al.*, 1998).

CONCLUSIONS

As with all other aqueous environments, drinking water treatment and distribution systems offer habitats suitable for biofilm growth. Biofilm accumulation in treatment, which has historically been viewed as extremely undesirable, is now being optimized as a component of water treatment. Biological treatment can then reduce the potential for biofilm formation in distribution systems where control is very difficult. Biofilm growth in distribution systems is governed by a host of variables that may be system specific. Some of the factors that appear to be important in biofilm control are the type and concentration of disinfectant, the organic content of the water, the type of pipe material, and the use of a corrosion control scheme. It is important to note that each system is unique and a biofilm control strategy that is successful for one utility may not be appropriate for another.

REFERENCES

Abernathy C.G. (1998). Effects of corrosion control treatments and biofilm disinfection on unlined ferrous pipes. PhD Thesis, Montana State University, USA.

Ahmad R., Amirtharajah A. (1998). Detachment of particles during biofilter backwashing. *J Am Water Works Assoc*, **90**, 74–85.

Ahmad R., Amirtharajah A., Al-Shawwa A., Huck P.M. (1998). Effects of backwashing on biological filters. *J Am Water Works Assoc*, **90**, 62–73.

Allen M.J., Geldreich E.E., Taylor R.H. (1980). The occurrence of microorganisms in water main encrustations. *J Am Water Works Assoc*, 72, 614–618.

Amirtharajah A., Westein D.P. (1980). Initial degradation of effluent quality during disinfection. *J Am Water Works Assoc*, 72, 518–524.

Bablon G.P., Ventresque C., Aim R.B. (1988). Developing a sand-GAC filter to achieve high-rate biological filtration. *J Am Water Works Assoc*, **80**, 47–53.

Block J.C. 1992. Biofilms in drinking water distribution systems. In: Bott T.R., Fletcher M., Capdeville B. (eds) *Biofilms – Science and Technology*. NATO Advanced Study Institute, Portugal, Kluwer Publishers, The Netherlands.

Bouwer E.J., Crowe P.B. (1988). Biological processes in drinking water treatment. *J Am Water Works Assoc*, **80**, 82–90.

Borbiogot M.M., Dodin A., Lherritier R. (1982). Limiting bacterial aftergrowth in distribution systems by removing biodegradable organics. *Proc AWWA Annual Conf*, Miami Beach, Fl. American Water Works Assoc, Denver, CO.

Briganti L.A., Wacker S.C. (1995). *Fatty Acid Profiling and the Identification of Environmental Bacteria for Drinking Water Utilities.* American Water Works Association, Denver, CO.

Bryant E.A., Fulton G.P., Budd G.C. 1992. *Disinfection Alternatives for Safe Drinking Water.* Van Nostrand Reinhold, New York.

Bucklin K.E., McFeters G.A., Amirtharajah A. (1991). Penetration of coliforms through municipal drinking water filters. *Water Res*, **25**, 1013–1017.

Burke V., Robinson J., Gracey M., Peterson D., Meyer N., Haley V. (1984). Isolation of *Aeromonas* spp. from an unchlorinated domestic water supply. *Appl Environ Microbiol*, **48**, 367–370.

Burr M., Camper A.K., DeLeon R., Hacker P. (2000). *Colonization of Biologically Active Filter Media with Pathogens*. American Water Works Association, Denver, CO (In press).

Buswell C.M., Herlihy Y.M., Lawrence L.M., McGuiggan J.T.M., Marsh P.D., Keevil C.W., Leach S.A. (1998). Extended survival and persistence of *Campylobacter* spp. in water and aquatic biofilms and their detection by immunofluorescent-antibody and rRNA staining. *Appl Environ Microbiol*, **64**, 733–741.

Camper A.K. (1996). *Factors Limiting Microbial Growth in the Distribution System: Pilot and Laboratory Studies.* American Water Works Association Research Foundation, Denver, CO.

Camper A., Jones W.L., Hayes J.T. (1996). Effect of growth conditions and substratum composition on the persistence of coliforms in mixed population biofilms. *Appl Environ Microbiol*, **62**, 4014–4018.

Camper A.K., LeChevallier M.W., Broadaway S.C., McFeters G.A. (1985). Growth and persistence of pathogens on granular activated carbon filters. *Appl Environ Microbiol*, **50**, 1378–1382.

Camper A.K., LeChevallier M.W., Broadaway S.C., McFeters G.A. (1986). Bacteria associated with granular activated carbon particles in drinking water. *Appl Environ Microbiol*, **52**, 434–438.

Camper A.K., Broadaway S.C., LeChevallier M.W., McFeters G.A. (1987). Operational variables and the release of granular activated carbon particles in drinking water. *J Am Water Works Assoc*, **79**, 70–74.

Camper A.K., Warnecke M., Jones W.L., McFeters G.A. (1998). *Pathogens in Model Distribution System Biofilms*. American Water Works Association, Denver, CO.

Camper A.K., Ellis B., Butterfield P., Jones W.L., Anderson W., Huck P.M., Volk C.A., LeChevallier M.W. (1999). *Biological Stability of Water in Treatment Plants and Distribution Systems*. American Water Works Association, Denver, CO.

Characklis W.G. (1988). *Bacterial Regrowth in Distribution Systems*. American Water Works Association, Denver, CO.

Centers for Disease Control (CDC). (1985). Detection of elevated levels of coliform bacteria in a public water supply – Connecticut. *Morbid Mortal Weekly Rpt*, **34**, 142–144.

Chen C.I., Griebe T., Srinivasan R., Stewart P.S. (1993). Effects of various metal substrata on accumulation of *Pseudomonas aeruginosa* biofilms and the efficacy of monochloramine as a biocide. *Biofouling*, 7, 241–251.

Cipparone L.A., Diehl A.C., Spietel G.E. Jr. (1997). Ozonation and BDOC removal: effect on water quality. *J Am Water Works Assoc*, **89**(2), 84–97.

Coffey B.M., Krasner S.W., Sclimenti M.J., Hacker P.A., Gramith J.T. (1995). Comparison of biologically active filters for the removal of ozone by-products, turbidity, and particles. *Proc Water Qual Technol Conf*. American Water Works Association, Denver, CO, pp. 357–389.

Collins C.H., Grange J.M., Yates M.D. (1984). Mycobacteria in water. *J Appl Bacteriol*, **57**, 193–211.

deBeer D., Srinivasan R., Stewart P.S. (1994). Direct measurement of chlorine penetration into biofilms during disinfection. *Appl Environ Microbiol*, **60**, 4339–4344.

Delanoue A., Holt D., Woodward J.C., McMath S.M., Smith S.E. (1997). Effect of pipe materials on biofilm growth and deposit formation in water distribution systems. *Proc Water Qual Technol Conf*. American Water Works Association, Denver, CO.

DeWaters J.E., DiGiano F.A. (1990). The influence of ozonated natural organic matter on the biodegradation of a micropollutant in a GAC bed. *J Am Water Works Assoc*, **82**, 69–75.

DiGiano F.A. (1992). *Microbial Activity on Filter-Adsorbers*. American Water Works Association, Denver, CO.

Donlan R.M., Pipes W.O. (1988). Selected drinking water characteristics and attached microbial population density. *J Am Water Works Assoc*, **80**, 70–76.

Eberhardt M., Madsen S., Sontheimer H. (1977). Investigations of the use of biologically effective activated carbon filters in the processing of drinking water. EPA-TR-77-503. US Environmental Protection Agency, Cincinnati, OH.

Ellis B.D, Butterfield P., Jones W.L., McFeters G.A., Camper A.K. (1999). Effects of carbon source, carbon concentration, and chlorination on growth related parameters of heterotrophic biofilm bacteria. *Microb Ecol* (In press).

Gilbert P., Brown M.R.W. (1985). Mechanisms of the protection of bacterial biofilms from antimicrobial agents. In: Lappin-Scott H.M., Costerton J.W. (eds) *Microbial Biofilms*. Cambridge University Press, Cambridge, UK, pp. 118–130.

Griebe T., Chen C.-I., Srinivasan R., Stewart P.S. (1993). Analysis of biofilm disinfection by monochloramine and free chlorine. In: Geesey G.G., Lewandowski Z., Flemming H.-C. (eds) *Biofouling and Biocorrosion in Industrial Water Systems*. Lewis Publishers, Boca Raton, pp. 151–161.

Goel S., Hozalski R. M., Bouwer E.J. (1995). Biodegradation of NOM: Effect of NOM source and ozone dose. *J Am Water Works Assoc*, **87**, 90–105.

Gu B., Mehlhorn T.L., Liang L., McCarthy J.F. (1996). Competitive adsorption, displacement, and transport of organic matter on iron oxide: I. Competitive adsorption. *Geochim Cosmochim Acta*, **60**, 1943–1950.

Gu B., Schmidt J., Chen Z., Liang L., McCarthy J. F. (1994). Adsorption and desorption of natural organic matter on iron oxide: mechanisms and models. *Environ Sci Technol*, **28**, 38–46.

Havelaar H.H., Versteegh J.F.M., During M. (1990). The presence of *Aeromonas* in drinking water supplies in the Netherlands. *Zbl Hyg*, **190**, 236–256.

Horsburgh C.R. (1991). *Mycobacterium avium* complex infection in the acquired immunodeficiency syndrome. *New Engl J Med*, **324**, 1332–1338.

Hozalski R.M., Bouwer E.J. (1998). Deposition and retention of bacteria in backwashed filters. *J Am Water Works Assoc*, **90**, 71–85.

Hozalski R.M., Goel S., Bouwer E.J. (1995). TOC removal in biological filters. *J Am Water Works Assoc*, **87**, 40–54.

Huck P.M,. Anderson W.B. (1992). Quantitative relationships between the removal of NVOC, chlorine demand, and AOX formation potential in biological water treatment. *Vom Wasser*, **78**, 281.

Hudson L.D., Hankins J.W., Battaglia M. (1983). Coliforms in a water distribution system: a remedial approach. *J Am Water Works Assoc*, **75**, 564–568.

Jacangelo J.G., DeMarco J., Owen D.M., Randtke S.J. (1995). Selected processes for removing NOM: An overview. *J Am Water Works Assoc*, **87**, 64–77.

Jones K., Bradshaw S.B. (1996). Biofilm formation by the Enterobacteriaceae: a comparison between *Salmonella enteriditis, Escherichia coli*, and a nitrogen-fixing strain of *Klebsiella pneumoniae*. *J Appl Bacteriol*, **80**, 458–464.

Joret J.C. (1994). Control of biodegradable organic matter during drinking water treatment. *Proc Int Sem Biodegradable Organic Matter*. Montreal, Quebec.

Kerneis A., Deguin A., Feinberg M. (1994). Modeling applications of the number of microorganisms according to the residence time of drinking water in a distribution system. *Proc Int Sem Biodegradable Organic Matter*. Montreal, Quebec.

Kerr C.J., Osborn K.S., Robson G.D, Handley P.S. (1999). The relationship between pipe material and biofilm formation in a laboratory model system. *J Appl Microbiol*, **85**, 29S–38S.

Klein P.D., Gastrointestinal Physiology Working Group, Graham D.Y., Galillor A., Opekun A.R, O'Brian E. (1991). Water source as a risk factor for *Helicobacter pylori* infection in Peruvian children. *Lancet*, **337**, 1503–1506.

Knocke W.R. (1988). Soluble manganese removal on oxide coated filter media. *J Am Water Works Assoc*, **80**, 65–70.

Knocke W.R., Shorney H.L., Bellamy J.D. (1994). Examining the reactions between soluble iron, DOC, and alterative oxidants during conventional treatment. *J Am Water Works Assoc*, **86**, 117–127.

Koudjonou B.K., Prevost M., Lafrance P. (1997). Assessing the impact of chlorination on the composition of a drinking water biofilm. *Proc Water Qual Technol Conf*. American Water Works Association, Denver, CO.

Krasner S.W., Sclimenti M.J., Coffey B.M. (1993). Testing biologically active filters for removing aldehydes formed during ozonation. *J Am Water Works Assoc*, **85**, 62–68.

LeChevallier M.W. (1990). Coliform regrowth in drinking water: a review. *J Am Water Works Assoc*, **82**, 74–86.

LeChevallier M.W., Babcock T.M., Lee R.G. (1987). Examination and characterization of distribution system biofilms. *Appl Environ Microbiol*, **53**, 2714–2717.

LeChevallier M.W., Cawthon C.D., Lee R.G. (1988). Factors promoting survival of bacteria in chlorinated water supplies. *Appl Environ Microbiol*, **54**, 649–654.

LeChevallier M.W., Lowry C.D., Lee R.G. (1990a). Disinfecting biofilms in a model distribution system biofilm. *J Am Water Works Assoc*, **82**, 87–99.

LeChevallier M.W., Olson B.H., McFeters G.A. (1990b) *Assessing and Controlling Bacterial Regrowth in Distribution Systems*. American Water Works Association, Denver, CO.

LeChevallier M.W., Schulz W., Lee R.G. (1991). Bacterial nutrients in drinking water. *Appl Environ Microbiol*, **57**, 857–862.

LeChevallier M.W., Welch N.J., Smith D.B. (1996a). *Factors Limiting Microbial Growth in the Distribution System: full Scale Experiments.* American Water Works Association, Denver, CO.

LeChevallier M.W., Welch N.J., Smith D.B. (1996b). Full-scale studies of factors related to coliform regrowth in drinking water. *Appl Environ Microbiol*, **62**, 2201–2211.

LeChevallier M.W., Hassenauer T.S., Camper A.K., McFeters G.A. (1984). Disinfection of bacteria attached to granular activated carbon. *Appl Environ Microbiol*, **48**, 918–923.

LeChevallier M.W., Becker W.C., Schorr P., Lee R.G. (1992). Evaluating the performance of biologically active rapid sand filters. *J Am Water Works Assoc*, **84**, 136–146.

LeChevallier M.W., Lowry C.D., Lee R.G., Gibbon D.L. (1993). Examining the relationship between iron corrosion and the disinfection of biofilm bacteria. *J Am Water Works Assoc*, **85**, 111–123.

LeChevallier M.W., Norton C.D., Camper A.K., Morin P., Ellis B., Jones W., Rompre A., Prevost M., Coallier J., Servais P., Holt D., Delanoue A., Colbourne J. (1998). *Microbial Impact of Biological Filtration*. American Water Works Association, Denver, CO.

Lowther E.D., Moser R.H. (1984). Detecting and eliminating coliform regrowth. *Proc Water Qual Technol Conf*. American Water Works Association, Denver, CO, pp. 323–336.

Mackay W.G., Gibbon L.T., Barer M.R., Reid D.C. (1999). Biofilms in drinking water systems: A possible reservoir for *Helicobacter pylori*. *J Appl Microbiol*, **85**, 52S–59S.

Martin R.S., Gates W.H., Tobin R.S., Sumarah R, Wolfe P., Forestall P. (1982). Factors affecting coliform bacterial growth in distribution systems. *J Am Water Works Assoc*, **74**, 34–37.

McCarthy J.F., Williams T., Liang L., Jardine P., Jolley L., Taylor D., Palumbo A., Cooper L. (1993). Mobility of natural organic matter in a sandy aquifer. *Environ Sci Technol*, **27**, 667–676.

McMeen C.R., Benjamin M.M. (1997). NOM removal by slow sand filtration through iron oxide-coated olivine. *J Am Water Works Assoc*, **89**, 57–71.

Miltner R.J., Rice E.W., Summers R.S. (1992). A pilot-scale study of biological treatment. *Proc AWWA Annual Conf*. American Water Works Association, Denver, CO.

Moran C.M., Moran D.C., Cushing R.S, Lawler D.F. (1993). Particle behavior in deep-bed filtration: Part 2 – particle detachment. *J Am Water Works Assoc*, **85**, 82–93.

Morin P., Camper A.K., Jones W., Gatel D., Goldman J.C. (1996). Colonization and disinfection of biofilms hosing coliform-colonized carbon fines. *Appl Environ Microbiol*, **62**, 4428–4432.

Neden D.G., Jones R.J., Smith J.R., Kirmeyer G.J., Foust G.W. (1992). Comparing chlorination and chloramination for controlling bacterial regrowth. *J Am Water Works Assoc*, **84**, 80–88.

Nightingale S.D., Byrd L.T., Southern P.M., Jockusch J.D., Cal S.X., Wynne B.A. (1992). *Mycobacterium avium-intracellulare* complex bacteremia in human immunodeficiency virus positive patients. *J Infect Dis*, **165**, 1082–1085.

Niquette P., Prevost M., Maclean R.G., Thibault D., Coallier J., Desjardins R., Lafrance P. (1998). Backwashing first-stage sand-BAC filters. *J Am Water Works Assoc*, **90**, 86–97.

O'Connor J.T., Hash L., Edwards A.B. (1975). Deterioration of water quality in distribution systems. *J Am Water Works Assoc*, **67**, 113–116.

Oliveri V.P., Bakalian A.E., Bossung K.W., Lower E.D. (1985). Recurrent coliforms in water distribution systems and the presence of free residual chlorine. In: Jolley R.L., Bull R.J., Davis W.P. Katz S., Roberts M.H., Jacobs V.A. (eds) *Water Chlorination, Chemistry, Environmental Impact, and Health Effects*. Lewis Publishers, Boca Raton, FL.

Ollos P.J., Slawson R.M., Huck P.M. (1997). Modeling biofilm accumulation in drinking water distribution systems. *Proc Water Qual Technol Conf*. American Water Works Association, Denver, CO.

Olson S.C. (1996). Phosphate based corrosion inhibitors effects on distribution system regrowth. *Nat Conf Integrating Corrosion Control & Other Water Quality Goals*. American Water Works Association, Denver, CO.

Olson B.H. (1982). Assessment and implications of bacterial regrowth in water distribution systems. Environmental Protection Agency project summary EPA-600/S2-82-072.

Owen D.M., Amy G.L., Chowdhury Z. K., Paode R, McCoy G., Viscosil K. (1995). NOM characterization and treatability. *J Am Water Works Assoc*, **87**, 46–63.

Parfitt R.L., Fraser A.R., Farmer V.C. (1977). Adsorption on hydrous oxides, III. Fulvic acid and humic acid on goethite, gibbsite and imogolite. *J Soil Sci*, **28**, 289–296.

Prevost M., Dejardins R., Duchesne D., Poirier C. (1990). Chlorine demand removal by biological activated carbon filtration. *Proc Water Qual Technol Conf*. American Water Works Association, Denver, CO, pp. 407–428.

Price M.L. (1994). *Ozone and Biological Treatment for DBP Control and Biological Stability.* American Water Works Association, Denver, CO.

Qui L. (1999). The surface interactions among iron oxides, humic substances and biofilms. MS Thesis, Montana State University, USA.

Reilly K.J., Kippen J.S. (1984). Relationship of bacterial counts with turbidity and free chlorine in two distribution systems. *J Am Water Works Assoc*, **75**, 309–314.

Rollinger Y., Dott W. (1987). Survival of selected bacterial species in sterilized activated carbon filters and biological activated carbon filters. *Appl Environ Microbiol*, **53**, 777–781.

Schreppel C.K., Fredericksen D.W., Geiss A.A. (1997). The positive effects of corrosion control on lead levels and biofilms. *Proc Water Qual Technol Conf*. American Water Works Association, Denver, CO.

Schulze-Robbecke R., Fischeder R. (1989). Mycobacteria in biofilms. *Zbl Hyg*, **188**, 385–390.

Servais P., Laurent P., Randon G. (1993). Impact of biodegradable dissolved organic carbon (BDOC) on bacterial dynamics in distribution systems. *Proc Water Qual Technol Conf*. American Water Works Association, Denver, CO.

Shahamat M., Mai U., Paszko-Kolva C., Kessel M., Colwell R.R. (1993). Use of autoradiography to assess viability of *Helicobacter pylori* in water. *App Environ Microbiol*, **59**, 1231–1235.

Smith D.B., Hess A.F., Hubbs S.A. (1990). Survey of distribution system coliform occurrences in the United States. *Proc Water Qual Technol Conf*. American Water Works Association, Denver, CO, pp. 1103–1107.

Sontheimer H. (1979a). Applying oxidation and adsorption techniques: a summary of progress. *J Am Water Works Assoc*, **71**, 612–617.

Sontheimer H. (1979b). Design criteria and process schemes for GAC filters. *J Am Water Works Assoc*, **71**, 618–622.

Sontheimer H., Hubel D. (1987). The use of ozone and granular activated carbon in drinking water treatment. In: Huck P.M., Toft P. (eds) *Treatment of Drinking Water for Organic Contaminants*. Pergamon Press, NY.

Sontheimer H., Heilker E., Jekel M.R., Nolte H., Vollmer F. (1978). The Mulheim process. *J Am Water Works Assoc*, **70**(7), 393–396.

Speitel G.E. Jr., Symons J.M., Diehl A.C., Sorensen H.W., Cipparone L.A. (1993). Effect of ozone dosage and subsequent biodegradation of DBP precursors. *J Am Water Works Assoc*, **85**, 86–95.

Srinivasan R. Stewart P., Griebe T., Chen C-I. (1995). Biofilm parameters influencing biocide efficacy. *Biotechnol Bioeng*, **46**, 553–560.

Stewart M.H., Wolfe R.L., Means E.G. (1990). Assessment of the bacteriological activity associated with granular activated carbon treatment of drinking water. *Appl Environ Microbiol*, **56**, 3822–3829.

Stringfellow W.T., Mallon K., DiGiano F.A. (1993). Enumerating and disinfecting bacteria associated with particles released from GAC filter-adsorbers. *J Am Water Works Assoc*, **85**, 70–80.

Tipping E. (1981). The adsorption of aquatic humic substances by iron oxides. *Geochim et Cosmochim Acta*, **45**, 191–199.

Urfer D., Huck P.M., Booth S.D.J., Coffey B. (1997). Biological filtration for BOM and particle removal: a critical review. *J Am Water Works Assoc*, **89**, 83–98.

USEPA. (1998). Announcement of the drinking water contaminant candidate list; notice. *Federal Register*, **63**, 10274–10287, March 2.

Warren T.M., Williams V., Fletcher M. (1992). Influence of solid surface, adhesive ability, and inoculum size on bacterial colonization in microcosm studies. *Appl Environ Microbiol*, **58**, 2954–2959.

van der Kooij D. (1988). Properties of aeromonads and their occurrence and hygienic significance in drinking water. *Zentralbl Bakteriol Mikrobiol Hygiene*, **187**, 1–17.

van der Kooij D. (1992). Assimilable organic carbon as an indicator of bacterial regrowth. *J Am Water Works Assoc*, **84**, 57–65.

van der Kooij D., Hijnen A.M. (1985). Measuring the concentration of easily assimilable organic carbon in water treatment as a tool for limiting regrowth of bacteria in distribution systems. *Proc Water Qual Technol Conf*. American Water Works Association, Denver, CO, pp. 35–52.

van der Kooij D., Oorhuizen W.A. (1997). Biofilm formation in a distribution system supplied with unchlorinated slow sand filtrate. *Proc Water Qual Technol Conf*. American Water Works Association, Denver, CO.

van der Kooij D., Visser A., Hijnen W.A.M. (1982). Determining the concentration of easily assimilable organic carbon in drinking water. *J Am Water Works Assoc*, 74, 540–545.

van der Kooij D., Hijnen W.A.M., Kruithof J.H.C. (1989). The effects of ozonation, biological filtration, and distribution on the concentration of easily assimilable organic carbon (AOC) in drinking water. *Ozone Sci & Engrg*, **11**, 297–311.

van der Wende E., Characklis W.G. (1990). Biofilms in potable water systems. In: McFeters G.A. (ed) *Drinking Water Microbiology*. Springer-Verlag, New York.

van der Wende E., Characklis W.G., Smith D.B. (1989). Biofilm and bacterial water quality. *Water Res*, **23**, 1313–1322.

Varadachari C., Chattopadhyay T., Gosh K. (1997). Complexation of humic substances with oxides of iron and aluminum. *Soil Sci*, **162**, 28–34.

Vasconcelos J.J., Boulos P.F., Grayman W.M. Laurent K., Wable O., Biswas P., Bari A. Rossman L.A., Clark R.J., Goodrich J.A. (1996). *Characterization and Modeling of Chorine Decay in Distribution Systems*. American Water Works Association, Denver, CO.

Volk C., LeChevallier M., Friedman M. (1998). *Interactions between Pipe Materials, Organics, Corrosion Inhibitors, and Disinfectants on Distribution Biofilms: Field Studies*. National Water Research Institute, Fountain Valley, CA.

Warren T.M., Williams V., Fletcher M. (1992). Influence of solid surface, adhesive ability and inoculum size on bacterial colonization in microcosm studies. *Appl Environ Microbiol*, **58**, 2954–2959.

West A.P., Millar M.R., Tompkins D.S. (1992). Effect of physical environment on survival of *Helicobacter pylori*. *J Clin Pathol*, **45**, 228–231.

Werner P., Hambsch B. (1986). Investigation of the growth of bacteria in drinking water. *Water Suppl*, **4**, 227–231.

Zhou J.L., Rowland S., Mantoura R., Braven J. (1994). The formation of humic coatings on mineral particles under simulated estuarine conditions – a mechanistic study. *Water Res*, **28**, 571–579.

19 Biofilm Control in Industrial Water Systems: Approaching an Old Problem in New Ways

Rodney M. Donlan

The process of biofilm formation in industrial water systems was documented and described over 25 years ago. Biofilms are now known to be ubiquitous and to result in a number of serious problems for industry. The plant operator responsible now must realize that good microbiological control hinges upon an understanding of biofilm processes, how and where biofilms will develop and how they can be controlled. A first step in developing a biofilm control program is adequate monitoring. This can range from collection and examination of pipe deposits to the use of an on-line electrochemical device interfaced with a computer. Laboratory testing of various candidate biofilm control strategies is also necessary prior to treatment of an industrial water system. Various test apparatus using different biofilm measurement techniques are presented and compared. The choice of method will depend upon cost and time constraints. Control strategies in the technical and patent literature are numerous. Because chlorine has been shown to be ineffective against biofilms and because there are safety concerns in the handling of gaseous chlorine, there is a trend to use other biocides for biofilm control. A number of non oxidizing biocides are available, though it is not clear how effective these are against biofilms. Other biocides which appear to have merit for biofilm control include bromine-based products and ozone. A wide variety of newer approaches include techniques which disrupt biofilms either mechanically or by using electrical fields, techniques which hydrolyze or destabilize the extracellular polymeric matrix, enhance penetration into the biofilms, or alter biofilm surface tension. A successful treatment will target the various biofilm components in a cost effective and environmentally friendly manner. A recognition by industry that biofilms are a major contributor to biofouling in industrial water systems has finally begun to occur. This is the first step in developing control strategies that will be broadly accepted by industry.

KEY WORDS: biofilm, industrial water system, biofilm monitoring, biofilm testing, biocides, biofilm control

INTRODUCTION

As early as 1940, Heukelekian and Heller (1940) noted that growth of microorganisms on surfaces provided a competitive advantage in aquatic systems. Jones *et al.* (1969) characterized the slime layers of aquatic bacteria, contributing important information on biofilm structure and new procedures for biofilm characterization. Characklis

(1973a), reviewed the role of slimes in natural and man-made aquatic systems, noting slimes to be the cause of marine fouling, to increase fluid frictional resistance in pipelines, to affect the heat transfer properties of heat exchangers, and to contribute to scale up problems in biological reactors. Characklis (1973b) also presented several case studies where pipelines in industrial water systems had suffered significant losses in capacity due to microbial slimes. He noted also that though chlorine had been used extensively in process water systems, it was often ineffective in destroying attached slime. Costerton *et al.* (1978) proposed that bacterially produced exopolymers were important in attaching bacteria to inert surfaces in aquatic and in many other systems (human and animal tissues, for example). They argued that attachment would be advantageous in providing a constant supply of nutrients and a mechanism that would resist removal from the system (adhesion). These and other early papers document the importance of biofilms in microbial growth and industrial fouling and understanding of biofilm processes has continued to expand significantly.

Figure 1 (a) Industrial cooling tower. Recirculating cooling water is distributed over the surface of packing (fill) as air is drawn in and over the packing surfaces by means of a fan at the top of the tower. The cooled water collects in the basin of the tower and is pumped back into the cooling water system. (Photograph courtesy of Tom Erdner, Calgon Corporation, Pittsburgh, PA.); (b) Deck of cooling tower showing buildup of algal biofilm. Cooling water is recirculated over the deck surface and flows down over the surfaces of fill material to achieve cooling. Note the small distribution holes in the deck surface which may become occluded due to biofilm formation. (Photograph courtesy of Tom Erdner, Calgon Corporation, Pittsburgh, PA.)

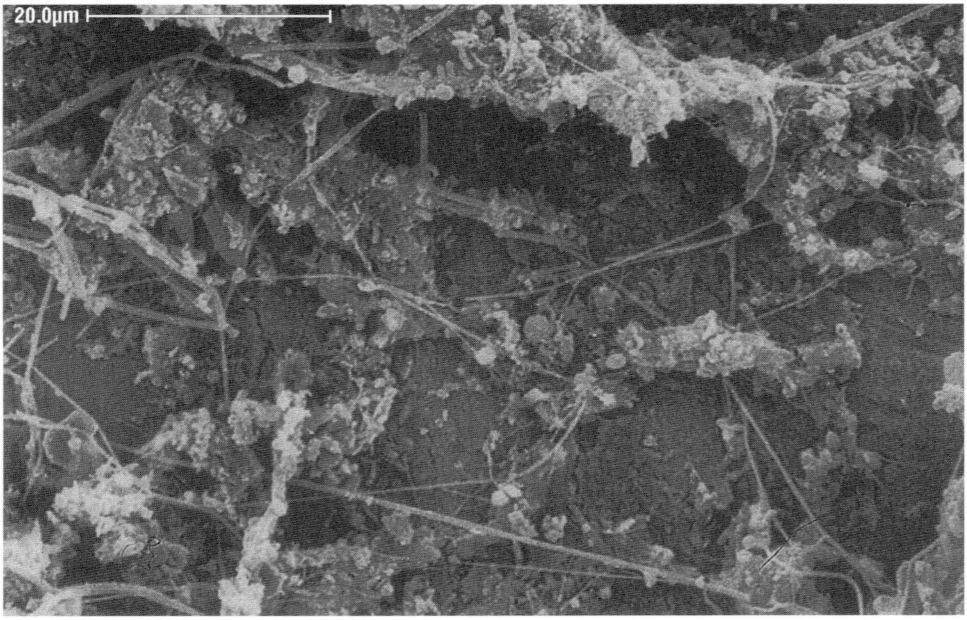

Figure 2 Scanning electron micrograph of biofilm on the surface of 90/10 copper nickel cylinders, exposed in a recirculating cooling water system of a gas fired power plant, and examined after fixation, dehydration, critical point drying, and palladium/gold coating using a Cambridge scanning electron microscope. Error bar = 20 μm. (Reproduced from Donlan *et al.*, 1997, with permission.)

The focus of this chapter is on biofilms in industrial water systems. The breadth and scope of the latter range from high purity water systems used in manufacturing processes to recirculating cooling water in power generation facilities, and from water systems used in a process stream (*e.g.* fiber finish plant) to cooling water. Figure 1a shows an industrial water system cooling tower, which contains many sites and components susceptible to biofilm formation (Figure 1b). The number and type of organisms, flow dynamics, temperature range, organic loading, suspended solids levels, general water chemistry, substratum physical and chemical characteristics, are all important in biofilm formation in industrial water systems, and may vary significantly. Therefore, it is difficult to generalize about the types of biofilms that form in these systems. Suffice to say, that any industrial water system subject to microbial contamination will probably be susceptible to biofilm formation. Nutrients and other conditions necessary for biofilm formation in these systems will usually be adequate for at least a portion of the microbial population. Biofilms are ubiquitous (Costerton *et al.*, 1995), so it would be expected that most industrial process or cooling water systems will be subject to biofilm formation. Figure 2 shows a biofilm on a metal surface collected from a recirculating cooling water system.

BIOFILM MONITORING

Characklis (1990) proposed an expert system approach to biofouling monitoring in industrial water systems, in which data collected from a fouling monitor(s) is relayed directly to a central computer. As an increase in biofouling above a preselected level was detected, treatment chemical could automatically be added to reduce biofilms to acceptable levels. However, this concept has not been implemented to any great extent by industry, partly because there is no real consensus on accepted biofilm monitoring techniques, and partly because there is a paucity of information on concentrations of biocides or other treatments necessary to control biofilm formation in dynamic industrial water systems (as opposed to laboratory data). That being said, there are a number of approaches in the published technical and patent literature regarding biofilm monitoring techniques. Each has its strengths and weakness, and these should be carefully considered before choosing one technique over another. These techniques can be grouped operationally into 1) *in situ* procedures, which would include sampling of deposits and slimes from piping and other surface, 2) in-place monitors, which allow installation of monitoring devices directly into the water system piping, 3) side stream monitors, which would include coupon racks and any other devices allowing installation and removal of test substrata from the device, 4) side stream monitors which generally measure total deposition onto exposed surfaces (such as delta "p" and delta "t"), and 5) electrochemical monitors, which can detect changes in the substratum chemistry as a function of biofilm growth in real time. The operator responsible for microbiological water quality is therefore faced with the question of which technique will provide the greatest and most accurate information regarding biofilm formation for the most reasonable cost/time expenditure. Those techniques which appear easiest to use may provide the least information.

In Situ Techniques

Rosser (personal communication) noted that routine biofouling monitoring of the large (60" ID) carbon steel seawater transmission lines of the Saudi Aramco seawater injection system is done by collecting and analyzing solids removed from pipe walls by routine scraping operations, which occurred every two weeks. Because the cleaning operation is performed biweekly, it is possible to establish baseline data on the volume/composition of the solids removed. The disadvantages of *in situ* sampling are that pipe surfaces must be accessible to be sampled, often necessitating expensive shutdowns, and that biofilms may be altered by desiccation during the process, prior to collection and analysis. The benefit is that the samples collected represent biofilms as they exist in nature. The program described may avoid some of the drawbacks and thus could provide interesting and useful information. LeChevallier *et al.* (1987) described an approach for collecting biofilms from drinking water systems. Sediment samples were collected from fire hydrants and examined. Donlan (1987) collected biofilm samples from the walls of water mains removed from a drinking water distribution system. Pipe sections were drained, cut out, and samples were immediately collected, returned to the laboratory, and processed. Goysich and McCoy (1989) described a technique for sampling and analyzing algal biofilms from cooling tower decks, measuring dry weight, chlorophyll *a*, protein, and optical density at 260 nm (nucleic acids).

These approaches might provide a reality check for more sophisticated side stream monitoring approaches to industrial water systems.

In-place Monitoring Devices

Yohe *et al.* (1986) and Donlan *et al.* (1994) described a device allowing installation of substrata directly into pressurized water mains, with the intent of collecting biofilms from these systems. By collecting water and biofilm samples from adjacent locations over designated exposure intervals, the relationship between biofilm formation rate and extent and water quality can be determined. Jones *et al.*,(1993) used a retractable bioprobe, installed at different locations in an oil production water injection system to collect monthly samples of sessile sulfate-reducing bacteria. Each of these techniques has the advantage of monitoring biofilm formation in the water system over designated time periods, and conditions that influence biofilm formation, such as flow dynamics, temperature, bulk water chemistry, microbial population density, and biocide concentration, can be closely duplicated. However, for studies investigating biofilm removal, these approaches, as with any other approach, will not mimic exactly what is occurring on the pipe wall, where biofilms may have developed for extended time periods (years).

Side Stream Monitoring Devices

There are many more published works on the use of side stream devices for biofilm monitoring. McCoy and Costerton (1982) described a device called the Robbins device, which has been used more than any other approach for biofilm monitoring. This device allows for the installation of plugs into the wall, being flush with the inner lumen. The chemistry and surface texture of the plugs can be varied, depending upon the specific requirements for the application. Donlan (1992) used a similar device for studies in cooling water. The substrata used were in the form of removable rings instead of plugs, providing a much greater surface area and eliminating potential problems with flow differences which might occur at the plug surface in a Robbins device. Chexal *et al.* (1997) used an annular reactor, of the type originally developed in Characklis' laboratory for monitoring biofilm formation. The literature should be consulted for other types of side stream monitors (Howsam and Tyrrel, 1989; Characklis, 1991; Donlan *et al.*, 1991; 1997; Wolfaardt *et al.*, 1991; Araujo Da Silva, 1994; Bakich, 1995; Jacobs *et al.*, 1996; Piriou *et al.*, 1997; Puckorius, 1997). For each of these devices, substrata are removed from the device and processed for biofilm removal. Like the in-place monitors, this approach provides a means of sampling biofilm formation in the water system as it occurs. Assuming substratum characteristics and flow dynamics are similar to the bulk water, this approach should provide useful information on the biofouling characteristics of the water system. It is important to note that devices should be installed in close proximity and with the same water supply to provide the most relevant data.

Real Time Monitors

Recently, a number of devices have been introduced and tested which provide real time information. If Characklis' vision was to provide an expert system for biofilm control in industrial water systems, then these approaches might provide the basis for such a system. Zuniga *et al.* (1990), defined an adequate fouling monitor as one which will provide a heated surface with precise control of heat transfer rate, control cooling water flow within 0.06 ft s^{-1}, by its geometry, be able to mimic flow through heat exchangers, be sensitive to minor system changes, and collect data continuously. The oldest, most commonly used approach is the delta "p" device and/or delta "t" device, designed to measure the pressure drop across the length of a tube section as a function of fouling or change in the thermal characteristics of a tube as a function of fouling. The Bridger Scientific Fouling Monitor, described by Bloch and DiFranco (1995), provides an example of such a device. Zuniga *et al.* (1990); Strauss (1992); Aylott *et al.* (1995); and Zisson *et al.* (1995) also provided examples of the use of this type device for biofouling monitoring. A concern with these devices is that they will measure total fouling, which may include clay/silt, corrosion and scale deposits, as well as biofilms. The fouling monitor therefore is not specific for biofilm formation, and data must be reviewed with this in mind. Licina *et al.* (1994) utilized the BioGeorge probe to monitor biofouling in water systems. This measures changes in electrochemical reactions produced by biofilms on stainless steel electrodes. Readings can be collected continuously and fed into a computer. Deposits that produce an increase in the current required to achieve the applied potential, such as biofilms, may be detected by monitoring the current that flows during the polarization cycle. Though laboratory and field studies have been run with this device, questions still remain as to whether the data produced are specific for biofilm formation. Wetegrove *et al.* (1997) described an optical fouling monitor and found that results using this device correlated with delta "p" measurements. However, no data were presented describing the nature of the fouling deposits for either measurement. Though questions arise regarding the meaning of data collected from these devices, they are nonetheless pointing in the right direction, getting away firstly from the incorrect notion that plate counts of water samples relate to biofouling extent, and secondly from the "scrape and plate" techniques commonly in use. More studies comparing biofilm levels using several different techniques could begin to address some of -these concerns and help to bring closer a true "expert system approach" to biofilm monitoring.

LABORATORY TESTING

To best determine the efficacy of biofilm control treatments, laboratory testing is necessary prior to in-field evaluation. Table 1 provides a sampling of apparatus which has been used by both industrial and academic research groups for biofilm testing. It is apparent that there is no real consensus in this area, in spite of the fact that there are two ASTM standards that address biofilm testing in industrial water systems (American Society for Testing Materials, 1991; 1997). Test apparatus range from simple batch reactors inoculated with pure bacterial cultures (Videla *et al.*, 1993; 1994;

Table 1 Biofilm laboratory testing apparatus.

Apparatus	*Inoculum*	*Substratum*	*Growth Medium*	*Reference*
Model cooling tower	Mixed consortia: algae, bacteria	Plastic and wooden slats	Low nutrient: tap water, N and P added	Williams and Holz, 1998
Model cooling tower	*P. aeruginosa* *K. pneumoniae* *E. aerogenes*	316 L stainless steel tubing	High nutrient: synthetic cooling water, 500 ppm glucose	Williams and Holz, 1998
Continuous flow heat exchanger loop	*Sphaerotilus natans*	316 L stainless steel	Tap water plus CYG medium	Ludensky, 1998
Recirculating test loop with Robbins device	North Sea consortia grown on "rock pile"	Mild steel sampling studs	Seawater salt-supplemented with T-Soy broth, sodium lactate	Tanner *et al.*, 1985
Recirculating water system with biofilm sampling device	Power plant recirculating cooling water	CPVC sampling rings	Power plant recirculating cooling water	Donlan *et al.*, 1997
Recirculating test loop	Fluids and metal coupons from gas production equipment	Mild steel coupons	Gas production fluids	Pope *et al.*, 1990
Continuous flow loop	Mixed cooling water consortia	PVC tubing	Synthetic cooling water	Dallmier *et al.*, 1997
Model cooling tower	Potable water	316L Stainless Steel coupons	Cycled up potable water	Coughlin and Steimel, 1997
Batch system	Consortium isolated from seawater corrosion deposit	AISI 304 stainless steel coupons	Postgate's Medium C	Tapper *et al.*, 1997
Batch system	Cooling tower water	Galvanized steel washers	Cooling tower water	Walter and Cooke, 1997
Batch system	*E. aerogenes*	Culture contained in alginate matrix	5 mM CaCl2	McIlwaine *et al.*, 1997
Batch system	*P. fluorescens*	1020 Carbon steel, AISI 304 stainless steel coupons	Synthetic cooling water	Videla *et al.*, 1994
Dynamic once-through reactor	*Klebsiella rubiacearum*	Silicone rubber	Complex medium: gluconse, yeast extract, peptone	Markx and Kell, 1990

Table 1 Continued.

Apparatus	*Inoculum*	*Substratum*	*Growth Medium*	*Reference*
Batch system	Wild type pure culture isolate	Culture immobilized in agar matrix contained on a slide/coupon	Any industrially relevant cooling or process water	Characklis, 1991
Recirculating continuous culture reactor	*Pseudomonass fluorescens*	Glass tubing and coupons	Water containing 2–4 mg l^{-1} glucose	Taylor, 1996
Batch system	*Vibrio alginolyticus*	AISI 304L stainless steel coupons	Seawater supplemented with yeast extract	Videla *et al.*, 1993
Batch system	*Klebsiella pneumoniae*	316 stainlesss steel coupons	1/10 Trypticase Soy Broth	Yu and McFeters, 1994
Annular reactor	Drinking water plus coliform bacteria	Mild steel and polycarbonate coupons	Tap water supplemented with N, P, and assimilable organic carbon	Camper *et al.*, 1996
Disc reactor continuous flow system	*Pseudomonas aeruginosa*	Polycarbonate discs	Dilute Trypticase Soy Broth	Heersink and Zelver, unpublished
Chemostat-fed flow cell	*Pseudomonas fluorescens*	Stainless steel coupons	Complex "Oligotrophic" culture medium	Mittelman *et al.*, 1992
Flow cell in combination with FTIR apparatus	Mixed culture isolated from human saliva	Germanium prism	Glucose-tap water	Suci *et al.*, 1997

Yu and McFeters, 1994) to dynamic model cooling towers or recirculating water systems inoculated with either mixed consortia or cooling water (Coughlin and Steimel, 1997; Donlan *et al.*, 1997; Williams and Holz, 1998). The advantage of batch systems with pure or mixed cultures is that fewer variables need be controlled to obtain reproducible results. Generally, investigators have used Gram-negative organisms common in cooling water (*Pseudomonas spp.*, *Klebsiella pneumoniae*, *Enterobacter spp.*). Continuous flow systems, on the other hand provide information more representative of the water system. In this case, investigators have either used deposits collected from a water system or the water as makeup, to test the apparatus. The obvious downside of these systems is greater variability and more difficulty in reproducing results. An example of a laboratory recirculating water system is shown in Figure 3.

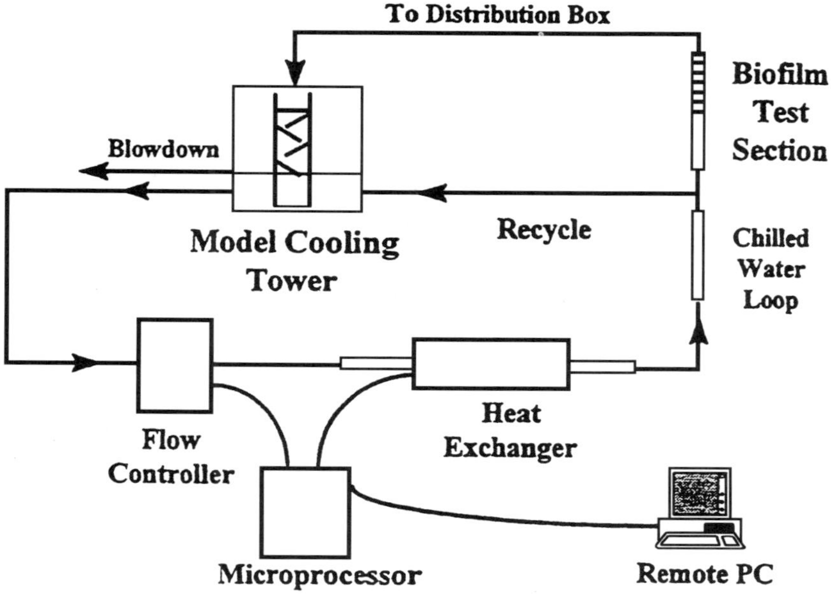

Figure 3 Schematic of model cooling tower with heat exchanger. (Reproduced from Williams and Holz, 1998, with permission.)

Table 2 provides a listing of different measurement techniques which have been used for biofilm measurement. Each method has been categorized by describing the component measured, ease of use, and ability to model an industrial water system biofilm. The purpose of this ranking is to provide the potential user with some level of guidance in choice of method. For example, the biofilm heterotrophic plate count technique is relatively simple to perform (ease of use = 5). However, its ability to model a biofilm in a water system ranks at only a 3. There are other techniques which may more accurately represent a water system biofilm (cryosectioning, confocal laser scanning microscopy, other biomass techniques shown), but most of these are either more difficult to use or require sophisticated/expensive equipment. Techniques which have relatively high scores in both criteria are absorbance at 550 nm, the Micro-Lowry protein assay, and direct cell count using epifluorescence. Also, techniques which are non-destructive are preferable since they enable examination of the intact biofilm. Destructive techniques will involve some kind of disruption to remove the cells (scraping, sonication). Biofilm biomass, thickness, or biovolume also are preferred since they will provide information on all the components of the biofilm, not merely the cellular component.

There are several needs in the area of biofilm testing and measurement. First, it is necessary to establish reproducibility for each particular method. Whenever a method is described, the variability of the results should be presented, allowing the reader to determine whether that is tolerable for their specific application. Secondly, it would be helpful for researchers to compare their method to older, more established

Table 2 Techniques for biofilm measurement.

Method	*Component Measured*	*Ease of Use*[1]	*Represents IW Biofilm*[2]	*Reference*
Biofilm heterotrophic plate count	Viable bacterial cells	5	3	Donlan *et al.*, 1997
Chlorophyll *a*	Algae and Cyanobacteria	3	2	Williams and Holz, 1998
Dry weight	Total biofilm deposit (ND)	5	3	Williams and Holz, 1998
Heat transfer resistance	Total biofilm deposit (ND)	3	3	Ludensky, 1998
Direct cell count using epifluorescence	Total microbial cells	3	4	Tapper *et al.*, 1997
ATP	Total cellular ATP	3	3	Walter and Cooke, 1997
Dielectric spectroscopy	Total cellular biovolume (ND)	2	4	Markx and Kell, 1990
Absorbance at 950 nm	Biofilm thickness (ND)	2	4	Taylor, 1996
Cryosectioning of biofilm	Total biofilm thickness and cell count (ND)	2	5	Yu *et al.*, 1994
Attenuated total reflection/FTIR	Biofilm biomass (cellular + EPS) (ND)	1	4	Nivens *et al.*, 1993b
Confocal scanning laser microscopy	Total microbial cells (ND)	1	5	Stewart *et al.*, 1997
Differential interference contrast microscopy	Total microbial cells (ND)	2	4	Rogers and Keevil, 1992
Quartz crystal microbalance	Total microbial cells (ND)	1	4	Nivens *et al.*, 1993a
Bioluminescence	Biomass of luminescent cells (ND)	2	3	Arrage *et al.*, 1995
Tryptophan fluorescence	Bacterial biomass (ND)	2	4	Angell *et al.*, 1993
Absorbance at 550 nm	Biofilm thickness (ND)	4	4	Jacobs *et al.*, 1996
Confocal scanning laser microscopy	Biofilm thickness (ND)	1	5	Stoodley *et al.*, 1997
Micro-Lowry protein assay	Biofilm biomass	3	4	Mittelman *et al.*, 1992

[1] Scale = 1-5, with 1 being the easiest; [2] Scale = 1-5, 5 being the most representative; ND = non-destructive

methods. This has been done in a number of studies, where biofilm heterotrophic plate counts were compared with other less commonly used methods. In the case of such plate counts, this type of comparison would only be useful if the time, temperature of incubation, and medium for the plate count were standardized, however. Another need in this area is to establish how well laboratory methods predict biofilm control efficacy in field application. For example, by testing a treatment on a small comfort cooling tower where biofilm was measured using some kind of side stream testing device and comparing the results with laboratory data, the latter could take on greater significance. Finally, the development of simple techniques for quantifying the non-cellular component of biofilms would be very helpful.

BIOFILM CONTROL USING BIOCIDES

Chen and Stewart (1996), and LeChevallier *et al.* (1988) demonstrated that biofilm organisms are significantly more resistant to biocides (and antibiotics) than their planktonic counterparts. There are several reasons for this. Biocide molecules tend to react with the extracellular component of biofilms, rendering them incapable of reaching the cells they are intended to inactivate. The EPS matrix also acts as a physical barrier and the molecules must diffuse through this material to reach embedded cells. It has also been shown (Huang *et al.*,1995) that biofilm cells are physiologically different; the biofilm state appears to confer a greater ability to resist biocidal treatment. For these reasons, treatment of industrial water systems for biological control must encompass not only reduction in planktonic organisms, but both removal and disinfection of biofilms.

Historically, bleach (5.25 or 12.5% sodium hypochlorite) or gaseous chlorine has been the treatment of choice for biological control in industrial water systems. This is because chlorine is a relatively inexpensive commodity chemical. Also, when fed at appropriate levels and frequencies it can be effective, at least in systems without an elevated chlorine demand. However, chlorine, especially gaseous chlorine, has come under increasing scrutiny. Recirculating cooling water systems tend to be operated at high cycles of concentration and at pH values above 8.0. At this pH range, chlorine, regardless of whether it is fed as bleach or gaseous chlorine, will dissociate to hypochlorite ion, which is much less biocidal than hypochlorous acid. Another concern, especially with gaseous chlorine, is the safety of working with a potentially hazardous compressed gas. In addition, chlorine will not provide adequate microbiological control in systems that are either heavily fouled initially or that have persistent system leaks (for example refineries). For these reasons, many power utilities and other industrial water users have either moved away from chlorine treatment alone in favor of a combination with nonoxidizing biocides or towards elimination of chlorination entirely.

There are number of nonoxidizing biocides currently in use for biological control. With the recognition that biofilms are the primary driving force behind biofouling of industrial water systems, biocide manufacturers and water treatment companies have begun testing of their products for efficacy against biofilms, using methods already discussed in this chapter. Because dosages and test conditions often vary from one study to the next, it is not clear which products are the most effective. Results may depend, for example, on system pH, contact time, and the types of organisms used. When choosing a product for biofilm control, it is advisable to evaluate the potential effect of system chemistry on the product, whether a fast acting or slower acting product is preferred, and the environmental fate. Specific papers should be consulted for guidelines for treatment of industrial water systems with nonoxidizing biocides (*e.g.* Eager *et al.*, 1986; Eager and Theis, 1987; Grab and Theis, 1992; Videla *et al.*, 1996; Boivin, 1997; Kleina *et al.*, 1997; Walter and Cooke, 1997). More studies that compare several products on either an equal active or equal cost basis, such as the paper by Walter and Cooke (1997) would be helpful. Also, it is essential that evaluations incorporate both laboratory and field evaluations. There is a trend in the industry towards combining nonoxidizing with anionic or nonionic surfactants to provide even better penetrating ability, a process that will be described in detail below.

Oxidizing Biocides Containing Bromine

In addition to the increased use of nonoxidizing biocides over the last few decades, there is also a trend toward the use of bromine based biocides. Oxidizing compounds which have been used as biocides in cooling water systems include bromine chloride (BrCl), 1-bromo-3-chloro-5,5-dimethylhydantoin (BCDMH) and other brominated hydantoins, sodium bromide blended in line with either sodium hypochlorite or chlorine gas, stabilized bromine products, and sodium bromide or potassium bromide with chlorinated isocyanurate solid mixtures or chlorinated hydantoins (Bartholomew, 1998). Because the pH of cooling water systems tends to run at or above pH 8.0, bromine compounds are more effective than chlorine. This is because the dissociation constant for hypobromous acid is higher than hypochlorous acid (Bartholomew, 1998), and the undissociated form of each is the most effective form. Assuming that hypobromous acid and hypochlorous acid are similar in terms of biocidal efficacy, the free halogen residual may be reduced yet achieve the same effect on planktonic bacteria. Another advantage of bromine over chlorine is that bromamines and free bromine have similar biocidal properties against planktonic bacteria (as compared to chlorine and chloramines) (Bartholomew,1998). Smith *et al.* (1993) also showed that bromine was less corrosive towards copper containing alloys, while maintaining good biological control. It has been documented that chloramines are more effective than free chlorine for biofilm disinfection (LeChevallier *et al.*, 1988). A study by Hunt and Blatchey (1998) found that bromamines were as effective as chloramines for biofilm control. A consideration with bromine based biocides, however, is that they may interact antagonistically with other water treatment chemicals (such as tolyltriazole and hydroxyethylidenediphosphonic acid (HEDP)) (Bartholomew, 1998). Dallmier *et al.* (1997), and McCoy *et al.* (1998) described a new, proprietary stabilized bromine product which provided approximately equivalent biocidal efficacy in laboratory systems but with a lower consumption of tolyltriazole, a copper corrosion inhibitor, apparently alleviating or at least minimizing this concern. They also provided evidence in laboratory systems that this product was significantly more effective in reducing biofilms, as measured by pressure drop, biofilm heterotrophic plate count, and an unexplained biofilm volumetric measurement than unstabilized bromine at equivalent concentrations. They theorized that the reason for the greater effect on biofilms was compatibility with scale and corrosion inhibitors and less reaction with organics in the water.

Table 3 Effect of different bromine based biocides for controlling biofilms.

Product (Reference)	*Biofilm Organism(s)*	*Test System*	*Measurement*	*Minimum Dosage Providing Biofilm Control*
BrMEH [a] (Ludyankskiy and Himpler, 1997)	*Sphaerotilus natans*	Continuous flow heat exchanger loop	Delta "T" % Dissolved oxygen	15 mg l^{-1} slug + continuous treatment for 3 h
BCDMH [b] (Walker *et al.*, 1994)	Consortia	Chemostat	Biofilm HPC [c]	>4 mg l^{-1} slug
Stabilized Bromine (Dallmier *et al.*, 1997)	Consortia	Laboratory recirculating water system	Biofilm HPC % Pressure drop	10 mg l^{-1} slug
NaOCl + NaBr (Franklin *et al.*, 1991)	Gram-negative and Gram-positive environmental isolates	Batch system	Biofilm HPC acetate uptake	16 mg l^{-1} slug + 2 mg l^{-1} for 24 h

[a] = bromomethylethylhydantoin; [b] = bromochlorodimethylhydantoin; [c] = biofilm heterotrophic plate count

Table 3 provides several of the currently used bromine releasing biocides and references documenting their performance in laboratory studies. Because these biocides are oxidants, they will tend, like the other halogen based biocides to be highly reactive with organic material. Biofilm EPS will react with these biocides, lessening their effectiveness. This means that they would not be expected to be as effective as nonoxidizing biocides against established biofilms and biofouling deposits. However, Kaucic (personal communication) found that in a full scale evaluation in a power plant cooling tower a combination of sodium dichloroisocyanurate and sodium bromide was effective in remediating biofouling on plate heat exchangers.

Ozone

Ozone is the strongest commercially available oxidizing agent (Pryor and Bukay, 1990) and has been used in Europe as a disinfectant for potable water systems for almost 100 years (Pryor and Bukay, 1990). It is produced on site by the corona discharge method, eliminating the drawbacks of both chlorine and bromine, storage and feeding of concentrated stock solutions. Ozone also has the advantage of leaving no residual products to be discharged and high compatibility with other treatment chemicals (Ditzler and McGrane, 1995). Akhlaq *et al.* (1990), studied the effects of ozone on alginic acid, a model compound for the polyuronic acids of bacterial extracellular polysaccharides. They found that ozone caused a reduction in the molecular weight of the alginic acid, as measured by a reduction in viscosity. Kauer *et al.* (1992), and Videla *et al.* (1993) examined the effect of ozone in laboratory experimental water systems. Videla *et al.* (1993) found that ozone was effective in killing aerobic organisms within biofilms by using 0.2–0.5 mg l^{-1} for a contact time of 10–15 min. They found no significant improvement when the contact time was increased under quiescent conditions. Kauer *et al.* (1992) found that biofilm removal by ozone was dependent upon age, initial biofilm mass, and preadaptation of the biofilm to chlorine. Older biofilms (45 d *vs* 19 d) were more resistant to a 0.08 mg l^{-1} dose. A biofilm mass increase of ten fold resulted in an approximate 50% reduction in removal rate. Biofilms that had been pre-exposed to >0.1 mg l^{-1} total chlorine required more than 10 times more ozone to achieve approximately equivalent removal. Use of ozone in recirculating cooling water systems to both disinfect and remove biofilms is worthy of consideration. Watanabe *et al.* (1996), patented ozonized water for peeling off slime from the walls of heat exchanger tubing. Ditzler and McGrane (1995) found ozone to be effective for both reducing planktonic bacterial counts and removing algal growth in a glycol contaminated cooling tower. Though limited in a recirculating water system application due to its very short half life in the water, ozone treatment could probably be combined successfully with other biofilm control strategies for both disinfecting and removing established biofilms.

NEWER APPROACHES FOR BIOFILM CONTROL

Because a biofilm by definition is a complex deposit consisting of cells, extracellular polymers, and entrapped particles, effective biofilm control will target each of these components. A variety of novel approaches have emerged, especially over the last decade. These rely either upon physical or chemical means for disrupting the biofilm EPS matrix and releasing immobilized cells or in more effectively penetrating these deposits to reach and kill biofilm associated organisms. Whittaker *et al.* (1984) evaluated a plethora of approaches and treatments for removing biofilms from reverse-osmosis membranes. These encompassed surfactants and detergents, chaotrophic agents which denature proteins and solubilize organic constituents, bactericides, enzymes which hydrolyze either proteinaceous or glycoprotein exopolymers, and chelants. The treatments were tested individually or in combinations, and efficacy was determined using scanning electron microscopy and by viable counting. They found that the anionic and chaotrophic agent combinations and combinations involving enzymes were most effective. Since this work was published, a number of approaches have been proposed and tested. They can be categorized as 1) mechanical/electrical disruption techniques, 2) techniques which hydrolyze or destabilize the extracellular polymeric matrix, 3) techniques which enhance biofilm penetration, and 4) techniques which alter the biofilm surface tension.

Mechanical/Electrical Disruption Techniques

Zips *et al.* (1990) found that ultrasound was effective in removing up to 95% of *Pseudomonas diminuta* cells attached to ultrafiltration membranes, and this value was dependent upon intensity, distance from the transducer, and exposure time. Huang *et al.* (1996) found that ultrasound treatment reduced *P. aeruginosa* biofilm levels on stainless steel slides by 1.5 logs, and also modestly improved the efficacy of gentamicin

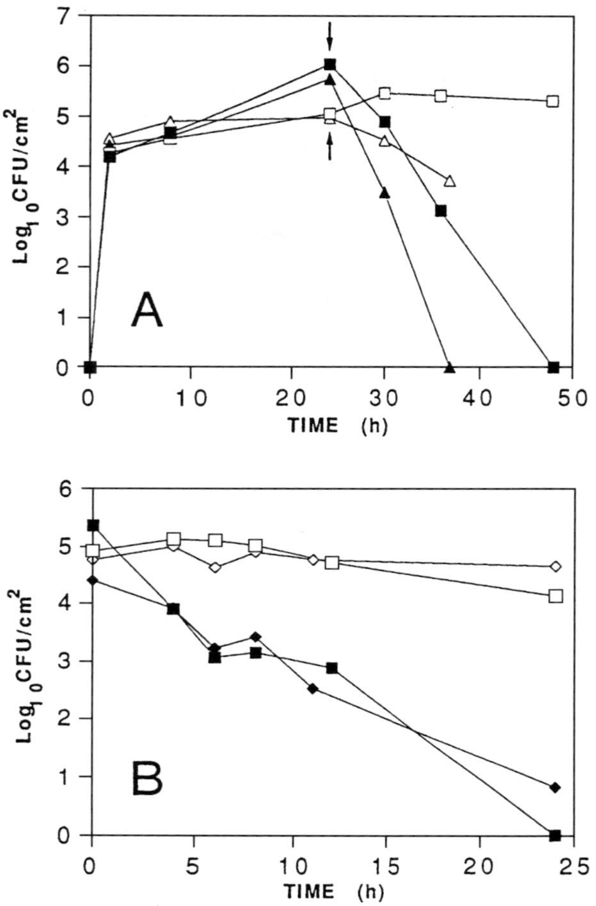

Figure 4 Effect of low strength electrical field combined with low current density followed by biocide application (arrows) on *P. aeruginosa* colonization ($\bar{x}$, n = 2). At 24 h, glutaraldehyde (5 ppm) (□, ■) or kathon (1 ppm) (△, ▲) was applied to both electrified and control devices. (B) Effect of biocides on an established (24-h) *P. aeruginosa* biofilm in the presence and absence of an EF-CD. Glutaraldehyde (5 ppm) (□, ■) or quaternary ammonium compound (10 ppm) (◇, ◆) was supplied to both electrified and control devices for 24 h ($\bar{x}$, n = 2). The electrified devices are represented by solid symbols (■, ▲, ◆). (Reproduced from Blenkinsopp *et al.*, 1992, with permission.)

against established biofilms. They postulated that ultrasound could have removed the upper layers of the *P. aeruginosa* biofilms, providing greater penetration of the gentamicin. Blenkinsopp *et al.* (1992), demonstrated that a low strength electrical field (± 12 V cm^{-1}) combined with a low current density (± 2.1 mA cm^{-2}) could enhance the biocidal efficacy of several commercial biocides against *P. aeruginosa* biofilms, at levels below the threshold normally required for efficacy against planktonic cells (Figure 4). They hypothesized that this treatment disrupted the charges on the exopolysaccharide matrix, allowing penetration of the biocide molecules to the biofilm cells.

Techniques which Hydrolyze or Destabilize the Extracellular Polymeric Matrix

Turakhia *et al.* (1983) found that addition of a calcium specific chelant, EGTA, caused detachment of a biofilm from A1-6X steel surfaces in a laboratory reactor supplied with secondary sewage sludge. The EGTA was fed at a concentration of 1 mM and detachment occurred within 5 min, as measured by an increase in suspended solids and total carbohydrate concentrations. They hypothesized that the calcium ions maintained the tertiary structure of the extracellular polymers. A drawback with using this type of treatment in industrial water systems would be high costs (due to high levels of divalent cations in recirculating cooling water systems), and environmental concerns (affect on aquatic organisms) (Vanderpool, personal communication).

More recently, enzymes have been investigated for biofilm control in industrial water systems. Johansen *et al.* (1997) examined the effect of several commercially available enzymes for their efficacy in removing laboratory-grown bacterial biofilms. They found that a multi component enzyme preparation (Pectinex Ultra SP, Novo Nordisk, Bagsvaerd, Denmark) containing a wide range of carbohydrases, including pectinase, arabanase, cellulase, hemicellulase, beta-glucanase, and xylanase activities, was most effective against biofilms of several different bacteria. This preparation targeted the extracellular polymers, resulting in a release of biofilms from surfaces while having no measurable bactericidal effect, as determined by CTC staining. The authors noted that economic considerations might limit acceptance of this technique for biofilm control in industrial water systems. Wiatr (1991) patented a combination of cellulase, alpha-amylase and protease for controlling slime formation in cooling towers, wastewater treatment, and paper making systems. The approach of using a suite of enzymes to target the diversity of polymers found in biofilms (Sutherland, 1983) may make this type of treatment more acceptable in industrial water systems, at least as a niche type treatment.

There are other approaches which target the biofilm extracellular polymer matrix. Christensen *et al.* (1990) found that 0.005% (50 mg l^{-1}) hydrogen peroxide in combination with either 10^{-4} M Fe^{2+} or 0.1 M abscorbate caused significant removal of *Pseudomonas sp.* biofilms in laboratory reactors. Depending upon the organism, this dosage/contact time could be considered subcidal (National Research Council, 1980). Kramer and Snow (1997) described an invention based on the use of an alkaline persalt and a positively charged phase transfer agent together with surfactant for cleaning and disinfecting materials and media. This technology was successfully used for cleaning biofilms in dental water unit lines and, though not published, may have application in the treatment of industrial water systems (Kramer, personal communication).

Enhanced Biopenetration

Because biofilm microorganisms are embedded in a matrix of extracellular polymers, treatment which could possibly penetrate this barrier to the target cells could be beneficial in biofilm control. Although total removal of the "dead biofilm" would be helpful, disinfection of the biofilm cells would significantly mitigate further biofilm growth, at least prior to recolonization. Clark and Langley (1990), patented a concept for biofilm control whereby stabilized chlorine dioxide followed by addition of a bacterial nutrient source was added to a test heat exchanger tube fouled with untreated fresh water. The basis for their invention was that the stabilized chlorine dioxide will generate reactive chlorine dioxide when activated by acidic metabolites of the biofilm bacteria after addition of the nutrient source, generally glucose or another readily metabolizable carbohydrate. They were able to demonstrate a reduction in heat transfer resistance and friction factor in test heat exchangers following treatment.

Mitchell and Nevo (1965) described a marine bacterium capable of degrading the capsular polysaccharides of several bacteria (*Flavobacterium*, *E. coli*, *Arthrobacter*) in culture, as demonstrated by growth on medium containing only bacterial capsular polysaccharide as the sole carbon source and by release of reducing sugars by the culture filtrates in the presence of capsular polysaccharides. These findings point out that there are organisms in nature that might be used to degrade biofilm exopolymers in industrial settings. Pollock and Yamazaki (1993) used *Lysobacter* species to clarify commercial microbial polysaccharides (Keltrol, produced by *Xanthomonas campestris*, and K1A96, produced by *Alcaligenes* sp.), as well as a number of exopolysaccharide producing bacteria embedded in agar. *Lysobacter* spp. are Gram-negative soil or aquatic organisms which produce proteolytic and polysaccharolytic enzymes, and have been shown to lyse Gram-negative and Gram-positive bacteria, and also actinomycetes, yeasts, and filamentous fungi, Cyanobacteria, and nematodes (Christensen, 1989). Because these organisms can hydrolyze both bacterial proteins and exopolymers, they might be good candidates for biofilm control, though there are no published data on this application.

Araki and Hosomi (1990) used bacteriophage specific for a *Pseudomonas* sp. to treat planktonic cultures of this slime forming organism. Though they demonstrated a reduction in optical density with treatment, plaque forming units of phage used were not given. Though tests against biofilms were not performed it could be speculated that addition of a mixture of phage isolated from cooling water bacteria might be effective in disinfecting biofilms.

Techniques which Alter Surface Tension

Of the new and emerging treatments for biofilm control discussed, probably none is as widely used as surfactant and surfactant/biocide combinations. Surfactants, or surface active agents, alter the surface tension within the biofilm and at the biofilm/substratum interface, allowing enhanced penetration by biocide molecules and also more effective removal of the biofilm deposits from surfaces. In this section, the use of synthetic surfactants, primarily anionic and nonionic materials, alone, and in combination with biocides for both biofilm inhibition and control will be discussed. Surfactants can be characterized according to their structure, solubility, adsorption and orientation at surfaces, micelle formation, and by other functional properties (Kirk and Othmer, 1983). Laboratory and field applications of surfactants for biofilm control demonstrate their efficacy for either inhibiting biofilm formation, or cleaning or disinfecting established biofilms. For inhibition, the surfactants may be administered alone (Lutey *et al.*, 1989; Blainey and Marshall, 1991; Whitekettle, 1991; Cloete and

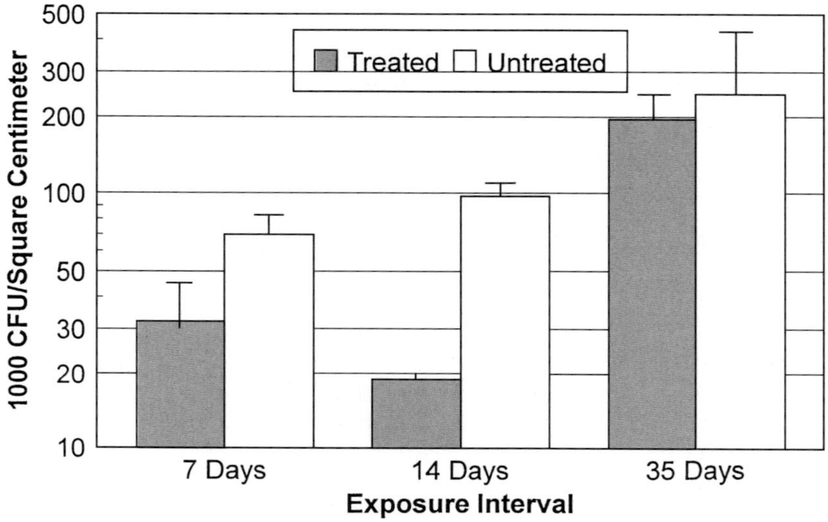

Figure 5 Effect of ethylene oxide/propylene oxide block copolymer surfactant on biofilm control on PVC in recirculating cooling water. (Reproduced from Donlan *et al.*, 1997, with permission.)

Brozel, 1992; Robertson, 1994; Robertson and Taylor, 1994; Wiatr, 1994; Wright and Michalopoulos, 1996b; Donlan *et al.*, 1997), or combined with biocides (Christensen and Zivtins, 1978; Koryu, 1980; Wright and Michalopoulos, 1996a; Donlan *et al.*, 1997; Meade *et al.*, 1997; Yu *et al.*, 1998). Most of the studies cited in the literature are laboratory investigations utilizing pure cultures of bacteria under controlled conditions. However, several used either mixed consortia or ran studies on a side stream of a cooling water system. For example, Donlan *et al.* (1997) investigated the effect of an ethylene oxide/propylene oxide block copolymer surfactant for inhibiting biofilm formation onto the surface of PVC in recirculating cooling water and untreated makeup water for a power plant. They found that feeding 250 mg l^{-1} of a 20% active solution of surfactant continuously, reduced biofilm and biofouling onto PVC for >14 d (Figure 5). Lutey *et al.* (1989) treated both closed loop and open recirculating cooling water systems with dimethylamide (DMATO) and found that it reduced fouling and prevented further plugging and fouling of high efficiency fill material. However, no data for these studies were presented. Wright and Michalopoulos (1996) showed that 48 mg l^{-1} dinonylsulfosuccinate fed continuously in an alkaline fine paper mill reduced biofouling onto exposed surfaces in a side stream test device for up to 8 d. Meade *et al.* (1997) patented the combination of peracetic acid with an ethylene oxide/propylene oxide nonionic surfactant for biological control in paper mill systems and demonstrated reduction in fouling when a combination of 20 mg l^{-1} peracetic acid combined with low levels of the surfactant were added continually to the water system. Treating large industrial water systems on a continuous basis with a surfactant may be cost prohibitive. Therefore, it may be more reasonable to use surfactants alone or in combination with other treatments for biofilm removal rather than inhibition.

Whittaker *et al.* (1984) found that sodium dodecyl sulfate in combination with urea was the most successful surfactant combination for cleaning reverse osmosis membranes. Lutey *et al.* (1989) found that treatment of biofilms on the surface of glass beads with DMATO caused the release of cells from the surfaces. The authors theorized that the DMATO depolymerized the biofilm and dispersed the cells. Fletcher *et al.* (1991) showed that Tween 20, a nonionic surfactant, caused biofilms to expand and cells to separate from glass surfaces to which they were attached. They postulated that the surfactant interfered with the hydrophobic interactions between biofilm cells and the substratum. Kanuth and Puckorius (1992) presented several different cleaning strategies for cooling tower fill material, including the use of oxidants in combination with dispersants and surfactants, noting one case study where 50–70% of the deposits comprised of silt and biomass were removed by a cleaning treatment in a natural draft cooling tower. Yu *et al.* (1998) found that daily treatment of cooling water systems at a geothermal power plant with 5 mg l^{-1} of an anionic surfactant (biodetergent) in combination with daily addition of bleach (200 gallons fed to a 400,000 gallon system) resulted in a 22% decrease in average fill weight and a 20% increase in thermal performance over the six week trial period. At the same time, the levels of ATP, protein, turbidity, and heterotrophic plate counts in the water increased. Ten Eyck (personal communication) has found that combinations of an ethylene oxide/propylene oxide surfactant (12 mg l^{-1} continuous) with continuous chlorination was effective in removing biofouling and silt deposits from cooling tower fill material in a recirculating cooling water system as measured by a significant decrease in fill weight and an increase in the total suspended solids in the treated system. Analyses showed that the deposit was approximately 23% organic (as measured by loss on ignition at 450°C) and comprised of silt and microorganisms. In this case, the surfactant material was added approximately 30 min prior to the oxidant, resulting in significant increases in total suspended solids from the fill material. Gill *et al.* (1995) showed that deposits on cooling tower high efficiency fill material are comprised primarily of biofilms and silt. The treatment is believed to work by oxidizing the exopolymer matrix of the biofilms, releasing the entrapped solids (primarily clay) from the surfaces.

It is clear from the literature that biofilm formation leading to biofouling of industrial water systems is a recalcitrant problem, not easily controlled, and even then, at significant cost. The old approach of increasing biocide dosage to remediate a biofouling problem appears to be misguided. Since biofouling is a surface-associated phenomenon, it must be treated as such, by targeting treatments for controlling surface-associated or sessile organisms. These sessile organisms reside in a matrix of extracellular polymers and commonly also abiotic materials. A successful treatment will be one which targets the multiple components of the biofilm cost effectively, with minimal effect on the environment and workers. The approach used by Whittaker *et al.* (1984) for cleaning reverse osmosis membranes is sound and might be used as a guide for future biofilm control strategies. In most cases, it will probably not be possible to prevent biofilm formation in industrial water systems. An approach which combines both biofilm inhibition and control (disinfecting and removing established biofilms) will probably be necessary. Control strategies which target specific components of biofilms; *viz.* the cells, exopolymers, and entrapped particles will have the greatest chance of success. In Table 4 the benefits and drawbacks of various biofilm control strategies are compared.

Table 4 Comparison of different treatments for biofilm control.

Treatment	*Benefits*	*Drawbacks*	*Reference*
Chlorination	Commodity; inexpensive	Ineffective against EPS; much less effective at higher pH	Lechavallier *et al.*, 1988; Hassan and Oh, 1989; Chen and Stewart, 1996
Bromine oxidizing biocides	Effective over a wide pH range; bromamines as effective as hypobromite	High levels required; ineffective against EPS	Batholomew, 1988; Dallmier *et al.*, 1997; Hunt and Blatchley, 1998; McCoy *et al.*, 1998
Ozone	Strong oxidant; hydrolyzes EPS; effective against cells and EPS	Short half life in cooling water; more expensive than other oxidizing biocides	Akhlaq *et al.*, 1990; Pryor and Bukay, 1990; Kaur *et al.*, 1992; Videla *et al.*, 1993; Ditzler and McGrane, 1995; Watanabe *et al.*, 1996
Ultrasound + biocide	Possibly good penetrant	Not demonstrated in the field	Zips *et al.*, 1990; Huang *et al.*, 1996
Electical current + biocide	Possibly good penetrant	Not demonstrated in the field	Blenkinsopp *et al.*, 1992
EGTA chelant	Destabilizes EPS	Environmental problems; effect on system piping; not demonstrated in the field	Turakhia *et al.*, 1983
Specific enzymes	Targets EPS and cells	Cost prohibitive; not demonstrated in the field	Wiatr, 1991; Johansen *et al.*, 1997
Hydrogen peroxide + reducing agents	Targets EPS at subcidal levels	Not demonstrated in the field; will not kill the organisms	Christensen *et al.*, 1990
Alkaline per salts + cationic surfactants	Cleans and disinfects biofilms	Not demostrated in cooling water	Kramer and Snow, 1990
Stabilized chlorine dioxide + bacterial nutrients	Environmental	Not demonstrated in the field; cost	Clark and Langley, 1990
Polysaccharidase producing bacteria	Might allow environmentally friendly biofilm control	Unproven	Mitchell and Nevo, 1965; Pollack and Yamazaki, 1993
Bacteriophage	Targets biofilm bacteria	Highly specific; no effect on EPS; not demonstrated in cooling water	Araki and Hosomi, 1990

Table 4 Continued.

Treatment	*Benefits*	*Drawbacks*	*Reference*
Surfactants alone	Proven in field applications	Cost; does not destroy biofilm components; may foam at higher levels	Lutey *et al.*, 1989; Whitekettle, 1991; Blainey and Marshall, 1991; Cloete and Brozel, 1992; Robertson, 1994; Robertson and Taylor, 1994; Wiatr, 1994; Wright and Michalopoulos, 1996b
Surfactants plus biocides	Proven in field applications; targets EPS and cells	Cost; environmental problems with biocides; foaming	Christensen and Zivtins, 1978; Koryu, 1980; Wright and Michalopoulos, 1996a; Donlan *et al.*, 1997; Meade *et al.*, 1997
Surfactant plus bleach	Proven in field applications; targets EPS and cells	Requires higher levels of bleach which may affect system metallurgy; Requires rapid system blowdown	Yu *et al.*, 1998

Future Trends in Biofilm Control

Over the last two decades, knowledge of biofilm processes has increased substantially and the scientific literature contains many reports of work on biofilm processes and biofilm control. Nevertheless, biofilms continue to cause serious problems for industry. Treatments which are cost effective, environmentally acceptable, and inert towards system materials and other water treatment products are not readily available for industrial use. The trend in the water treatment arena is likely to be towards stronger environmental regulations. Chlorine, though historically the workhorse for biological control is not only ineffective towards biofilms but also of concern from a safety and environmental perspective. Henley (1992) surveyed representatives from several industrial water treatment companies and found that most felt gaseous chlorine was likely to be replaced by other forms of chlorine or different biocides. It is generally believed that biocides which are non-toxic and degrade quickly will be used more frequently, including ozone and bromine. If adjuvants such as surfactants are found to enhance biocidal efficacy, and especially to have an effect on biofilms then these are likely to be used more in the industry. This would depend on the environmental profile of the surfactants under consideration, such as biodegradability and foaming tendencies. Bromine based biocides, as already discussed, are equally effective at higher pH values, and will allow the user to avoid the use of chlorine gas cylinders, a direct safety benefit. Bromine would also have a limited half life in the environment due to reaction with organic materials. Alhough cost is often a driving force in acceptance of new approaches, it is probable that environmental concerns will play an even greater role, potentially minimizing cost issues. It is also likely that more research will focus on natural control strategies, developed in nature. McCoy *et al.* (1998) suggest that certain marine organisms produce natural antifoulants. These types of materials may eventually gain acceptance, especially as chlorine and other commercial biocides lose favor, either because of environmental concerns or because they are ineffective against biofilms. Several other novel, and relatively untested approaches, have been referred to in this communication which need to be examined further. Davies *et al.*, (1998) showed that cell-to-cell communication is important in bacterial biofilms, and that biofilms comprised of organisms lacking specific molecular signals formed biofilms that were much more susceptible to biocide treatments such as sodium dodecyl sulfate. These authors point out that strategies which could interfere with these cell-to-cell signals might effectively produce biofilms much more susceptible to various biofilm control treatments.

Another trend in the industry will likely be to utilize water reuse to a greater extent (Vaska and Lee, 1994). Market forces driving this trend are the need by industry for reduced chemical consumption and reduced environmental discharge, in addition to a shortage of freshwater sources in certain areas. Reuse brings its own problems, including increased levels of ammonia and phosphate, total dissolved solids, and organics, all of which may affect microbiological growth and biofilm formation. Treatments which can control biofilms under these more demanding conditions will be needed.

It is clear that a paradigm shift has already occurred in industry regarding biofilms. It is rare to find anyone in the water industry, whether that individual is involved as an end user, service representative, or supplier, who is not familiar with the term biofilm. Most will recognize the importance of biofilms and their potential effect on the industrial process. This shift in thinking is the first step towards developing and implementing effective biofilm control strategies.

ACKNOWLEDGEMENTS

The author would like to thank the following individuals for providing published methods for biofilm monitoring and testing: Richard Walter and Lisa Cooke (Dow Chemical), Larry Grab (Ecolab), John Diemer (Union Carbide), Mark Weincek and Terry Williams (Rohm and Haas), Mike Winter (Amoco), Rick Rosser (Saudi Aramco), Gary Horocek, Bill McCoy, and Tony Dallmier (Nalco), Michael Ludensky (Lonza), Nick Zelver, Bill Costerton, and Becky Pitts (Center for Biofilm Engineering, Montana State University), Cecilia McGough and Peter Ten Eyck (Calgon), David Taylor (English China Clays), David White (University of Tennessee), and Bill Keevil (CAMR). The author also acknowledges the support of the Calgon Corporation and especially Betty Schwarz for providing the literature searches. Much of the planning and a portion of the writing was done while the author was employed by the Calgon Corporation.

REFERENCES

Akhlaq M.S., Schuchmann H.-P., von Sonntag C. (1990). Degradation of the polysaccharide alginic acid: a comparison of the effects of UV light and ozone. *Environ Sci Technol*, 24, 379–383.

American Society for Testing Materials. (1991). Standard guide for selecting test methods to determine the effectiveness of antimicrobial agents and other chemicals for prevention, inactivation, and removal of biofilm. Standard E1427–91.

American Society for Testing Materials. (1997). Standard test method for efficacy of microbiocides used in cooling systems. Standard E-645-97.

Angell P., Arrage A.A., Mittelman M.W., White D.C. (1993). On line, non-destructive biomass determination of bacterial biofilms by fluorometry. *J Microbiol Methods*, **18**, 317–327.

Araki M., Hosomi M. (1990). Using bacteriophage for slime control in the paper mill. *Tappi J*, **73**, 155–158.

Araujo Da Silva, R. (1994). Multipurpose apparatus for monitoring biofilms and microbiological corrosion in industrial water systems. Brazilian Patent No. BR9204970.

Arrage A.A., Vasishtha N., Sundberg D., Bausch G., Vincent H.L., White D.C. (1995). On-line monitoring of antifouling and fouling-release surfaces using bioluminescence and fluorescence measurements during laminar flow. *J Indust Microbiol*, **15**, 277–282.

Aylott P.J., Stott J.F.D., Eden R.D., Grover H.K. (1995). Monitoring of marine biofouling of titanium tubed heat exchanger using a remote controlled thermal resistance method. National Association of Corrosion Engineers Annual Conference, Paper No. 195.

Bakich, S.G. (1995). Application of the biofilm coupon as a direct measure of the *in situ* growth potential of water. MS Thesis, Montana State University, Bozeman, USA.

Batholomew R.D. (1998). Bromine-based biocides for cooling water systems: a literature review. International Water Conference Annual Conference, Paper No. 98-74.

Blainey B.L., Marshall K.C. (1991). The use of block copolymers to inhibit bacterial adhesion and biofilm formation on hydrophobic surfaces in marine habitats. *Biofouling*, 4, 309–318.

Blenkinsopp S.A., Khoury A.E., Costerton J.W. (1992). Electrical enhancement of biocide efficacy against *Pseudomonas aeruginosa* biofilms. *Appl Environ Microbiol*, **58**, 3770–3773.

Bloch K.P., DiFranco P. (1995). Preventing MIC through experimental, on-line fouling monitoring. National Association of Corrosion Engineers Annual Conference, Paper No. 257.

Boivin J. (1997). Control of microbiologically influenced corrosion in oil transmission pipelines. *Mater Perf*, **36**, 53–55.

Camper A.K., Jones W.L., Hayes J.T. (1996). Effect of growth conditions and substratum composition on the persistence of coliforms in mixed-population biofilms. *Appl Environ Microbiol*, **62**, 4014–4018.

Characklis W.G. (1973a). Attached microbial growths-I.Attachment and growth. *Water Res*, 7, 1113–1127.

Characklis, W.G. (1973b). Attached microbial growths-II. Frictional resistance due to microbial slimes. *Water Res*, 7, 1249–1258.

Characklis W.G. (1990). Microbial biofouling control. In: Characklis, W.G., Marshall, K.C. (eds) *Biofilms*. John Wiley and Sons, NY, pp. 585–633.

Characklis, W.G. (1991). Method for determining the quality of a medium. US Patent No. 5,051,359.

Chen X., Stewart P.S. (1996). Chlorine penetration into artificial biofilm is limited by reaction-diffusion interaction. *Environ Sci Technol*, **30**, 2078–2083.

Chexal B., Horowitz J., Munson D., Spalaris C., Angell P., Gendron T., Ruscak, M. (1997). Biofilm monitoring in power plant waters for use in prediction and control of MIC. EPRI Tenth Service Water Reliability Improvement Seminar, Denver, CO.

Christensen, P. (1989). Lysobacteraceae In: Staley J.T., Bryant M.P., Pfennig N., Holt J.G. (eds) *Bergey's Manual of Systematic Bacteriology* Vol. 3. Williams and Wilkins, Baltimore, USA, pp. 2083–2089.

Christensen R.J., Zivtins G.J. (1978). Removal and prevention of biological sludge and slime formation. German Patent No. DE2758540.

Christensen B.E., Trønnes H.N.,Vollan K., Smidsrød O., Bakke R. (1990). Biofilm removal by low concentrations of hydrogen peroxide. *Biofouling*, **2**, 165–175.

Clark J.B., Langley D.E. (1990). Biofilm Control. US Patent No. 4929365.

Cloete T.E., Brozel V.S. (1992). Practical aspects of biofouling control in industrial water systems. *Int Biodeterior Biodegr*, **29**, 299–341.

Costerton J.W., Geesey G.G., Cheng K.-J. (1978). How bacteria stick. *Sci Am*, **238**, 86–95.

Costerton J.W., Lewandowski Z., Caldwell D.E., Korber D.R., Lappin-Scott H.M. (1995). Microbial biofilms. *Annu Rev Microbiol*, **49**, 711–45.

Coughlin M.F., Steimel L. (1997). Performance of hydrogen peroxide as a cooling water biocide and its compatibility with other cooling water inhibitors. National Association of Corrosion Engineers Annual Conference, Paper No. 397.

Dallmier A.W., Martens J.D., McCoy W.F. (1997). Performance of stabilized halogen biocides in cooling water. National Association of Corrosion Engineers Annual Conference, Paper No. 398.

Davies D.G., Parsek M.R., Pearson J.P., Iglewski B.H., Costerton J.W., Greenberg E.P. (1998). The involvement of cell-to-cell signals in the development of a bacterial biofilm. *Science*, **280**, 295–298.

Ditzler L.C., McGrane W.K. (1995). Guidelines and examples of ozone in cooling tower applications. *J Cooling T Inst*, **16**, 10–15.

Donlan R.M. (1987). Evaluation of attached microbial communities in drinking water pipes. PhD Thesis, Drexel University, Philadelphia, USA.

Donlan R.M. (1992). Correlation between Sulfate Reducing Bacterial colonization and metabolic activity on selected metals in a recirculating cooling water system. National Association of Corrosion Engineers Annual Conference, Paper No. 183.

Donlan R.M., Muia R.A., Smith A.L. (1991). Relationship of biofilm formation on stainless steel in untreated once through cooling water to seasonal water quality trends. Cooling Tower Institute Annual Meeting, Paper No. TP91-05.

Donlan R.M., Pipes W.O., Yohe T.L. (1994). Biofilm formation on cast iron substrata in water distribution systems. *Water Res*, **28**, 1497–1503.

Donlan R.M., Elliott D.L., Gibbon D.L. (1997). Use of surfactants to control silt and biofilm deposition onto PVC fill in cooling water systems. International Water Conference, Paper No. IWC-97-73.

Eager R.G. Jr., Theis A.B. (1987). Control of microbiological fouling with glutaraldehyde. Cooling Tower Institute Annual Meeting, Paper No. TP87-1.

Eager R.C., Theis A.B., Turakhia M.H., Characklis W.G. (1986). Glutaraldehyde: impact on corrosion causing biofilms. National Association of Corrosion Engineers Annual Conference, Paper No. 125.

Fletcher M., Lessmann J.M., Loeb G.I. (1991). Bacterial surface adhesives and biofilm matrix polymers of marine and freshwater bacteria. *Biofouling*, **4**, 129–140.

Franklin M.J., Nivens D.E., Vass A.A., Mittelman M.W., Jack R.F., Dowling N.J.E., White D.C. (1991). Effect of chlorine and chlorine/bromine biocide treatments on the number and activity of biofilm bacteria and on carbon steel corrosion. *Corrosion*, **47**, 128–134.

Gill J.S., Yorke M.A., Donlan R.M., Gibbon D.L. (1995). Fouling of film forming cooling tower fills-a mechanistic approach. *J Cooling T Inst*, **16**, 10–19.

Goysich M.J., McCoy, W.F. (1989). A quantitative method for determining the efficacy of algicides in industrial cooling towers. *J Ind Microbiol*, 4, 429–434.

Grab L.A., Theis A.B. (1992). Comparative biocidal efficacy vs. Sulfate-Reducing Bacteria. National Association of Corrosion Engineers Annual Conference, Paper No. 184.

Hassan R.S., Oh L.C.P. (1989). Effect of sodium hypochlorite (Clorox) and its mode of application on biofilm development. *Biofouling*, **1**, 353–361.

Henley M. (1992). Environmental update, stronger regulations bringing change to water treatment. *Ind Water Treat*, November/December, pp. 13–15.

Heukelekian H., Heller A. (1940). Relation between food concentration and surface for bacterial growth. *J Bacteriol*, **40**, 547–558.

Howsam P., Tyrrel S. (1989). Diagnosis and monitoring of biofouling in enclosed flow systems – experience in groundwater systems. *Biofouling*, **1**, 343–351.

Huang C.-T, Yu F.P., McFeters G.A., Stewart P.S. (1995). Nonuniform spatial patterns of respiratory activity within biofilms during disinfection. *Appl Environ Microbiol*, **61**, 2252–2256.

Huang C.-T., James G., Pitt W.G., Stewart P.S. (1996). Effects of ultrasonic treatment on the efficacy of gentamicin against established Pseudomonas aeruginosa biofilms. *Colloids Surf B: Biointerfaces*, **6**, 235–242.

Hunt B.A., Blatchley E.R. (1998). Optimization of halogen dose for biofilm control. Proc Disinfection '98, The Latest trends in wastewater disinfection: chlorination vs U.V. disinfection. *Water Environ Fed*, 99–108.

Jacobs L., DeBruyn E.E., Cloete T.E. (1996). Spectrophotometric monitoring of biofouling. *Water Sci Technol*, **34**, 533–540.

Johansen C., Falholt P., Gram L. (1997). Enzymatic removal and disinfection of bacterial biofilm. *Appl Environ Microbiol*, **63**, 3724–3728.

Jones H.C., Roth I.L., Saunders W.M. III. (1969). Electron microscopic study of a slime layer. *J Bacteriol*, **99**, 316–325.

Jones D.S., O'Rourke P.C., Caine C.W. (1993). Detection and control of microbiologically-influenced corrosion in a Rocky Mountain oil production system – a case history. National Association of Corrosion Engineers Annual Conference, Paper No. 311.

Kanuth J.G., Puckorius P.R. (1992). Cooling tower film fill water quality operations guidelines for successful utilization. *J Cooling T Inst*, **13**, 12–18.

Kaur K., Bott T.R., Leadbeater B.S.C. (1992). Effects of ozone as a biocide in an experimental cooling water system. *Ozone Sci Eng*, **14**, 517–530.

Kirk I., Othmer D.F. (1983). Surfactants and detersive systems, In: *Encyclopedia of Chemical Technology*, 3rd edition, Vol. 22. Wiley Interscience, John Wiley and Sons, NY, pp. 332–432.

Kleina L.G., Czechowski M.H., Clavin J.S., Whitekettle W.K., Ascolese C.R. (1997). Performance and monitoring of a new nonoxidizing biocide: the study of BNPD/ISO and ATP. National Association of Corrosion Engineers Annual Conference, Paper No. 403.

Koryu K.K.K. (1980). Slime removal. Japanese Patent No. JP55059896.

Kramer D.N., Snow P.A. (1997). Cleansing and disinfecting method. US Patent Patent No. 4941989.

LeChevallier M.W., Babcock T.M., Lee R.G. (1987). Examination and characterization of distribution system biofilms. *Appl Environ Microbiol*, **53**, 2714–2724.

LeChevallier M.W., Cawthon C.D., Lee R.G. (1988). Inactivation of biofilm bacteria. *Appl Environ Microbiol*, **54**, 2492–2499.

Licina G.J., Nekoksa G., Howard R.L. (1994). An electrochemical method for on-line monitoring of biofilm activity in cooling water using the BioGeorge Probe. American Society for Testing Materials, Standard Technical Publication 1232.

Ludensky M.L. (1998). An automated system for biocide testing of biofilms. *J Ind Microbiol*, **20**, 109–115.

Ludyanskiy M.L., Himpler F.J. (1997). The effect of halogenated hydantoins on biofilms. National Association of Corrosion Engineers Annual Conference, Paper No. 405.

Lutey R.W., King V.M., Gleghorn M. (1989). Mechanisms of action of dimethylamides as a penetrant/dispersant in cooling water systems. International Water Conference Annual Conference, Paper No. IWC-89-33.

Markx G.H., Kell D.B. (1990). Dielectric spectroscopy as a tool for the measurement of the formation of biofilms and of their removal by electrolytic cleaning pulses and biocides. *Biofouling*, 2, 211–227.

McCoy W.F., Costerton J.W. (1982). Fouling biofilm development in tubular flow systems. *Dev Ind Microbiol*, 23, 551–558.

McCoy W.F., Allain E.J., Yang S., Dallmier A.W. (1998). Strategies used in nature for microbial fouling control: applications for industrial water treatment. National Association of Corrosion Engineers Annual Conference, Paper No. 520.

McIlwaine, Diemer J., Grab L. (1997). Determining the biofilm penetrating ability of various biocides utilizing an artificial biofilm matrix. National Association of Corrosion Engineers Annual Conference, Paper No. 400.

Meade R.J., Robertson L.R., Taylor N.R., LaZomby J.G. (1997). Method for preventing microbial deposits in the papermaking process with ethylene oxide/propylene oxide copolymers. US Patent No. 5,624,575.

Mitchell R., Nevo Z. (1965). Decomposition of structural polysaccharides of bacteria by marine micro-organisms. *Nature (Lond)*, **205**, 1007–1008.

Mittelman M.W., Kohring L.L., White D.C. (1992). Multipurpose laminar-flow adhesion cells for the study of bacterial colonization and biofilm formation. *Biofouling*, 6, 39–51.

National Research Council. (1980). Drinking Water and Health, Vol. 2, 93.

Nivens D.E., Chambers J.Q., Anderson T.R., White D.C. (1993a). Long-term, on-line monitoring of microbial biofilms using a quartz crystal microbalance. *Anal Chem*, **65**, 65–69.

Nivens D.E., Chambers J.Q., Anderson T.R., Tunlid A., Smit J., White D.C. (1993b). Monitoring microbial adhesion and biofilm formation by attenuated total reflection/Fourier transform infrared spectroscopy. *J Microbiol Methods*, **17**, 199–213.

Piriou P., Jousset M., Levi Y. (1997). Direct biofilm measurement in potable water: an evaluation of the bacterial regrowth risk in distribution systems. Proc American Water Works Association Water Quality Technology Conference, 5D4/1–14.

Pollack T.J., Yamazaki M. (1993). Clarification of microbial polysaccharides with enzymes secreted from *Lysobacter* species. *J Indust Microbiol*, **11**, 187–192.

Pope D.H., Zintel T.P., Aldrich H., Duquette D. (1990). Laboratory and field tests of efficacy of biocides and corrosion inhibitors in the control of microbiologically influenced corrosion. National Association of Corrosion Engineers Annual Conference, Paper No. 34.

Pryor A., Bukay M. (1990). Historical perspective of cooling tower ozonation. *Ind Water Treat*, October, 26–32.

Puckorius P.R. (1997). Monitoring requirements for refinery cooling system reuse water. *Mater Perf*, **36**, 42–47.

Robertson L.R. (1994). Prevention of microbial adhesion. Tappi 1994 Biological Sciences Symposium, pp. 225–232.

Robertson L.R., Taylor N.R. (1994). Biofilms and dispersants: a less toxic approach to deposit control. *Tappi J*, **77**, 99–103.

Rogers J., Keevil C.W. (1992). Immunogold and fluorescein immunolabelling of *Legionella pneumophila* within an aquatic biofilm visualized by using episcopic differential interference contrast microscopy. *Appl Environ Microbiol*, **58**, 2326–2330.

Smith A.L., Breckenridge R., Clay V., Swidle D. (1993). Bromine vs. gaseous chlorine, a comprehensive review of case histories. National Association of Corrosion Engineers Annual Conference, Paper No. 637.

Stewart P.S., Camper A.K., Handran S.D., Huang C.-T., Warnecke M. (1997). Spatial distribution and coexistence of *Klebsiella pneumoniae* and *Pseudomonas aeruginosa* in biofilms. *Microb Ecol*, **33**, 2–10

Stoodley P., Yang S., Lappin-Scott H., Lewandowski Z. (1997). Relationship between mass transfer coefficient and liquid flow velocity in heterogenous biofilms using microelectrodes and confocal microscopy. *Biotechnol Bioeng*, **56**, 681–688.

Strauss S.D. (1992). Instrumentation advances improve fouling, corrosion monitoring. *Power*, September 1992, pp. 17–20.

Suci P.A., Siedlecki K.J., Palmer R.J. Jr., White D.C., Geesey G.G. (1997). Combined light microscopy and attenuated total reflection fourier transform infrared spectroscopy for integration of biofilm structure, distribution, and chemistry at solid-liquid interfaces. *Appl Environ Microbiol*, **63**, 4600–4603.

Sutherland I.W. (1983). Microbial exopolysaccharides-their role in microbial adhesion in aqueous systems. *Crit Rev Microbiol*, **10**, 173–201.

Tanner R.S., Haack T.K., Semet R.F., Greenley D.E. (1985). A mild steel tubular flow system for biofilm monitoring. UK Corrosion '85, Harrogate, UK.

Tapper R.C., Smith J.R., Beech I.B., Viera M.R., Guiamet P.S., Videla H., Swords C.L., Edyvean R.G.J. (1997). The effect of glutaraldehyde on the development of marine biofilms formed on surfaces of AISI 304 Stainless steel. National Association of Corrosion Engineers Annual Conference, Paper No. 205.

Taylor R.J. (1996). Efficacy of industrial biocides against bacterial biofilms. PhD Thesis, University of Birmingham, Birmingham, UK.

Turakhia M.H., Cooksey K.E., Characklis W.G. (1983). Influence of a calcium-specific chelant on biofilm removal. *Appl Environ Microbiol*, **46**, 1236–1238.

Vaska M., Lee B. (1994). Successful water reuse in open recirculating cooling systems. National Association of Corrosion Engineers Annual Conference, Paper No. 455. 124.

Videla H.A., Viera M.R., Guiamet P.S., Staibano Alais J.C. (1994). Combined action of oxidizing biocides for controlling biofilms and MIC. National Association of Corrosion Engineers Annual Conference, Paper No. 260.

Videla H.A., Gomez de Saravia S.G., deMele M.F.L., Hernandez G., Hartt W. (1993). The influence of microbial biofilms on cathodic protection at different temperatures. National Association of Corrosion Engineers Annual Conference, Paper No. 298.

Videla H.A., Guiamet P.S., Viera M.R., Gomez de Saravia S.G., Gaylarde C.C. (1996). A comparison of the action of various biocides on corrosive biofilms. National Association of Corrosion Engineers Annual Conference, Paper No. 286.

Videla H.A., Viera M.R., Guiarnet P.S., deMele M.F.L., Bianchi F., Canales C.G. (1993). Laboratory studies on the effect of ozone on the passivity of steel and mixed bacterial biofilms. National Association of Corrosion Engineers Annual Conference, Paper No. 486.

Walker J.T., Rogers J., Keevil C.W. (1994). An investigation of the efficacy of a bromine containing biocide on an aquatic consortium of planktonic and biofilm micro-organisms including *Legionella pneumophila*. *Biofouling*, **8**, 47–54

Walter R.W., Cooke L.M. (1997). 2-(Decylthio)ethanamine hydrochloride: a new multifunctional biocide which enhances corrosion inhibition. National Association of Corrosion Engineers Annual Conference, Paper No. 410.

Watanabe S., Onda K., Matsuda T., Hayashi A. (1996). Ozonized water for peeling off slime in the wall of heat exchangers. Japanese Patent No. JP08166197 A2.

Wetegrove R.L., Banks R.H., Hermiller M.R. (1997). Optical monitor for improved fouling control in cooling systems. *J Cooling T Inst*, **18**, 52–59.

Whitekettle W.K. (1991). Effects of surface-active chemicals on microbial adhesion. *J Ind Microbiol*, **7**, 105–116.

Whittaker C., Ridgeway H.R., Olson B.H. (1984). Evaluation of cleaning strategies for removal of biofilms from reverse-osmosis membranes. *Appl Environ Microbiol*, **48**, 395–403.

Wiatr C.L. (1991). Enzyme blend containing cellulase to control industrial slime. US Patent No. 4,994,390.

Wiatr C.L. (1994). Development of biofilms. Tappi 1994 Biological Sciences Symposium, pp. 203–223.

Williams T.M., Holz J.W. Jr. (1998). Biofouling studies with methylchloro/methylisothiazolone in model cooling systems. National Association of Corrosion Engineers Annual Conference, Paper No. 298.

Wolfaardt G.M., Archibald R.E.M., Cloete T.E. (1991). The use of DAPI in the quantification of sessile bacteria on submerged surfaces. *Biofouling*, **4**, 265–274.

Wright J.B., Michalopoulos D.L. (1996a). Method and composition for enhancing biocidal activity. European Patent No. EP0741109A2.

Wright B.J., Michalopoulos D.L. (1996b). Method for inhibiting microbial adhesion on surfaces. US Patent No. 5,512,186

Yohe T.L., Donlan R.M., Kyriss K.K. (1986). Sampling device for determining conditions on the interior surface of a water main. US Patent No. 4,631,961.

Yu F.P., McFeters G.A. (1994). Physiological response of bacteria in biofilms to disinfection. *Appl Environ Microbiol*, **60**, 2462–2466.

Yu F.P., Ginn D., McCoy W.F., Castanieto H. (1998). Cooling tower fill fouling control in a geothermal power plant. National Association of Corrosion Engineers Annual Conference, Paper No. 529.

Yu F.P., Callis G.M., Stewart P.S., Griebe T., McFeters G.A. (1994). Cryosectioning of biofilms for microscopic examination. *Biofouling*, **8**, 85–91.

Zips A., Schaule G., Flemming H.C. (1990). Ultrasound as a means of detaching biofilms. *Biofouling*, **2**, 323–333.

Zisson P.S., Whitaker J.M., Neilson H.L., Mayne L.L. (1995). Experiences with monitoring and control of microbiological growth in the standby service water system of a BWR nuclear power plant. National Association of Corrosion Engineers Annual Conference, Paper No. 264.

Zuniga P.O., Miller K., Winters M.A. (1990). Cooling water fouling monitor senses upsets, evaluates changes. *Chem Proc*, April, 1990.

20 Environmentally Acceptable Control of Microbial Biofilms

Manfred Zinn, Richard C. Zimmerman and David C. White

Surfaces that are exposed to liquids in the environment become covered by microorganisms at a rapid rate. Under good growth conditions these cells divide and form a layer of cells called a biofilm. This chapter will give a survey of the methods that are used to date to prevent the formation of biofilms (biofouling), and consider what kind of approaches might be used for future protection due to their environmental friendliness.

KEY WORDS: biofouling, biofilm control, antifoulant, biocide, minimal effective release rate, coating

DEFINITION OF BIOFOULING

The undesired deposition of cells and the subsequent formation of a cell layer (biofilm) on a surface are called biofouling (Flemming *et al.*, 1996). Biofouling occurs between liquid-solid, gas-solid or even liquid-liquid interfaces. The largest problems with biofilm growth are encountered at the liquid-solid interface. Thus, discussion will be limited to this particular interaction.

PROBLEMS CAUSED BY BIOFOULING

Biofilms contribute to a range of costly problems in daily life. They may be responsible for the biodeterioration of materials, *e.g.* corrosion of metals (Ford and Mitchell, 1990; Little *et al.*, 1991), degradation of polyester-polyurethane (Gu *et al.*, 1998b), and deterioration of concrete (Dierks *et al.*, 1991; Gu *et al.*, 1998a). Water treatment plants are especially sensitive to biofouling (van der Wende *et al.*, 1989; Byrd *et al.*, 1991; Block *et al.*, 1993; Speth *et al.*, 1998; Donlan, 2000) because a continuously high quality of the drinking water has to be maintained. However, limited biofouling is often observed in drinking water distribution systems (Camper *et al.*, 1999) and can reduce the transport of freshwater through tubes (Munson *et al.*, 1990; Lewandowski and Stoodley, 1995). Analogous observations were reported for biofouling of ships' hulls; biofilms increase the drag of ships (Bohlander, 1991) and therefore significantly raise the consumption of fuel (Alberte *et al.*, 1992). Biofilms also may impose an imminent danger or even a life threatening problem to people that have a deficient immunosystem. The settlement of opportunistic pathogens on catheters and medical implants (Mittelman, 1996; Bayston, 2000) can lead to severe complications. Moreover, the integration of opportunistic pathogens, *e.g.* coliform bacteria (Camper *et al.*, 1996) or *Helicobacter pylori* (Mackay *et al.*, 1999), into natural and non-harmful biofilms poses a threat to public health.

All these examples represent inherently difficult problems. In many cases the effect can be limited only through the frequent cleaning or in some cases by the replacement of the fouled surface.

STRATEGIES TO REDUCE BIOFOULING

The biofouling of a surface occurs in a sequence of distinguishable events. In a first step planktonic cells attach to an inert surface mainly through hydrophobic interactions (Rosenberg and Kjelleberg, 1986). The avoidance of this event is considered as most important in the reduction of biofouling (McEldowney and Fletcher, 1986; Palmer *et al.*, 1997). However, the best method of avoidance is strongly dependent on a number of factors such as the physiology of the attaching cells (Marshall *et al.*, 1971; Fletcher, 1991; Korber *et al.*, 1994; Jucker *et al.*, 1996; Neu, 1996; Mohamed *et al.*, 1998), the general nutrition in the system (Cowell *et al.*, 1999), the size of the system (see below), the fluid dynamics in the liquid (Korber *et al.*, 1989; Mittelman *et al.*, 1990; Lau and Liu, 1993; Brading *et al.*, 1995), and the property of the surface (Wardell *et al.*, 1980; Caldwell, 1984; Fletcher and Pringle, 1985, Lappin-Scott and Costerton, 1993). Thus, the modification of the material surface appears to be a good strategy to limit cell attachment.

When cells attach successfully to a surface, genes are activated which lead to the synthesis of various compounds (Angles and Goodman, 2000). An important category of compounds for biofilm formation is the extracellular polymeric substances (EPSs), which consist of proteins (adhesins) and extracellular polysaccharides (Sutherland, 1980; Allison and Sutherland, 1987; Davies *et al.*, 1993) and are typically found in biofilms at concentrations of 1–2% w/v (Christensen and Characklis, 1990). Polysaccharides are typically composed of repeating sugar units, usually glucose, galactose, mannose, rhamnose, N-acetylglucosamine, glucuronic acid and galacturonic acid. They are assembled intracellularly into a polymer from sugar-nucleotides (*e.g.* UDP-galacturonic acid) *via* lipid-linked intermediates (Sutherland, 1982). Three basic types of extracellular polysaccharides are observed, *viz.* lipopolysaccharides (LPS), capsular polysaccharides (CPS) both located on the outer membrane of Gram-negative cells (Whitfield and Valvano, 1993), and slime polysaccharides (SPS). SPS are completely released from the cell and differ from LPS and CPSs with respect to molecular structure and sugar composition (Hughes, 1995). CPS and SPS often have extremely high molecular weights, in the range of millions. Aqueous solutions of SPS are viscous and behave in a non-Newtonion way, *i.e.* the viscosity is dependent on the shear rate. The presence of uronic acid and pyruvate in the polymer influences the physical properties significantly, since charged polymers are more soluble in water (Hart *et al.*,

1999) and can react as cation exchangers (Christensen and Characklis,1990; Linton *et al.*, 1999). Thus, in special cases the exopolysaccharides can also take an important role in corroding the surface to which the bacterium is attached. For instance, it was found that *Thiobacillus ferrooxidans* is not able to leach non-ferrous sulphide (synthetic covellite, CuS) when the exopolysaccharides were removed (Pogliani and Donati, 1999). However, the addition of isolated exopolysaccharides restored their ability to leach covellite immediately.

Further, it was concluded that the growth condition determines the composition and amount of exopolysaccharides produced (Christensen and Characklis, 1990). Thus, the oxidation state and the specific rate of production of SPS appear to be inversely related to the growth efficiency of the producing organism (Linton, 1991). EPS synthesis may therefore be seen as a kind of overflow metabolism. Consequently, it can be assumed that attached cells have similar EPS. However, when the biofilm thickness increases, nutrient gradients are detectable (Rittman and Dovantzis, 1983) and heterogeneity in EPS production can be expected (*e.g.* production rate). The effect of a nutrient gradient may be studied with planktonic cells, since there is evidence that SPS produced by a bacterium in a biofilm is identical to that produced by the same bacterium in the planktonic state differing only in their physical properties (Sutherland, 1995; Skillman *et al.*, 1999). Thus, planktonic *Streptococcus thermophilus* produces two types of exopolysaccharides with an identical monomer composition (galactose : glucose = 4 : 1) in batch cultures with broth milk and MRS broth (Skillman *et al.*, 1999). The polymers differed significantly in molecular weight (4.1×10^5 Da and 1.8×10^6 Da). When the carbon/nitrogen ratio of the broth was low (high N-content) the low mass polymer was favored, whereas at a high carbon/nitrogen ratio the smaller molecule was predominant.

It can be summarized that environmental conditions have an imminent influence on the establishment of a primary cell layer on the substratum. Since in many cases the environmental conditions cannot be modified, an appropriate strategy is to remove fixed cells with agents that preferentially destroy EPS.

In the last stage of biofilm formation successfully attached cells further divide and form microcolonies with a complex structure (Lewandowsky, 2000). Allison and Sutherland (1987) concluded that without the synthesis of EPS no microcolonies were formed. An indirect confirmation of this observation was only reported recently. Biofilms that were exposed simultaneously to low nutrient concentrations, high flow rates (up to 0.72 m s^{-1}), and turbulent flow conditions ($4200 < Re < 12000$) showed an increased physical stability (Melo and Vieira, 1999). The substrate consumption flux at high flow rates (0.62 m s^{-1}) was calculated to be considerably smaller than at slower flow (0.28 m s^{-1}). Further investigations revealed that biofilm structure is highly heterogeneous and after a certain time reaches a constant thickness due to biomass loss and cell growth (van Losdrecht *et al.*, 1995; Huang *et al.*, 1998; Xu *et al.*, 1998). Consequently, biofilm growth and steady-state conditions are difficult to describe mathematically (Wanner, 1996).

The system seems to be more complicated because it has been reported that cells in biofilms are able to communicate (Davies *et al.*, 1998) and are able to exchange genes (Angles and Goodman, 2000; Holmström and Kjelleberg, 2000). Moreover, a mixed culture of *Enterobacter agglomerans* and *Klebsiella pneumoniae G1* showed more successful biofilm formation than in isolation (Skillman *et al.*, 1999). This was explained by the affinity of the EPSs of both strains; a 2:1 mixture of *E. agglomerans*-EPS : *K. pneumoniae*-EPS reached a higher viscosity than a 1:1 mixture.

To date, the strategy of controlling heavily fouled surfaces is to use strong chemicals to kill the microorganisms. However, biofilms are more difficult to remove than previously thought (Muraca *et al.*, 1990; Callow, 1993; Williams *et al.*, 1997). Extensive research has revealed that cells of a biofilm are generally more resistant to biocides than cells in suspension (Chevallier *et al.*, 1988; Anderson *et al.*, 1990; Hoyle *et al.*, 1992; Skillman *et al.*, 1997). The problem is more complex because material of dead organisms will remain on the surface and may serve as nutrient and new anchor points for new cells. Moreover, the frequent application of toxic chemicals has caused problems in places that were not subjected to disinfection. For example, it was found that biocides were accumulated through food chain amplification. Thus, to avoid this negative effect efficient and environmentally friendly antifoulants are needed for future biofouling control. What are the tasks of such an antifoulant? Houghton (1984) formulated the required characteristics, as follows: a) active at low concentrations; b) economical; c) harmless to human beings and other non-target organisms; d) unaffected by inclusion in suitable matrix; e) non-polluting, and f) biodegradable.

METHODS TO INHIBIT FOULING

The previously mentioned strategies for the active reduction of biofouling all depend on the system being protected (closed, semi-open, and open). In the following, the kinds of techniques used in combination with which antifoulants are reviewed.

Closed And Semi-Open Systems

Depending on the size of the system, frequent biocide additions may become a very expensive way to control biofilms. First, the costs of the required chemicals may be large over time. Usually expensive oxidizing (bromine, chlorine, iodine, peracetic acid, hydrogen peroxide) and non-oxidizing agents (benzoate, bisulfite, formaldehyde, glutaraldehyde, quaternary amines) are applied in various processes. Second, without some mechanism for monitoring biofilm thickness or formation rates, the frequency of biocide addition is difficult to titrate for maximum effectiveness. Over-dosing can waste costly biocides while under-dosing may not provide effective control. Third, oxidizing biocides may have an additional negative side effect, *viz.* the housing of the system may corrode. This is an important problem frequently encountered in drinking water distribution systems. Electrochemical reactions at the pipe surface may cause the formation of pits which can originate larger nodules composed of ferric hydroxide. Growth of the corrosion nodules increases turbulent resistance to flow within the pipe and reduces the shear stress on the surface. The increase in turbulent boundary layer favors the adhesion of nutrients to the biofilm, improving the nutrition of the habitat for microbial cells. Recently, it has been shown with glass capillary flow cells that cell growth under such turbulent flow conditions was not as fast as under laminar flow conditions. However, a denser packing and higher cell numbers were detected when steady-states were obtained (Stoodley *et al.*, 1999). This confirms the observation

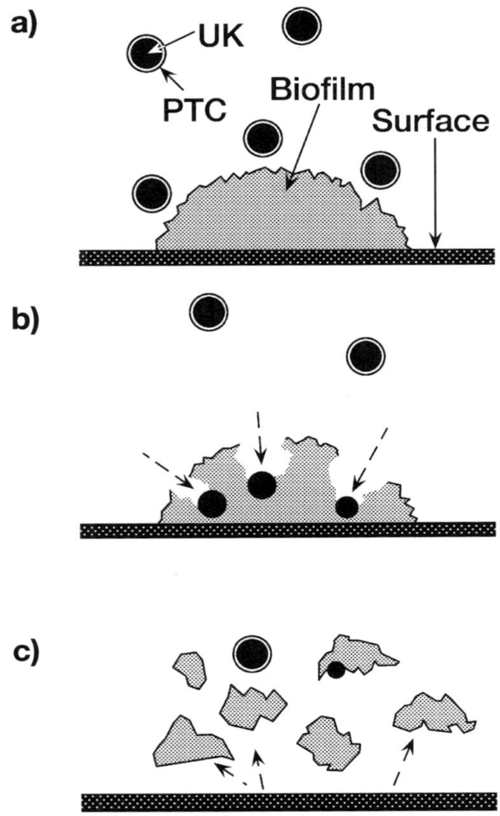

Figure 1 A phase transfer catalyst (PTC) and an integrated biocide increase the efficiency of the oxidizing agent (UK). a = contact with the biofilm; b = penetration of the outer layer and disintegration of biofilm matrix; c = destruction and elimination of biofilm through oxidation, hydrolysis and solubilization. Oxidizing agent = Ultra-Kleen® (hydroperoxide as active chemical). (Courtesy of Sterilex, Baltimore, MD, USA.)

that samples of iron tubercles in drinking water distribution pipes showed higher counts of coliform bacteria than not corroded systems (LeChevallier *et al.*, 1987).

Once a biofilm is formed, removal requires increased biocide concentrations (Brown and Gauthier, 1993; Chen *et al.*, 1993; Xu *et al.*, 1996).

A new method was developed by Sterilex (Baltimore, MD, USA) using a phase-transfer catalyst (Figure 1) to transport non-corrosive hydroperoxide to the cell membrane. The anion of hydroperoxide reacts with acetyl esters and forms peracetic acid, which produces strong hydroxyl radicals that are very reactive and decompose cell material efficiently. The advantage of peroxide as a biocide is that no corrosion is induced. First positive results with this design were reported on the disinfection of *Listeria monocytogenes* attached to latex gloves (McCarthy, 1996).

Non-oxidizing compounds such as quaternary ammonium compounds (QACs) bind by adsorption to the negatively charged cell surface (Schott and Young, 1972). Frequent usage of QACs for other applications such as fabric softeners in households and antistatic agents in industry has been reported to lead to contamination of river water by up to 25 mg l^{-1} (Neu, 1996). There are also reports of resistance towards QACs (Jones *et al.*, 1989; Volkering *et al.*, 1995), *e.g. P. aeruginosa* became resistant because of a change in the outer membrane composition (Jones *et al.*, 1989). This indicates that treatment with large amounts of biocides may cause new resistant microorganisms.

An environmentally friendly method to sanitize a closed system represents the application of an enzyme cocktail. Johansen *et al.* (1997) used glucose oxidase and lactoperoxidase in combination with polysaccharide hydrolyzing enzymes, which removed biofilms of *Staphylococcus aureus*, *Staphylococcus epidermidis*, *Pseudomonas fluorescens*, and *Pseudomonas aeruginosa*. However, these enzymes are not cheap to produce and therefore the application may be limited to small systems only.

The research on antifoulants needs accurate and appropriate test systems. Ludensky (1998) suggested a method that enables the on-line monitoring of the antifoulant efficiency for a heat exchanger. Parameters such as heat transfer, dissolved oxygen, and the pH of the solution were measured continuously, enabling the documentation of the efficiency of biocides.

Open Systems

The treatment of biofilms is much more difficult in open systems than in closed or semi-open systems. Surfaces exposed to seawater or lake water are quickly covered by a bacterial layer (Kerr *et al.*, 1998). This may enhance the attachment and fouling by larger organisms such as mussels and algae (Evans, 1981; Kirchman *et al.*, 1982; Holmström and Kjelleberg, 1994; Gu *et al.*, 1997). Thus, the best approach is to reduce the formation of bacterial biofilms as much as possible. As previously mentioned, this can be performed through frequent and consequently expensive, cleaning of the surface. This is usually done with water jets (Swain and Schultz, 1996), steam (Flemming *et al.*, 1996), ultrasound (Zips *et al.*, 1990) or acid and base baths (Speth *et al.*, 1998). A frequently encountered problem with these methods is that the surface may be harmed by strong chemicals and treatments. The incomplete removal of organic compounds such as exopolysaccharides may propagate new fouling. In addition, these methods are labor intense and time consuming and may not be applicable to all fouling prone systems, *e.g.* large ships and oilrigs.

As previously mentioned, researchers have concluded that the avoidance of biofilm has to occur at the very beginning, with reduction of microbial attachment (Gerhardt *et al.*, 1988; Holmström and Kjelleberg, 1994). This seems reasonable since bacterial biofilms are considered to be the first step towards macrofouling by mussels and algae. Consequently, new methods had to be developed.

A plausible approach is to modify the surface. Specifically designed coatings (Cooksey and Wigglesworth-Cooksey, 1992) that can be applied to surfaces should prevent biofouling, and to date, three different approaches exist, *viz.* 'non-stick' coatings, ablative and eroding coatings, and leaching coatings.

'Non-stick' surfaces

The measurement of the contact angle of a liquid on a surface can be used to describe the wettability and the adhesive property of that surface (Zisman, 1964; Baier and Meyer, 1992). Fluorinated polymers (Milne and Callow, 1984) and silicone elastomers (Callow *et al.*, 1986; Arrage *et al.*, 1995) have low-energy surfaces and low wettabilities. As a consequence, the interaction between surface and organism is weaker and a cleaning step can be more efficient (Meyer *et al.*, 1988; Arrage *et al.*, 1995). However, the use of silicone elastomers (*e.g.* RTV-11, Mera *et al.*, 1997) and modified siloxane resins (Santos and Bott, 1992) are limited to only a few applications due to poor abrasion resistance (Evans and Clarkson, 1993).

The polymer poly(ethylene glycol) has been investigated for its fouling resistance by several investigators (Blainey and Marshall, 1991; Desai *et al.*, 1992; Prime and Whitesides, 1993; Ista *et al.*, 1996). Self-assembled monolayers formed from $HS(CH_2)_{11}(OCH_2CH_2)_6OH$ were very efficient in repelling, attachment was reduced by 99.7% in the case of a medical (*S. epidermidis*) or a marine organism (*Deleya marina*) (Ista *et al.*, 1996).

"Smart" polymers represent a new alternative material and have gained much interest in biotechnology (Galaev, 1995). They undergo rapid and reversible changes of phase in response to a triggering signal, such as temperature, ionic strength, pH, light, or an electrical field (Galaev, 1995). Ista and Lopez (1998) showed a first application in cleaning of fouled surfaces. They used poly(*N*-isopropylarylamide) (PNIPAAM) which has a lower critical solubility temperature of 32°C. Test surfaces were fouled with *Halomonas marina* in natural bay water above 32°C. More than 90% of the attached fouling material could be removed when a phase change in the PNIPAAM was provoked by the temperature decrease. These results indicate that such materials have great potential and more research should be performed in this field.

The construction of an active and biocidal surface is another approach to reducing cell attachment. Toxic metals like copper, zinc, and silver appeared to have an antifouling effect when used as surface material or when integrated into paint. However, this effect is observed for only a short time because a few bacteria are able to overcome the toxicity by the production of protective exopolymers (Silver and Misra, 1988; Babycos *et al.*, 1993; Geesey, 1994; Rogers *et al.*, 1995; Srivastava *et al.*, 1995; Flemming *et al.*, 1996; Tang and Cooney, 1998). In addition, some surfaces such as copper, rapidly oxidize to insoluble, non-toxic salts. Thus, mechanical cleaning of such surfaces remains a necessity. Depending on the antifouling agents incorporated into the surface coating, cleaning can produce highly toxic waste that must be treated with care, adding to the costs and difficulty of treatment.

The application of DC5700 (3-trimethoxysilyl)-propyloctadecyldimethyl ammonium chloride), a quaternary ammonium compound covalently bound to a silicone matrix, showed promising results (Evans and Clarkson, 1993). However, the application was found to be restricted, since the intramolecular bonding was weakened in presence of sea water. The antifouling effect was lost after 7 d and octadecane, a degradation product of DC5700 was detected in the leachates (Evans and Clarkson, 1993). The covalent binding of other, more environmentally friendly antifoulants (see above) might give more acceptable results.

Ablative and eroding surfaces

Ablative or self-polishing marine paints containing toxic metals are designed to erode physically over time to liberate a new layer of paint that still has the original antifouling effect (Smith and Smith, 1975). The most efficient additive to marine paints is organotin, and ships of the US Navy did not show significant macrofouling over 5 years (Bohlander, 1991). However, the organotin antifouling agent proved to be refractory. Once eroded from the paint, it accumulated in the environment. Microorganisms degrade tributyltin (TBT) at a slow rate (Callow and Willingham, 1996). The half-life of sediment-bound TBT was determined to be 160 d (Langston and Burt, 1991). Investigations showed that low levels of TBT (0.05 $\mu g\ l^{-1}$) lead to growth of male sex organs in the female dog-whelk *Nucella lapillus* (Gibbs and Bryan, 1986). The use of TBT was prohibited in the US on ships smaller than 25 m and on non-aluminum hulled vessels (Dalley, 1989). Other countries, including most of Western Europe, and Scandinavia, Australia and New Zealand have similarly banned TBT. As a result of past widespread use, the concentration of total organotin compounds in the river Elbe in Germany has been found to range between 30 and 96 $ng\ l^{-1}$ (Shawky and Emons, 1998). Fish contained between 54 and 223 $ng\ g^{-1}$ (fresh mass) in their liver and 27–202 $ng\ g^{-1}$ in their muscles (Shawky and Emons, 1998). In addition, sea ducks from British Columbia were found to contain concentrations as high as 1100 $ng\ g^{-1}$ (Kannan *et al.*, 1998). Recently, Tang and Cooney (1998) showed with a Robbins device (a flow cell with laminar flow conditions) that TBT-resistant *P. aeruginosa* PAO-1 made up to 50% of a biofilm culture.

The replacement of organotin with high levels of cuprous oxide appeared to reduce biofouling also. The use of such coating material (*e.g.* ABC®, Ameron International, Brea, CA, USA) in combination with an anticorrosive coating (epoxy or coal-tar in epoxy) showed acceptable protection for 2 years, but did not attain the initially good performance of organotin (Bohlander, 1997).

Leaching coatings

The diffusion of antifouling compounds from a surface, called leaching, offers another field of applications (Swain and Schultz, 1996). This method of surface protection was copied from nature, since marine organisms like sponges (Thompson, 1985; Thompson *et al.*, 1985; Sears *et al.*, 1990), Gorgonian corals (Keifer *et al.*, 1986; Vrolijk *et al.*, 1990), and the eelgrass *Zostera marina* (Harrison and Chan, 1980; Todd *et al.*, 1993) contain naturally occurring compounds shown to inhibit a wide range of fouling organisms. There are now more than 90 antifouling compounds of natural origin described in literature (see survey by Clare, 1996). The structure of natural antifoulants (see Figure 2) is usually complex and therefore the production by chemical synthesis is in many cases very difficult (Clare, 1998). In most cases, the chemical reaction responsible for the antifouling activity of these compounds is unknown.

Zosteric acid (ZA), however, is readily synthesized chemically from readily available precursors (sulfonation reaction, Todd *et al.*, 1993). The antifouling property of ZA results from binding to attachment sites on cell surfaces, preventing cell adhesion (Sundberg *et al.*, 1997). Whether other natural antifoulants work in a similar way is unknown.

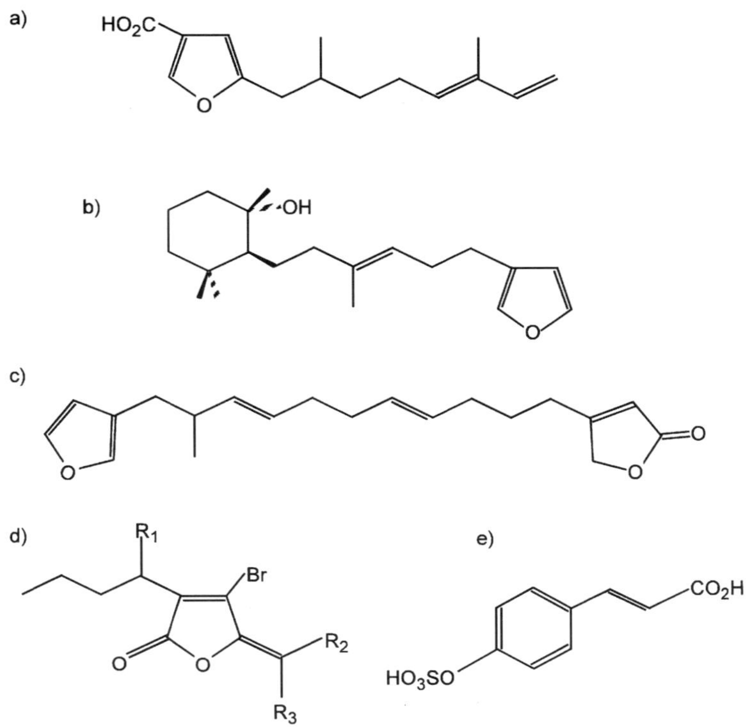

Figure 2 Natural antifouling agents of marine organisms a = (9E)–4-(6,10-Dimethylocta–9,11-dienyl) furan–2-carboxylic acid; b = Ambliol-A; c = Furospongolide; d: Halogenated furanoses; e = *p*-coumaric acid sulphate (zosteric acid). (Adapted from Clare, 1998, and Todd *et al.*, 1993.)

Several antifouling compounds have been successfully commercialized, including Sea-Nine 211 (4,5-dichloro-2-n-octyl-4-isothiayolin-3-one) by Rohm & Haas, USA, Irgarol 1051 (2-methyltio-4-ter-butylamino-6-cyclopropylamino-s-triazine) by Novartis, Switzerland, Diuron (2-(3,4-dichlorophenyl) 1-1-dimethylurea) by Bayer, UK, and Chlorothalonil (2,4,5,6 tetrachloro 1,3 benzendicarbonitrile) by ISK Biotech, USA (Callow and Willingham, 1996). Although these compounds possess significant and broad toxicity, they tend to be biodegradable and have short half-lives in sea water (Callow and Willingham, 1996). The mode of action of all commercially available antifoulants is based on their toxicity.

Methods to determine the minimal effective release rate of a leaching antifoulant experimentally

The efficiency of the antifoulants is concentration dependent. Below a certain concentration antifouling efficacy is reduced, and in the worst case, the compound will be used as a carbon source (Eighmy *et al.*, 1992). Consequently, it is essential to determine the minimal effective release rate (MERR) of the antifoulant. This can be

Table 1 Typical minimal efficient release rates (MERR) of antifoulants.

Antifoulant	*MERR* {$\mu g\ cm^{-2}\ d^{-1}$}	*Reference*
Benzoic acid	>10	(Eighmy *et al.*, 1992)
Copper oxide	10	(Fischer *et al.*, 1984)
2,4-dinitrophenol	>15	(Eighmy *et al.*, 1992)
Sea-Nine 211	15	(Vasishtha *et al.*, 1995)
Tributyltin chloride	5	(Fischer *et al.*, 1984)
Zosteric acid	>50	(Haslebeck *et al.*, 1996)

done with a MERR device, consisting of a peristaltic pump delivering different test solutions from small liquid containers to submerged tubes closed with a porous membrane. The extent of the fouling on the membranes is then analyzed and the MERR for a particular antifoulant can be determined. Typical release rates are summarized in Table 1.

Although simple in principle, the MERR system has been difficult to carry out in practice. Mechanical problems have involved membrane clogging, deterioration of soluble agents in the feed tubes and unrealistically high mass flow of solvent at the membrane surface. Thus, while MERR systems may provide general ranges of effective concentrations, they do not simulate the leaching of antifouling agents from hard coatings.

Optimal leaching rates

Antifoulants that are integrated into a coating leach at decreasing flux rates as the reservoir is drained. The dynamics are derived from the Fickian law and can be calculated according to Higuchi (1963):

$$Q = (D(2A_0 - C_S)C_S t)^{1/2} \tag{1}$$

where Q is the amount of the diffusant released per unit area over time t, A_0 is the initial diffusant concentration in the matrix, C_S is the solubility, and D is the diffusion coefficient of the diffusant out of the matrix. The flux, or release rate F_{dc} is then given as (Weisman *et al.*, 1992):

$$F_{dc} = dQ/dt = (D(2A_0 - C_S)C_S)/(2t^{1/2}) \tag{2}$$

The time course of an assumed coating is shown in Figure 3. When the flux falls below the MERR of the antifoulant, then biofouling can be expected. At the worst the antifoulant is far below the working concentration and may be used as a nutrient source, resulting in extensive fouling.

The efficacy of coatings with antifoulants has to be verified experimentally. Flow cells have been found to be very useful for screening potential antifoulants and coating matrices (Angell *et al.*, 1993; Mittelman *et al.*, 1993; Rijnaarts *et al.*, 1993; Arrage and White, 1997). Good results can generally be obtained under laminar flow conditions ($1 < Re < 1000$; Zinn *et al.*, 1999). The coated metal coupons are integrated into a flow channel flush with the surface. Microorganisms are then inoculated at a

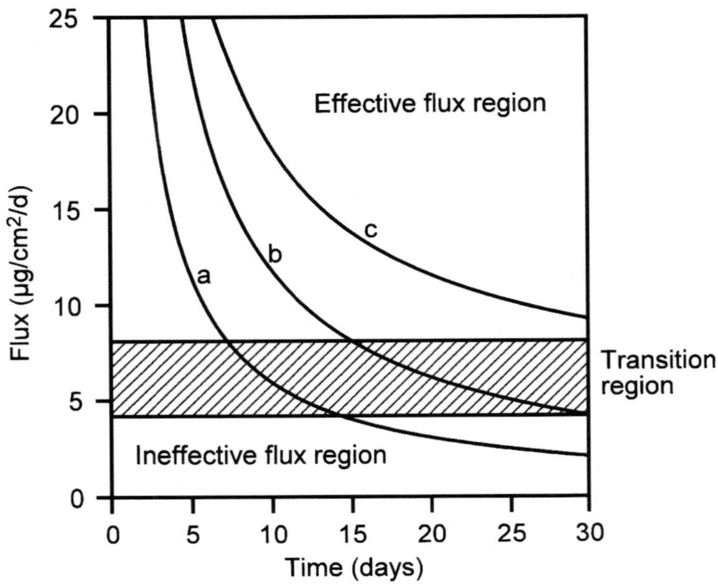

Figure 3 Conceptual plot showing flux vs time for three different coatings (a, b, and c). (Adapted from Vasishtha *et al.*, 1995.)

concentration of about 10^{-7} cells ml^{-1} over a period of 2 h under laminar flow conditions. The efficiency of a coating is determined through microscopic counting of the acridine orange stained cells after isolation from the surface. The dual comparison of the coated coupon with an uncoated coupon in the same flow chamber and with a reference flow cell allow clear statements about their effectiveness (Arrage *et al.*, 1995).

The most effective combinations of antifoulant and matrix have to be tested in natural systems using rafts (Evans and Clarkson, 1993; Swain and Schultz, 1996). Several coated test panels are exposed to freshwater or seawater for a longer period (>2 months). To date, quantification of antifouling effectiveness is difficult and has been limited to the description of species and counts of larger foulants like mussels. The development of optical sensors that are able to detect biological matter at a low concentration on test surfaces is under development (HOBI Labs, Watsonville, CA, USA). This would considerably help in making comparisons of different test panels and moreover, biofouling could be detected at a much earlier state, *e.g.* through the detection of microorganisms and microalgae.

NEW AND ENVIRONMENTALLY FRIENDLY WAYS TO PROTECT SURFACES FROM FOULING

Studies have shown that bacterial biofilms can inhibit the attachment of larvae of marine organisms (Maki *et al.*, 1988; Holmström and Kjelleberg, 1994). Burchard and Sorongon (1998) showed that this is also the case for the interaction between gliding bacteria. The cells were isolated from a marine biofilm and identified as being members of the genus *Cytophaga*. One strain (RB1057) produced an extracellular glycoprotein with a mass of about 60 kDa and this inhibited the other strain (RB1058) from adhering and gliding on substrata. It was found by the same group that this inhibitor was not effective against other aquatic gliding bacteria. However, a modification of the protein might broaden the impact on other species. Ideally, this antifoulant would be placed covalently bound to a suitable matrix.

The screening for potential microorganisms and their products can be carried out with hydrogels (Gatenholm *et al.*, 1995), where the bacteria are embedded into a gel and are able to live and produce the potential antifoulants. Subsequent attachment tests with larvae or microorganisms will indicate which strains produce effective antifoulants. The responsible chemical can then be isolated and integrated into a leaching or eroding coating.

To date, there are only a few reports that mention the biological inactivation of attached bacteria by bacteriophages (viruses). Roy and Ackermann (1993) showed that *Listeria monocytogenes*, a bacterium often found in food-processing plants, can be successfully inactivated with listeriphages. The combination with other types of phages or other compounds enhanced the sanitizing effect (Roy and Ackermann, 1993).

In a more recent study (Hughes *et al.*, 1998) found that *Enterobacter agglomerans* 53b grown in a biofilm culture was susceptible to the bacteriophage SF153b that was isolated from a sewage treatment plant. Infected *E. agglomerans* lysed and the exopolysaccharide was degraded by a viral glucanase. The reduction of biofilm cells occurred at a rapid rate, reducing cell density 30-fold in about 30 min. The virus was very specific because a close relative strain *Serratia marcescens* is not contaminated nor was its exopolysaccharides degraded.

Experiments with electric currents showed that biofilms of *P. aeruginosa* were more sensitive to an antibiotic treatment with tobramycin ("bioelectric effect", Jass *et al.*, 1995). The best results were observed for currents up to 20 mA cm^{-2}, and further current increases did not enhance the susceptibility. Studies with a mixed biofilm showed that biofilm thickness varied with the polarity of the surface material (Stoodley *et al.*, 1998). A positive charge resulted in a compression of the biofilm to 74% of the original thickness, whereas a negative charge resulted only in an expansion of approximately 4%. A different approach was chosen by Cho and Choi (1999). They investigated whether alternate currents (0.2 A, 500 Hz) had an antifouling effect in a single-tube heat exchanger. A fouling resistance of 20–38% could be found at flow velocities between 0.52 and 0.78 m s^{-1}. However, for slower flow rates the treatment was not successful. Calcium carbonate scales were shaped in clusters and were elliptic crystals, whereas under no current conditions the scale organized in needle-shaped crystals and much more difficult to be removed. Whether their observation was related to the biological activity was not given.

Bioerodible materials are also under investigation for the *in situ* delivery of anti cancer drugs (Heller, 1988; Mathiowitz *et al.*, 1989). Here, the erosion of the surface releases the drug that is integrated in the matrix. Successful studies have been reported with poly(3-hydroxybutyrate) (PHB) as a bioerodible matrix (Akhtar *et al.*, 1992; Gopferich, 1996; Pouton and Akhtar, 1996). Further, biodegradable rubber made of poly(3-hydroxalkanoates) (PHA) (de Koning *et al.*, 1994) seems to have potential for use as an environmentally friendly matrix. The antifouling agent could be incorporated by cross-linking with the polymer through irradiation. The antifoulant would then be released through the bioerosion of the matrix. This could theoretically enhance the efficiency of the antifoulant, since it is postulated that growing cells are more sensitive to toxic compounds (Brown *et al.*, 1988; Williams *et al.*, 1997). However, this has to be confirmed experimentally.

FUTURE CHALLENGES

The reviewed strategies suggest that the inhibition of microbial fouling by highly toxic chemicals is somewhat ambiguous since they may significantly harm organisms. Biological methods represent an alternative, since they are, in general, more environmentally friendly due to their biodegradability. However, it is important to note that rate of biodegradation vary significantly among so-called natural antifouling agents. These agents are not necessarily inherently less toxic to non-target organisms. Such biological approaches may have a commercial advantage because the public perception of natural compounds in the environment is, at this time, positive. The ideal antifoulant agent has get to be developed or discovered. Based upon the informant presented in this review, it may be stated that the ideal antifouling agent will provide extended activity against a narrow range of taret organisms. More emphasis should be placed on delivery methods that enable controlled release of agents in direct response to the presence of fouling organisms.

ACKNOWLEDGEMENTS

The authors thank Marc Mittelman for his critical comments and for useful discussions.

REFERENCES

Akhtar S., Pouton C.W., Notarianni L.J. (1992). Crystallization behaviour and drug release from bacterial polyhydroxyalkanoates. *Polymer*, **33**, 117–126.

Alberte R.S., Snyder S., Zahuranec B.J., Whetstone M. (1992). Biofouling research needs for the United States NAVY: Program history and goals. *Biofouling*, **6**, 91–95.

Allison D.G., Sutherland I.W. (1987). The role of exopolysaccharides in adhesion of freshwater bacteria. *J Gen Microbiol*, **133**, 1319–1327.

Anderson R.L., Holland B.W., Carr J.K., Bond W.W., Favero M.S. (1990). Effect of disinfectants on pseudomonads colonized on the interior surface of PVC pipes. *Am J Public Health*, **80**, 17–21.

Angles, M.L., Goodman, A.E. (2000). Plasmid transfer between bacteria in biofilms. In: Evans, L.V. (ed) *Biofilms: Recent Advances in their Study and Control*. Harwood Academic Publishers, pp. 81–99.

Angell P., Arrage A.A., Mittelman M.W., White D.C. (1993). On-line, non-destructive bio -mass determination of bacterial biofilms by fluorometry. *J Microbiol Methods*, **18**, 317–327.

Arrage A.A., White D.C. (1997). Monitoring biofilm-induced persistence of Mycobacterium in drinking water systems using GFP fluorescence. In: Hastins J.W., Kricka L.J. Stanley P.E. (eds) *Bioluminescence and Chemiluminescence: Molecular Reporting with Photons*. John Wiley & Sons, New York, pp. 383–386.

Arrage A.A., Vasishtha N., Sundberg D., Bausch G., Vincent H.L., White D.C. (1995). On-line monitoring of antifouling and fouling-release surfaces using bioluminescence and fluorescence measurements during laminar flow. *J Ind Microbiol*, **15**, 277–282.

Babycos C.R., Barrocas A., Webb W.R. (1993). A prospective randomized trial comparing the silver-impregnated collagen cuff with the bedside tunneled subclavian catheter. *J Parenter Enteral Nutr*, 17, 61–63.

Baier R.E., Meyer A.E. (1992). Surface analysis of fouling-resistant marine coatings. *Biofouling*, 6, 165–180.

Bayston, R. (2000). Biofilm infections on implant surfaces. In: Evans, L.V. (ed) *Biofilms: Recent Advances in their Study and Control*. Harwood Academic Publishers, pp. 117–131.

Blainey B.L., Marshall K.C. (1991). The use of copolymers to inhibit bacterial adhesion and biofilm formation on hydrophobic surfaces in marine habitats. *Biofouling*, 4, 309–318.

Block J.C., Houdidier K., Paquin J.L., Miazga J., Levi Y. (1993). Biofilm accumulation in drinking water distribution systems. *Biofouling*, 6, 333–343.

Bohlander G.S. (1991). Biofilm effects on drag: Measurements on ships. *Trans I Mar E (C)*, Conference 3, **103**, 135–138.

Bohlander G.S. (1997). Biofouling, fleet maintenance and operational needs. *Nav Res Rev*, 44, 9–17.

Brading M.G., Boyle J., Lappin-Scott H.M. (1995). Biofilm formation in laminar flow using *Pseudomonas fluorescens* EX101. *J Ind Microbiol*, **15**, 297–304.

Brown M.L., Gauthier J.J. (1993). Cell density and growth phase as factors in the resistance of a biofilm of *Pseudomonas aeruginosa* (ATCC 27853) to iodine. *Appl Environ Microbiol*, **59**, 2320–2322.

Brown M.R.W., Allison D.G., Gilbert P. (1988). Resistance of bacterial biofilms to antibiotics: A growth-rate related effect? *Antimicrob Chemother*, **22**, 777–783.

Burchard R.P., Sorongon M.L. (1998). A gliding bacterium strain inhibits adhesion and motility of another gliding bacterium strain in a marine biofilm. *Appl Environ Microbiol*, 64, 4079–4083.

Byrd J.J., Xu H.-S., Colwell R.R. (1991). Viable but nonculturable bacteria in drinking water. *Appl Environ Microbiol*, **57**, 875–878.

Callow M.E. (1993). A review of fouling in freshwaters. *Biofouling*, 7, 313–327.

Callow M.E., Willingham G.L. (1996). Degradation of antifouling biocides. *Biofouling*, **10**, 239–249.

Camper A.K., Jones W.L., Hayes J.T. (1996). Effect of growth conditions and substratum composition on the persistence of coliforms in mixed-population biofilms. *Appl Environ Microbiol*, **62**, 4014–4018.

Camper A., Burr M., Ellis B., Butterfield P., Albernathy C. (1999). Development and structure of drinking water biofilms and techniques for their study. *J Appl Microbiol*, **85**, 1S–12S.

Cho Y.I., Choi B.G. (1999) Validation of an electronic anti-fouling technology in a single-tube heat exchanger. *Int Comm Heat Mass Transfer*, 42, 1491–1499.

Chen C.-I., Griebe T., Characklis W.G. (1993). Biocide action of monochloramine on biofilm systems of *Pseudomonas aeruginosa*. *Biofouling*, 7, 1–17.

Christensen B.E., Characklis W.G. (1990). Physical and chemical properties of biofilms. In: Characklis W.G., Marshall K.C. (eds) *Biofilms*. John Wiley & Sons, New York, pp. 93–130.

Clare A.S. (1996). Marine natural product antifoulants: status and potential. *Biofouling*, **19**, 211–229.

Clare A.S. (1998). Towards nontoxic antifouling. *J Mar Biotechnol*, **6**, 3–6.

Cowell B.A., Willcox M.D.P., Herbert B., Schneider R.P. (1999). Effect of nutrient limitation on adhesion characteristics of *Pseudomonas aeruginosa*. *J Appl Microbiol*, **86**, 944–954.

Dalley R. (1989). Legislation affecting tributyltin antifoulings. *Biofouling*, **1**, 363–366.

Davies D.G., Chakrabarty A.M., Geesey G.G. (1993). Exopolysaccharide production in biofilms: substratum activation of alginate gene expression by *Pseudomonas aeruginosa*. *Appl Environ Microbiol*, **59**, 1181–1186.

Davies D.G., Parsek M.R., Pearson J.P., Iglewsi B.H., Costerton J.W., Greenberg E.P. (1998). The involvement of cell-to-cell signals in the development of a bacterial biofilm. *Science*, **280**, 295–298.

de Koning G.J.M., van Bilsen H.H.M., Lemstra P.J., Hazenberg W., Witholt B., Preusting H., van der Gali J.G. In: Schirmer A., Jendrossek D. (1994). A biodegradable rubber by crosslinking poly(hydroxyalkanoate) from *Pseudomonas oleovorans*. *Polymer*, **35**, 2090–2097.

Desai N.P., Hossainy F.A., Hubbel J.A. (1992). Surface immobilized polyethylene oxide for bacterial repellence. *Biomaterials*, **13**, 417–420.

Dierks M., Sand W., Bock E. (1991). Microbial corrosion of concrete. *Experientia*, **47**, 514–516.

Donlan, R.M. (2000) Biofilm control in industrial water systems: approaching an old problem in new ways. In: Evans, L.V. (ed) *Biofilms: Recent Advances in their Study and Control*. Harwood Academic Publishers, pp. 333–360.

Eighmy T.T., Arwa J., De Rome L., Brown M.G., Cimini R.A., Sundberg D.C., Weisman G.R. (1992). Controlled release antifouling coatings. II. The effects of controlled release of 2,4-dinitrophenolate and benzoate on marine biofilm development and metabolic activity. *Biofouling*, **6**, 147–163.

Evans L.V. (1981). Marine algae and fouling; a review with particular reference to ship fouling. *Bot Mar*, **24**, 167–171.

Evans L.V., Clarkson N. (1993). Antifouling strategies in the marine environment. *J Appl Bacteriol*, **74**, 119S–124S.

Fischer E.C., Castelli V.J., Rodgers S.D., Bleile H.R. (1984). Technology for control of marine biofouling — a review. In: Costlow J.D., Tipper R.C. (eds) *Marine Biodeterioration: an Interdisciplinary Study*. US Naval Institute Press, Annapolis, MD, pp. 261–299.

Flemming H.-C., Griebe T., Schaule G. (1996). Antifouling strategies in technical systems — a short review. *Water Sci Technol*, **34**, 517–524.

Fletcher M. (1991). Physiological activity of bacteria attached to solid surfaces. *Adv Microb Physiol*, **32**, 53–85.

Fletcher M., Pringle J.H. (1985). The effect of surface free energy and medium surface on bacterial attachment to solid surfaces. *J Coll Interfac Sci*, **104**, 5–14.

Ford T., Mitchell R. (1990). The ecology of microbial corrosion. In: Marshall K.C. (ed) *Advances in Microbial Ecology*. Plenum Press, New York, pp. 231–262.

Galaev J.Y. (1995). 'Smart' polymers in biotechnology and medicine. *Russ Chem Rev*, **64**, 471–489.

Gatenholm P., Holmström C., Maki J.S., Kjelleberg S. (1995). Toward biological antifouling surface coatings: marine bacteria immobilized in hydrogel inhibit barnacle larvae. *Biofouling*, **8**, 293–301.

Geesey G.G. (1994). Biofouling/biocorrosion in industrial water systems. In: Geesey G.G., Lewandowski Z., Flemming H.-C. (eds) *Biofouling and Biocorrosion in Industrial Water Systems*. Lewis Publishers, Boca Raton, pp. i-iv.

Gerhardt D.J., Rittschof D., Mayo S.W. (1988). Chemical ecology and the search for antifoulants. *J Chem Ecol*, **14**, 1903–1915.

Gibbs P.E., Bryan G.W. (1986). Reproductive failure in populations of the dog-whelk, *Nucella lapillus*, caused by imposex induced by tributyltin from antifouling paints. *J Marine Biol Assoc UK*, **66**, 767–777.

Gopferich A. (1996). Mechanisms of polymer degradation and erosion. *Biomaterials* 17, 103–114.

Gu J.-D., Maki J.S., Mitchell R. (1997). Microbial biofilms and their role in the induction and inhibition of invertebrate settlement. In: D'Itri F.M. (ed) *Zebra Mussels and Aquatic Nuisance Species*. Ann Arbor Press, Chelsea, pp. 343–357.

Gu J.-D., Ford T.E., Berke N.S., Mitchell R. (1998a). Biodeterioration of concrete by fungus *Fusarium*. *Int Biodeter Biodegrad*, **41**, 101–109.

Gu J.-D., Roman M., Fesselman T., Mitchell R. (1998b). The role of microbial biofilms in deterioration of space station candidate materials. *Int Biodeter Biodegrad*, **41**, 25–33.

Harrison P.G., Chan A.T. (1980). Inhibition of the growth of micro-algae and bacteria by extracts of eelgrass (*Zostera marina*) leaves. *Mar Biol*, **61**, 21–26.

Hart T.D., Chamberlain A.H.L., Lynch J.M., Newling B., McDonald P.J. (1999) A stray field magnetic resonance study of water diffusion in bacterial exopolysaccharides. *Enzyme Microb Technol*, **24**, 339–347.

Haslebeck E., Kavanagh C.J., Shin H.-W., Banta W.C., Song P., Loeb G.I. (1996). Minimum effective release rate of antifoulants (2): measurement of the effect of TBT and zosteric acid on hard fouling. *Biofouling*, **10**, 175–186.

Heller J. (1988). Chemically self-regulated drug delivery-systems. *J Controlled Release*, **8**, 111.

Higuchi T. (1963). Mechanism of sustained-action medication. Theoretical analysis of rate of release of solid drugs dispersed in solid matrices. *J Pharm Sci*, **52**, 1145.

Holmström C., Kjelleberg S. (1994). The effect of external biological factors on settlement of marine invertebrate and new antifouling technology. *Biofouling*, **8**, 147–160.

Holmström C., Kjelleberg S. (2000). Bacterial interactions with marine fouling organisms. In: Evans, L.V. (ed) *Biofilm: Recent Advances in their Study and Control*. Harwood Academic Publishers, pp. 101–115.

Houghton D.R. (1984). Toxicity testing of candidate antifouling agents and accelerated antifouling paint testing. In: Costlow J.D., Tipper D.G. (eds) *Marine Biodeterioration: an Interdisciplinary Study*. US Naval Institute, Annapolis, pp. 256–258.

Hoyle B.D., Alcantara J., Costerton J.W. (1992). *Pseudomonas aeruginosa* biofilm as a diffusion barrier to piperacillin. *Antimicrob Agents Chemother*, **36**, 2054–2056.

Huang C.-T., Xu K.D., McFeters G.A., Steart P.S. (1998). Spatial patterns of alkaline phosphatase expression within bacterial colonies and biofilms in response to phosphate starvation. *Appl Environ Microbiol*, **64**, 1526–1531.

Hughes, K.A. (1995). Bacterial polysaccharides from industrial environments. In: Wimpenny J., Handley P., Gilbert P., Lappin-Scott H. (eds) *The Life and Death of Biofilm*. Bioline, Cardiff, pp. 107–108.

Hughes K.A., Sutherland I.W., Jones M.V. (1998). Biofilm susceptibility to bacteriophage attack: the role of phage-born polysaccharide depolymerase. *Microbiology*, **144**, 3039–3047.

Ista L.K., Lopez G.P. (1998). Lower critical solubility temperature materials as biofouling release agents. *J Ind Microbiol*, **20**, 121–125.

Ista L.K., Fan H., Baca O., Lopez G.P. (1996). Attachment of bacteria to model solid surfaces: oligo(ethylene glycol) surfaces inhibit bacterial attachment. *FEMS Microbiol Lett*, **142**, 59–63.

Jass J., Costerton J.W., Lappin-Scott H.M. (1995). The effect of electrical currents and trombamycin on *Pseudomonas aeruginosa* biofilms. *J Ind Microbiol*, **15**, 234–242.

Johansen C., Falholt P., Gram L. (1997). Enzymatic removal and disinfection of bacterial biofilms. *Appl Environ Microbiol*, **63**, 3724–3728.

Jones M.V., Herd T.M., Christie H.J. (1989). Resistance of *Pseudomonas aeruginosa to amphoteric and quaternary ammonium biocides*. *Microbios*, **58**, 49–61.

Jucker B.A., Harms H., Zehnder A.J.B. (1996). Adhesion of the positively charged bacterium *Stenotrophomonas (Xanthomonas) maltophilia* 70401 to glass and teflon. *J Bacteriol*, **178**, 5472–5479.

Kannan K., Senthilkumar K., Elliott J.E., Feyk L.A., Giesy J.P. (1998). Occurence of butyltin compounds in tissues of water birds and seaducks from the United States and Canada. *Arch Environ Contam Toxicol*, **35**, 64–69.

Keifer P.A., Rhinehart K.L., Hooper I.R. (1986). *Renilla foulins* antifouling diterpenes from the sea pansy *Renilla reniformis* (Octocorallia). *J Org Chem*, **5**, 4450–4454.

Kerr A., Cowling M.J., Beveridge C.M., Smith M.J., Parr A.C.S., Head R.M., Davenport J., Hodgkiess T. (1998). The early stages of marine biofouling and its effect on two types of optical sensors. *Environ Int*, **24**, 331–343.

Kirchman D., Graham S., Reish D., Mitchell R. (1982). Bacteria induce settlement and metamorphosis of *Janua (Dexiospora) brasiliensis* (Grube). *J Exp Mar Biol Ecol*, **56**, 153–163.

Korber D.R., Lawrence J.R., Caldwell D.E. (1994). Effect of motility on surface colonization and reproductive success of *Pseudomonas fluorescens* in dual-dilution continuous culture and batch culture systems. *Appl Environ Microbiol*, **60**, 1421–1429.

Korber D.R., Lawrence F.R., Sutton B., Caldwell D.E. (1989). The effect of laminar flow on the kinetics of surface recolonization by mot^+ and mot^-, *Pseudomonas fluorescens*. *Microb Ecol*, **18**, 1–19.

Langston W.J., Burt G.R. (1991). Bioavailability and effects of sediment-bound TBT in deposit-feeding clams, *Scrobiculuria plana*. *Mar Environ Res*, **32**, 61–77.

Lappin-Scott H.M., Costerton J.M. (1993). Bacterial biofilms and surface fouling. *Biofouling*, **2**, 323–342.

Lau Y.L., Liu D. (1993). Effect of flow rate on biofilm accumulation in open channels. *Water Res*, **27**, 355–360.

LeChevallier M.W., Babock T.S., Lee R.G. (1987). Examination and characterization of distribution system biofilms. *Appl Environ Microbiol*, **53**, 2714–2724.

LeChevallier M.W., Cawthon C.D., Lee R.G. (1988). Inactivation of biofilm bacteria. *Appl Environ Microbiol* 54, 2492–2499.

Lewandowski, Z. (2000). Structure and function of biofilms. In: Evans L.V. (ed) *Biofilms: Recent Advances in their Study and Control*. Harwood Academic Publishers, pp. 1–17.

Lewandowski Z., Stoodley P. (1995). Flow induced vibrations, drag force, and pressure drop in conduits covered with biofilm. *Water Sci Technol*, **32**, 19–26.

Linton C.J., Sherriff A., Millar M.R. (1999). Use of a modified Robbins device to directly compare the adhesion of *Staphylococcus epidermis* RP62A to surfaces. *J Appl Microbiol*, **86**, 194–202.

Linton J.D. (1991). Metabolite production and growth efficiency. *Antonie van Leeuwenhoek*, **60**, 293–311.

Little B., Wagner P., Mansfield F. (1991). Microbiologically influenced corrosion of metals and alloys. *Int Mater Rev*, **36**, 253–272.

Ludensky M.L. (1998). An automated system for biocide testing on biofilms. *J Ind Microbiol*, **20**, 109–115.

Mackay W.G., Gribben L.T., Barer M.R., Reid D.C. (1999). Biofilms in drinking water systems: a possible reservoir for *Helicobacter pylori*. *J Appl Microbiol*, **85**, 52S–59S.

Maki J.S., Rittschof D., Costlow J.D., Mitchell R. (1988). Inhibition of attachment of larval barnacles, *Belanus amphitrite*, by bacterial surface films. *Mar Biol*, **97**, 199–206.

Marshall K.C., Stout R., Mitchell R. (1971). Mechanisms of the initial events in the sorption of marine bacteria to surfaces. *J Gen Microbiol*, **68**, 337–348.

Mathiowitz E., Ron E., Mathiowitz G., Langer R. (1989). Surface morphology of bioerodible poly(anhydrides). *Polym Prepr*, **30**, 40.

McCarthy S.A. (1996). Effect of sanitizers on *Listeria monocytogenes* attached to latex gloves. *J Food Saf*, **16**, 231–237.

McEldowney S., Fletcher M. (1986). Effect of growth conditions and surface characteristics of aquatic bacteria on their attachment to solid surfaces. *J Gen Microbiol*, **132**, 513–523.

Melo L.F., Vieira M.J. (1999). Physical stability and biological activity of biofilms under turbulent flow and low substrate concentration. *Bioprocess Engin*, **20**, 363–368.

Mera A.E., Fox R.B., Bullock S., Swain G.W., Schultz M.P., Gatenholm P., Wynne K.J. (1997). Toward minimally adhesive surfaces utilizing siloxanes. *Nav Res Rev*, **49**, 4–8.

Meyer A.E., Baier R.E., King R.W. (1988). Initial fouling of nontoxic coatings in fresh, brackish, and sea water. *Can J Chem Eng*, **66**, 55–62.

Milne A., and Callow M.E. (1984). Non-biocidal processes. *Trans I Mar E 97*, Conference 1, 229–233.

Mittelman M.W., Nivens D.E., Low C., White D.C. (1990). Differential adhesion, activity, and carbohydrate: protein ratios of *Pseudomonas atlantica* monocultures attaching to stainless steel in a linear shear gradient. *Microb Ecol* 19, 269–278.

Mittelman M.W., Packard J., Arrage A.A., Bean S.L., Angell P., White D.C. (1993). Test systems for determining antifouling coating efficacy using on-line detection of bioluminescence and fluorescence in a laminar-flow environment. *J Microbiol Methods*, **18**, 51–60.

Mittelman M.W. (1996). Adhesion to biomaterials. In: Fletcher M. (ed) *Bacterial Adhesion: Molecular and Ecological Diversity*, Wiley-Liss, New York, pp. 89–127.

Mohamed M.N., Lawrence J.R., Robarts R.D. (1998). Phosphorus limitation of heterotrophic biofilms from the Fraser river, British Columbia, and the effect of pulp mill effluent. *Microb Ecol*, **36**, 121–130.

Munson B.R., Young D.F., Okiishi T.H. (1990). *Viscous Flow in Pipes*. John Wiley & Sons, New York, pp. 465–559.

Muraca P.W., Yu V.L., Goetz A. (1990). Disinfection of water distribution systems for *Legionella*: a review of application procedures and methodologies. *Infect Control Hosp Epidemiol*, **11**, 79–88.

Neu T.R. (1996). Significance of bacterial surface active compounds in interaction of bacteria with surfaces. *Microbiol Rev*, **60**, 151–166.

Palmer R.J., Phiefer C., Burlage R., Sayler G., White D.C. (1997). Monitoring biofilm-induced persistence of *Mycobacterium* in drinking water systems using GFP fluorescence. In: Hastins J.W., Kricka L.J. Stanley P.E. (eds) *Bioluminescence and Chemiluminescence: Molecular Reporting with Photons*. John Wiley & Sons, New York, pp. 445–450.

Pogliani C., Donati E. (1999). The role of exopolymers in the bioleaching of a non-ferrous metal sulphide. *J Ind Microbiol Biotechnol*, 22, 88–92.

Pouton C.W., Akhtar S. (1996). Biosynthetic polyhydroxyalkanoates and their potential in drug delivery. *Adv Drug Delivery Rev*, **18**, 133–162.

Prime K.L., Whitesides G.M. (1993). Adsorption of proteins onto surfaces containing end-attached oligo(ethylene oxide): a model system using self-assembled monolayers. *J Am Chem Soc*, **115**, 10714–10721.

Rijnaarts H.H.M., Norde W., Bouwer E.J., Lyklema J., Zehnder A.J.B. (1993). Bacterial adhesion under static and dynamic conditions. *Appl Environ Microbiol*, **59**, 3255–3265.

Rittman B.E., Dovantzis K. (1983). Dual limitation of biofilm kinetics. *Water Res*, **17**, 1727–1734.

Rogers J., Dowsett A.B., Keevil C.W. (1995). A paint incorporating silver to control mixed biofilms containing *Legionella pneumophila*. *J Ind Microbiol*, **15**, 377–383.

Rosenberg, M., Kjelleberg S. (1986). Hydrophobic interactions; role in bacterial adhesion. *Adv Microb Ecol*, **9**, 353–393.

Roy B., Ackermann H.W. (1993). Biological inactivation of adhering *Listeria monocytogenes* by listeriaphages and a quaternary ammonium compound. *Appl Environ Microbiol*, **59**, 2914–2917.

Santos R, Bott T.R. (1992). Coated surfaces in relation to biofilm formation. In: Melo L.F., Bott T.R., Fletcher, M., Capedeville, B. (eds) *Biofilms – Science and Technology*. Kluwer Academic Publishers, Dordrecht, pp. 105–110.

Schott H., Young C.Y. (1972). Electrokinetic studies of bacteria. II. Effect of cetyltrimethylammonium bromide on *Streptococcus faecalis*. *J Pharm Sci*, **61**, 762–765.

Sears M.A., Gearhart D.J., Tittschof D. (1990). Antifouling agents from marine sponge *Lissodendoryx isodictyalis* Carter. *J Chem Ecol*, **16**, 791–799.

Shawky S., Emons H. (1998). Distribution pattern of organotin compounds at different trophic levels of aquatic ecosystems. *Chemosphere*, **36**, 523–535.

Silver S., Misra T.K. (1988). Plasmid-mediated heavy metal resistances. *Ann Rev Microbiol*, **42**, 717–743.

Skillman L.C., Sutherland I.W., Jones M.V. (1997). Co-operative biofilm formation between two species of Enterobacteriaceae. In: Wimpenny J.W., Handley P. Gilbert P., Lappin-Scott H.M., Hones M.V. (eds) *Biofilms Community Interactions and Control*. Bioline Publications, Chippenham, pp. 119–129.

Skillman L.C., Sutherland, I.W. Jones M.V. (1999). The role of exopolysaccharides in dual species biofilm development. *J Appl Microbiol*, **85**, 13S–18S.

Smith P., and Smith L. (1975). Organotin compounds and applications. *Chem Brit*, **11**, 208–213.

Speth T.F., Summers R.S., Gusses A.M. (1998). Nanofiltration foulants from a treated surface water. *Environ Sci Technol*, **32**, 3612–3617.

Srivastava R.B., Chongdar S., Karande A.A. (1995). Observations on two copper tolerant marine biofilm bacteria. *Mater Org (Ber)*, **29**, 311–318.

Stoodley P., Lewandowski Z., Boyle J.D., Lappin-Scott H.M. (1998). Oscillation characteristics of biofilm streamers in turbulent flowing water as related to drag and pressure drop. *Biotechnol Bioeng*, **57**, 536–544.

Stoodley P., Dodds I., Boyle J.D., Lappin-Scott H.M. (1999). Influence of hydrodynamics and nutrients on biofilm structure. *J Appl Microbiol*, **85**, 19S–28S.

Sundberg D.C., Vasishtha N., Zimmerman R.C., Smith C.M. (1997). Selection, design and delivery of environmentally benign antifouling agents. *Nav Res Rev*, **49**, 51–59.

Sutherland I.W. (1980). Polysaccharides in the adhesion of marine and freshwater bacteria. In: Berkeley R.C.W., Lynch J.M., Melling J., Rutter P.R., Vincent B. (eds) *Microbial Adhesion to Surfaces*. Ellis Horwood, London, pp. 329–338.

Sutherland I.W. (1982). Biosynthesis of microbial exopolysaccharides. *Adv Microb Physiol*, **23**, 79–150.

Sutherland, I.W. (1995). Biofilm-specific polysaccharides – do they exist? In: Wimpenny, J., Handley, P., Gilbert, P., Lappin-Scott, H. (eds) *The Life and Death of Biofilm*. Bioline Publications, Cardiff, pp. 103–106.

Swain G.W., Schultz M.P. (1996). The testing and evaluation of non-toxic antifouling coatings. *Biofouling*, **10**, 187–197.

Tang R.J., Cooney J.J. (1998). Effects of marine paints on microbial biofilm development on three materials. *J Ind Microbiol Biotechnol*, **20**, 275–280.

Thompson J.E. (1985). Exudation of biologically-active metabolites in the sponge *Aplysinia fistularis* 1. Biological evidence. *Mar Biol*, **88**, 23–26.

Thompson J.E., Walker R.P., Faulkner D.J. (1985). Screening and bioassays for biologically-active substances from forty marine sponge species from San Diego, California, U.S.A. *Mar Biol*, **88**, 11–21.

Todd J.S., Zimmerman R.C., Crews P., Alberte R.S. (1993). The antifouling activity of natural and synthetic phenolic acid sulphate esters. *Phytochemistry* 34, 401–404.

van der Wende E., Characklis W.G., Smith D.B. (1989). Biofilms and bacterial drinking water quality. *Nat Res*, **23**, 1313–1322.

van Losdrecht M.C.M., Eikelboom D., Gjaltema A., Mulder A., Tijhuis L., Heijnen J.J. (1995). Biofilm structures. *Water Sci Technol*, **32**, 35–43.

Vasishtha N., Sundberg D.C., Rittschof D. (1995). Evaluation of release rates and control of biofouling using monolithic coatings containing an isothiazolone. *Biofouling*, **9**, 1–16.

Volkering F., Breure A.M., van Andel J.G., Rulkens W.H. (1995). Influence of nonionic surfactants on bioavailability and biodegradation of polycyclic aromatic hydrocarbons. *Appl Environ Microbiol*, **61**, 356–361.

Vrolijk N.H., Targett N.M., Baier R.E., Meyer A.E. (1990). Surface characterization of two gorgonian coral species: implications for a natural antifouling defense. *Biofouling*, **2**, 39–54.

Wanner O. (1996). Modelling of biofilms. *Biofouling* 10, 31–41.

Wardell J.N, Brown C.M., Ellwood D.C. (1980). A continuous culture study of the attachment of bacteria to surfaces. In: Berkeley R.C.W., Lynch J.M., Melling J., Rutter P.R., Vincent B. (eds) *Microbial Adhesion to Surfaces*. Ellis Horwood, Chichester, pp. 221–230.

Weisman G.R., Sundberg D.C., Cimini R.A., Brown M.G., Beno B.R., Eighmy T.T. (1992). Controlled release antifouling coatings. I. Approaches for controlled release of 2,4-dinitrophenolate and benzoate into seawater. *Biofouling*, **6**, 123–146.

Williams I., Venables W.A., Lloyd D., Paul F., Critchley I. (1997). The effects of adherence to silicone surfaces on antibiotic susceptibilty in *Staphylococcus aureus*. *Microbiology*, **143**, 2407–2413.

Whitfield C., Valvano M.A. (1993). Biosynthesis and expression of cell-surface polysaccharides in Gram-negative bacteria. *Adv Microbiol Physiol*, **35**, 135–246.

Xu K.D., Stewart P.S., Xia F., Huang C.-T., McFeters G.A. (1998). Spatial physiological heterogeneity in *Pseudomonas aeruginosa* biofilm is determined by oxygen availability. *Appl Environ Microbiol*, **64**, 4035–4039.

Xu X., Stewart P.S., Chen X. (1996). Transport limitation of chlorine disinfection of *Pseudomonas aeruginosa* entrapped in alginate beads. *Biotechnol Bioeng*, **49**, 93–100.

Zinn M., Kirkegaard R.D., Palmer R.J., White D.C. (1999). Laminar flow chamber for continuous monitoring of biofilm formation and succession. *Methods Enzymol*, **310**, pp. 224–232.

Zips A.G., Schaule G., Flemming H.-C. (1990). Ultrasound as a means for detachment of biofilms. *Biofouling*, **2**, 323–336.

Zisman W.A. (1964). Relation of the equilibrium contact angle to liquid and solid constitution. In: Gould R.F. (ed) *Contact Angle, Wettability and Adhesion*. American Chemical Society, Washington D.C., pp. 1–51.

21 Towards Environmentally Acceptable Control of Biofilms in the Pulp and Paper Industry

J. Barry Wright

One of the largest industries in the world is the manufacture of different types of paper, from tissues to writing paper to heavy cardboard. This is also an industry that suffers, to a significant degree, from the development of biofilm on the process equipment. The most significant of the biofilm-related problems result in diminution of product quality and lost production time. In spite of the size and the economic importance of biofilm-related problems in this industry, relatively little understanding of these problems exists outside of the industry. To add to the difficulties experienced as a result of the growth of biofilm-forming microorganisms within paper mills, modern environmental and health concerns are driving significant changes to the manner in which biofilm control occurs in paper mills. Currently, the most common method is based on the use of significant quantities of biocides. However, as these compounds become increasingly heavily regulated and, in some cases, banned, novel means of control are being sought. Two of the approaches currently being investigated are the use of novel surfactants and enzymes. These approaches have shown varying degrees of success and applicability owing to the diversity of conditions within the mills. A review of the problems associated with paper mill biofilms and the methods currently employed to control them is presented along with a discussion of the uses and limitations of these methods.

INTRODUCTION

The production of paper, from tissue to heavy-weight cardboard, is one of the largest industries in the world. The estimated 3500 paper machines operating in North America and Europe have an annual production of about 160 million metric tonnes of product. Each production machine typically consumes 10 to 100 m^3 of water per tonne of paper produced (Väisänen *et al.*, 1994). Given the volume of water used and the operating conditions, paper mills are near perfect environments for the growth of microorganisms, resulting in significant problems in paper production.

In recent years, these problems have been exacerbated by the conversion of many papermaking systems from acidic to more neutral or alkaline process conditions (Gudlauski, 1996). As the majority of paper mills convert in this way, fungi have become less important in the mill environment, hence the focus on bacterial problems in this review. However, comments made specifically about bacteria also generally apply to fungal problems. The conversion to alkaline papermaking conditions has favoured the growth of bacteria in mills and increased the costs of their control (Farkas *et al.*, 1990; Sorrelle and Belgard, 1991). The conversion has been prompted by a number of concerns including, increased product opacity and whiteness, improved production rates, increased paper strength due to better bonding between fibres, and greater permanence associated with the end product (Camp, 1989). These factors enhance economic returns and generally outweigh the increased costs of microbial control. Nonetheless, the problems (most notably, bacterial fouling) cause considerable difficulties and have resulted in a continuing search for better methods to minimize the impact of microbial growth in mill process waters.

Environmental concerns assist in the creation of conditions that favour the growth of bacteria in paper mills. One such concern is the consumption of large quantities of timber, resulting in a concerted move towards increased recycling of paper. Sorrelle and Belgard (1991) estimated that such use of recycled fibre has resulted in the numbers of bacteria in papermaking process waters increasing by about 1000-fold. This is because of the recycled fibre used (*i.e.* post-consumer waste) and the storage of recycled paper bales under conditions that allow for microbial contamination and proliferation.

Related to the increased use of recycled fibre is an effort to reduce the water used in the papermaking process. In many machines, the 1000 m^3 of water used for each tonne of paper produced has been reduced 100-fold (Väisänen *et al.*, 1994). In addition, the increased use of recycled water results in an increase in the concentration of bacterial cells, as well as in the nutrients for their growth. Therefore, the combination of recycled fibre use, decreased water consumption, and the move towards alkaline paper making processes have contributed to the creation of conditions that favour the proliferation of bacterial cells, as planktonic populations and as biofilms. The present discussion will illuminate the problems associated with this, and document some of the newer methods being employed or investigated to minimize these problems.

PAPER MILL AREAS PRONE TO MICROBIAL PROBLEMS

There are a variety of problems that microorganisms may cause in paper mills, and these may be partially defined by the area of the mill where the problem is occuring as well as by whether or not the problematic organisms are predominantly sessile or planktonic. Biofilm will form, to some degree, on any site in the mill where water comes into contact with the surfaces of equipment. These problems may be examined by tracing the papermaking process (see Figure 1) and focussing on problems specific to particular areas/functions of the mill.

Plugging

Incoming pulp (thick stock) is mixed with process water (white water) and passes through a variety of screens and cleaners to remove debris, bark and large fibre bundles. The screens and centrifugal cleaners are among the first areas to be affected and biofilm is the major biological challenge. The biofilm may be formed *de novo* or it may be

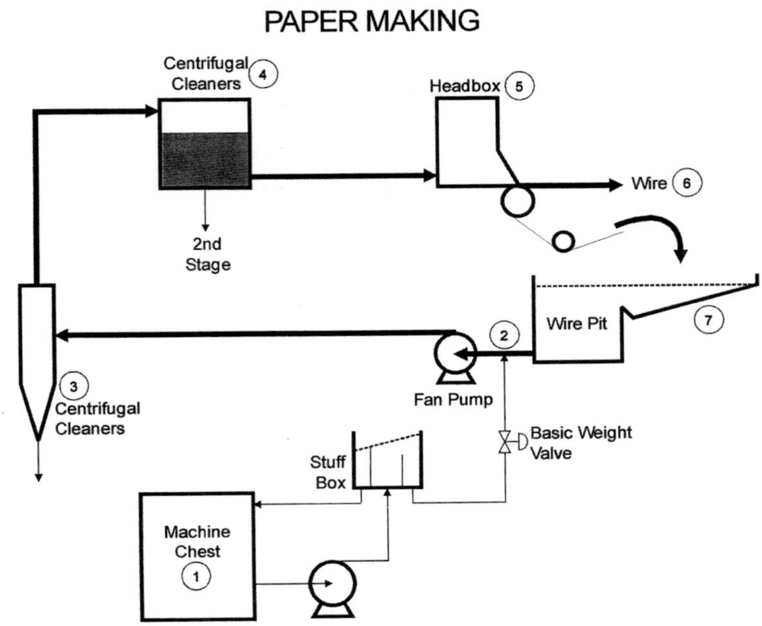

Figure 1 Representation of the steps involved in papermaking. Blended wood fibres are resident in the machine chest ① until needed for paper production. In many instances various chemical additives, including biocides, may be added into the process stream at this point. During paper making, pulp is transferred from the machine chest through the stuff box (if present) to the fan pump ② where it is mixed with process water (called "white water") to lower the consistency of the process stream. From the fan pump the stock (white water and pulp) passes through a series of centrifugal cleaners ③ and pressure screens ④ which remove large particulates (*e.g.* large fibre bundles, bark) from the process stream. The cleaned pulp is then piped to the headbox of the paper machine ⑤ where paper formation begins. Pulp in the headbox is spread across a moving belt called the "wire" ⑥ which supports the nascent paper. Paper formation occurs as the process water is removed from the nascent paper. The water is collected in a large reservoir or "wire pit" ⑦ under the paper machine for subsequent use or to be diverted to the sewer. The paper then travels through a drier to further reduce the moisture content.

"chunks" of biological matter sloughed from remote sites throughout the wet end of the paper machine. The former is probably the most problematic because "chunks" of biofilm carried in the process water tend to be removed along with other debris. When biofilm forms *de novo* on the screens and cleaners, they become occluded and unable to perform their functions effectively with a resultant reduction in device efficiency.

Corrosion

In most of the mill pipework, the most significant microbiological problem is microbially induced corrosion (Blanco *et al.*, 1996) associated with biofilm. In areas where biofilm accumulates on the walls of various supply lines, microbial activity may be sufficient to cause the formation of electrochemical cells. These form between areas of active metabolism (*i.e.* under a microcolony) and areas where there is little metabolic activity (*i.e.* a clean part of the pipe). Once electrochemical cells (Characklis and Cooksey, 1983; Cloete *et al.*, 1992) form, continued bacterial metabolism promotes corrosion of the metal substrate, the end result of which is that the pipes eventually permit water leakage. The corroded areas then need to be replaced and that typically entails a major overhaul and a resultant loss of production if the flow cannot be rerouted.

Storage Chest Fouling

Any of the chests where pulp (or white water) is stored are subject to biofouling problems. These tend not to be of serious concern with respect to functional aspects of the process. However, the development of biofilm in any area leads to the establishment of a "nursery" population that will foster rapid repopulation of a particular environment once conditions permit. In addition, in many storage chests the turnover of the contents may not be very rapid, or there may be poor mixing within the chest, resulting in the development of areas of depleted oxygen concentration where anaerobes may proliferate and cause potentially serious problems.

Anaerobic problems

The problems caused by anaerobes are typically two-fold. The most common of these is the production of volatile fatty acids, which impart to the paper a lingering, foul smell that may substantially reduce the desirability of the produced paper. Secondly, anaerobes have been cited as the cause of the production of explosive gasses in paper mills, and on occasion, explosion of these gasses has been identified as the cause of the destruction of parts of the mill, as well as killing or injuring mill workers (Sorrelle and Belgard, 1992).

Additive Spoilage

Throughout the process of paper production various additives are used to give the paper particular properties related, for example, to its final colour, opacity, brightness and wet strength. Many of the additives are organic in nature and are susceptible to microbial degradation, as is the wood pulp that may also be stored in chests. Two additives that are highly susceptible to microbial spoilage are starches and clays, which are typically received by the mill in very large quantities and stored in a variety of chests or silos. Since no area of the paper mill is sterile, it is necessary to preserve these materials. Many of the additives (*e.g.* clay), are shipped from the supplier already dosed with large amounts of biocides (preservatives) to control bacterial growth during transport and to help preserve the material upon arrival at the mill. Storage of these materials on site may allow for microbial spoilage to commence, causing a direct economic problem. In addition, once some of these materials are added to the process stream, they may become nutrient sources for both planktonic and sessile bacterial populations.

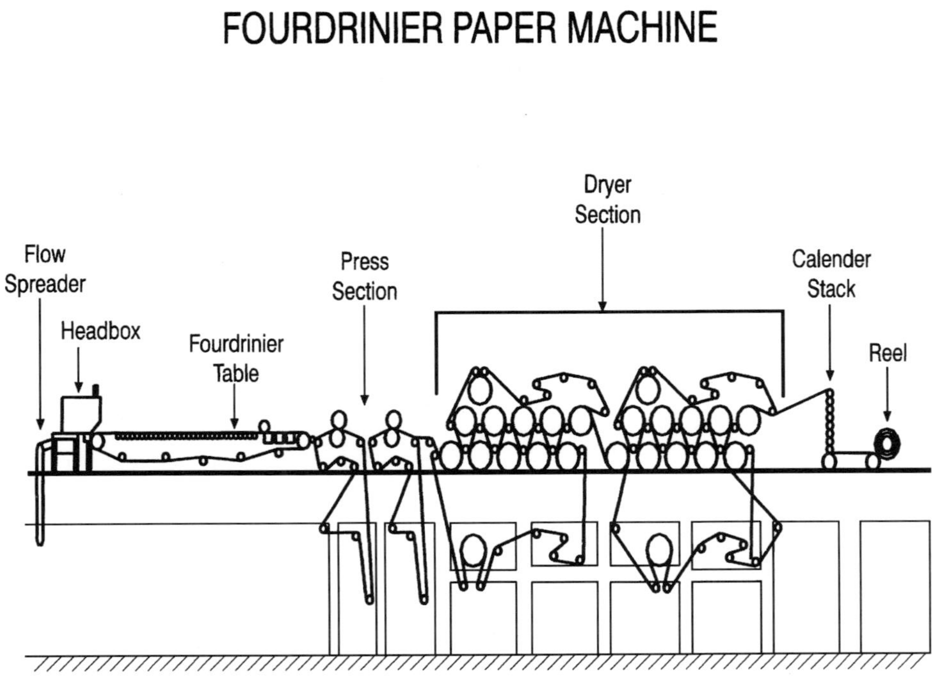

Figure 2 Sections of a Fourdrinier paper machine.

Machine Fouling

Once an examination of the paper formation area of the paper machine (see Figure 2) is undertaken, the problems of microbial biofilm development become apparent. The headbox is the area where incoming pulp is evenly distributed across the width of the "wire" upon which paper formation takes place, and this is an area that is particularly sensitive to the growth of biofilm. The interior walls of the headbox may easily be seen (or felt) to develop significant amounts of slime in a relatively short time, depending upon the microbial control programme in place in the mill. Similarly, the headbox lip (the area where the pulp leaves the headbox and is accepted by the wire) is also very prone to rapid development of microbial slime.

Headbox

The headbox and the headbox lip are perceived to be the sites of the most damaging build-up of biofilm, which eventually reaches a critical size at which a portion is sloughed from the surface. The sloughed material may be transported throughout the white water system and serve as an inoculum for a variety of chests as well as providing boli of cells, enmeshed in exopolysaccharide, that pose potential problems for the screens and centrifugal cleaners. Additionally, the sloughed biomass may also become enveloped in the nascent sheet, causing more immediate problems, discussed in detail in a subsequent section, recognized as the primary problems associated with the formation of biofilm on paper machines.

Paper machine table

Another area that is very prone to the development of slime is the table of the paper machine. Slime development here occurs on the foil boxes, suction boxes and table supports over which the wire, carrying the nascent sheet, passes and water is removed from the sheet. These areas receive a continuous bathing in process water that is relatively free of large paper fibers and that is somewhat cooler than the majority of the system as well as being rich in nutrients. The water, as it bathes these areas, deposits microorganisms on the surfaces of the table structures, and this results in the rapid and abundant growth of biofilm. This is particularly true since the scrubbing action of paper fines flowing past the surface in high velocity water is considerably reduced in these areas compared to conditions in the piping that delivers the stock to the headbox. Similar statements can be made for surfaces in the wire pit (where water removed from the nascent paper sheet is collected), the save all (a device that recovers paper fibres from the wire pit process water) and the process water storage chests. All these areas experience conditions similar to those on the table and are, likewise, very prone to the development of extensive biofilm "deposits". The problems caused by the formation of biofilm in these areas are similar to those caused by biofilm formed on various parts of the table and headbox and are discussed in a subsequent section.

Spray jets

The various spray jets used in the wet end of the paper machine are also adversely affected by the formation of biofilm. Some jets are positioned above the table of the paper machine and are used to control the rate of drying of the nascent sheet as it passes along the table on the wire. Trim jets are also positioned at the edges above the wire and produce a forceful stream of water which trims the edges of the nascent sheet to ensure a uniform width as the sheet completes formation and enters the dryer section of the paper machine. Additionally, spray nozzles may be placed above open white water silos and these spray water to disperse foam caused by large amounts of entrained air in the white water. The different spray jets may experience a substantial build up of slime if they are fed by the clear white water from the save-all, this water being particularly supportive of biofilm formation. Alternatively, depending upon the design of the machine, some of the spray jets may have fresh water as their water source, frequently surface water from either a lake or river. If the water is not adequately treated, contamination with freshwater organisms may cause a severe problem due to the presence of large numbers of filamentous organisms. Fouling caused by filamentous freshwater organisms, *e.g. Beggiatoa* spp. or *Sphaerotilus natans*, characteristically results in the formation of long "stringers" of bacterial cells that may extend a substantial distance. Biofilm in these areas forms as a pendulous mass, eventually reaching a critical length, sloughing from the jets, and potentially being incorporated into the paper sheet. Alternatively, if the biofilm mass accumulates inside the spray nozzles, plugging of the spray jets may occur.

Misted Areas

Other areas of the paper machine may be within the mist generated by the spray jets (or by the wire movement), and may also develop biofilm in significant amounts over time. One of the problems associated with slime formation in these areas is the difficulty of ensuring that they receive adequate amounts of biocide to control the growth of the biofilm population. Even though the areas may be getting well misted, they do not necessarily experience more than a very high relative humidity on a continuous basis, so that water droplets containing biocides may not reach them. Alternatively, if biocide-containing water reaches these areas, the biofilm may repel the majority of the liquid. In some of these areas, the biofilm may slough to some extent and cause problems similar to those described for the table of the machine. However, biofilm in these areas may also cause significant safety hazards particularly on walkways and stairs. In addition, there is often concern about the poor aesthetics associated with significant amounts of slime on structures within the mill.

PRODUCT QUALITY PROBLEMS CAUSED BY BIOFILM

Biofilm in the paper mill causes a variety of different problems depending upon the area/structure of the mill in/on which the biofilm occurs. A number of the problems associated with parts of the mill were described previously. However, the most severe biofilm-associated problems occur in the wet end of the machine as a result of slime build-up on various parts of the headbox, machine table, wire pit, spray nozzles, etc. When biofilm forms in one or more of these areas and eventually sloughs off, it may eventually end up being incorporated into the nascent sheet (Glazer, 1991). This may cause a variety of problems for the mill, all of which are very costly.

Holes/Spots

The most common result of the incorporation of a biofilm mass in the sheet is the formation of holes and spots in the paper (Camp, 1989; Glazer, 1991). When small amounts of biofilm are incorporated, particularly when accompanied by some inorganic and/or organic material (not of microbial origin), they may show up as coloured flecks in the paper. Holes are caused by dehydration of larger incorporated slime masses. The extracellular polymer that surrounds bacterial cells is 99% water (Sutherland, 1977), and assuming that the bacterial cell mass is also approximately 90% water, it is possible to understand how a hole could form as the sheet dries. The biofilm mass takes up space in the sheet as it is forming on the wire. However, as the paper enters the dryer, the majority of the water content of the sheet and the biofilm mass is removed. Whereas the paper sheet is composed mainly of fibres that have definite structure and volume, the microbiological mass does not. Therefore, as the biofilm mass dehydrates, it contracts with the resultant formation of a hole in the sheet. Holes (or spots) decrease the quality of the finished product, and lower its monetary value.

This brief discussion of holes and spots has focused only on the potential microbial origin of these problems, excluding other potential causes of sheet defects. In reality, the cause of specific defects in a paper sheet requires a well-trained eye and, frequently, a series of chemical tests to conclusively determine.

Sheet Breaks

When a larger amount of biofilm becomes incorporated into a sheet, a localized weakness in the forming paper web may develop. This may be manifest by the sheet ripping (a sheet break) as it enters the dryer section of the machine. An occasional break is not normally a severe problem and may be caused by problems other than incorporation of slime. However, when the nascent sheet breaks, a large amount of production is lost because, rather than feeding through the dryer section, it ends up in the "broke" chest to be recycled. Although machine operators are very skilled at "re-threading" the paper machine rapidly, numerous occurrences waste a significant amount of production. Again, other process variables may also result in breaks in the paper web as it enters the dryer section and cause similar problems.

CHEMICAL CONDITIONS EXTANT IN PAPER MILLS

In order to fully appreciate the complexity of the paper mill environment in which biofilm develops a rudimentary understanding of the chemical environment extant in the mill is necessary. Paper mill process waters vary greatly depending on the location of the mill, its product, and its proprietary operating conditions, even after minimizing the range of observed conditions by examining only one mill type (*i.e.* alkaline papermaking processes). The conditions extant in the process stream vary significantly between mills and influence the type(s) of biofilm control programmes that may be instituted. Only broad generalizations of the process conditions may be made. For example, pH can vary broadly in paper mills of different types; the normal pH in mills ranges between approximately 3.5 and 8.5–9.0. However, the pH variance in alkaline mills is somewhat less broad and a realistic pH range for mills of this type is 6.5–8.5.

The temperatures at which various mills operate also vary widely, from approximately 25–30°C to 60°C. Alkaline fine paper mills typically operate with white water temperatures between 35–50°C. Thus the observed conditions of temperature and pH for most mills nearly optimal for survival and proliferation of bacteria.

Generalizations with respect to the chemical composition of the process stream are difficult to make. There is a wide variation in the types of water conditions in various mills, to some extent related to the types of products being produced. However, even within a given genre of mill there can be very significant differences in the overall water chemistry, related to a variety of potential factors including the location of the mill. Location influences a mill's water source and that may greatly affect a number of its process water properties, including water hardness. Another factor that may play affect the overall water chemistry is the extent of water recycling which occurs. A mill that recycles large amounts of water may experience a significant change in water

Table 1 Water chemistries from 21 alkaline fine paper mills.

Component/Parameter	*Minimum value*	*Maximum value*
Cl, ppm	34	546
SO_4, ppm	8.4	1530
Al, total, ppm	0.2	8.8
Ba, total, ppm	0.0	0.2
Ca, total, ppm	54.0	986
Cu, total, ppm	<0.05	0.01
Hardness, total, ppm	78.4	1090
Fe, total, ppm	<0.05	0.62
Mg, total, ppm	6.2	249
Mn, total, ppm	0.01	1.29
P, total, ppm	<0.4	68.3
Na, ppm	0.2	538
Conductivity, umhos	340	2820
pH	6.0	7.7
SiO_2, total, ppm	3.9	36.8

conditions from the beginning of a cycle to the end. This change is generally modulated and minimized by alterations in chemical feed rates. Table 1 demonstrates the differences in water chemistries from 21 mills producing similar products from the United States, Canada and Europe.

From this discussion it is obvious that it is difficult to specify the conditions under which an agent added to the process stream must function. As a result, one or two conditions must apply to any product that may be added to control bacterial proliferation and biofilm formation, *viz.* it must be capable of functioning over a wide range of possible water conditions, or the conditions under which the product functions must be well-defined and a range of products that cover as broad of range of potential conditions as possible must be available.

TRADITIONAL BIOFILM CONTROL

Boil-outs

The development of biofilm on various parts of a paper machine can have a number of far-reaching effects, as described earlier. Once a significant amount of slime is present on the machine and it has begun to slough off and cause operational problems, the most cost-effective action may be to interrupt production and perform a process referred to as a boil-out. This involves the use of a highly alkaline, high temperature flush of the paper machine and is effective in removing various biological and chemical deposits. The cost of the chemicals used to conduct boil-outs is relatively low. However, even the substantial costs associated with lost production and operator productivity (ranging into tens of thousands of dollars per occurrence) during the boil-out, are frequently less than those associated with an on-going loss of production or diminution in the quality, and therefore price, of the finished product.

In order to minimize the frequency of boil-outs and to maximize the quality of the finished product, various types of biofilm control programmes are instituted. Generally, the better the control programme in place, the longer the period of time between boil-outs. In the majority of paper mills, the main concern, microbiologically, (and most often measured parameter) is the level of bacterial contamination in the white water and various stock chests. Although this is usually an indicator of the efficacy of the current biocide programme, it can be very misleading. In particular, planktonic bacterial counts can be especially misleading when these are relied upon to determine the frequency and dosing of the biocide programme. There is an abundance of research demonstrating that sessile populations are significantly more resistant to biocides than their planktonic counterparts (Ruseska *et al.*, 1982; Costerton and Lashen, 1984; Wright *et al.*, 1991). The reasons behind this difference (see, for example, Costerton *et al.*, 1985; Evans *et al.*, 1991; Gilbert and Brown, 1995) are still under investigation and are beyond the scope of the present discussion. If, as previously alluded to, biocides are used to control the planktonic population without controlling the sessile population, the latter may serve as a "nursery" for bacterial cells, able to rapidly repopulate the system as soon as conditions become favourable for growth. Regardless of the level of the planktonic counts, they may not be indicative of the efficacy of the biocide programme against the sessile population. The less efficient the biocide programme is against biofilm organisms, the more regularly boil-outs are required.

In addition to the aforementioned factors, other concerns may influence the frequency of boil-outs. One of these is a machine that is not operating under ideal chemical conditions. As a result of unbalanced water chemistries within the process stream, chemical deposition (*e.g.* chemical slime) may also build up on the machine. Although this type of problem will not be affected by a microbiological control programme, a variety of observations indicate that biological slime can exacerbate the problem of chemical, and fines, deposition on the machine (Glazer, 1991; Marmo *et al.*, 1991; Wright, 1997). In mills operating within normal chemical parameters, the deposition of slime appears to cause accumulation of chemical deposits on the paper machine, due, in large measure, to the charged nature of the exopolymer that surrounds bacterial cells (Luft, 1971; Fassel *et al.*, 1992). The exopolymer binds a variety of organic and inorganic moieties and serves as a nidus for the formation of multi-component deposits. Evidence for the combined role of microbes and non-microbial organic and inorganic materials in the formation of some types of deposits was provided by Väisänen *et al.* (1994).

Biocide Programmes

Water treatment companies must supply products that function over a broad range of conditions or a range of products with well-defined ranges of optimal performance. The first of these criteria describes how biocides are usually marketed, *i.e.* they normally function under a broad range of conditions. A variety of concerns, other than water chemistry, influence the choice of biocide in a particular application. These include the type of organism(s) to be controlled, cost-constraints of the mill, regulatory issues surrounding the type of chemicals that can be used in the mill or its product(s), and the range of different products available from the water treatment company. A number of different types of biocides exist in water treatment companies' product lines. Broadly speaking, these can be divided into two groups; *viz.* oxidizing and non-oxidizing biocides.

Oxidizing biocides

Oxidizing biocides function by degrading the cell walls of bacteria and eventually disrupting the cells' metabolic processes. Among these agents are the hypohalous acids, chlorine dioxide, and the peracids. Oxidizing biocides are very effective, acting rapidly at relatively low concentrations (Alasri *et al.*, 1992; Bueckner and Post, 1998). Another feature is that they are relatively inexpensive and some can be generated in-house from various mill processes, particularly if the plant is bleaching its own wood fibres. However, these agents also have significant drawbacks. Many are dangerous to store and handle and therefore pose significant safety risks. In addition, their use in the mill is hampered by rapid reduction in their activity caused by the large quantities of organic materials present in the process stream (Fiessinger *et al.*, 1981). Regular or heavy use of many of these agents also causes corrosion of the metal piping in the mill (Bueckner and Post, 1998). Although many are, themselves, relatively environmentally benign, the use of some of these materials has been heavily criticised because of their propensity to form halogenated organic molecules (*e.g.* dioxins) that are carcinogenic or environmentally unacceptable for other reasons (Travis and Hattemer-Frey, 1991; Mukerjee, 1998).

Non-oxidizing biocides

Non-oxidizing biocides are a diverse group that are very common in microbiological control programmes in paper mills. A number of different compounds are utilized as biocides, including bromonitropropanediol, carbamates, chlorosulfone, dibromonitrilopropionamide, dodecylguanidine hydrochloride, glutaraldehyde, isothiazolinone, and methylene bisthiocyanate. Most of these are believed to be metabolic inhibitors. Water treatment companies that supply these agents do not typically own patents on the microbicidal compounds in their products. However, most of the companies hold patents on combinations of these agents that are reported to act synergistically in the control of microbial populations. Alternatively, the companies may hold patents on combinations of the biocidal agents with other agents that increase their solubility, effectiveness, or ease of handling, *i.e.* features that increase the value of the product. Therefore, water treatment companies conduct research to discover better ways of delivering a relatively limited number of microbicidally active ingredients. This has led to the large number of products available from the different suppliers along with associated claims as to the efficacy of a given product for a particular type of problem.

Surface active biocides

A relatively small number of surfactants are utilized as biocides in the pulp and paper industry. Primarily, these are quaternary ammonium chlorides or "quats." At the present time, the most popular of these cationic surfactants are the alkyl dimethylbenzyl ammonium chlorides (Buecker and Post, 1998), which are regularly blended in with other biocides to yield a synergistic product. Their mode of action is believed to be a combination of their membrane disrupting properties (Wright and Gilbert, 1987) and a surface cleaning (detergency) functionality attributed to their surface-active nature (Davis, 1990). Therefore, these products are often marketed as agents with enhanced surface-cleaning abilities, in addition to their microbicidal activities.

Microbial Resistance to Biocides

In bacteria, an ability to acquire resistance to a variety of microbicides is common, and the acquisition of antibiotic resistance is becoming a major challenge for physicians and drug companies around the world (Mulligan *et al.*, 1993; Gaynes, 1997; Strausbaugh, 1997). However, the paper industry has not experienced a problem with microbial populations developing resistance to biocides and paper mills appear to be able to utilize the same biocides for years at a time without having to change due to the development of a biocide-resistant mill flora.

Concerns Surrounding the Use of Biocides

In spite of calls for reduced packaging and moves towards "paperless" offices, the demand for paper and paper products remains strong, and a variety of new problems are emerging for paper companies as they compete in today's market. These centre on environmental concerns of specific groups and, increasingly, the general population. The increasing environmental awareness of the general populace is resulting in an increasing number of regulations being legislated that require major changes in the treatment of paper process streams (Hanley, 1990). As a result, biocidal agents, as well as many of their adjuvants and solvents, are coming under increased scrutiny. In some cases, this scrutiny is resulting in the banning of their use in paper mills either by direct legislation or by the requirement for onerous use documentation that makes their use cost-prohibitive. This results in a very strong demand for alternative methods of combating microbial problems, particularly fouling, in paper mills.

NOVEL APPROACHES TO SLIME CONTROL

Generally, satisfaction with biocides, in terms of their ability to enhance machine runnability, is high, largely because of their cost-effectiveness. Many industrial biocides are highly cost-effective, assuming they are applied correctly. This, combined with a cyclical demand for paper that generally results in the relatively low price of the end product, makes these agents very attractive to the producer and the associated financial community. However, as discussed above, legislation and environmental pressures are forcing mills to adopt strategies to minimize their reliance on chemical biocides. These pressures combined with a comparative dearth of new agents coming to the market are resulting in companies looking for novel methods to control slime.

Surface Attachment Inhibitors

One of the ways of controlling biofilm formation on industrial surfaces is the utilization of methods that prevent it from forming. Biofilm typically forms after the deposition of a conditioning layer on the surface. This is followed by bacterial cell deposition and further recruitment of additional cells as well as multiplication of the initially adhering cells (Marshall *et al.*, 1971; van Hoogmoed *et al.*, 2000). Once bacteria are attached to a surface, they promote the attachment of additional cells by elaborating an exopolymer that is more conducive to colonization than the original surface. Dead bacteria that still possess their exopolymer, as well as exopolymer left adherent to a surface after the bacterial cells have been removed remain effective in promoting further colonization of the surface (Nickels *et al.*, 1981; Costerton *et al.*, 1995). Therefore, if bacterial cells could be prevented from attaching to a surface, the difficulties associated with biofilm formation could potentially be eliminated.

Cationic polymers

A variety of methods have been investigated to provide a mechanism whereby initial biofilm formation can be inhibited. One of these involves the use of particular cationic polymers (Jaquess and Hollis, 1991) that can be added to industrial process streams. These are generally described as ionene polymers (*e.g.* tetramethyl ethanediamine polymer with oxybischloroethane) and are claimed to be effective at inhibiting slime formation at concentrations below which they are bactericidal (Jaquess and Hollis, 1991). These authors demonstrated that such polymers were effective at inhibiting the formation of biofilms derived from a biculture of *Klebsiella oxytoca* and *Pseudomonas aeruginosa*. However, the lack of published reports regarding the *in situ* performance of these molecules prevents any discussion of their relative efficacy in operating paper mills.

A variety of polymers and surfactants have been utilized in an attempt to control biofilm formation in the laboratory with a defined bacterial culture. However, it is much more difficult to control a more diverse population of bacteria in situations that mimic the "natural" environment. Biofilm formation appears to be such a preferred mode of growth (Davies *et al.*, 1998) that even when one species of bacteria is prevented from attaching to a given surface, another species in the population may colonize even if it is not normally the initial colonizer of a system. This propensity towards biofilm formation and the ability of multiple species to serve as potential initial colonizers increases the difficulty in preventing biofilm formation in many industrial settings.

Surfactant-based slime control

A novel method of controlling slime formation in mills, recently commercialized, involves the use of a surfactant system that appears to effectively inhibit bacterial attachment to surfaces (Wright, 1997). A series of laboratory and mill experiments demonstrated that this proprietary sulfosuccinate-based substance (Wright and Michalopoulos, 1997) has functions not normally associated with surfactants. Surfactants are generally thought to alter interfacial tension and change surface wettability,

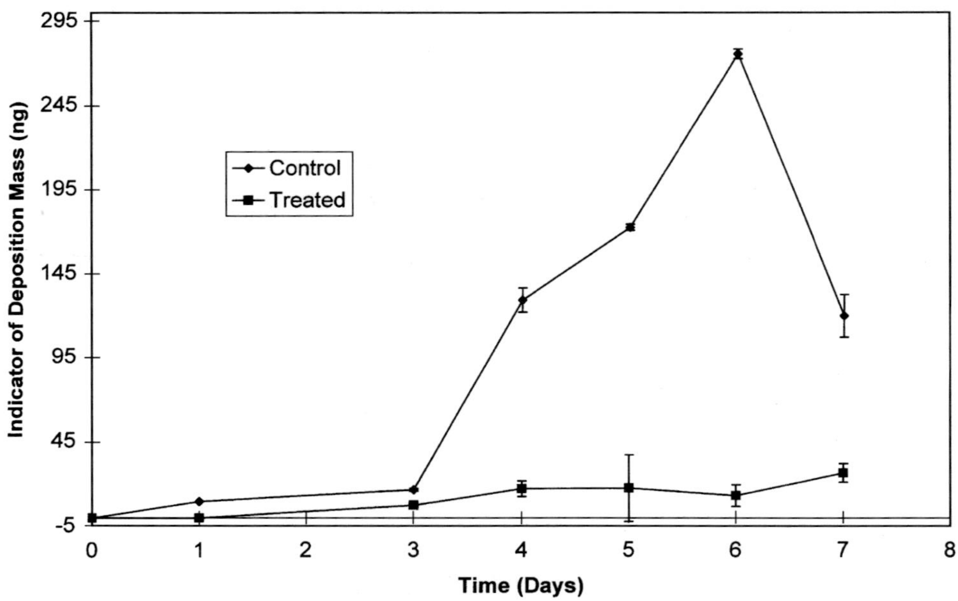

Figure 3 Biological deposition on a paper machine at an American alkaline tissue mill when treated with a sulfosuccinate-based antifouling product (◆) compared to deposition on the machine with a conventional (carbamate-based) biocide programme in place (■). n = 4 (Reproduced from Wright, 1997, with permission).

thereby affecting the ability of bacteria to adhere (Blainey and Marshall, 1991; Quirynen *et al.*, 1994; Busscher *et al.*, 1997; van Hoogmoed *et al.*, 1999). Although some evidence for this mechanism exists, evidence of its success in a complex environment remains inconclusive. The activity of the sulfosuccinate-based product seems to be different in that its primary mode of action does not appear to be the result of an alteration of surface energies (Wright, 1997).

Figure 3 shows an example of the efficacy of the sulfosuccinate-based surfactant in preventing microbial deposition on test surfaces placed in the process water system of an American alkaline tissue machine. Bacterial biomass was measured using a probe containing removable sampling coupons that allowed for the formation of biofilm that was representative of that observed on surfaces throughout the wet end of the paper machine. Addition of the surfactant to the process stream, resulted in surfaces that remained essentially devoid of bacterial and non-bacterial fouling over a 3-week period in this and other mills (results shown are from the first 7 d). The control observations (Figure 3), were made as the machine produced a similar grade of paper (over a similar 7-day period) with a conventional (carbamate-based) biocide programme in place. Each run was started immediately following a boil-out, when the system was as clean as possible. (The sharp decline in the amount of mass attached to the coupons at the final control sampling point was an artefact caused by the loss of a significant portion of the deposit from one coupon during sampling).

In addition to demonstrating good efficacy, the sulfosuccinate-based product is reasonably biodegradable (studies with a very similar molecule indicated that biodegradation of the sulfosuccinate was 94% (Hales, 1993)), lacks significant aquatic toxicity, and is approved for use in paper with direct aqueous and fatty food contact (Wright, 1997).

Enzymes

Enzymes have been considered as valuable slime control agents in the pulp and paper industry in large part because of their attractive environmental and safety profiles (Moor and Hach, 1984). Enzymes are generally regarded as being safe to utilize in a process stream, and their environmental profile is favourable, particularly since they are biodegradable and have low environmental toxicity. In general, enzymes are also safe to handle, although some irritation may result from contact with them, and some people may develop sensitivities to them.

Since enzymes have been seen as desirable slime control products in the pulp and paper industry, as well as in other industries that utilize large quantities of process waters, a significant number have been patented for use as slime control agents. A number of these have been single enzyme products designed to attack a specific component of the bacterial biofilm, for example the bacterial exopolymer. Owing to its role in binding bacterial cells together and to various substrates, this has been regarded as a convenient and relatively well-defined structure against which to examine enzyme activity. Since *Pseudomonas* spp. are common problematic organisms in paper process streams (Väätänen and Niemelä, 1983; Blanco *et al.*, 1996), the alginate elaborated by these bacteria has been targeted for enzymatic control. LaMarre (1990) demonstrated that the use of a crude alginase preparation was effective in minimizing the amount of bacterial slime produced in a laboratory test apparatus over a 4-d period, by preventing the attachment of bacteria and promoting the removal of any attached cells.

The complexity of most industrial process streams, particularly in paper mills requires that a more robust product be developed that is effective against a much larger proportion of potential colonizers. Therefore, additional products have been developed that combine the effective properties of a number of enzymes (*e.g.* Wiatr, 1990). However, given the potential variety of organisms that may exist in a paper mill, the most common target for enzymes remains the bacterial exopolymer.

One of the main problems with the use of an enzyme-based product is the wide range of (sometimes inhospitable) conditions under which it would be required to function. These include wide fluctuations in the temperature, pH and chemical composition of the process water. In addition, it is necessary to determine what type(s) of enzyme(s) should be incorporated into a product, which has to be effective in a wide range of environmental conditions but must not cause other process-related problems. For example, a protease would need to be effective against the bacterial biofilm but not auto-degrade during storage and not interfere with the casein that may be a process additive in some mills; a carbohydrase would need to be effective against a range of exopolymers produced by a variety of bacteria, but not hydrolyze starches and wood fibres.

The cost-effectiveness of an enzyme-based slime control protocol also needs to be taken into consideration, although this is not unique to enzyme products. Although formulating and providing an industrial grade enzyme is frequently more expensive than working with bulk synthetic chemicals, this problem could be overcome by demonstrating a highly effective product produced by an efficient methodology. In addition, a small premium in price could be expected for a product that functions well and reliably and is environmentally sensitive. Associated with the cost of the enzyme is the cost of transporting it. Most mills prefer not to work with solid products that need to be fed into the process stream because the feeding method is typically more costly, difficult to accomplish, and more difficult to automate. Alternatively, enzymes may be distributed as solutions but the ease-of-use benefit would be offset, at least to some degree, by the added cost of shipping the solvent.

One of the greatest benefits of enzymes, their biodegradability, is also a detriment with respect to storage and shipping. An enzyme solution could be subject to bacterial contamination and spoilage and although this could be overcome by the addition of a biocide, that would detract from the purpose the enzyme-based product. It is also increasingly difficult to supply spoilage-prone products to mills because of concerns over sensitization reactions to isothiazolinones (Toren *et al.*, 1997), one of the most widely used preservatives. Mills are also cautious about purchasing products that claim to be environmentally acceptable and are simply "preserved" by the addition of a biocide, since a number of these products have been demonstrated to be only as effective as the biocide added to "preserve" the so-called active ingredient.

CONCLUSION

One of the most difficult balances that industries, such as pulp and paper, face is that to be cost-competitive, the fouling control methodologies they adopt must be affordable. Therefore, as new technologies are developed they must be designed to be cost-competitive. An increased understanding of the problems faced by industries that treat millions of litres of water on an annual basis, combined with an improved ability to monitor biofouling in these systems, will further efforts to find an economically and environmentally sound method for dealing with the problems of biofouling in these industries. Certainly, steps have been taken that are pointing the industry in the right direction. However, research is certain to find better mechanisms to accomplish these goals. Other technologies lie on the horizon that suggest that novel methods will be found that will be more effective, and eventually more economical than the ones presented. One of these areas of investigation will doubtlessly encompass a better understanding (and exploitation) of the mechanisms utilized by various organisms to prevent (or encourage) fouling (see for example Clare, 1996). Examples of antifouling mechanisms employed by organisms include chemicals produced in by sea squirts (Wahl *et al.*, 1994) and seaweeds (Steinberg *et al.*, 1997) that inhibit microbial fouling of their surfaces. In addition, a better understanding of the molecular mechanisms of fouling, particularly related to quorum sensing, suggests that products that capitalize on these understandings will be forthcoming (Davies *et al.*, 1998). Thus, it appears that the progress made to date in finding an environmentally benign method of controlling fouling may eventually be surpassed by emerging technologies.

REFERENCES

Alasri A., Roques C., Michel G., Cabassud C., Aptel P. (1992). Bactericidal properties of peracetic acid and hydrogen peroxide, alone and in combination, and chlorine and formaldehyde against bacterial water strains. *Can J Microbiol*, **38**, 635–642.

Blainey B.L., Marshall K.C. (1991). The use of block copolymers to inhibit bacterial adhesion and biofilm formation on hydrophobic surfaces in marine habitats. *Biofouling*, **4**, 309–318.

Blanco M.A., Negro C., Gaspar I., Tijero J. (1996). Slime problems in the paper and board industry. *Appl Microbiol Biotechnol*, **46**, 203–208.

Bueckner B., Post R. (1998). Control of biofouling in evaporative cooling systems. *Chem Eng Prog*, Sept, 45–50.

Busscher H.J., van Hoogmoed C.G., Geertsema-Doornbusch G.I., van der Kuijl-Booij M., van der Mei H.C. (1997). *Streptococcus thermophilus* and its biosurfactants inhibit adhesion by *Candida* spp. on silicone rubber. *Appl Environ Microbiol*, **63**, 3810–3817.

Camp V. (1989). Microbiology in alkaline fine paper machine systems and the control of slime and deposits using pretreated sewage water as a fresh water source. *Paper S Afr*, **Nov/Dec**, 12–17.

Characklis W.G., Cooksey K.E. (1983). Biofilms and microbial fouling. *Adv Appl Microbiol*, **29**, 93–138.

Clare A.S. (1996). Marine natural product antifoulants: status and potential. *Biofouling*, **9**, 211–229.

Cloete T.E., Brözel V.S., Von Holy A. (1992). Practical aspects of biofouling control in industrial water systems. *Int Biodeterior Biodegr*, **29**, 299–341.

Costerton J.W., Lashen E.S. (1984). Influence of biofilm on efficacy of biocides on corrosion-causing bacteria. *Mater Perf*, **23**, 34–37.

Costerton J.W., Marrie T.J., Cheng K.-J. (1985). Phenomena of bacterial adhesion. In: Savage D.C., Fletcher M. (eds) *Bacterial Adhesion: Mechanisms and Physiological Significance*. Plenum Press, New York, pp. 1–43.

Costerton J.W., Lewandowski Z., Caldwell D.E., Korber D.R., Lappin-Scott H.M. (1995). Microbial biofilms. *Ann Rev Microbiol*, **49**, 711–745.

Davis B. (1990). Surfactant-biocide interactions. In: Porter M.R. (ed) *Recent Developments in the Technology of Surfactants*. Elsevier, New York, pp. 65–131.

Davies D.G., Parsek M.R., Pearson J.P., Iglewski B.H., Costerton J.W., Greenberg E.P. (1998). The involvement of cell-to-cell signals in the development of a bacterial biofilm. *Science*, **280**, 295–298.

Evans D.J., Allison D.G., Brown M.R.W., Gilbert P. (1991). Susceptibility of *Pseudomonas aeruginosa* and *Escherichia coli* biofilms towards ciprofloxacin: Effect of specific growth rate. *J Antimicrob Chemother*, **27**, 177–184.

Farkas J.P., Johns K.A., Maszkiewicz A.C., Ormeron D.L. (1990). Alkaline papermaking and biological control. *PIMA Mag*, **Jul**, 24–26.

Fassel T.A., van Over J.E., Hauser C.C., Buchholz L.E., Edmiston C.E., Sanger J.R., Remsen C.C. (1992). Evaluation of bacterial glycocalyx preservation and staining by ruthenium red, ruthenium red-lysine, and alcian blue for several methanotroph and staphylococcal species. *Cells Mater*, 2, 37–48.

Fiessinger F., Richard Y., Montiel A., Musquere P. (1981). Advantages and disadvantages of chemical oxidation and disinfection by ozone and chlorine oxidation. *Sci Total Environ*, **18**, 245–261.

Gaynes R. (1997). The impact of antimicrobial use on the emergence of antimicrobial-resistant bacteria in hospitals. *Infect Dis Clin N Am*, **11**, 757–765.

Gilbert P., Brown M.R.W. (1995). Mechanisms of the protection of bacterial biofilms from antimicrobial agents. *Plant Microb Biotechnol Res*, **5**, 118–130.

Glazer J.A. (1991). Overview of deposit control. *TAPPI J*, **74**, 72–77.

Gudlauski D.G. (1996). Whitewater system closure means managing microbiological buildup. *Pulp Paper*, **70**, 161–165.

Hales S.G. (1993) Biodegradation of the anionic surfactant dialkyl sulphosuccinate. *Environ Toxicol Chem*, **12**, 1821–1828.

Hanley R.W. (1990). Setting the scene for environmental compliance in the last decade of the twentieth century. *TAPPI J*, **73**,155–176.

Jaquess P.A., Hollis C.G. (1991). Process for inhibiting bacterial adhesion and controlling fouling in aqueous systems. Australian Patent AU-A-63940/90, issued April 18, 1991.

LaMarre T.M. (1990). Slime control in industrial waters using enzymes. Canadian Patent 1,274,442, issued September 25, 1990.

Luft, J.H. (1971). Ruthenium red and violet. I. Chemistry, purification, methods of use for electron microscopy and mechanism of action. *Anat Rec*, **171**, 347–368.

Marmo S.A., Nurmiaho-Lassila E.L., Varjonen O., Salkinoja-Salonen M.S. (1991) Biofouling and microbially-induced corrosion on paper machines. In: Dowling N.J., Danko J.C. (eds) *Microbially Influenced Corrosion and Biodeterioration*. University of Tennessee Press, Knoxville, pp. 433–449.

Marshall, K.C., Stout R., Mitchell R. (1971). Mechanism of the initial events in the sorption of marine bacteria to surfaces. *J Gen Microbiol*, **68**, 337–348.

Moor A.H., Hach M.J. (1984). Enzymes to control microbiological deposits. In: *PIRA Conf, Slime and its Control*. Pira International, London, SPB/3.

Mukerjee D. (1998). Health impact of polychlorinated dibenzo-p-dioxins: a critical review. *J Air Waste Manag Assoc*, **48**, 157–165.

Mulligan M.E., Murray-Leisure K.A., Ribner B.S., Standiford H.C., John J.F., Korvick J.A., Kaufmann C.A., Yu V.L. (1993). Methicillin-resistant *Staphylococcus aureus*: a consensus review of the microbiology, pathogenesis, and epidemiology with implications for prevention and management. *Am J Med*, **94**, 313–328.

Nickels J.S., Bobbie R.J., Lott D.F., Martz R.F., Benson P.H., White D.C. (1981). Effect of manual brush cleaning on biomass and community structure of microfouling film formed on aluminium and titanium surfaces exposed to rapidly flowing seawater. *Appl Environ Microbiol*, **41**, 1442–1453.

Ruseska I., Robbins J., Lashen E.S., Costerton J.W. (1982). Biocide testing against corrosion-causing oilfield bacteria help control plugging. *Oil Gas J*, **8**, 253–264.

Quirynen M., van der Mei H.C., Boller C.M.L., Geertsema-Doornbusch G.I., Busscher H.J., van Steenberghe D. (1994). Clinical relevance of the influence of surface free energy and roughness on the supergingival and subgingival plaque formation in man. *Colloids Surf B*, **2**, 25–31.

Sorrelle P.H., Belgard W. (1991). The effect of recycled fiber use on paper machine biological control. In: *TAPPI Papermakers Conf.* TAPPI Press, Atlanta, pp. 569–575.

Sorrelle P.H., Belgard W. (1992). Growth in recycling escalates costs for paper machine biological control. *Pulp Paper*, **May**, 57–64.

Steinberg P.D., Schneider R., Kjelleberg S. (1997). Chemical defenses of seaweeds against microbial colonization. *Biodegradation*, **8**, 211–220.

Strausbaugh L. (1997). Antimicrobial resistance: problems, laments, and hopes. *Am J Infect Control*, **25**, 294–296.

Sutherland I.W. (1977). Bacterial exopolysaccharides: their nature and composition. In: Sutherland I.W. (ed). *Surface Carbohydrates of the Prokaryotic Cell*. Academic Press, London, pp. 27–96.

Toren K, Brisman J., Meding B. (1997). Sensitization and exposure to methylisothiazoloinones (Kathon) in the pulp and paper industry: a report of two cases. *Am J Ind Med*, **31**, 551–553.

Travis C.C., Hattemer-Frey H.A. (1991). Human exposure to dioxin. *Sci Total Environ*, **104**, 97–127.

Väätänen P., Niemelä S.I. (1983). Factors regulating the density of bacteria in process waters of a paper mill. *J Appl Bacteriol*, **54**, 367–371.

Väisänen O.M., Nurmiaho-Lassila E-L., Marmo S.A., Salkinoija-Salonen M.S. (1994). Structure and composition of biological slimes on paper and board machines. *Appl Environ Microbiol*, **60**, 641–653.

van Hoogmoed C.G., van der Mei H.C., Busscher H.J. (2000). The role of biosurfactants in affecting initial microbial adhesion mechanisms. In: Evans L.V. (ed) *Biofilms: Recent Advances in their Study and Control*. Harwood Academic Publishers, pp. 237–251.

Wahl M., Jensen P.R., Fenical W. (1994). Chemical control of bacterial epibiosis on ascidians. *Mar Ecol Prog Ser*, **110**, 45–57.

Wiatr C. (1990). Application of cellulase to control industrial slime. United States patent 4,936,994, issued June 26, 1990.

Wright J.B. (1997). Significantly reduced toxicity approach to paper machine deposit control. In: *1997 Engineering & Papermakers Conf Proc.* TAPPI Press, Atlanta, pp. 1083–1088.

Wright N.E., Gilbert P. (1987). Antimicrobial activity of n-alkyltrimethylammonium bromides: influence of specific growth rate and nutrient limitation. *J Pharm Pharmacol*, **39**, 685–690.

Wright J.B., Michalopoulos D.L. (1997). Method for inhibiting microbial adhesion on surfaces. United States Patent 5,593,599, issued January 14, 1997.

Wright J.B., Ruseska I., Costerton J.W. (1991). Decreased biocide susceptibility of adherent *Legionella pneumophila*. *J Appl Bacteriol*, **71**, 531–538.

22 Study of Biofouling Control with Fluorescent Probes and Image Analysis

F. Philip Yu and Gordon A. McFeters

This chapter discusses the use of various fluorescent probes in biofilm study. As fluorescence microscopy and digital imaging technology progress rapidly due to the development of high-speed personal computer processors, it will no longer be necessary to rely on a bulky super computer to do the complicated algorithmic calculations. Micrographs can be digitized, enhanced, and visualized beyond the limit of the human eye. Digital image processing has brought light microscopy into three-dimensional resolution, which is crucial if biofilm researchers are to understand the spatial distribution and heterogeneity among biofilms. Current progress in biofouling control, with the combination of fluorochromes, microscopy and image analysis is reviewed here.

INTRODUCTION

The existence of biofilms was first recognized in soil (Söhngen, 1913), and later studied in aquatic environments (Henrici, 1933; Zobell, 1943; Mekalanos, 1992). Since then much progress has been made in understanding the ecology and physiology of adherent microorganisms (Characklis and Marshall, 1990; Lappin-Scott *et al.*, 1992). The adhesion event exerts a profound effect on bacteria. Attachment alters their physiological processes (Costerton *et al.*, 1987; Davies and McFeters, 1988; Kölbel-Boelke and Hirsch, 1989), their surface structures (Rosenberg *et al.*, 1967; Costerton *et al.*, 1981; Costerton and Lappin-Scott, 1989; Marshall, 1992; Wolfaardt and Cloete, 1992) and their relationships to the bulk fluid (Wolfaardt *et al.*, 1992; Lau and Liu, 1993). Both biofouling and microbiologically influenced corrosion are phenomena that are linked to the existence, properties and activities of biofilms. Biofilm processes are manifested in many forms and are studied by researchers from a wide variety of disciplines. There are numerous industrial environments where corrosion and biofouling processes are potentially troublesome, including cooling water systems, storage tanks, water and wastewater treatment facilities, filters, piping, and drinking water distribution systems (Flemming and Geesey, 1990).

Control of biofilm problems has usually been attempted by the application of biocides in water systems. It was initially assumed that the disinfection kinetics of attached bacteria would be similar to those of their planktonic counterparts. Many industrial systems have experienced the inevitable problems caused by biofilms, even in the presence of an effective disinfectant residual. It is now recognized that strategies to control attached bacteria must be based on data generated from biofilm studies. Difficulties associated with studying surface-associated cells have hindered work on characterizing the activities of adherent bacteria, compared to the progress made with free-living bacteria. For instance, it is necessary to remove the cells from the substratum prior to the enumeration of viable bacteria attached to surfaces. Differences in physiological activity between attached and free-living bacteria (Fletcher, 1984) may explain the diverse susceptibilities and growth requirements after cells have been removed from the substratum. In addition, enumeration of viable bacteria by plate count (PC) methods may not detect all viable cells, particularly those injured by environmental stress (Camper and McFeters, 1979; McFeters *et al.* 1982; Roszak and Colwell, 1987). Also, detached bacteria that are aggregated may be problematic in the plate counting technique.

The conventional approach to biofilm study involves physically removing samples from the substratum, followed by biological and chemical analyses. Using this approach, a great deal of information on biofilm heterogeneity has been revealed from years of research; however, spatial patterns were neglected in the process. However, recent advances in optic technologies have enabled researchers to conduct non-destructive study on the spatial distribution, thickness and the physiological activities of biofilms and their responses to biofouling control agents. Furthermore, digital image processing has given light microscopy a three-dimensional resolution, which has lead to major advances in biofilm study and biofouling control. With the aid of high speed CPU chips, complicated logarithm calculations can be achieved with personal computer within seconds.

The purpose of this chapter is to review progress in combining fluorescent probes, microscopy and image analysis in biofouling control.

FLUORESCENT PROBES

Accurate detection and enumeration of bacteria is an important task in many areas of microbiological investigation. However, most of the methods commonly used to count bacteria in samples taken from natural environments have limitations (Daley, 1982). One example is the widely acknowledged failure of bacteria to form colonies on specific media due to the taxonomic (Atlas, 1984) and physiological (Roszak and Colwell, 1987; McFeters, 1989) diversity of microorganisms in most environments. Viable counts generated with plate enumeration from aquatic environments yield as little as 1 % of the total microscopic counts for bacterial isolates from freshwater (Servais and Menon, 1991). Microscopic enumeration showing various physiological activities within a microbial population has significant advantages over conventional methods.

An ideal fluorescent probe should allow the selective microscopic examination of particular activities or physiological properties of bacteria. Fluorochromes may bind specifically to components of the cell envelope or contents, and some are modified by cellular activity. After binding or metabolism, the fluorochrome can be observed directly with epifluorescence microscopy using appropriate excitation and emission filters.

Acridine Orange

Acridine orange (AO), one of the nucleic acid intercalating dyes, was originally applied to stain bacterial cells for fluorescence microscopy (Rigler, 1973) in order to detect and enumerate bacteria. The method commonly employed by microbial ecologists is the AO direct count (AODC) technique, also known as the direct epifluorescent technique (DEFT). The unique metachromatic property of AO (Bitton *et al.*, 1993) led to the widespread application of AO to vital staining. AO fluoresces either green or red, depending on the nature of the binding reaction with nucleic acid. It was assumed that the color of AO-stained bacteria could be used to discriminate between active and inactive cells (van Es and Meyer-Reil, 1982; Lopez-Torres *et al.*, 1988; Menezes *et al.*, 1995), because AO bound to double stranded DNA fluoresces green while single stranded RNA-bound AO fluoresces orange. Actively metabolising cells would be expected to have higher RNA content and thus appear orange with AO staining. However, studies (McFeters *et al.*, 1991) have shown that the AO staining reaction may be suggestive of physiological activity only under defined conditions. Variables in staining and fixation procedures as well as uncertainties associated with mixed bacterial populations in environmental samples may produce results that are not consistent with the classical interpretation of this reaction. Hence, AO has limited application as a vital stain.

The direct microscopic enumeration method using AO (AODC) usually shows a reasonable correlation with viable counts when applied to exponentially growing cultures. However, AODC data can exceed PC enumeration by several orders of magnitude (Roszak and Colwell, 1987) when used to examine bacterial populations in natural environments. The direct viable count (DVC) method developed by Kogure *et al.* (1979; 1984), has been employed successfully in enumerating bacteria within environmental samples (Maki and Remsen, 1981; Xu *et al.*, 1982; Rollins and Colwell, 1986; Liebert and Barkay, 1988; Singh *et al.*, 1990). The DVC method was later applied as a direct *in situ* enumeration method for thin biofilms (Yu *et al.*, 1993), where the results indicated that this adaptation of the method can provide rapid (4h) as well as more accurate information regarding bacterial number and viability within biofilms.

However, using the DVC method on biofilm bacteria it would be difficult to enumerate the elongated cells without removing and dispersing the biofilm aggregate. Therefore, an *in situ* biofilm activity assessment might not be possible using the DVC method.

Fluorescein Diacetate

Fluorescein diacetate (FDA) is a fluorochrome conjugated to two acetate radicals. After FDA enters the cell membrane *via* active metabolism (Brunius, 1980), it is hydrolyzed by intracellular esterases then fluorescein is released from the compound. Fluorescein has an absorbance at 490 nm, and fluoresces green when excited with blue light. The use of FDA has been suggested for the detection of microbial activity in bacterial suspensions (Jarnagin *et al.*, 1980; Chrzanowski *et al.*, 1984) and biofilms (Pawley, 1990; Safferman and Bishop, 1996; Battin, 1997). However, FDA hydrolysis might be limited to environments rich in eukaryotes and Gram-negative cells (Chrzanowski *et al.*, 1984). It has also been observed that fluorescein derived from cleaved FDA in bacteria tends to leak out under some physiological conditions.

Tetrazolium Salts

The reduction of tetrazolium salts to formazan has been used for many years in histo-, cyto-, and biochemical determinations of oxidase and dehydrogenase activities. Zimmermann *et al.* (1978) used the redox dye, 2-(*p*-iodophenyl)-3-(*p*-nitrophenyl)-5-phenyltetrazolium chloride (INT) to study respiratory activity in aquatic bacteria. However, the reduced form of INT (INT-formazan) can only be observed within bacteria by light microscopy, which can not be applied to the study of biofilm bacteria on opaque substrata without removal of the cells. Although the combination of fluorescent-antibody (FA) and INT reduction has been applied successfully to the study of cellular activity (Baker and Mills, 1982), the tedious procedures of preparing and examining FA has limited this approach as a general application. Another redox dye, 5-cyano-2,3-ditolyl tetrazolium chloride (CTC), has been applied successfully to the study of physiological activity within eukaryotic (Stellmach, 1984) and prokaryotic (Rodriguez *et al.*, 1992; Kaprelyants and Kell, 1993) cells. The *in situ* performance of CTC reduction was compared with the *in situ* DVC and PC methods in the determination of bacterial viability within pure culture biofilms. Both the CTC and DVC methods showed comparable numbers, and the results were two-fold higher than PC enumeration (Yu and McFeters, 1994a).

The CTC method has also been applied successfully to other biofilm research concerned with disinfection, culturability, and enzymatic removal (Stewart *et al.*, 1994; Yu and McFeters, 1994b; Pyle *et al.*, 1995; Johansson *et al.*, 1997; Kalmbach *et al.*, 1997).

Rhodamine 123

Rhodamine (Rh) 123 is a cationic fluorescent dye which is found to be concentrated in mitochondria by the relatively high negative potential across the energized mitochondrial membrane (Johnson *et al.*, 1981). Rh 123 is a proton motive force (PMF)-driven dye, which is only taken up by viable cells. In bacterial cells Rh 123 is accumulated in an uncoupler-sensitive fashion *via* transmembrane potential (Haugland, 1996). This fluorochrome has been utilized to assess the physiological states of *Micrococcus luteus* and *Escherichia coli* (Kaprelyants and Kell, 1992), *Salmonella typhimurium* (Mason *et al.*, 1995; Lopez-Amoros *et al.*, 1995) and *Aeromonas salmonicida* (Morgan *et al.*, 1993) by flow cytometry.

Rh 123 is not readily absorbed by Gram-negative bacteria because of a permeability limitation in their outer membrane (Nikaido and Vaara, 1985). However, treatment with Tris and EDTA at alkaline pH (Kaprelyants *et al.*, 1992) eliminates

(a)

(b)

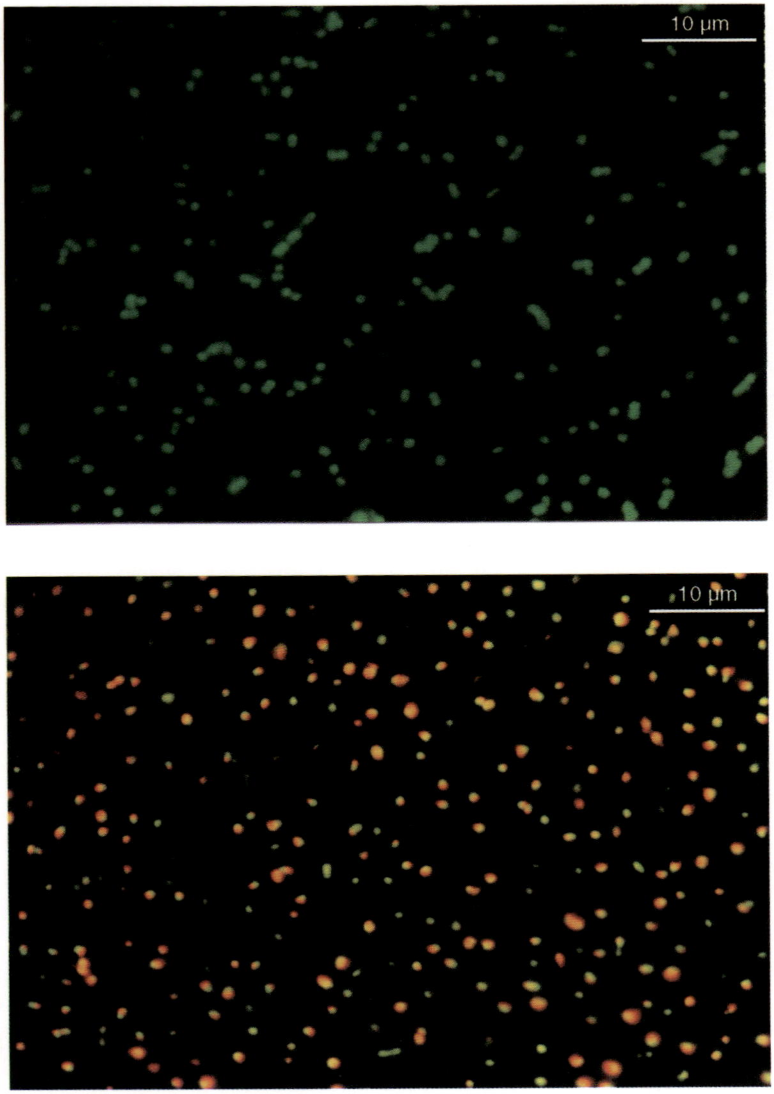

Figure 1 Epifluorescence micrographs of a *K. pneumoniae* biofilm on 314 stainless steel stained with (a) Rh 123 and (b) CTC-formazan and DAPI (green), using a Leitz filter block H.

this barrier and achieves optimal staining within 2h. Both the Rh 123 and the CTC methods gave comparable enumeration results on biofilm bacteria (Yu *et al.*, 1994a). Figure 1 is an example of Rh 123 and CTC staining of a monolayer *Klebsiella pneumoniae* biofilm. Active biofilm bacteria appear green under epifluorescence microscopy using a Leitz filter block H (Figure 1a). The "H" filter block has a combination of excitation filter (BP420-490), dichromatic mirror (RKP 455) and suppression filter (LP515). Figure 1b shows good color contrast for both DAPI-stained green non-respiring cells and respiring cells that contain red CTC-formazan crystals.

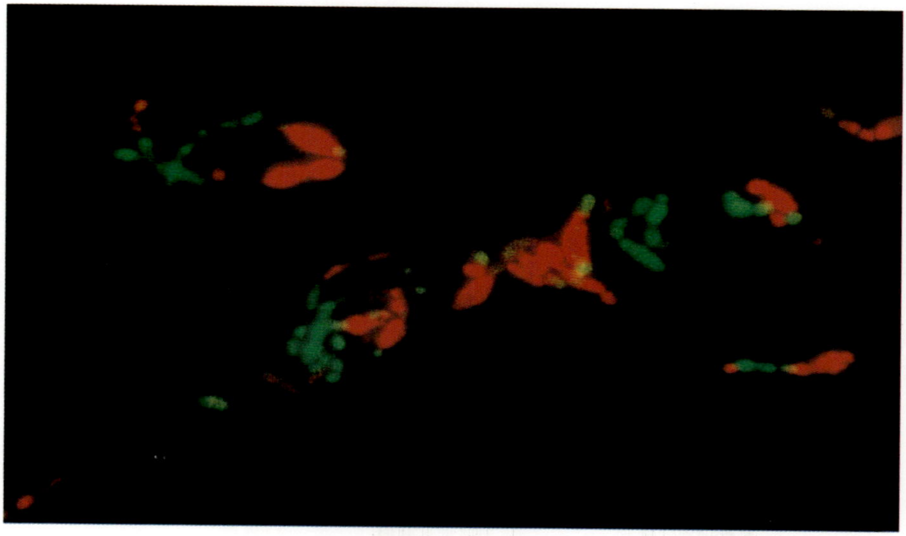

Figure 2 *Bifidobacterium sp.* stained with the *Bac*Light probe. When incubated with the SYTO 9 and propidium iodide nucleic acid stains, live bacteria with intact cell membranes fluoresce green and dead bacteria with compromised membranes fluoresce red. (Image contributed by Bruce Roth and Paul Millard, Molecular Probes Incorporated.)

Live/Dead Viability Assay

The Live/Dead *Bac*Light viability assay (Molecular Probes Incoporated, Eugene, Oregon) has been successfully applied to study the susceptibility of bacterial biofilms (Korber *et al.*, 1997; Wood *et al.*, 1998) treated with antimicrobials. The *Bac*Light probe distinguishes live bacteria with intact plasma membranes from dead cells with compromised membranes (Haugland, 1996). Figure 2 shows *Bifidobacterium sp.* stained with the SYTO 9 and propidium iodide, where live bacteria with intact cell membranes fluoresce green and dead bacteria with compromised membranes fluoresce red. Two different nucleic acid stains are included; live bacteria fluoresce green by taking up the stain SYTO 9, and propidium iodine causes the dead cells to show red fluorescence. The *Bac*light assay also allows the end-user to vary the ratio of SYTO 9 and propidium iodine to give balanced staining of most samples. The bacterial fluorescence at each emission wavelength can be calibrated in quantitative assays using equipment such as a fluorometer, fluorescence microplate reader or flow cytometer.

Other dual staining approaches to distinguish viable from non-viable bacteria are the oxonol dye and calcolfluor white (Mason *et al.*, 1995) combination. The dye bis(1,3-dibutylbarbituric acid) trinmethine oxonol ($DiBAC_4(3)$) is a lipophilic anion that responds by decreasing fluorescence with increasing membrane potential. Calcafluor white (CFW) is the disodium salt of 4′,4′-bis(4 anilino-bis-diethyl amino-s-triazin-2-ylamino)-2,2′-stilbene dissulphonic acid); it is used as fluorescence brighteners in the dye industry. The viable cell can exclude CFW whereas non-viable cells have bright fluorescence.

Other Vital Stains

There are several proprietary fluorescent probes that have been applied to industrial cooling water biofilm study without revealing detailed mechanisms. The proprietary stain PRB is reported to be visualized under both light and epifluorescence microscopy (Chalut *et al.*, 1994). Active cells stained pink with a dark spot; inactive cells do not have a spot under bright field illumination. Both active and inactive cells appear red under epifluorescence illumination (Kogure *et al.*, 1980). Fluorochrome APY is reported to stain RNA but not DNA (Chalut *et al.*, 1994). Cells killed with chemical biocide would not be stained or else appeared distorted (Kogure *et al.*, 1980). Another fluorochrome CEB is specific for bacterial exopolymeric substances (EPS) (Chalut *et al.*, 1994). The slime layer containing EPS fluoresces neon blue, whereas the bacterial cells are yellow under the epifluorescence microscope (Chalut *et al.*, 1994).

Fluorochromes also have various applications in fluorescence conjugates and oligonucleotide probes. These research topics are not covered in this chapter.

IMAGE ANALYSIS

Image analysis is the process of manipulation by converting images through digital processing with computer, then analyzing with customized software (Figure 3). The process involves multidisciplinary elements of optics, electronics, mathematics and computer science. Image analysis contains two main components, *viz.* image acquisition and object analysis.

Image Acquisition

An image under the microscope is an object with two-dimensional spatial representation. The analog image can be acquired using a camera and processed by a frame grabber within the computer, then converted into a digital form. The digitized image is usually represented as a mathematical expression of $f(x,y)$, which is composed of numerous small rectangular picture elements called pixels. The dynamic range of pixel intensities varies, and the quality of the digitized form is dependent on this. For an 8-bit image, each pixel will have grey levels ranging from 0 to 255, which is an indication of the signal intensity or brightness. For a modern 32-bit operating system, such as Microsoft Windows® 98, the range will be from 0 to 1023.

Video Camera

A video camera, which is different from conventional still frame single-lens reflex film cameras, produces a continuous analog output of video images. These output signals conform to a specific video standard that varies around the world. The three main video types are 525-line NTSC (National Television System Committee) in the United States, 625-line PAL (Phase Alternating Line) in Europe and 625-line SECAM (Sequential Color a Memoire) in France. These standards vary in field rate (Hz), format (H × V) and pixel rate (MHz). The most recent video cameras have replaced the tube

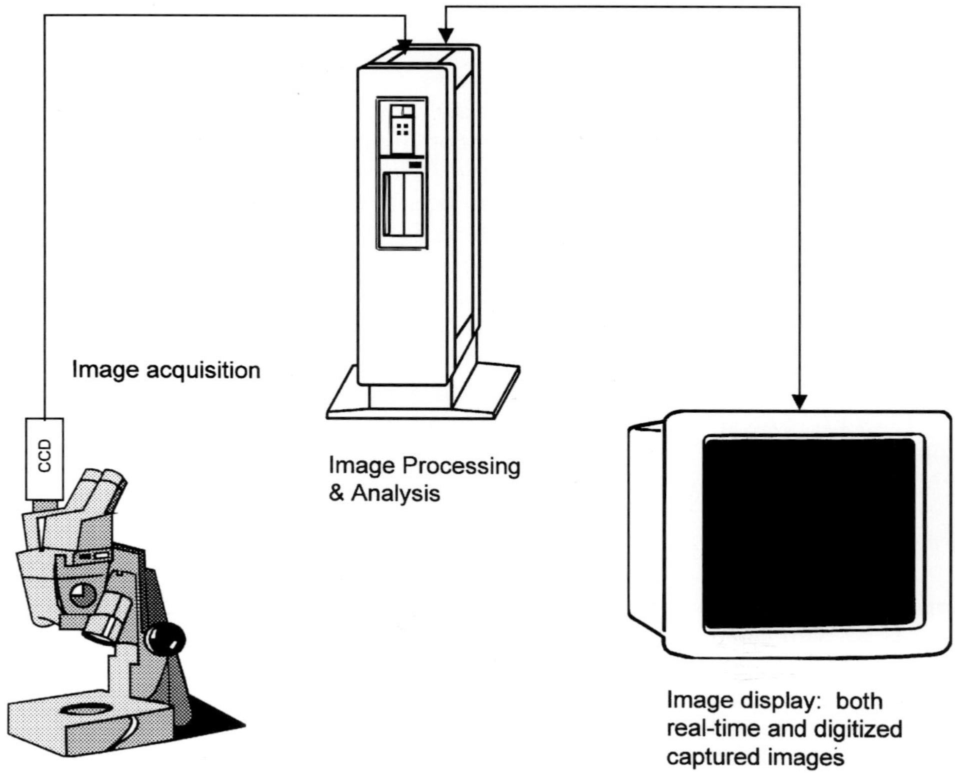

Figure 3 An image analysis system.

with a solid-state image sensor, called a charge-coupled device (CCD), that changes light energy into electrical pulses that can be recorded on videotape. The clarity and resolution of the video images depend on the number of pixels that the CCD can create. CCD devices can be divided into two groups, *viz.* video camcorder and scientific cameras. Scientific CCD cameras employ full frame CCD elements, the readout is processed through an on-chip pre-amplifier and an off-chip analog-to-digital converter to deliver a single high-quality digital image.

The CCD sensor is a monochrome imaging device. One way to create a color image is reconfiguration through three single color filters; a more advanced technology employs three CCDs to separate three color channels red, green and blue.

There is no standard for scientific cameras. They come with different options including pixel sizes, image dimensions, readout rate, integration time, cooling, quality grade and dynamic range (Castleman, 1996). These options determine the quality of image outputs. Since the computer and software can regulate shutter control, the common bleaching problem with fluorescence microscopy can be reduced to a minimum. It is not unusual for the captured images show higher resolution due to the image enhancement features of high-end cameras.

Image Analysis

Image analysis is accomplished by computer software to give quantitative data, such as the number of cells within a microscopic image, or the diameter or length of a selected bacterium. The output is the numerical data of the digitized image. The rapid development of the personal computer (PC) has allowed scientists today to process and analyze complicated images using a PC with specialized image processing boards. There are many sophisticated image processing packages in the market, which can be used without computer programming. Software such as IBAS (Kontron, Germany), Quantimet (Cambridge, UK), Visilog (Noesis, Canada) and Image-Pro Plus (Silver Spring, MD, USA) often provide similar functionality as image processing libraries, but in a much more user-friendly style.

The processed image in consecutive pixels can be stored as bytes in binary files. For example a 512 × 512 pixel image, stored at one byte per pixel, will take up 262 kilobytes (KB) of disk space. The most common output portable file formats are GIF (Graphics Interchangeable Format), TIFF (Tag Image File Format), JPEG (Joint Photographic Expert Group), BMP (standard Windows bitmap image), PCX (PC Paintbrush® file format), PDF (Adobe Acrobat® file format) and PICT (Macintosh graphics format). Of these formats, only TIFF images can be extended to include three dimensions.

BIOFOULING STUDY WITH IMAGE ANALYSIS

A great deal of work on biofilms has been done by removing bacteria from substrata and replacing them on glass slides for examination. Most structural studies have relied on light and electron microscopy (Kinner *et al.*, 1983; Robinson *et al.*, 1984; Costerton *et al.*, 1987; Eighmy *et al.*, 1983; Ganczarczyk *et al.*, 1992; Lappin-Scott *et al.*, 1992; Stewart *et al.*, 1993). Problems associated with these techniques include disruption of biofilm structure during removal from the substratum, laborious preparation, and extensive sample processing that may introduce artifacts. Light microscopy used in combination with computer-enhanced microscopy is an effective tool, but it is best applied during the early phases of biofilm development (Lawrence *et al.*, 1989). Visualization of monolayer bacterial biofilms can also be accomplished easily by either light or fluorescence microscopy (Yu *et al.*, 1993; 1994). Due to the resolution limits of optical microscopy, studies on thicker biofilms require mechanical removal of biofilms from the substratum prior to further analysis or optical sectioning by scanning laser confocal microscopy (SLCM).

Biofilms are highly heterogeneous in structure and physiological activity. Without understanding the spatial structure, composition and metabolic activity within biofilms, it is difficult to understand and control the processes of biofouling.

Scanning Laser Confocal Microscopy (SLCM)

The concept of confocal microscopy was first filed as a US patent in 1957 by Minsky (1961). In Minsky's embodiment of the confocal microscope, the conventional condenser is replaced by a lens identical to the objective lens. The field of illumination is limited by a pinhole, positioned on the microscope axis (Pawley, 1990). The pinhole blocks the light from the out-of-focus planes above and below the plane of focus, which makes only the light from the plane of focus visible. This pinhole based confocal microscopy is applicable to both trans-illuminating and epi-illuminating modes. With the development of lasers in microscopy, the biological application of SLCM technology (Carlsson *et al.*, 1985; Amos and White, 1987; Carlsson and Åslund, 1987) was first published in 1985 and later introduced commercially by Srasstro, Biorad, Olympus, Zeiss and Leitz in the late 1980's. SLCM is also referred to as laser scanning confocal microscopy (LSCM), confocal laser scanning microscopy (CLSM) and scanning confocal laser microscopy (SCLM) in various publications.

SLCM technology has permitted major advances in biofilm research. The SLCM eliminates out-of-focus haze and allows horizontal and vertical optical sectioning. The reconstruction of images is based on optical sections that can be applied nondestructively to specimens in a matter of minutes. The systems are equipped with highly sophisticated image analysis capabilities that enable researchers to visualize reconstructed 2-D and 3-D images without physically disrupting the biofilm (Lawrence *et al.*, 1991; Caldwell *et al.*, 1992a; 1992b, Dalton *et al.*, 1994; de Beer *et al.*, 1994; Korber *et al.*, 1994; Stoodley *et al.*, 1994; Wolfaardt *et al.*, 1994; de Beer and Stoodley, 1995; Stewart *et al.*, 1995; Doolittle *et al.*, 1996, Müller *et al.*, 1996; Sanford *et al.*, 1996; Swope and Flickinger, 1996; de Beer *et al.*, 1997; Jayaraman *et al.*, 1997; Korber *et al.*, 1997; Neu and Lawrence, 1997; Okabe *et al.*, 1997; Lawrence *et al.*, 1998; Wolfaardt *et al.*, 1998). However, this technology involves relatively expensive instrumentation and has limited resolution when applied to thicker biofilms. These aspects restrict it from general application.

Cryoembedding and Cryosectioning

Cryoembedding and cryosectioning of human and animal tissues for light and fluorescence microscopy are well-established histological techniques (Troyer, 1980; Bancroft, 1982; Carson, 1990; Elias, 1990). Cryosectioning techniques have been utilized successfully in biofilm research to visualize the structure of bacterial biofilms(Yu *et al.*, 1994). Cryoembedding is performed with Tissue-Tek® OCT compound (Miles Incorporated, Elkhart, IN, USA) by placing the biofilm coupon on top of a dry ice slab. The embedded biofilm within the frozen OCT block can then be sectioned with a cryostat to yield 5 μm slices of biofilm cross-section.

Different embedding techniques have been applied to biofilms for measurement and evaluation of morphological parameters. Embedding media including paraffin (Li and Ganczarczyk, 1990), plastic resin (glycol methacrylate) (Ganczarczyk *et al.*, 1992; Stewart *et al.*, 1993) and agar (Ganczarczyk *et al.*, 1992) have been utilized. When combined with image analysis, the thin (2.5 μm) sections reveal quantitative data on biofilm heterogeneity such as thickness and size (Stewart *et al.*, 1993).

The cryoembedding technique involves less sample processing and is more rapid than any of other procedures. The whole process can be completed in less than 24 h. This method preserves the biofilm with minimal preparative artifacts. The advantage of the approach is that the biofilm cryosections can be examined with conventional

(a)

(b)

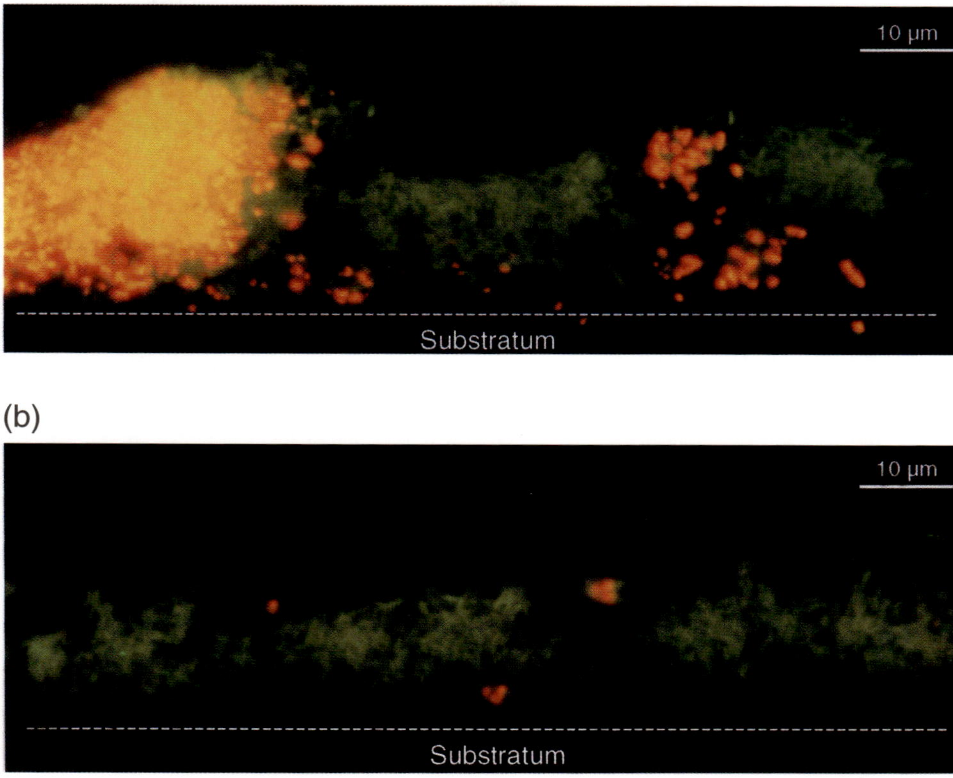

Figure 4 Epifluorescence micrographs of a frozen sections of a mixed *K. pneumoniae* and *P. aeruginosa* biofilm grown on 314 stainless steel and treated with monochloramine (4 mg l^{-1}). Disinfection intervals are (a) 30 min and (b) 60 min. The dashed lines indicate the position of the substratum. Biofilm bacteria were stained with CTC (red) and DAPI (green).

light and epifluorescence microscopy. The cryoembedding procedure is compatible with specialized staining or labeling techniques, such as fluorochromes (Yu *et al.*, 1994), immunofluorescence staining, oligonucleotide probing and radioisotope labeling. Fixatives and stains may also be applied to frozen sections after air drying (Bancroft, 1982; Elias, 1990). The cryosectioning technique offers an alternative, minimally disruptive approach to the study of biofilms by microscopy. Individual cells, microcolonies, void areas and biofilms thickness can be quantitatively determined by image analysis software (Murga *et al.*, 1995; Huang *et al.*, 1996; Wentland *et al.*, 1996). Figure 4 shows cryosections of a mixed *K. pneumoniae* and *Pseudomonas aeruginosa* biofilm. The biofilm bacteria showed decreased respiratory activity and thickness after treating with 4 mg l^{-1} of monochloramine. This approach also provides an unique opportunity to study the spatial response of biofilm bacteria to antimicrobial agents, and enables the mechanisms explaining their comparative resistance to disinfection to be addressed in a way that has not been possible using traditional techniques (Huang *et al.*, 1995; McFeters *et al.*, 1995).

In a study comparing scanning electron microscopy (SEM), SLCM and cryosectioning techniques to visualize biofilm structural heterogeneity (Stewart *et al.*, 1995), SLCM and cryosectioning were shown to be superior to SEM in their ability to image the biofilm interior, and in their potential to provide quantitative information.

Digital Confocal Microscopy

Digital confocal microscopy (DCM) is a recent advancement in which digital signal processing (DSP) chips and software technology are combined to produce an image processing system capable of removing out-of-focus haze. The pinhole based hardware approach produces the analog confocal microscopy. Analog confocal microscopy, such as SLCM, produces sharp and clear images with a theoretical 1.4x improvement in resolution. However, the pinhole blocks 85% to 99% of the light, which limits its application under low-light conditions. In addition, SLCM systems are delicate and complex instruments that are often subject to alignment and laser stability problems (Beckwith and Margerum, 1997). The drawbacks of SLCM include the costs of the instrument and additional laser light sources.

DCM employs deconvolution algorithms to calculate and remove out-of-focus haze images captured with a standard research microscope and video camera. The process can be processed quickly with a PC under either IBM or Macintosh platforms. With a Z-axis stage motor and controller, DCM can de-blur a stack of microscope images and reconstruct a 3-D image of high quality (Kesterson and Richardson, 1991). The results are comparable to those produced by SLCM, but at a much more affordable cost.

The deconvolution approach of DCM is versatile. It can be used with a wide variety of specimens, such as paper fibers, computer chips and sandstone (Brading *et al.*, 1996). It is also valuable in applications that require a specific wavelength of light not available with the SLCM system, such as ultraviolet. It is gaining popularity in biological and microbiological research (Gorby, 1994; Richardson, 1997; Kunkier *et al.*, 1998) for visualizing structures.

With the availability of DSP chips, faster PC processors and image processing software, the DCM approach provides a powerful and useful tool with great versatility. However, DCM technology can only be used for epifluorescence applications in visualizing biofilm spatial distribution due to the difficulties involved with optical sectioning under bright ground illumination. The biofilm researcher may benefit from DCM technology since it is an affordable alternative to image analysis or SLCM. However, during recent years, with the increasing numbers of manufacturers of SLCM systems and image analysis software, these have become highly competitive.

In the future, biofilm research is moving towards the use of biosensors, chemical sensors, physical sensors and controllers that are characterized by merging concepts in classical disciplines like chemistry, biology, physics, medicine, computer and electrical engineering. The rapid development of these technologies will help in the better understanding of biofilms, and may lead to more effective biofouling control.

REFERENCES

Amos W.B, White J.G. (1987). Use of a confocal imaging in the study of biological structures. *Appl Opt*, **26**, 3239–3243.

Atlas R.M. (1984). Use of microbial diversity measurements to assess environmental stress In: Klug M.J., Reddy C.A. (eds) *Current Perspectives in Microbial Ecology*. American Society for Microbiology, Washington D.C., pp. 450–454.

Baker K.H., Mills A.L. (1982). Determination of the number of respiring *Thiobacillus ferrooxidans* cells in water samples by using combined fluorescent antibody-2-(*p*-iodophenyl)-3-(*p*-nitrophenyl)-5-phenyltetrazolium chloride staining. *Appl Environ Microbiol*, **43**, 338–344.

Bancroft J.D. (1982). Frozen and related sections. In: Bancroft J.D., Stevens A. (eds) *Theory and Practice of Histological Techniques*. Churchill Livingston, New York, p. 82.

Battin T.J. (1997). Assessment of fluorescein diacetate hydrolysis as a measure of total esterase activity in natural stream sediment biofilms. *Sci Total Environ*, **198**, 51–60.

Beckwith R.C., Margerum D.W. (1997). Kinetics of hypobromous acid disproportionation. *Inorg Chem*, **36**, 3754–3760.

Bitton G., Koopman B., Jung K., Voiland G., Kotob M. (1993). Modification of the standard epifluorescence microscopic method for total bacterial counts in environmental samples. *Water Res*, **27**, 1109–1112.

Brading M.G., Boyle J., Lappin-Scott H.M. (1996). The influence of laminar flow and surface type on bacterial adhesion. IChemE Res Event, Eur Conf Young Res Chem Eng, Volume I, 25–27.

Brunius G. (1980). Technical aspects of 3′,6′-diacetyl fluorescein for vital fluorescent staining of bacteria. *Curr Microbiol*, **4**, 321–323.

Caldwell D.E., Korber D.R., Lawrence J.R. (1992a). Confocal laser microscopy and digital image analysis in microbial ecology. In: Marshall K.C. (ed) *Advances in Microbial Ecology*, Volume 12, Plenum Press, New York, pp. 1–67.

Caldwell D.E., Korber D.R., Lawrence J.R. (1992b). Imaging of bacterial cells by fluorescence exclusion using scanning confocal laser microscopy. *J Microbiol Methods*, **15**, 249–261.

Camper A.K., McFeters G.A. (1979). Chlorine injury and enumeration of waterborne coliform bacteria. *Appl Environ Microbiol*, **37**, 633–641.

Carlsson K., Åslund N. (1987). Confocal imaging for 3-D digital microscopy. *Appl Opt*, **26**, 3232–3238.

Carlsson K., Danielsson P., Lenz R., Liljeborg A., Majlöf L., Åslund N. (1985). Three-dimensional microscopy using a confocal laser scanning microscope. *Opt Lett*, **10**, 53–55.

Carson F.L. (1990). *Histotechnology, A Self Instructional Text*. ASCP Press, Chicago, IL.

Castleman K.R. (1996). *Digital Image Processing*. Prentice Hall, Englewood Cliffs, NJ.

Chalut J., Cairns J., Korkorian N. (1994). Identification and quantification of cooling water biofilms using fluorescent staining and ATP monitoring techniques. *CORROSION '94*, Paper No. 272, NACE Internarional, Houston, TX.

Characklis W.G., Marshall K.C. (1990). *Biofilms*. John Wiley & Sons Incorporated, New York.

Chrzanowski T.H., Crotty R.D., Hubbard J.G., Welchi R.P. (1984). Applicability of the fluorescein diacetate method of detecting active bacteriain freshwater. *Microb Ecol*, **10**, 179–185.

Costerton J.W., Lappin-Scott H.M. (1989). Behavior of bacteria in biofilms. *Am Soc Microbial News*, **55**, 650–654.

Costerton J.W., Irvin R.T., Cheng K.-J. (1981). The bacterial glycocalyx in nature and disease. *Ann Rev Microbiol*, **35**, 299–324.

Costerton J.W., Cheng K.-J., Geesey G.G., Ladd T.I., Nickel J.C., Dasgupta M., Marrie T.J. (1987). Bacterial biofilms in nature and disease. *Ann Rev Microbiol*, **41**, 435–464.

Daley R.J. (1982). Direct epifluorescence enumeration of native aquatic bacteria: uses, limitations and comparative accuracy. In: Costerton J.W., Colwell R.R. (eds) *Native Aquatic Bacteria: Enumeration, Activity and Ecology*. ASTM, Philadelphia, pp. 29–45.

Dalton H.M., Poulsen L.K., Halasz P., Angles M.L., Goodman A.E., Marshall K.C. (1994). Substratum-induced morphological changes in a marine bacterium and their relevance to biofilm structure. *J Bacteriol*, **176**, 6900–6906.

Davies D.G., McFeters G.A. (1988). Growth and comparative physiology of *Klebsiella oxytoca* attached to granular activated carbon particles and in liquid media. *Microb Ecol*, **15**, 165–175.

de Beer D., Stoodley P. (1995). Relation between the structure of an aerobic biofilm and transport phenomena. *Water Sci Technol*, **32**, 11–18.

de Beer D., Stoodley P., Lewandowski Z. (1997). Measurement of local diffusion coefficients in biofilms by microinjection and confocal microscopy. *Biotechnol Bioeng*, **53**, 151–158.

de Beer D., Stoodley P. Roe F., Lewandowski Z. (1994). Effects of biofilm structures on oxygen distribution and mass transport. *Biotechnol Bioeng*, **43**, 1131–1138.

Doolittle M.M., Cooney J.J., Caldwell D.E. (1996). Tracing the interaction of bacteriophage with bacterial biofilms using fluorescent and chromogenic probes. *J Ind Microbiol*, **16**, 331–341.

Eighmy T.T., Maratea D., Bishop P.L. (1983). Electron microscopic examination of wastewater biofilm formation and structural components. *Appl Environ Microbiol*, **45**, 1921–1931.

Elias J.M. (1990). *Immunohistopathology, A Practical Approach to Diagnosis*. ASCP Press, Chicago, IL.

Flemmir H.-C., Geesey G.G. (1990). *Biofouling and Biocorrosion in Industrial Water Systems*. Springer-Verlag, Berlin.

Fletcher M. (1984). Comparative physiology of attached and free-living bacteria. In: Marshall K.C. (ed) *Microbial adhesion and aggregation.* Springer-Verlag, New York, pp. 223–232.

Ganczarczyk J.J., Zahid W.M., Li D.-H. (1992). Physical stabilization and embedding of microbial aggregates for light microscopy. *Water Res*, **26**, 1695–1699.

Gorby G.L. (1994). Digital confocal microscopy allows measurement and three-dimensional multiple spectral reconstruction of *Neisseria gonorrhoeae*/epithelial cell interactions in the human fallopian tube organ culture model. *J Histochem Cytochem*, 42, 297–306.

Haugland R.P. (1996). *Handbook of fluorescent Probes and Research Chemicals*. Molecular Probes, Incorporated, Eugene, OR.

Henrici A.T. (1933). Studies of freshwater bacteria I. A direct microscopic technique. *J Bacteriol*, **25**, 277–286.

Huang C.-T., McFeters G.A., Stewart P.S., (1996). Evaluation of physiological staining, cryoembedding and autofluorescence quenching techniques on fouling biofilms. *Biofouling*, **9**, 269–277.

Huang C.-T., Yu F.P., McFeters G.A., Stewart P.S. (1995). Nonuniform spatial patterns of respiratory activity within biofilm during disinfection. *Appl Environ Microbiol*, **61**, 2252–2256.

Jarnagin J.-L., Luchsinger D.W., Petes T.D. (1980). The use of fluorescein diacetate and ethidium bromide as a stain for evaluating viability of mycobacteria. *Stain Technol*, **553**, 253–258.

Jayaraman A., Cheng E.T., Earthman J.C., Wood T.K. (1997). Importance of biofilm formation for corrosion inhibition of SAE 1018 steel by axenic aerobic biofilms. *J Ind Microbiol Biotechnol*, **18**, 396–401.

Johansson C., Falholt P., Gram L., (1997). Enzymatic removal and disinfection of bacterial biofilms. *Appl Environ Microbiol*, **63**, 3724–3728.

Johnson L.V., Walsh M.L., Bockus B.J., Chen L.B. (1981). Monitoring of relative mitochondrial membrane potential in living cells by fluorescence microscopy. *J Cell Biol*, **88**, 526–535.

Kalmbach S., Manz W., Szewzyk I.U. (1997). Dynamics of biofilm formation in drinking water: phylogenetic affiliation and metabolic potential of single cells assessed by formazan reduction and *in situ* hybridization. *FEMS Microbiol Ecol*, **22**, 265–279.

Kaprelyants A.S., Kell D.B. (1992). Rapid assessment of bacterial viability and vitality by rhodamine 123 and flow cytometry. *J Appl Bacteriol*, **72**, 410–422.

Kaprelyants A.S., Kell D.B. (1993). The use of 5-cyano-2,3-ditolyl tetrazolium chloride and flow cytometry for the visualization of respiratory activity in individual cells of *Micrococcus luteus*. *J Microbiol Methods*, **17**, 115–122.

Kinner N.E., Balkwill D.L., Bishop L. (1983). Light and electron microscopic studies of microorganisms growing in rotating biological contactor biofilms. *Appl Environ Microbiol*, **45**, 1659–1669.

Kogure K., Simidu U., Taga N. (1979). A tentative direct microscopic method for counting living marine bacteria. *Can J Microbiol*, **25**, 415–420.

Kogure K., Simidu U., Taga N. (1980). Distribution of viable marine bacteria in neritic seawater around Japan. *Can J Microbiol*, **26**, 318–323.

Kogure K., Simidu U., Taga N. (1984). An improved direct viable count method for aquatic bacteria. *Arch Hydrobiol*, **102**, 117–122.

Korber D.R., James G.A., Costerton J.W. (1994). Evaluation of fleroxacin activity against established *Pseudomonas fluorescence* biofilms. *Appl Environ Microbiol*, **60**, 1663–1669.

Korber D.R., Choi A., Wolfaardt G.M., Ingham S.C., Caldwell D.E. (1997). Substratum topography influences susceptibility of *Salmonella enteritidis* biofilms to trisodium phosphate. *Appl Environ Microbiol*, **63**, 3352–3358.

Kölbel-Boelke J.M., Hirsch P. (1989). Comparative physiology of biofilm and suspended organisms in the groundwater environment. In: Characklis W.G., Wilderer P.A. (eds) *Structure and Function of Biofilms*, John Wiley & Sons Ltd, New York, pp. 221–238.

Kunkier P.E., Kraig R.P. (1998). Calcium waves precede electrophysiological changes of spreading depression in hippocampal organ cultures. *J Neurosci*, **18**, 3416–3425.

Lappin-Scott H.M., Costerton J.W., Marrie T.J. (1992). Biofilms and biofouling. In: Lenderberg J. (ed) *Encyclopedia of Microbiology*, Academic Press Incorporated, New York, pp. 277–284.

Lau Y.L., Liu D. (1993). Effect of flow rate on biofilm accumulation in open channels. *Water Res*, **27**, 355–360.

Lawrence J.R., Korber D.R., Caldwell D.E. (1989). Computer-enhanced darkfield microscopy for the quantitative analysis of bacterial growth and behavior on surfaces. *J Microbiol Methods*, **10**, 123–138.

Lawrence J.R., Neu T.R., Swerhone G.D.W. (1998). Application of multiple parameter imaging for the quantification of algal, bacterial and exopolymer components of microbial biofilms. *J Microbiol Methods*, **32**, 253–261.

Lawrence J.R., Korber D.R., Hoyle B.D., Costerton J.W., Caldwell D.E. (1991). Optical sectioning of microbial biofilms. *J Bacteriol*, **173**, 6558–6567.

Li D.-H., Ganczarczyk J.J. (1990). Structure of activated sludge flocs. *Biotechnol Bioeng*, **35**, 57–65.

Liebert C., Barkay T. (1988). A direct viable counting method for measuring tolerance of aquatic microbial communities to Hg^{2+}. *Can J Microbiol*, **34**, 1090–1095.

Lopez-Amoros R., Comas J., Vives-Rego J. (1995). Flow cytometric asessment of *Escherichia coli* and *Salmonella typhimurium* starvation-survival in seawater using rhodamine 123, propidium iodide, and oxonol. *Appl Environ Microbiol*, **61**, 2521–2526.

Lopez-Torres A.J., Prieto L., Hazen T.C. (1988). Comparison of the *in situ* survival and activity of *Klebsiella pneumoniae* and *Escherichia coli* in tropical marine environments. *Microb Ecol*, **15**, 41–57.

Maki J.S., Remsen C.C. (1981). Comparison of two direct count methods for determining metabolizing bacteria in freshwater. *Appl Environ Microbiol*, **41**, 1132–1138.

Marshall K.C. (1992). Biofilms. An overview of bacterial adhesion, activity, and control at surfaces. *Am Soc Microbiol News*, **58**, 202–207.

Mason D.J., Lopez-Amoros R., Allman R., Stark J.M., Lloyd D. (1995). The ability of membrane potential dyes and Calcofluor White to distinguish viable and non-viable bacteria. *J Appl Bacteriol*, **78**, 309–315.

McFeters G.A. (1989). Detection and significance of injured indicator and pathogenic bacteria in water. In: Ray B., (ed) *Injured Index and Pathogenic Bacteria: Occurrence and Detection in Foods, Water and Feeds*. CRC Press Incorporated, Boca Raton, FL, pp. 179–210.

McFeters G.A., Cameron S.C., LeChevallier M.W., (1982). Influence of diluents, media, and membrane filters on the detection of injured waterborne coliform bacteria. *Appl Environ Microbiol*, **43**, 97–102.

McFeters G.A., Yu F.P., Pyle B.H., Stewart P.S. (1995). Physiological methods to study biofilm disinfection. *J Ind Microbiol*, **15**, 333–338.

McFeters G.A., Singh A., Byun S., Callis P.R., Williams S. (1991). Acridine orange staining reaction as an index of physiological activity in *Escherichia coli*. *J Microbiol Methods*, **13**, 87–97.

Mekalanos J.J. (1992). Environmental signals controlling expression of virulence determinants in bacteria. *J Bacteriol*, **174**, 1–7.

Menezes T.M., Band D.E., Gaylarde C.C. (1995). Biofilm and biocide assessment using epifluorescence microscopy. *9th Int Biodeterior Biodegrad Symp*, pp. 144–149.

Minsky M. (1961). Microscopy apparatus. US Patent 3,013,467.

Morgan J.A.W, Rhodes G., Pickup R.W. (1993). Survival of nonculturable *Aeromonas salmonicida* in lake water. *Appl Environ Microbiol*, **59**, 874–880.

Murga R., Stewart P.S., Daly D. (1995). Quantitative analysis of biofilm thickness variability. *Biotechnol Bioeng*, **45**, 503–510.

Müller S., Pedersen A.R., Poulsen L.K., Arvin E., Molin S. (1996). Activity and three-dimensional distribution of toluene-degrading *Pseudomonas putida* in a multispecies biofilm assessed by quantitative *in situ* hybridization and scanning confocal laser microscopy. *Appl Environ Microbiol*, **62**, 4632–4640.

Neu T.R., Lawrence J.R. (1997). Development and structure of microbial biofilms in river water studied by confocal laser scanning microscopy. *FEMS Microbiol Ecol*, **24**, 11–25.

Nikaido H., Vaara M. (1985). Molecular basis of bacterial outer membrane permeability. *Microbiol Rev*, **49**, 1–32.

Okabe S., Yasuda T., Watanabe Y. (1997). Uptake and release of inert fluorescence particles by mixed population biofilms. *Biotechnol Bioeng*, **53**, 459–469.

Pawley J.B. (ed) (1990). *Handbook of Biological Confocal Microscopy*. Plenum Press, New York.

Pyle B.H., Broadaway S.C., McFeters G.A. (1995). Factors affecting the determination of respiratory activity on the basis of cyanoditolyl tetrazolium chloride reduction with membrane filtration. *Appl Environ Microbiol*, **61**, 4304–4309.

Richardson M. (1997). 3D decovolution of microscope data. *Sci Technol J*, **20**, 20–21.

Rigler R. (1973). Staining of DNA with acridine orange. *Nobel Symp*, **23**, 335–341.

Robinson R.W., Akin D.E., Nordstedt R.A., Thomas M.V. (1984). Light and electron microscopy examinations of methane-producing biofilms from anaerobic fixed-bed reactor. *Appl Environ Microbiol*, **480**, 127–136.

Rodriguez G.G., Phipps D., Ishiguro K., Ridgway H.F. (1992). Use of fluorescent redox probe for direct visualization of actively respiring bacteria. *Appl Environ Microbiol*, **58**, 1801–1808.

Rollins D.M., Colwell R.R. (1986). Viable but nonculturable stage of *Campylobacter jejuni* and its roles in survival in the natural aquatic environment. *Appl Environ Microbiol*, **52**, 531–538.

Rosenberg B., Renshaw E., Vancamp L., Hartwicki J., Drobnik J. (1967). Platinum-induced filamentous growth in *Escherichia coli. J Bacteriol*, **93**, 316–721.

Roszak D.B., Colwell R.R. (1987). Survival strategies of bacteria in the natural environment. *Microbiol Rev*, **51**, 365–379.

Safferman S.I., Bishop P.L. (1996). Aerobic fluidized bed reactor with internal media cleaning. *J Environ Eng*, **122**, 284–291.

Sanford B.A., de Feijter A.W., Wade M.H., Thomas V.L. (1996). A dual fluorescence technique for visualization of *Staphylococcus epidermidis* biofilm using scanning confocal laser microscopy. *J Ind Microbiol*, **16**, 48–56.

Servais P., Menon P. (1991). Fate of autochthonous and fecal bacteria in marine ecosystems. *Kiel Meeresforsch*, **8**, 290–296.

Singh A., Yu F.P., McFeters G.A. (1990). Rapid detection of chlorine-induced bacterial injury by the direct viable count method using image analysis. *Appl Environ Microbiol*, **56**, 389–394.

Söhngen N.L. (1913). Influence of colloids on microbiological processes. *Zentralbl Bakteriol Parasitenk Infektionskr Abt 1*, **238**, 622–647.

Stellmach J. (1984). Fluorescent redox dye. *Histochemistry*, **80**, 137–143.

Stewart P.S., Peyton B.M., Drury W.J., Murga R. (1993). Quantitative observation of heterogeneities in *Pseudomonas aeruginosa* biofilms. *Appl Environ Microbiol*, **59**, 327–329.

Stewart P.S., Murga R., Srinivasan R., de Beer D. (1995). Biofilm structural heterogeneity visualized by three microscopic methods. *Water Res*, **29**, 2006–2009.

Stewart P.S., Griebe T., Srinivasan R., Yu F.P., de Beer D., McFeters G.A. (1994). Comparison of respiratory activity and culturability during monochloramine disinfection of binary population biofilms. *Appl Environ Microbiol*, **60**, 1690–1692.

Stoodley P., de Beer D., Lewandowski Z. (1994). Liquid flow in biofilm systems. *Appl Environ Microbiol*, **60**, 2711–2716.

Swope K., Flickinger M.C. (1996). The use of confocal scanning laser microscopy and other tools to characterize *Escherichia coli* in a high-cell-density synthetic biofilm. *Biotechnol Bioeng*, **52**, 340–356.

Troyer H. (1980). *Principles and Techniques for Histochemistry*. Little, Brown and Co Boston, MA.

van Es F.B., Meyer-Reil L.A. (1982). Biomass and metabolic activity of heterotrophic marine bacteria. *Adv Microb Ecol*, **6**, 111–170.

Wentland E.J., Stewart P.S., Huang, C-T., McFeters G.A. (1996). Spatial variations in growth rate within *Klebsiella pneumoniae* colonies and biofilm. *Biotechnol Prog*, **12**, 316–321.

Wolfaardt G.M., Cloete T.E. (1992). The effect of some environmental parameters on surface colonization by microorganism. *Water Res*, **26**, 527–537.

Wolfaardt G.M., Lawrence J.R., Robarts R.D., Caldwell D.E. (1998). *In situ* characterization of biofilm exopolymers involved in the accumulation of chlorinated organics. *Microb Ecol*, **35**, 213–233.

Wolfaardt G.M., Lawrence J.R., Headley J.V., Robarts R.D., Caldwell D.E. (1994). Microbial exopolymers provide a mechanism for bioaccumulation of contaminants. *Microb Ecol*, **27**, 279–291.

Wood P., Caldwell D.E., Evans E., Jones M., Korber D.R., Wolfhaardt G.M., Wilson M., Gilbert P. (1998). Surface-catalysed disinfection of thick *Pseudomonas aeruginosa* biofilms. *J Appl Microbiol*, **84**, 1092–1098.

Xu H-S., Roberts N., Singleton F.L., Attwell R.W., Grimes D.J., Colwell R.R. (1982). Survival and viability of nonculturable *Escherichia coli* and *Vibrio cholerae* in the estuarine and marine environment. *Microb Ecol*, **8**, 313–323.

Yu F.P., McFeters G.A. (1994a). Rapid *in situ* assessment of physiological activities in bacterial biofilms using fluorescent probes. *J Microbiol Methods*, **20**, 1–10.

Yu F.P., McFeters G.A. (1994b). Physiological responses of bacteria in biofilms to disinfection. *Appl Environ Microbiol*, **60**, 2462–2466.

Yu F.P., Pyle R.H., McFeters G.A. (1993). A direct viable count method for the enumeration of attached bacteria and assessment of biofilm disinfection. *J Microbiol Methods*, **17**, 167–180.

Yu F.P., Callis G., Stewart P.S., Griebe T., McFeters G.A. (1994). Cryosectioning of biofilms for microscopic examination. *Biofouling*, **8**, 85–91.

Zimmermann R., Iturriaga R., Becker-Birck J. (1978). Simultaneous determination of the total number of aquatic bacteria and the number thereof involved in respiration. *Appl Environ Microbiol*, **36**, 926–935.

Zobell C.E. (1943). The effect of solid surfaces upon bacterial activity. *J Bacteriol*, **46**, 39–56.

23 Microbially Influenced Corrosion in the Context of Metal Microbe Interactions

W. Allan Hamilton

Microbially influenced corrosion of both mild and stainless steels is considered as being one particular example of the general phenomena of metal microbe interactions. The essential redox character of the electrochemical reactions of corrosive metal loss is stressed and parallels are drawn with the redox reactions of microbial cellular energetics. An electron transfer hypothesis is proposed which both rationalises present understanding of the mechanisms involved, and provides a theoretical framework for the design and interpretation of future studies of microbial corrosion.

KEY WORDS: redox reactions; biomineralisation; sulphate, Fe(III), MnO_2 reduction

INTRODUCTION

Microbially influenced corrosion (MIC) is generally thought of as being a subject with high technical content but limited intellectual focus, and of interest to only a small band of specialists. It is now possible, however, to mount a vigorous case countering such a narrow view. This arises not only from new experimental findings but, more particularly, from the development of a theoretical framework that incorporates the essential elements of electrochemistry and materials science, and places MIC firmly within mainstream microbial ecophysiology. The essence of this altered perspective lies in the appreciation that the electrochemical redox reactions that are fundamental to all corrosion processes are only one manifestation of a whole series of electron transfer reactions associated with metal microbe interactions that are themselves central issues in microbial energetics.

Seen from the point of view of the altered redox status, and often also solubility of the metals, such interactions have been given the generic name, biomineralisation. For example, the aerobic corrosion of ferrous metals gives rise to mixed ferric oxide/hydroxide which is precipitated as a corrosion product. Further, the combined action of iron and sulphur oxidising bacteria can be harnessed to the solubilisation and leaching of metals such as copper and uranium from their insoluble ores (Ehrlich,

1990; Schippers *et al.*, 1996). In bioremediation the metallic component, again most usually iron, can act as either electron donor or acceptor leading, respectively, to stimulated reductive dechlorination or enhanced anaerobic degradation of organic pollutants (Lovley, 1995a,b).

This more open view of MIC stems from the elucidation of the key elements in the mechanism of anaerobic corrosion of mild steel caused by the sulphate-reducing bacteria. Reinforcement is given to the hypothesis proposed by the recent findings of the role of manganese oxidising bacteria in corrosive attack on stainless steel. There remain, however, a number of experimental findings still to be accommodated within the model. It will be the aim of this chapter to place knowledge of microbial corrosive mechanisms within the wider context of microbial physiology, and to identify those elements that may yet bring further elucidation to understanding of the processes involved.

ANAEROBIC CORROSION BY SULPHATE-REDUCING BACTERIA

MIC of mild steel by sulphate-reducing bacteria (SRB) has recently been the subject of a series of review papers. These have considered the subject within the broader context of electrochemical reactions in a wide range of biotic and abiotic corrosive processes (Hamilton and Lee, 1995; Lee *et al.*, 1995; Lewandowski *et al.*, 1997), and against the background of the basic physiology of the SRB (Hamilton, 1998; 1999). The features of primary importance have been identified, and are presented below.

SRB stimulate corrosion as a consequence of their activities within mixed species microbial consortia in the form of biofilms adherent to the mild steel substratum. Maximal corrosive activity is demonstrated where there is access to oxygen and the biofilm develops a characteristic aerobic-anaerobic (O_2/AnO_2) interface.

The primary role of the SRB is generic and comprises the production of sulphide. At least a proportion of this sulphide arises directly from the oxidation of hydrogen formed at the cathode of the electrochemical corrosion cell.

Iron dissolution takes place at the electrochemical anode and combines with the biogenic sulphide to give precipitated iron sulphides as corrosion products. The chemical nature and physical form of the iron sulphides are dependent upon a number of factors, *viz.* the relative concentrations of soluble iron and sulphide, access of oxygen, the presence of other bacteria within the biofilm, *e.g.* iron and/or sulphur oxidising species, and the time course of development of biofilm and associated corrosion processes.

Iron sulphide corrosion products may be either protective or corrosive. Where they are in the form of a tightly adherent thin film or tarnish, they protect the underlying unreacted steel in a manner directly analogous to the oxide film that constitutes the corrosion resistance of stainless steel. In the case of iron sulphides, however, such films are inherently unstable and their rupture gives rise to extremely active corrosion cells between the iron sulphide (cathode) and the exposed steel surface (anode). Also, whereas oxide films on stainless steel have the capacity to repassivate after any perturbation, this property is not evident with the iron sulphides.

Thick, loosely adherent iron sulphide deposits act to stimulate corrosion in the same manner as outlined for ruptured protective films. In order for iron sulphide corrosion products to function by this mechanism, they must be in direct electrical contact with the underlying steel substratum.

A major feature of corroding biofilms is the nature and extent of the heterogeneities they demonstrate, *viz.* consortia of mixed microbial species, almost certainly located at separate sites within the biofilm, O_2/AnO_2 interface, a range of iron sulphide corrosion products overlaid, in the presence of oxygen, with ferric oxide/hydroxide, separate foci of electrochemical activity with iron dissolution and pit formation at anodic sites.

Despite the extent of present knowledge of MIC caused by the sulphate reducers, there remain areas where there are significant uncertainties, however. The quantitative importance of the oxidation of cathodic hydrogen has not been established. It is still not possible to relate unambiguously protective and corrosive effects to the nature of the corrosion product(s) Fe_xS_y deposited. Perhaps most importantly, it remains unclear what is the nature of the relationships, direct and indirect, between the heterogeneity of the electrochemical corrosion reactions and the microbial, chemical and physical heterogeneities within the biofilm itself.

Model for SRB Corrosion

The most full and convincing exposition of the current state of understanding of the mechanism of biocorrosion of mild steel caused by the SRB is given in the papers by Lee *et al.* (1993a, b) and Nielsen *et al.* (1993).

The dissolution of ferrous iron is an anodic process involving an oxidative reaction with the loss of electrons.

$$Fe^{o} \Rightarrow Fe^{2+} + 2e^{-}$$

In order for this reaction to proceed it must be coupled to a parallel reductive reaction. This occurs at a cathodic region on the metal surface where, under anaerobic conditions, protons or hydrogen sulphide (Costello, 1974) may act as electron acceptor.

$$2H^{+} + 2e^{-} \Rightarrow 2H \Rightarrow H_2$$

or

$$2H_2S + 2e^{-} \Rightarrow 2HS^{-} + H_2$$

These anodic and cathodic reactions, and their location at separate sites on the metal surface, constitute an electrochemical corrosion cell. Each partial reaction is thermodynamically characterised by its redox potential, and they are coupled at a potential value midway between the more positive cathodic reduction and the more negative anodic oxidation. For mild steel at neutral pH values this corrosion potential (E_{corr}) is around –700mV, as measured against the standard calomel electrode (SCE). The value of E_{corr}, and the rate of any subsequent corrosion in a given experimental situation however, is determined by environmental and kinetic factors affecting each of the individual partial reactions, as indicated below.

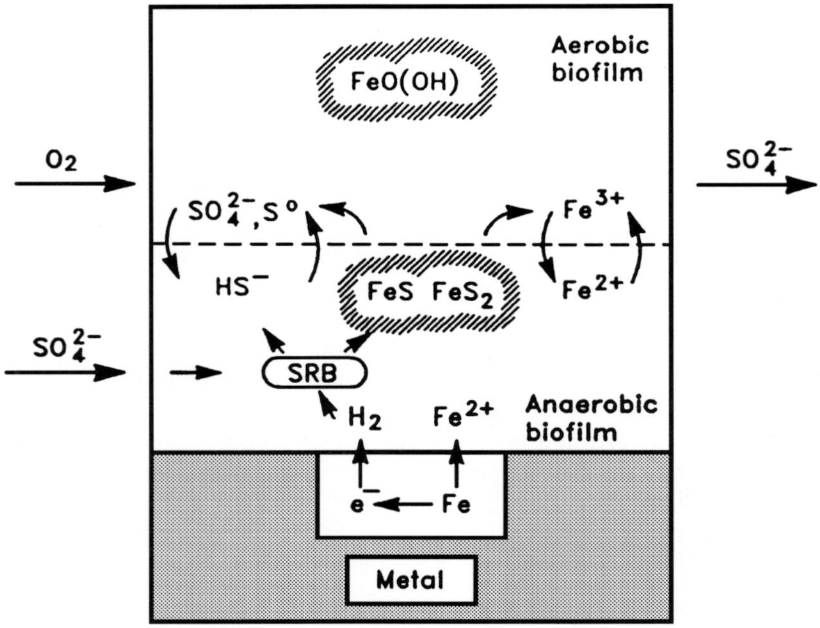

Figure 1 A descriptive model of the corrosion of mild steel resulting from the action of sulphate-reducing bacteria in a mixed anaerobic/aerobic system in which oxygen acts as the terminal electron acceptor. (Reproduced from Nielsen *et al.*, 1993, with permission).

The molecular hydrogen resulting from these coupled reactions is subject to biological oxidation by SRB. This removal of hydrogen from its site of production is termed cathodic depolarisation as it facilitates the electrochemical cathodic reaction which is generally considered to be the controlling step in the overall corrosion process. It should be mentioned at this point, however, that a number of authors have taken a contrary view in which it is propsed that SRB stimulate corrosion by a mechanism of anodic depolarisation (Hamilton and Lee, 1995). A key element in this hypothesis is localised acidification at the anode which results from the formation of iron sulphide corrosion products (Crolet, 1992; Daumas *et al.*, 1993).

$$Fe^{2+} + HS^- \Rightarrow FeS + H^+$$

According to both hypotheses, the production of sulphide results in the precipitation of ferrous sulphide corrosion products within the anaerobic regions of the biofilm. Across the O_2/AnO_2 interface however, both reduced ferrous and sulphide ions are subject to oxidation, abiotic and/or biotic, with oxygen as terminal electron acceptor. This, in turn, generates the ferric oxide/hydroxide and elemental sulphur as corrosion products which are recognised as being diagnostic for active SRB corrosion. That is to say, the apparent paradox of the key involvement of oxygen in so-called anaerobic SRB corrosion is solved, and a proper theoretical framework is established for the underlying mechanisms. This reaction scheme is illustrated in Figure 1.

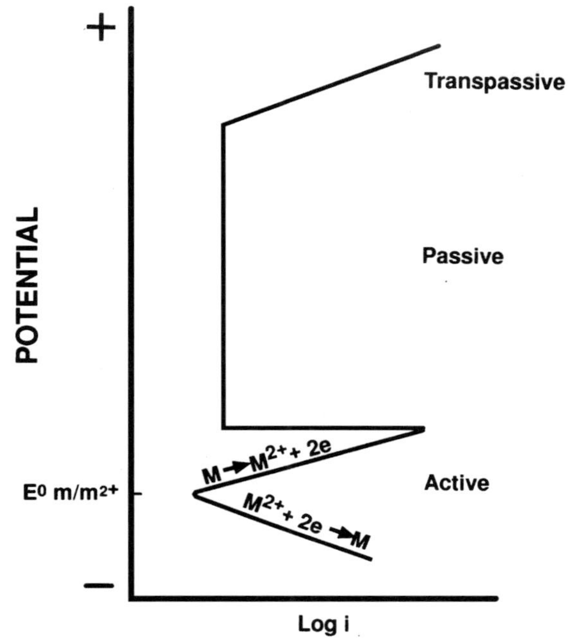

Figure 2 An idealised anodic curve obtained by direct current polarisation of a metal subject to passivation. (Reproduced from Dexter *et al.*, 1991, with permission).

Hypothesis

SRB corrosion of mild steel occurs by a process of electron transfer from the base metal to oxygen as ultimate electron acceptor, through a series of coupled redox reactions of electrochemical, biotic and abiotic character respectively.

The strength of any such hypothesis lies with the demonstration of its general validity through the confirmation of its basic tenets from experimental analysis of another system. Also, it must be able, possibly through modification or development, to accommodate new and even apparently contradictory data. Both of these circumstances can be seen to apply to the coupled electrochemical and biotic electron transfer hypothesis of microbially influenced corrosion.

MANGANESE OXIDISING BACTERIA AND THE CORROSION OF STAINLESS STEEL

Stainless steels owe their overall corrosion resistance to the inclusion in their metallurgical formulation of chromium, nickel, manganese and, possibly, molybdenum. Putting it at its simplest, these elements react with oxygen to form a stable oxide film which passivates the surface of the steel and constitutes the mechanism of its corrosion resistance. In electrochemical terms, this is best explained with reference to Figure 2.

The Figure gives an idealised representation of the anodic polarisation curve for a metal, such as stainless steel, which is subject to passivation. Using a technique known as direct current polarisation (Dexter *et al.*, 1991), the relationship is established between potential and the logarithm of the current density for the anodic metal dissolution reaction.

$$M \Rightarrow M^{2+} + 2e^-$$

For a so-called active metal such as mild steel which is not normally subject to passivation, at potentials close to and above the reversible metal potential, $E^o{}_{M/M}{}^{2+}$, there is a strong tendency to corrode, provided only that there is a suitable coupled cathodic reductive reaction. This is the situation referred to above, with protons or H_2S as electron acceptor and an active corrosion potential, E_{corr}, at around $-700mV_{sce}$.

With stainless steel however, the oxide film determines that with increasing potential a passive phase is reached where there is no parallel increase in current density; *i.e.* there is only a minimal so-called passive current flow, neither anodic nor cathodic reactions occur, and there is no corrosive metal loss. The resting potential for stainless steel is about $-150mV_{sce}$ which falls within this passive phase, as identified in Figure 2.

When the potential is further raised however, the anodic polarisation curve enters a new active, or transpassive phase which is once again characterised by parallel increases in potential and current density. In the presence therefore, of a suitable cathodic reactant with a sufficiently positive redox potential, an active electrochemical corrosion cell can be established. Where this occurs the effects are usually severe with the onset of rapid and extensive crevice or pitting corrosion.

Ennoblement

Ennoblement is the term used to describe the raising of the potential for a passivated metal surface with the attendent danger of inducing a corrosion reaction, particularly in the presence of high chloride content. Ennoblement has been described in a number of papers as resulting from the presence of a microbial biofilm (Dexter, 1995). There have been a number of differing hypotheses as to the possible mechanism of the effect, however. It was the work of Dickinson *et al.* (1996a) which first identified the deposition of MnO_2 as being the likely causative agent. These authors noted ennoblement of 316L stainless steel coupons after exposure for periods up to 35 d in a natural fresh water stream. Potentials increased from $-150mV_{sce}$ and held steady at values close to $+350mV_{sce}$, with an associated increase in cathodic current density. The biofilm formed during the exposure included characteristic 10–20 μm diameter annular deposits which were shown to be rich in precipitated MnO_2. It was found that the ennoblement effect on the stainless steel could be reproduced by coating the coupons with a MnO_2 paste. One or two electron transfer reactions were discussed as possible alternatives for the cathodic reduction of MnO_2.

$$MnO_2 + H_2O + e^- \Rightarrow MnOOH + OH^-$$

$$MnO_2 + 2H_2O + 2e^- \Rightarrow Mn^{2+} + 4OH^-$$

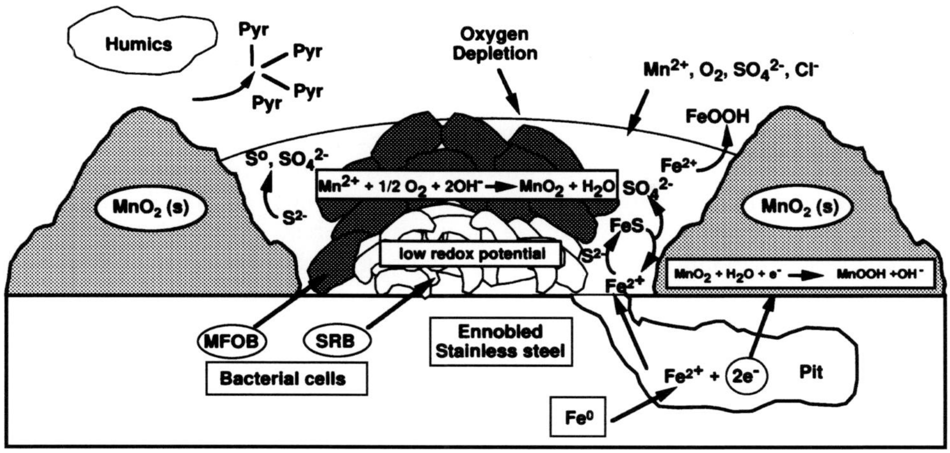

Figure 3 A hypothetical model for the ennoblement and corrosion of stainless steel consequent upon the biomineralisation of MnO_2. (Reproduced from Dickinson and Lewandowski, 1996, with permission).

Similar involvement of MnO_2 in the ennoblement of stainless steels has also been reported by Linhardt (1994) and Renner (1996).

In a series of papers, the Bozeman group have explored further the MnO_2 effect (Dickinson *et al.*, 1996b; 1997; Olesen *et al.*, 1999). Firstly, the biological nature of the process was established by the demonstration of ennoblement when stainless steel coupons were suspended in pure cultures of the manganese-oxidising bacterium *Leptothrix discophora*. The increase in potential was shown to correlate with the loss of soluble manganese from the medium and with the increase in the oxide precipitated on the steel surface.

With regard to the cathodic reduction of the manganese oxide deposits, X ray photoelectron spectroscopic (XPS) analysis of electrochemical reduction of plated MnO_2 has established that the reaction proceeds to Mn^{2+}, but with MnOOH as an intermediary reactant. Most recently, it has been shown that when 304L stainless steel is exposed to 0.35% NaCl subsequent to chemical or microbiological MnO_2 ennoblement, severe pitting corrosion develops, whereas unreacted steel remains corrosion-resistant at this concentration of NaCl (Olesen, Beyenal and Lewandowski, personal communication).

The implications of these findings, both in terms of corrosion mechanisms affecting passivated metals such as stainless steel, and with regard to the wider involvement of manganese-oxidising and sulphate-reducing bacteria in corrosion processes in general, have been considered in two important papers (Dickinson and Lewandowski, 1996; Lewandowski *et al.*, 1997). An illustrative model for the proposed reaction scheme and species interactions has been put forward, and is reproduced here as Figure 3. It is important to recognise that this model is predictive and to some degree hypothetical, whereas that for MIC with mild steel (Figure 1) is descriptive of what are known to be the processes occuring.

Figure 3 includes the possibility of anaerobic zones being generated within the annular MnO_2 deposits and beneath the actively metabolising manganese-oxidising bacteria; these would favour the growth of sulphate reducers whose activity at low values of electrochemical potential might further stimulate the corrosion process initiated by the MnO_2 ennoblement. Also, reference is made in the model to the widely recognised surface activity of MnO_2 which here may catalyse the breakdown of humic acids to low molecular weight metabolites, and the oxidation of reduced iron and sulphide to ferric oxide/hydroxide and elemental sulphur, repectively.

The essential point, however, of these findings on the ennoblement of stainless steel by the biomineralised deposition of MnO_2 is that it represents another clear demonstration of redox reactions in microbial physiology coupling with, and in this case making possible, the electrochemical redox reactions of metal corrosion.

There are, of course, major phenomenological differences between the two microbiologically influenced corrosion processes in mild and stainless steels. The SRB stimulate a pre-existing active corrosion in mild steel, and the biogenic sulphide reacts with ferrous ions already solubilised from the metal substratum. The biomineralised MnO_2 arises from the oxidation of soluble manganese in the bulk medium and makes possible a corrosion that would not otherwise occur. The SRB serve to initiate a series of reactions which are thereafter autocatalytic. In the case of MnO_2, it is reduced as the cathodic reductant and must, therefore, be continually replenished by ongoing microbial activity. The two microbially influenced cathodic reactions occur at very different electrochemical potentials.

The essential redox characteristics remain, however, and in so doing considerably strengthen the electrochemical/physiological electron transfer hypothesis of microbially influenced corrosion by extending its validity to both active mild and passivated stainless steels.

ENVIRONMENTAL AND PHYSIOLOGICAL FACTORS

Since the principal thesis of this discussion is that microbial corrosion should be seen as one particular expression of metal microbe interactions which occur also in many other forms within natural environments, it is worthwhile looking at some of these other processes in order to consider what facets they demonstrate which may add to, or challenge, the electron transfer hypothesis of MIC.

Alternative Electron Acceptors in Marine Sediments

It has been widely recognised for many years that carbon flux in sediment systems is coupled to the reduction of a series of electron acceptors of increasingly reductive redox potential, and located at increasing depth within the sediment. In marine and estuarine environments, although methanogenesis may be present at some greater depths, the principal terminal electron acceptor is clearly identified as sulphate, along with the associated sulphidogenic activity. Particularly in more recent studies, however, the role of iron, Fe(III), and manganese, Mn(IV), reduction has also been stressed (Nealson and Myers, 1992; Canfield *et al.*, 1993; Aguilar and Nealson, 1994; Aller,

1994; Nealson and Saffarini, 1994; Thamdrup *et al.*, 1994; Ramsing *et al.*, 1996; Thamdrup and Canfield, 1996).

This extensive body of work shows that across a wide range of marine habitats there is consistent evidence for the presence of Fe(III) and Mn(IV) reduction linked to organic carbon oxidation. In several instances the data suggest that the major role of iron and manganese takes the form of Fe(II)/Fe(III) and Mn(II)/Mn(IV) redox cycling, lying between and functionally coupling the quantitatively more important aerobic oxygen and anaerobic sulphate reduction zones. Although it is known that a broad spectrum of microbial species have the capability of manganese and/or iron reduction (Gounot, 1994; Lovley, 1993; 1995a), it appears likely that in natural sediment ecosystems much of the iron and virtually all of the manganese reduction occurs abiotically. Thamdrup *et al.* (1994), for example, have calculated that in their coastal sediment samples 100% of Mn(IV) reduction is coupled to Fe(II) and HS^- oxidation, while 63% of Fe(III) reduction is coupled to HS^- oxidation. These processes lead to the formation of elemental sulphur and of pyrite (FeS_2), and show minimal involvement of metal-reducing bacteria. Similar findings were reported by Nealson and Myers (1992).

Although they make no specific mention of corroding biofilms, Ramsing *et al.* (1996) stress that with regard to the vertical zonation of electron donors and acceptors in stratified water columns, marine sediments and microbial mats, the only significant differences are those of dimension or scale. It is clear, therefore, that these sediment data have major relevance to the closely parallel activities underpinning the involvement of MnO_2 in the ennoblement and pitting corrosion of stainless steel. Whereas one group of bacteria are responsible for the Mn(II) oxidation and precipitation of MnO_2, in corrosion the balancing reduction of Mn(IV) is the electrochemical cathodic reaction. What environmental factors, therefore, determine that in this particular circumstance Mn(IV) reduction is not carried out harmlessly by manganese-reducing bacteria or abiotically by reduced sulphide? This question, which remains unanswered at the present time, is even more pressing in situations where the presence of SRB has been reported in cases of ennobled stainless steel, and where it is suggested that there may be a synergy between the two microbial processes of manganese oxidation and sulphate reduction (Lewandowski *et al.*, 1997). Clearly for this to be the case such activities would have to be physically separated on the steel surface, and/or all the biogenic sulphide would require to be removed as precipitated iron sulphides in order to prevent its competing in the reduction of MnO_2, thereby inhibiting rather than enhancing the ennoblement.

Iron Reducing Bacteria

Although the microbial reduction of Fe(III) has been recognised for some time, even as late as 1993, it was possible to say "*Geobacter metallireducens* and *Shewanella putrefaciens* are the only well characterized organisms known to couple the oxidation of fermentation acids and/or hydrogen to the reduction of Fe(III)" (Coleman *et al.*, 1993). It is now known that the capacity for Fe(III) reduction is distributed widely within both the Archaeal and Bacterial kingdoms. The majority of bacterial species with this property are found amongst the γ and δ Proteobacteria, but it is also present in certain

Gram-positive organisms and in separate lineages such as *Geovibrio* and *Wolinella* (Caccavo *et al.*, 1996; Lovley, 1997). Fe(III) reduction has also now been found in both thermophilic bacteria (Bridge and Johnson, 1998) and hyperthermophilic Archaea (Vargas *et al.*, 1998).

It is also evident from these studies that Fe(III) is only one of several possible electron acceptors capable of being utilised by these organisms, although the identity of the alternatives shows a considerable degree of species-specificity. For example, *Geovibrio ferrireducens* and those *Pelobacter* spp. so far studied are also sulphur reducers, as are some but not all strains of *Desulfuromonas* and *Geobacter* (Coates *et al.*, 1995; Lovley *et al.*, 1995; Caccavo *et al.*, 1996; Lovley, 1997). On the other hand, sulphate reducing species such as *Desulfovibrio vulgaris*, *Desulfovibrio desulfuricans*, *Desulfobulbus propionicus*, *Desulfobacterium autotrophicum* and *Desulfobacter hydrogenophilus*, but not *Desulfobacter postgatei*, have been now shown to be capable also of Fe(III) reduction (Coleman *et al.*, 1993; Lovley, 1997). Moreover the affinity of *D. desulfuricans* for H_2 with Fe(III) as electron acceptor is an order of magnitude higher than with sulphate, which suggests that in many natural environments it is possible that this so-called sulphate-reducing bacterium might behave rather as a Fe(III)-reducing organism (Coleman *et al.*, 1993).

In contrast to these findings, *Shewanella putrefaciens*, probably the most studied Fe(III) reducer at the present time, can reduce tetrathionate, thiosulphate and sulphite, but not either sulphate or sulphur itself. Whereas another Fe(III) reducer *Bacillus infernus*, isolated from a deep subsurface environment and the only known anaerobic *Bacillus* sp., can utilize Mn(IV), trimethylamine oxide and nitrate, but none of the oxidised forms of sulphur as electron acceptor (Boone *et al.*, 1995).

The physiology, ecology, evolutionary development and phylogenetic relationships of the iron-reducing bacteria are at an early stage of concerted scientific study, but already there is a body of evidence which calls into question certain of the assumptions that presently determine the interpretation of respiratory mechanisms in anaerobic ecosystems, (see, for example, Lovley *et al.* (1995); Vargas *et al.* (1998)). In terms of the focus on MIC in this presentation, the suggestion that SRB might not always function as sulphate reducers with concomitant suphidogenesis, must at least give pause for thought in view of the centrality of that mechanism to the model for the corrosion of mild steel. Also, the capacity for Fe(III) reduction has the potential to influence the corrosion process directly. Obuekwe *et al.* (1981a; b; c), for example, have shown that an Fe(III)-reducing *Pseudomonas* sp. (subsequently reclassified as a *Shewanella* sp.) caused anodic depolarisation in mild steel and stimulated corrosion by the reductive removal of a protective, insoluble Fe(III) surface film.

Extracellular Polymeric Substances

In addition to the redox properties that dominate metal microbe interactions, it is significant that many of the products of biomineralisation are of altered, and often reduced solubility. In the present context, the insolubility of ferric oxide/hydroxide and MnO_2, and indeed of mild and stainless steels, requires that metal microbe interactions involving these materials incorporate some mechanism of solubilisation or direct cell-substrate contact.

In his review, Lovley (1997) cites an example of solubilisation where the addition of chelators such as nitrilotriacetic acid to sediments from the Fe(III) reduction zone of a petroleum-contaminated aquifer greatly accelerated Fe(III) reduction and its coupled oxidation of aromatic hydrocarbons. The addition of humic substances had a similar effect although in this case it was hypothesised that the humics were acting by an electron carrier mechanism in which chemical Fe(III) reduction coupled to humic oxidation was followed by re-reduction of the humics by the so-called iron-reducing bacteria.

In most instances, however, extracellular polymeric substances (EPS) have been proposed as providing the critical functional linkage between microbial cell and insoluble substrate. In the case of pyrite, FeS_2, oxidation (bioleaching) by *Thiobacillus ferrooxidans* and *Leptospirillum ferrooxidans*, Schippers *et al.* (1996) and Gehrke *et al.* (1998) have shown that the process is mediated by the formation of a lipopolysaccharide-Fe(III) complex. However, there is a considerable degree of specificity. When *T. ferrooxidans* is grown on elemental sulphur, although it again synthesises EPS, the polymer in this instance is more hydrophobic and does not bind to FeS_2. Another example of such substrate-specificity is the copious EPS production when *Shewanella putrefaciens* is grown on MnO_2 as compared to the absence of polymer when the same organism is grown with ferric oxide as terminal electron acceptor (Little *et al.*, 1998). A possible explanation of this effect is to be found in the work of Caccavo *et al.* (1997) who showed that the adhesion of *Shewanella alga* to amorphous Fe(III) oxide was dependent upon cell surface proteins and hydrophobic interactions. An adhesion-deficient mutant produced copious exopolysaccharide that possibly sterically hindered the protein-Fe(III) oxide interaction. Perhaps surprising, however, was the fact that both the wild type and the mutant had similar Fe(III) reducing activity.

In the case of microbial biofilms associated with corrosive weight loss from their metal substrata, there are many examples where the biofilm EPS has been claimed to play a significant role in the process (see, for example, Geesey *et al.* (1988), Ford and Mitchell (1990); Beech *et al.* (1991)). Beech and colleagues have examined the possible role of EPS from SRB in the corrosion of mild steel in an extended study. A particular isolate from a corroding ship's hull has been identified as a new species, *Desulfovibrio indonensis* (Feio *et al.*, 1998) and shown to be particularly aggressive in comparison with other strains isolated from non-corrosive environments. Comparison of EPS composition from aggressive and non-aggressive strains, both with and without the addition of mild steel coupons, demonstrated that the presence of the metal induced qualitative differences in the polymers, most noticeably the appearance of new protein bands after SDS-PAGE electrophoresis (Zinkevich *et al.*, 1996). Further analyses showed that the crude EPS from the aggressive *D. indonensis* is more active than that from a non-aggressive organism in an iron binding assay, and that the bound iron was in the form Fe(III) (Beech *et al.*, 1999). Furthermore, it was noted that iron binding was only found with EPS from cells grown in the presence of mild steel; soluble ferrous salts in the medium did not induce this effect. Partial purification of the EPS from *D. indonensis* has allowed identification of a thermostable polysaccharide-protein complex of molecular mass greater than 200kD which both binds iron from a steel surface and, when added to such a surface causes a pitting corrosion reaction (Beech *et al.*, 1998). The hypothesis put forward by these authors is that *D. indonensis* demonstrates species-specificity in its corrosive activity and that the basis of its aggressiveness lies with the high Fe(III)-binding affinity of a polysaccharide-protein fraction within the EPS produced in the presence of a susceptible metal substratum.

It is important to appreciate, however, that the presence of biological polymers at a metal surface may also demonstrate a number of possible non-specific effects through their capacity to change the near-surface physicochemical environment. Roe *et al.* (1996), for example, examined the effects of spotting pure polymer samples onto coupons of mild steel in an experiment designed to compare alginate with agarose. Whereas the former polymer has a net negative charge due to its carboxylic uronic acid groups and is known to bind positively charged metal ions, the neutral agarose possesses neither property. In terms of the pitting corrosion induced, however, there was no evidence for an increased effect with alginate. The authors deduce that the corrosion measured with both polymer additions was due to the formation of an oxygen concentration cell. In instances where there is an incomplete or patchy distribution of polymer, or of biofilm, corrosion can arise by this mechanism in which the lower concentration of oxygen under the deposit gives rise to an anodic region, as compared to that portion of the metal with direct access to oxygen which then acts as a cathode. In discussing their findings, the authors also draw attention to instances where a complete polymer coating is associated with a decrease in corrosion due to the shielding of the underlying metal from oxygen or other possible cathodic reactant.

These concerns have a particular relevance to laboratory studies of SRB corrosion of mild steel. It has been clearly demonstrated that during the initial stages of biofilm development patchy microbial growth gives rise to a primary phase of oxygen-dependent corrosion with associated tubercle formation and localised regions of low pH (Lee *et al.* 1993a; 1993b; Lee and de Beer, 1995). Subsequently, more contiguous growth then shields the substratum from oxygen and allows the development of sufficient sulphidogenic activity to precipitate iron sulphide corrosion products and establish the sulphide/metal galvanic couple which is responsible for the major, and longer term, SRB-dependent corrosion.

Whereas the more generic consequences of EPS additions to metal substrata can be differentiated from such other effects as are identified as being dependent on the presence of specific chemical components of the EPS, it is clear that care must be taken in the interpretation of such effects.

Species Specificity and Biofilm Heterogeneities

Passing reference has been made above to the capacity for SRB to utilise Fe(III) as electron acceptor, and for sulphur and thiosulphate to act as alternative substrates for sulphidogenesis. None the less it remains true to say that this presentation has focussed very firmly on the sulphidogenic activity of the SRB consequent upon their ustilisation of sulphate as electron acceptor. There are a number of papers, however, that have stressed the widespread occurrence of strains capable of thiosulphate reduction and have pointed out the potential corrosivity of such organisms (Ravot *et al.*, 1995; Magot *et al.*, 1997; Dawood and Brözel, 1998; Dawood *et al.*, 1998; McLeod *et al.*, 1998). One organism, in particular, has recently figured prominently in the corrosion literature. *Shewanella putrefaciens* is a facultative anaerobe with the capacity to utilise either oxygen or thiosulphate, amongst a number of other possible electron acceptors; that is to say, it has increased versatility compared to the strictly anaerobic SRB in that it can both generate anaerobic conditions from its own metabolic activity and then switch to a sulphidogenic mode of energy metabolism. It also has the capability of H_2 oxidation and has been clearly identified as a potentially corroding species (Dawood and Brözel, 1998; Dawood *et al.*, 1998; McLeod *et al.*, 1998). As noted above, the corrosive *Pseudomonas* isolated by Obuekwe (1981b) has now been reclassified as a *Shewanella*. In their review, Nealson and Little (1997) report further studies with this organism that show that it has corrosive capability from either or both Fe(III) and thiosulphate reduction. They also cite their own pure culture studies with *S. putrefaciens* designed to clarify the nature of the corrosion processes involved and, in particular, to examine the effect of Fe(III) reduction on the removal of potential passivating films on mild steel. The importance of these studies is considerable as it sheds new light on previously unrecognised complexity within the process of sulphidogenic corrosion of mild steel. It is also highly significant in terms of the case being argued in this present article that the reactions involved are again entirely compatible with the electrochemical/physiological electron transfer hypothesis.

CONCLUSION

The versatility of *S. putrefaciens* has made it possible to carry out controlled laboratory experiments with some reasonable confidence that natural environmental conditions are being adequately simulated. That key condition, however, for meaningful analyses and interpretation remains an assumption. There is now a major requirement for an experimental programme, using the molecular techniques that have so revolutionised other areas of microbial ecology, and designed to obtain a true, three dimensional picture of what organisms and activities are actually present within corroding microbial biofilms. Only from such a study will it be possible to answer the two major questions still outstanding. What are the true physiological activities and physicochemical microenvironments responsible for naturally occurring MIC? What is the relationship between the heterogeneities within the biofilm, and the heterogeneities on the corroding metal surface? It is proposed that the electrochemical/physiological electron transfer hypothesis presented here constitutes a suitable theoretical framework for such analyses.

REFERENCES

Aller R.C. (1994). The sedimentary Mn cycle in Long Island Sound – its role as intermediate oxidant and the influence of bioturbation, CO_2, and C (org) flux on diagenetic reaction balances. *J Mar Res*, **52**, 259–295.

Aguilar C., Nealson K.H. (1994). Manganese reduction in Oneida Lake, New York – estimates of spatial and temporal manganese flux. *J Bacteriol*, **175**, 7594–7603.

Beech I.B., Gaylarde C.C., Smith J.J., Geesey G.G. (1991). Extracellular polysaccharides from *Desulfovibrio desulfuricans* and *Pseudomonas fluorescens* in the presence of mild and stainless steel. *Appl Microbiol Technol*, **35**, 65–71.

Beech I.B., Zinkevich V., Tapper R., Gubner R. (1998). Direct involvement of an extracellular complex produced by a marine sulfate-reducing bacterium in deterioration of steel. *Geomicrobiol J*, **15**, 121–134.

Beech I.B., Zinkevich V., Tapper R., Gubner R., Avci R. (1999). Study of the interaction of sulphate-reducing bacteria exopolymers with iron using X-ray photoelectron spectroscopy and time-of-flight secondary ionisation mass spectrometry. *J Microbiol Methods*, **36**, 3–10.

Boone D.R., Liu Y., Zhao Z-J., Balkwill D.L., Drake G.R., Stevens T.O., Aldrich H.C. (1995). *Bacillus infernus* sp. nov., an Fe(III)- and Mn(IV)-reducing anaerobe from the deep terrestrial subsurface. *Int J Syst Bacteriol*, **45**, 441–448.

Bridge T.A.M., Johnson D.B. (1998). Reduction of soluble iron and reductive dissolution of ferric iron-containing minerals by moderately thermophilic iron-oxidizing bacteria. *Appl Environ Microbiol*, **64**, 2181–2186.

Caccavo F., Schamberger P.C., Keiding K., Nielsen P.H. (1997). Role of hydrophobicity in adhesion of the dissimilatory Fe(III)-reducing bacterium *Shewanella alga* to amorphous Fe(III) oxide. *Appl Environ Microbiol*, **63**, 3837–3843.

Caccavo F., Coates J.D., Rossello-Mora R.A., Ludwig W., Schleifer K.H., Lovley D.R., McInerney M.J. (1996). *Geovibrio ferrireducens*, a phylogenetically distinct dissimilatory Fe(III)-reducing bacterium. *Arch Microbiol*, **165**, 370–376.

Canfield D.E., Thamdrup B., Hansen J.W. (1993). The anaerobic degradation of organic matter in Danish coastal sediments: iron reduction, manganese reduction, sulfate reduction. *Geochim Cosmochim Acta*, **57**, 3867–3883.

Coates J.D., Lonergan D.J., Philips E.J.P., Jenter H., Lovley D.R. (1995). *Desulfuromonas palmitatis* sp. nov., a marine dissimilatory Fe(III) reducer that can oxidize long-chain fatty acids. *Arch Microbiol*, **164**, 406–413.

Coleman M.L., Hedrick D.B., Lovley D.R., White D.C., Pye K. (1993). Reduction of Fe(III) in sediments by sulphate-reducing bacteria. *Nature (Lond)*, **361**, 436–438.

Costello J.A. (1974). Cathodic depolarisation by sulfate-reducing bacteria. *S Africa J Sci*, **70**, 202–204.

Crolet J-L. (1992). From biology and corrosion to biocorrosion. *Oceanol Acta*, **15**, 87–94.

Daumas S., Magot M., Crolet J-L.(1993). Measurement of the net production of acidity by a sulfate-reducing bacterium: experimental checking of theoretical models of microbially influenced corrosion. *Res Microbiol*, **144**, 327–332.

Dawood Z., Brözel V.S. (1998). Corrosion-enhancing potential of *Shewanella putrefaciens* isolated from industrial cooling waters. *J Appl Microbiol*, **84**, 929–936.

Dawood Z., Ehrenreich L., Brözel V.S. (1998).The effect of molecular oxygen on sulfite reduction by *Shewanella putrefaciens*. *FEMS Microbiol Lett*, **164**, 383–387.

Dexter S.C. (1995). Effect of biofilms on marine corrosion of passive alloys. In: Gaylarde C., Videla H.A. (eds) *Bioextraction and Biodeterioration of Metals*. Cambridge University Press, Cambridge, pp. 129–168.

Dexter S.C., Duquette D.J., Siebert O.W., Videla H.A. (1991). Use and limitations of electrochemical techniques for investigating microbiological corrosion. *Corrosion*, **47**, 308–318.

Dickinson W.H., Lewandowski Z. (1996). Manganese biofouling and the corrosion behaviour of stainless steel. *Biofouling*, **10**, 79–93.

Dickinson W.H., Caccavo F., Lewandowski Z. (1996a). The ennoblement of stainless steel by manganic oxide biofouling. *Corrosion Sci*, **38**, 1407–1422.

Dickinson W.H., Lewandowski Z., Geer R.D. (1996b). Evidence for surface changes during ennoblement of type 316L stainless steel: dissolved oxidant and capacitance measurements. *Corrosion Sci*, **52**, 910–920.

Dickinson W.H., Caccavo F., Olesen B.H., Lewandowski Z. (1997). Ennoblement of stainless steel by the manganese-depositing bacterium *Leptothrix discophora*. *Appl Environ Microbiol*, **63**, 2502–2506.

Ehrlich H.L. (1990). *Geomicrobiology*. Marcel Dekker, New York.

Feio M.J., Beech I.B., Carepo M., Lopes J.M., Cheung C.W.S., Franco R., Guezennec J., Smith J.R., Mitchell J.I., Moura J.J.G., Lino A.R. (1998). Isolation and characterisation of a novel sulphate-reducing bacterium of the *Desulfovibrio* genus. *Anaerobe*, 4, 117–130.

Ford T.E., Mitchell R. (1990). The ecology of microbial corrosion. *Adv Microbial Ecol*, **11**, 231–262.

Geesey G.G., Jang L., Jolley J.G., Hankins M.R., Iwaoka T., Griffiths P.R. (1988). Binding of metal ions by extracellular polymers of biofilm bacteria. *Water Sci Technol*, **20**, 161–165.

Gehrke T., Telegdi J., Thierry D., Sand W. (1998). Importance of extracellular polymeric substances from *Thiobacillus ferrooxidans* for bioleaching. *Appl Environ Microbiol*, **64**, 2743–2747.

Gounot A. (1994). Microbial oxidation and reduction of manganese: consequences in groundwater and applications. *FEMS Microbiol Rev*, 14, 339–350.

Hamilton W.A. (1998). Sulfate-reducing bacteria: physiology determines their environmental impact. *Geomicrobiol J*, **15**, 19–28.

Hamilton W.A. (1999). Bioenergetics of sulphate-reducing bacteria in relation to their environmental impact. *Biodegradation*, **9**, 201–212.

Hamilton W.A., Lee W. (1995). Biocorrosion In: Barton L.L. (ed) *Sulfate Reducing Bacteria*. Plenum Press, New York, pp. 243–264.

Lee W., de Beer D. (1995). Oxygen and pH microprofiles above corroding mild steel covered with a biofilm. *Biofouling*, **8**, 273–280.

Lee W., Lewandowski Z., Nielsen P.H., Hamilton W.A. (1995). Role of sulfate-reducing bacteria in corrosion of mild steel: a review. *Biofouling*, **8**, 165–194.

Lee W., Lewandowski Z., Okabe S., Characklis W.G., Avci R. (1993a). Corrosion of mild steel underneath aerobic biofilms containing sulfate-reducing bacteria. Part I: at low dissolved oxygen concentration. *Biofouling*, 7, 197–216.

Lee W., Lewandowski Z., Morrison M., Characklis W.G., Avci R., Nielsen P.H. (1993b). Corrosion of mild steel underneath aerobic biofilms containing sulfate-reducing bacteria. Part II: at high bulk oxygen concentrations. *Biofouling* 7, 217–239.

Lewandowski Z., Dickinson W., Lee W. (1997). Electrochemical interactions of biofilms with the metal surfaces. *Water Sci Technol*, **36**, 295–302.

Linhardt P. (1994). Manganese oxidizing bacteria and pitting of turbine components made of Cr Ni steel in a hydroelectric plant. *Werkst Korr*, 45, 79–83.

Little B.J., Wagner P.A., Lewandowski Z. (1998). Spatial relationships between bacteria and mineral surfaces. *Rev Mineralogy*, **35**, 123–159.

Lovley D.R. (1993). Dissimilatory metal reduction. *Ann Rev Microbiol*, 47, 263–290.

Lovley D.R. (1995a). Microbial reduction of iron, manganese, and other metals. *Adv Agronomy*, 54, 175–231.

Lovley D.R. (1995b). Bioremediation of organic and metal contaminents with dissimilatory metal reduction. *J Ind Microbiol*, 14, 85–93.

Lovley D.R. (1997). Microbial Fe(III) reduction in subsurface environments. *FEMS Mircobiol Rev*, **20**, 305–313.

Lovley D.R., Phillips E.J.P., Lonergan D.J., Widman P.K. (1995). Fe(III) and S^o reduction by *Pelobacter carbinolicus*. *Appl Environ Microbiol*, **61**, 2132–2138.

Magot M., Ravot G., Campaignolle X., Ollivier B., Patel P.K.C., Fardeau M.L., Thomas P., Crolet J-L., Garcia J.L. (1997). *Dethiosulfovibrio peptidovorans* gen. nov, sp. nov, a new anaerobic slightly halophilic thiosulphate-reducing bacterium from corroding offshore oil wells. *Int J Syst Bacteriol*, 47, 818–824.

McLeod E.S., Dawood Z., MacDonald R., Oosthuizen M.C., Graf J., Steyn P.L., Brözel V.S. (1998). Isolation and identification of sulphite- and iron-reducing hydrogenase positive facultative anaerobes from cooling water systems. *Syst Appl Microbiol*, **21**, 297–305.

Nealson K.H., Myers C.R. (1992). Microbial reduction of manganese and iron: new approaches to carbon cycling. *Appl Environ Microbiol*, **58**, 439–443.

Nealson K.H., Saffarini D. (1994). Iron and manganese in anaerobic respiration: environmental significance, physiology, and regulation. *Annu Rev Microbiol*, **48**, 311–343.

Nealson K.H., Little B.J. (1997). Breathing manganese and iron: solid state respiration. *Adv Appl Microbiol*, **45**, 213–239.

Nielsen P.H., Lee W., Lewandowski Z., Morrison M., Characklis W.G. (1993). Corrosion of mild steel in an alternating oxic and anoxic biofilm system. *Biofouling*, **7**, 267–284.

Obuekwe C.O., Westlake D.W.S., Cook F.D., Costerton J.W. (1981a). Surface changes in mild steel coupons from the action of corrosion-causing bacteria. *Appl Environ Microbiol*, **41**, 766–774.

Obuekwe C.O., Westlake D.W.S., Plambeck J.A., Cook F.D. (1981b). Corrosion of mild steel in cultures of ferric iron reducing bacterium isolated from crude oil: polarization characteristics. *Corrosion*, **37**, 461–467.

Obuekwe C.O., Westlake D.W.S., Plambeck J.A., Cook F.D. (1981c). Corrosion of mild steel in cultures of ferric iron reducing bacterium isolated from crude oil: mechanism of anodic depolarization. *Corrosion*, **37**, 632–637.

Olesen B.H., Avci R., Lewandowski Z. (1999). Manganese dioxide as a potential cathodic reactant in corrosion of stainless steels. *Corrosion Sci* (In press).

Ramsing N.B., Fossing H., Ferdelman T.G., Andersen F., Thamdrup B. (1996). Distribution of bacterial populations in a stratified fjord (Mariager Fjord, Denmark) quatified by in situ hybridization and related to chemical gradients in the water column. *Appl Environ Microbiol*, **62**, 1391–1404.

Ravot G., Ollivier B., Magot M., Patel B.K.C. Crolet J-L. Fardeau M.L., Garcia J.L. (1995). Thiosulphate reduction an important physiological feature shared by members of the order Thermotogales. *Appl Environ Microbiol*, **61**, 2053–2055.

Renner M. (1996). Scientific, engineering, and economic aspects of MIC on stainless steel applications in the chemical process industry. *Dechema Monogr*, **133**, 59–70.

Roe F.L., Lewandowski Z., Funk T. (1996). Simulating microbiologically influenced corrosion by depositing extracellular biopolymers on mild steel surfaces. *Corrosion*, **52**, 744–752.

Schippers A., Jozsa P-G., Sand W. (1996). Sulfur chemistry in bacterial leaching of pyrite. *Appl Environ Microbiol*, **62**, 3424–3431.

Thamdrup B., Canfield D.E. (1996). Pathways of carbon oxidation in continental margin sediments off central Chile. *Limnol Oceanogr*, **41**, 1629–1650.

Thamdrup B., Fossing H., Jörgensen B.B. (1994). Manganese, iron and sulfur cycling in a coastal marine sediment, Aarhus Bay, Denmark. *Geochim Cosmochim Acta*, **58**, 5115–5129.

Vargas M., Kashefi K., Blunt-Harris E.L., Lovley D.R. (1948). Microbiological evidence for Fe(III) reduction on early earth. *Nature*, **395**, 65–67.

Zinkevich V., Bogdarina I., Kang H., Hill M.A.W., Tapper R., Beech I.B. (1996). Characterisation of exopolymers produced by different isolates of marine sulphate-reducing bacteria. *Int Biodeterior Biodegr*, **37**, 163–172.

24 Biofilms Without a Substratum: Flocs and Microbial Communities

Linda L. Blackall and Per Halkjær Nielsen

Agglomerations of microbial cells are a common phenomenon in wastewater treatment processes as well as the growth of microbes as these large communities are heavily relied upon for the successful implementation of the processes. The aggregations are known as flocs or granules depending upon the density of the mass. For example, the activated sludge process which classically is used to treat wastewater by largely aerobic mechanisms is colonised by aggregates called flocs, while the upflow anaerobic sludge blanket process, an anaerobic treatment option is dominated by granules. This chapter examines aspects of floc composition, chemistry, and microbiology in relation to wastewater treatment.

KEY WORDS: activated sludge, floc structure, microbial ecology

WASTEWATER TREATMENT BY THE ACTIVATED SLUDGE PROCESS

The treatment of domestic wastewater in the developed world today is taken for granted. However, the impetus for treating this wastewater only arose in the mid 1800s and was solely related to public health concerns due to the prevalence of water borne diseases like cholera. The contemporary goal for wastewater treatment is to reduce pathogens and to reduce environmental pollutants like nutrients. Most biological wastewater treatment processes rely upon the selective accumulation of pollutant-degrading microorganisms. Pathogens in the wastewater survive poorly in the process, leading to their reduction. The products of the process are more microbes (since they have grown on the pollutants), dissolved compounds of negligible pollutant impact and volatile compounds which are released into the atmosphere. Pollutants in domestic wastewater are typically measured as biochemical oxygen demand (BOD) or chemical oxygen demand (COD) and as some form of nitrogen (*e.g.* total nitrogen) and phosphorus (*e.g.* PO_4-P). Since the process is aqueous, the microorganisms must be able to be separated from the treated water. Suspensions of individual microorganisms are extremely difficult to separate from aqueous media and would be completely dependent on expensive additional processes like filtration or centrifugation. Therefore, numerous methods which facilitate contact between the microbes and the wastewater and expedite separation of the microbes from the treated water have been devised. Biofilms of the microorganisms growing on either suspended carriers (moving bed reactor) (*e.g.* Hem *et al.*, 1994), rotating discs (*e.g.* Klemetson and Lang, 1984) or solid media like rocks or plastic media (trickling filter reactor) (*e.g.* Daigger *et al.*, 1993; Henze *et al.*, 1995) facilitate easy separation of the biomass since it is well attached to a support which the water easily flows away from. Membrane reactors (*e.g.* Chiemchaisri and Yamamoto, 1993; Delanghe *et al.*, 1994) also rely on the growth of microorganisms as biofilms, and can be used in certain specialised situations.

One popular method for treating wastewater, particularly domestic wastewater, is the activated sludge process (Figure 1A and B) which was first devised at the beginning of the 20th century (Ardern and Lockett, 1914). Although initially devised as a "fill and draw" process, due to perceived operational constraints, it was implemented in full-scale as a continuous process with the water moving from one zone to another as depicted in Figure 1A.

The current activated sludge process is based on suspended growth in which the majority of the nutrient transformations are mediated in the aeration tank (Figure 1A and B) by a complex enrichment culture of a mixture of generalist and specialist microorganisms. The microorganisms comprising largely bacteria, protozoa and metazoa are contained within discrete three-dimensional structures known as "flocs" which range from about 70 μm up to hundreds of μm in diameter. Aeration provides two functions, *viz.* it keeps the mixed liquor in suspension allowing good contact between flocs and the nutrients in the wastewater, and it provides dissolved oxygen for the aerobic microorganisms.

The secondary sedimentation tank (Figure 1A) provides a quiescent zone for gravity sedimentation of the flocs leaving the water clear supernatant which is disinfected (*e.g.* by chlorination or ozonation) before being discharged to the receiving water body (*e.g.* a river or the ocean). This supernatant can be treated further by tertiary procedures such as filtration or centrifugation to remove even more biomass or suspended solids (SS) and BOD. The settled sludge comprises the flocs which are partially wasted (to accommodate the growth of microorganisms on the wastewater nutrients) but mostly recycled as Return Activated Sludge (RAS) to the aeration tank. Figure 1B shows the aeration tank of an activated sludge plant employing surface aerators and a rubber skirt to reduce aerosol dispersion.

NUTRIENT REMOVAL PROCESS DESIGN

Figure 1A describes the process that would be designed for carbon removal from wastewater. However, the removal of the nutrients nitrogen and phosphorus is also required during treatment. Detail on the design configurations for activated sludge systems can be found elsewhere (Henze *et al.*, 1995; Seviour *et al.*, 1999) but a summary will be given here. Most activated sludge plants are either of the plug-flow design with tapered aeration, step aeration or step feed, or operated as complete mixed processes. The sequencing batch reactor is also an option for wastewater treatment but currently, is less common than continuous processes. Organic carbon is typically used by aerobic microorganisms in their normal metabolism and converted to more microorganisms and CO_2. This typically occurs in single sludge processes where all the biomass circulates from the aeration tank to the clarifier and back to the aeration

A

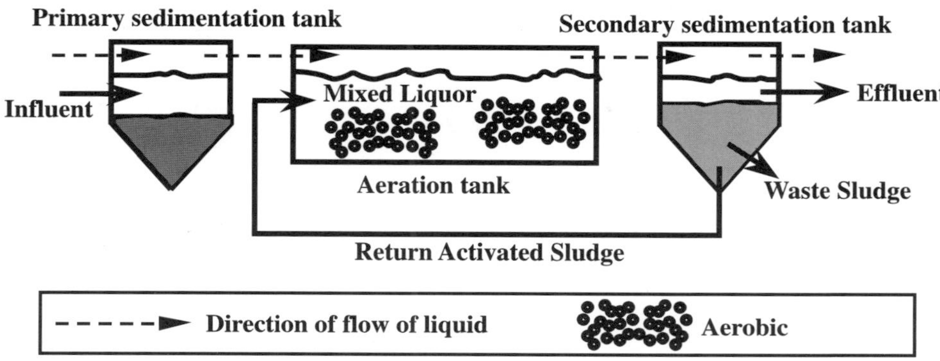

B

Figure 1 An activated sludge plant for carbon removal from domestic wastewater. A = diagram of the layout of the plant; B = photograph of the aeration tank section of a surface aerated plant, showing a rubber skirt to reduce the aerosol produced by vigorous mixing.

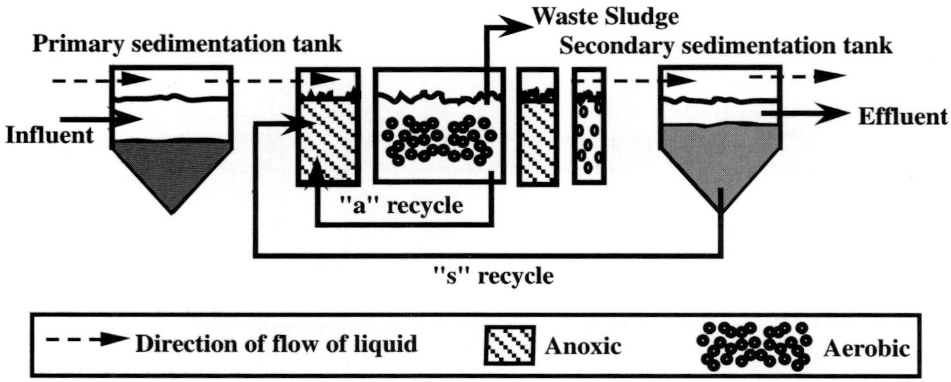

Figure 2 The Bardenpho process for removing nitrogen and carbon from domestic wastewater. Nitrified water is brought in the "a" recycle from the aeration tank to the first anoxic tank.

tank as shown in Figure 1A. In the last 20 years or so, there have been major advances in process design for nitrogen and phosphorus removal from wastewater. The microbial transformations required to achieve nitrogen removal are relatively well understood, but understanding of those for phosphorus removal is still at an elementary stage. This lack of knowledge in the fundamentals for phosphorus removal makes the design and operation of P removal processes very difficult.

Most nitrogen comes to the activated sludge plant in the influent in its reduced form and the method of choice for removal involves initial nitrification, followed by denitrification of the produced nitrate and nitrite to nitrogen gas which goes into the atmosphere. The influent also contains the organic carbon that the denitrifiers require so it requires flow recycles to satisfy the needs of both the nitrifiers and the denitrifiers. In early plant designs, the aerobic nitrification zone preceded the anoxic denitrification zone. Consequently, the organic carbon in the plant influent was partially used by heterotrophic organisms in the aerobic zone, leaving little or none for the denitrifiers in the subsequent anoxic zone. Carbon for denitrification could come from endogenous death and lysis of the active biomass or be added exogenously, typically as methanol. Later designs put the anoxic zone in front of and separate from the aerobic zone and installed a second recycle stream taking liquid from the end of the aerobic zone to the front of the anoxic zone. This modification ensured the denitrifiers were provided with both carbon and oxidised nitrogen and ensured the nitrifiers in the aerobic zone would have electron donors for metabolism in the form of reduced nitrogen coming through the anoxic zone from the influent. The final modification came when a second anoxic zone was installed after the aerobic zone to ensure that nitrate present in the flow to the clarifier was removed before it could contaminate the effluent. Figure 2 shows the so-called Bardenpho process for nitrogen removal from wastewater.

Phosphorus removal from wastewater can be carried out by passing the mixed liquor through an anaerobic zone followed by an aerobic zone and then to the clarifier with recycle of the sludge from the clarifier going to the anaerobic zone. If biomass is subjected to this pressure, it acquires the capacity to release phosphorus in the anaerobic zone and then to aerobically take up all the anaerobically released P and

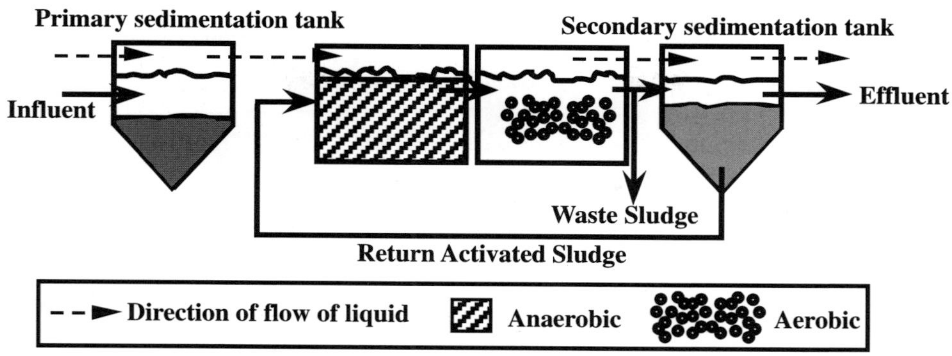

Figure 3 The A/O process for removal of phosphorus and carbon from domestic wastewater. Stored polyphosphate is released as PO_4 by the polyphosphate accumulating organisms in the anaerobic zone and taken up in the aerobic zone. Sludge at the end of the aerobic period loaded with polyphosphate is wasted to effect P removal.

the P from the influent and store it as polyphosphate. Concomitantly with P release in the anaerobic zone, the biomass takes up short chain fatty acids and produces intracellular poly-β-hydroxy alkanoates (PHA). In the aerobic zone, the PHA are utilised and intracellular glycogen is produced, which is used in the anaerobic zone. Therefore, both carbon and phosphorus transformations occur in the process and a stoichiometric relationship exists between the amount of carbon taken up and the P released. There are many empirically determined features about this process, but currently the full suite of microorganisms able to perform P release and uptake is not known, nor is the mechanism by which this occurs. More detail on P removal can be found in Bond and Rees (1999) and on the processes in Seviour *et al.* (1999). Features known to be important in biological phosphorus removal include the requirement for volatile fatty acids as the carbon source to drive P release in the anaerobic zone and the near complete absence of nitrates from the anaerobic zone. Figure 3 describes a summary of the process for P removal from wastewater.

Processes for both nitrogen and phosphorus removal were developed in parallel with those for P removal alone. Seviour *et al.* (1999) give details on the processes but a good process that achieves the required low levels of nitrogen and phosphorus in the effluent is the Modified University of Cape Town (MUCT) Process. Figure 4 describes the UCT process while in the MUCT process, the anoxic zone is divided into two compartments which allows more complete control over the recycle streams. The features important to biological phosphorus removal are provided by both configurations, while allowing simultaneous nitrogen removal and carbon removal.

AGGREGATION OF MICROORGANISMS

The elaboration of microorganisms as flocs in the activated sludge process is "automatic", *i.e.* when a process is initiated in the activated sludge configuration, microorganisms naturally grow in flocculated structures. There is no specific method for

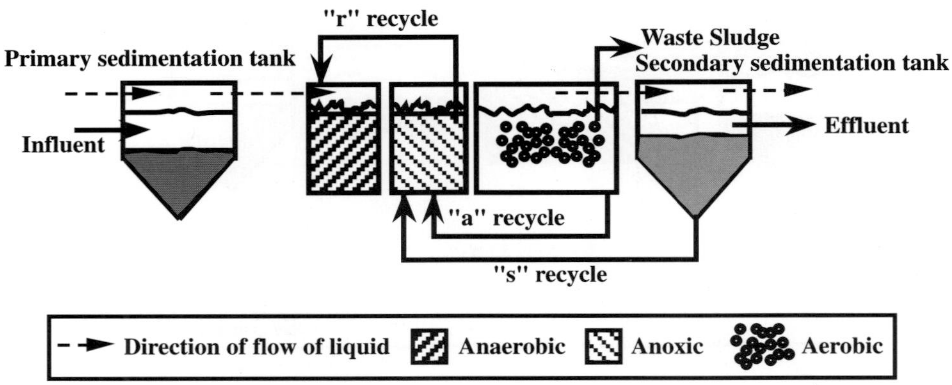

Figure 4 The University of Cape Town process for removal of phosphorus, nitrogen and carbon from domestic wastewater.

"forcing" microorganisms to grow in this manner. If they cannot grow at a rate that prevents their washout, they will not survive in the process. Therefore growth as flocs might be percieved to be a survival mechanism. However, efficient aggregation of microbes into flocs is crucial for process success since the sedimentation, compaction and dewaterability of the biomass in the clarifier relies upon the flocculation. Therefore, the study of fundamental aspects of flocculation is very important and can be exploited to improve sludge settleability, a major problem of the activated sludge process.

Types of Flocs and Filamentous Organisms

The activated sludge microbial population is initially investigated by microscopic investigation, an important method despite its apparent simplicity. Numerous early publications by wastewater researchers used this method in attempts to understand problems in floc settleability. Using simple phase-contrast microscopy, the flocs can be placed into different categories ranging from pinpoint flocs (very small, poorly compacting) through the ideal size that sediment quickly and compact well to large hydrated flocs with poor compaction in the sedimentation tank (Jenkins *et al.*, 1993). A major consideration in settleability became obvious from the early microscopic monitoring. This was related to the competitive growth of filamentous microorganisms (almost exclusively bacteria) in relation to so-called floc-forming bacteria. It was clear that in some situations, filamentous microorganisms dominated the microbial community in the activated sludge mixed liquor and it was hypothesised the filaments formed bridges between flocs preventing their compaction in the clarifier. There are many early publications on the observation and microscopic categorisation of filaments (Farquhar and Boyle, 1971a; 1971b; van Veen, 1973; Eikelboom, 1975; 1977) and more recently by Eikelboom and van Buijsen (1983) and Jenkins *et al.* (1993). At plant start-up, a form of microorganism growth known as "zooloeal fingers" can be commonly observed. These types of flocs are less common during stable later plant operation.

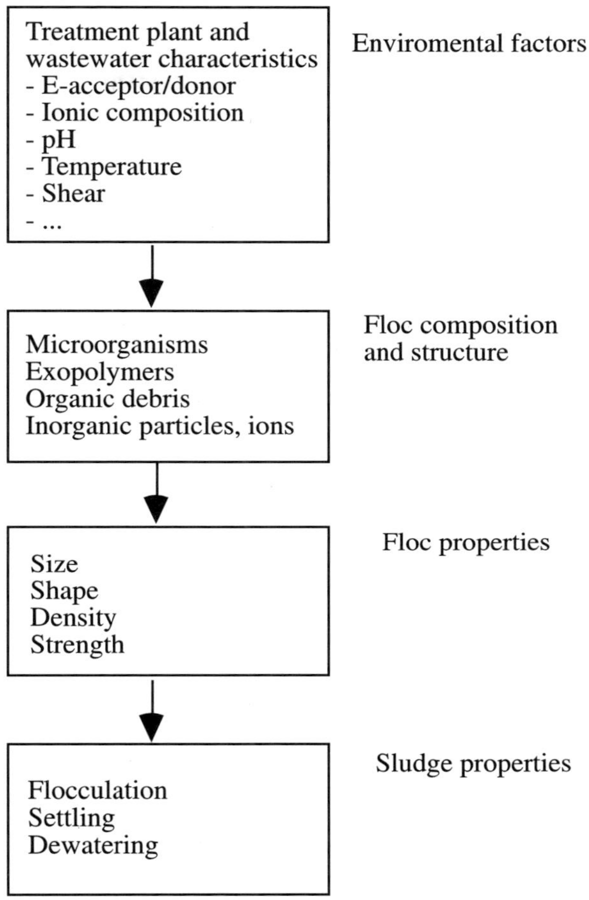

Figure 5 Possible interrelationships between treatment plant characteristics, floc composition, structure and properties, and sludge properties related to solid-liquid separation.

The parameters that lead to floc formation with certain properties important for the treatment plant performance are tentatively listed in Figure 5. Various external factors determine the microbial population as well as other components present in the floc (*e.g.* particles coming from the wastewater). All these components, with their inherent properties, build up the flocs, determining size distribution, shape, density, and strength. Ultimately, these floc properties determine the properties of functional sludge that in turn determine the operation of the plant, namely clarification (flocculation), settling, and dewatering.

FLOC STRUCTURE AND COMPOSITION

Activated sludge flocs are very heterogeneous, irregular structures consisting of microorganisms (as single cells, microcolonies and filaments), organic and inorganic adsorbed particles, extracellular polymeric substances (EPS) and organic fibers (Li and Ganczarczyk, 1993; Urbain *et al.*, 1993a; Jorand *et al.*, 1995b; Wagner *et al.*, 1995; Frølund *et al.*, 1996; Liss *et al.*, 1996; Zartarian *et al.*, 1997). The components may arise from wastewater, be produced by the bacteria, or be microbial lysis products. The size (diameter) of a floc is typically 40–100 μm, but in the mixed liquor in the process tanks they are usually aggregated loosely to much larger floc-aggregates with a size of several hundred micrometres to a few millimetres. These large flocs, which can be discerned with the naked eye, are easy to shear apart. The large structural heterogeneity of the flocs with regions of low and high density and with pores of water has been described by the fractal dimension, and typical values of 2.2–2.5 have been found (Snidaro *et al.*, 1997; Guan *et al.*, 1998).

The bacteria, both active and inactive, form a significant part of the floc, and the cell biomass is reported in the range of 10–20% of the organic matter in activated sludge flocs (*e.g.* Wanner, 1994b). Besides exopolymers and bacterial cells, organic debris (*e.g.* fibers) can be present, but its extent has not been well described. Additionally, inorganic particles and cations are present in the flocs, and these are important for linking the negatively charged macromolecules in the floc matrix.

EPS in flocs are considered very important for forming the network structure (or gel) that keeps the floc together. Thus, the amount and the composition of the EPS influence floc and sludge properties such as floc strength (Eriksson *et al.*, 1992), sludge settling (Becari *et al.*, 1980; Urbain *et al.*, 1993a), and sludge dewatering (Eriksson *et al.*, 1992; Nielsen *et al.*, 1996).

Extracellular Polymeric Substances

In activated sludge, EPS are usually defined as the amount of organic matter obtained by a certain extraction and purification procedure. Thus, EPS should be regarded as a very broad term including exopolymers produced by the cells, components arising from cell lysis and adsorbed matter from the wastewater (Nielsen and Jahn, 1999). Although of great interest, it is not possible to separate the different fractions by existing methods. The amount of extracted EPS is reported to vary from a few percent to *ca* 40% of the organic matter, and also the composition of EPS shows large variations (Gehr and Henry, 1983; Rudd *et al.*, 1983; Goodwin and Forster, 1985; Horan and Eccles, 1986; Karapanagiotis *et al.*, 1989; Morgan *et al.*, 1990; Urbain *et al.*, 1993a; 1993b; Frølund *et al.*, 1996). The reason for these varying results is likely to be the use of different extraction and purification procedures and different analytical techniques rather than very large differences among the sludges. Furthermore, much of the EPS probably cannot be extracted by the existing methods that mainly extract water soluble compounds (Nielsen and Jahn, 1999).

The composition of the EPS matrix of activated sludge flocs has been reported to be very complex, containing protein, polysaccharides, nucleic acids, lipids, various heteropolymers and humic substances. In most recent studies protein has been found to be the largest fraction in both total sludge and in the extracted EPS fractions (*e.g.* Frølund *et al.*, 1996) despite the use of different extraction and analysis methods. A typical amount is 50% of the organic matter of the extracted EPS. The function of the proteins in the floc matrix is not well investigated, but exoenzymes are present

(Frølund *et al.*, 1995), and some proteins may have structural functions (Higgins and Novak, 1997a).

Polysaccharides are usually considered the most important exopolymer in floc formation (*e.g.* Wanner, 1994a) and have in some studies been used as an indication of the amount of total exopolymers in the flocs (*e.g.* Andreadakis, 1993). However, a relatively small fraction, typically 15–20% of the total organic matter, is usually found as polysaccharide, so the amount in the EPS matrix is likely to be smaller. Lipids account for 25–30% of the organic matter in typical domestic wastewater (Raunkjær *et al.*, 1994), but their presence and function in activated sludge flocs has not been well studied. A typical content is 3–6% of the organic matter (Nielsen, unpublished results). Nucleic acids accumulate in the EPS matrix of flocs (Brown and Lester, 1980; Horan and Eccles, 1986; Urbain *et al.*, 1993a; Frølund *et al.*, 1996), and as polyelectrolytes, they are likely to be important for the floc structure.

Humic substances have been reported to amount to around 20% of the organic matter in activated sludge (Riffaldi *et al.*, 1982; Frølund *et al.*, 1996). However, the amount will depend on the content in wastewater and of the age of the sludge in the treatment plant, so can vary considerably (Nielsen, unpublished results). Humic substances as polyelectrolytes are well known for their ability to bind cations and are thus important for the floc structure and stability. They may also bind to and inhibit the activity of extracellular enzymes (Frølund *et al.*, 1995), and they are relatively easy to desorb from the floc by changing the ionic equilibrium of the system (Keiding and Nielsen, 1997).

FLOC PROPERTIES

As indicated in Figure 5, all the components due to their physico-chemical properties build up flocs and determine the size distribution, shape, density, and strength, which ultimately determine the functional properties of sludge.

The strength of activated sludge flocs is, like other biological aggregates, dependent on the interparticle forces between the different constituents (microorganisms, EPS, organic fibres, organic particles adsorbed from the wastewater and inorganic components). Common colloidal chemical interactions are assumed to be involved in the binding of the various entities, such as DLVO-type interactions (mainly electrostatic interactions and van der Waal interactions, Zita and Hermansson, 1994) and bridging of the EPS by means of divalent cations, particularly magnesium and calcium (Eriksson *et al.*, 1992; Keiding and Nielsen, 1997). However, if trivalent cations such as Fe(III) are present, they are more important than divalent ions (Nielsen and Keiding, 1998; Wilén *et al.*, personal observations). Hydrophobic interactions have been suggested to be of general importance for adhesion and flocculation of bacteria (*e.g.* van Loosdrecht *et al.*, 1987), and are likely to be of significance also in determining the structure and properties of activated sludge flocs (Urbain *et al.*, 1993a; Jorand *et al.*, 1995a). Hydrophobic bacteria have been shown to adsorb to activated sludge flocs more easily than hydrophilic bacteria (Zita and Hermansson, 1997; Olofsson *et al.*, 1998). A special phenomenon in activated sludge is foam formation, where the hydrophobicity of foaming sludge appears to be higher than for non-foaming sludge (Blackall and Marshall, 1989; Kocianova *et al.*, 1992). In addition, the physical entanglement of the long EPS macromolecules is important in the formation of the network structure of the floc (Eriksson *et al.*, 1992; Legrand *et al.*, 1998).

Floc Stability

When changes occur in the major intermolecular forces involved in keeping the floc together, a change in floc stability or floc strength can be observed (Eriksson *et al.*, 1992; Mikkelsen *et al.*, 1996; Keiding and Nielsen, 1997). The changes might be induced by changes in environmental parameters, *e.g.* ionic composition or pH, leading to changes in electrostatic interactions among exopolymers and particles in the floc. The result may be a strengthening or weakening of the floc, depending on type of change. Typical changes observed are decreased floc strength when pH increases and ionic strength is lowered, both due to increased electrostatic repulsion. Changes can also be induced by microbial activity, *e.g.* Fe(III)-reducing bacteria, where Fe(III) is reduced to a poorer flocculant, Fe(II) (Caccavo *et al.*, 1996). Recently, it has been reported that active aerobic metabolism of the bacteria is essential to maintain strong aggregates on a short term basis (Wilén *et al.*, personal observations), possibly due to an effect on EPS production in terms of quantity or properties.

Changes in floc strength and floc stability can be measured by recently developed methods (Eriksson *et al.*, 1992; Zita and Hermansson, 1994; Mikkelsen *et al.*, 1996; Mikkelsen and Keiding, 1999). The basic principle is that weakening of floc strength will lead to the release of small particles to the bulk water, where it can be measured as turbidity, the number of cells, total organic carbon or a change in filtration characteristics. It has become clear that a change in colloidal stability does not necessarily lead to deflocculation unless the flocs are exposed to some shear. Therefore, shear forces above a certain threshold level must be applied to the system to determine whether flocs break up more easily when exposed to a factor causing a potential stability change. The development of turbidity with time, or a related parameter, can be quantified using the initial slope (Eriksson *et al.*, 1992). Using a more fundamental approach and more comprehensive modelling, Mikkelsen *et al.* (1996) were able to determine the interaction energies and enthalpies for flocculation-deflocculation of activated sludge flocs, making it possible to compare the "shear sensitivity" of different sludges.

Floc Properties and Treatment Performance

The different separation steps in wastewater treatment plants are largely determined by the floc properties. In general, relatively-large, strong flocs are preferred, but it is difficult to obtain an ideal floc optimal for both clarification, settling and dewatering. It is even more difficult to find the optimal floc if the biological transformations are also included in the design, as small flocs with a low diffusion resistance would be preferred. Clarification is favoured by large flocs with an irregular shape that can collect minor flocs during the initial flocculation and formation of floc-aggregates. The floc must be strong to prevent break-up and production of many small particles.

Good settling flocs are large, spherical without many filamentous bacteria and with a high density. Small flocs can, for example, appear due to disintegration of larger flocs by high shear, anaerobic conditions or the presence of toxic compounds (*e.g.*

Eikelboom and van Buijsen, 1983; Wilén *et al.*, personal observations). A low density of flocs can appear due to extensive exopolymer production by bacteria such as *Zoogloea* spp. (Novak *et al.*, 1993), or due to a lack of inorganic ions (Higgins and Novak, 1997b).

In the dewatering process, large strong flocs are also preferable. Sludge with good dewatering properties is characterised by providing a high dry matter content, a low consumption of conditioning agents (organic polymers or iron/lime) and the dewatering process is fast. Poor dewatering is observed when the floc size distribution is broad with many small particles, for example, if the floc strength is weak so the flocs are sheared apart during dewatering (*e.g.* Mikkelsen *et al.*, 1996), or if the water binding capacity of the floc matrix is large. The latter is probably affected mainly by the amount and type of EPS in the floc.

MICROBIAL COMMUNITY STRUCTURE AND FUNCTION OF SLUDGE FLOCS

The complex microbial community in the activated sludge process has been investigated by classical microbiological approaches such as staining and isolation and identification procedures. From the types of transformations that are known to occur like nitrification, denitrification, carbon and sulfur oxidation, it can be safely presumed that nitrifiers, denitrifiers, heterotrophic carbon oxidisers, and sulfur oxidisers are present in the mixed liquor flocs. Indeed, it is easy to isolate many different physiological groups of microorganisms from the flocs. Probably the most well studied microbial group in activated sludge is the filamentous bacteria known to be responsible for sludge bulking and foaming (Blackall, 1999; Soddell, 1999) (see also *e.g.* Eikelboom and van Buijsen, 1983; Jenkins *et al.*, 1993; Wanner, 1994a). Bulking occurs where the filaments overgrow the flocs and prevent their settling and compaction in the secondary clarifier of the plant. Eventually, the sludge settleability is so impaired that biomass contaminates the normally clear supernatant, leading to violation of the plant licence for effluent suspended solids. Foaming describes the phenomenon where hydrophobic filaments adhere at the air-liquid interface of gas bubbles rising through the mixed liquor. Once the bubbles reach the surface of the mixed liquor, the filaments remain adhered to the bubble lamellae, where they preclude liquid drainage to the Gibbs borders of the bubble, thus stabilising the bubbles and leading to quite thick dense foams.

The nitrifiers in activated sludge, divided into the ammonia and the nitrite oxidisers, are typically reported to be *Nitrosomonas* and *Nitrobacter*, respectively (Halling-Sørensen and Jørgensen, 1993). Much recent research using molecular biological methods has helped to clarify the identity of both ammonia oxidisers (Wagner *et al.*, 1995; Juretschko *et al.*, 1998) and the nitrite oxidisers (Burrell *et al.*, 1998; Juretschko *et al.*, 1998) in activated sludges.

Bond and Rees (1999) have prepared a summary of the microbiological aspects of phosphorus removal in activated sludge processes, updating an earlier one by Jenkins and Tandoi (1991). Despite all the knowledge accrued in the last 20–30 years on the operation of enhanced biological phosphorus removal (EBPR) processes, the definitive identification of any one organism responsible for the process has not occurred. Early literature on the identity of the polyphosphate accumulating organisms

(PAOs) focussed on *Acinetobacter*, but chemotaxonomic and molecular biological methods like quinone profiling and gene probing have cast serious doubt on this. Recently, Bond *et al.* (1999) have identified one likely candidate PAO as a member of the beta-2 Proteobacteria, but it is likely that the diversity of PAOs is more broad than this.

The microbiology of activated sludge has been an intense area of research and development to the stage that this engineered process, reliant on the functioning of complex microbial communities, is employed throughout the developed world to treat the majority of sewage. However, the true *in situ* structure and function of the activated sludge microbial community has until recently eluded workers in this field.

Community Structure

The activated sludge process comprises a complex enrichment culture of a mixture of generalist and specialist microorganisms. Physical, chemical and biological factors interact to select the microbial community capable of degrading the pollutant material in the influent. Problems associated with EBPR processes and lack of fundamental understanding of these as well as of aspects of processes optimised to remove nitrogen from the influent have compelled researchers to examine and attempt to optimise the biological component of the mixed liquor of activated sludge plants. Since it is unlikely dilution and spread plate inoculation will allow the required insight into the true microbial diversity of activated sludge, molecular biological methods like cloning, DNA sequencing, probe design and gene probing (Wagner *et al.*, 1993; 1994a; 1994b; 1994c; 1995; Erhart *et al.*, 1997) in concert with community function methods such as microelectrodes (Schramm *et al.*, 1996; 1998; 1999) and *in situ* microautoradiography (Andreasen and Nielsen, 1997; 1998) are being used in an attempt to gain a better understanding and consequently the ability to manipulate the microbial communities. Other methods such as denaturing gradient gel electrophoresis (Santegoeds *et al.*, 1998) are likely to provide insight into some specific applications of activated sludge. It is paramount that in an investigation of the microbial community structure of an activated sludge plant, as much detail about the operation of the plant, including its operating data prior to and including the sampling time, is available. In the past, this has not been done with sufficient rigour and elegant molecular studies of microbial community structure and function cannot be linked with the operation of the plant.

In addition to the bacteria, protozoa are present in the mixed liquor milieu (Seviour, 1999). They play a vital role in keeping free swimming bacteria between flocs to a minimum and substantially contribute to effluent clarity. Some workers hypothesise that filaments make a "backbone" for flocs and the attachment of bacteria to the filaments is a common phenomenon. Indeed, the presence of attached bacterial growth on some filaments, such as Type 0041, is "diagnostic" in their identification.

Cloning for Microbial Community Structure Determination

Several studies into the microbial communities of activated sludge using cloning of 16S rDNA have been reported (Bond *et al.*, 1995; 1999; Schuppler *et al.*, 1995;

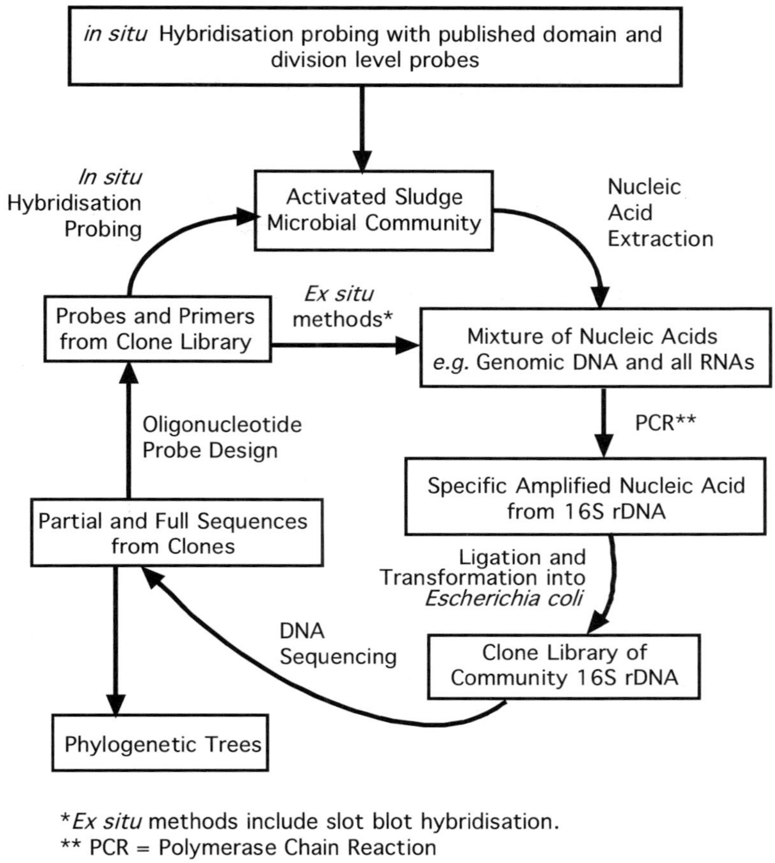

Figure 6 The Molecular Ecology Methods Wheel for investigating the microbial community structure of complex ecosystems such as activated sludge.

Blackall *et al.*, 1996; Burrell *et al.*, 1998; Juretschko *et al.*, 1998). The method has allowed the description of candidates for nitrification (Mobarry *et al.*, 1996; Burrell *et al.*, 1998; Schramm *et al.*, 1998), for competition with PAOs in EBPR (Bond *et al.*, 1999; Nielsen *et al.*, 1999), and for description of candidate PAOs (Bond *et al.*, 1995; 1999). The RNA approach, fundamental to the methods outlined in the papers referred to above, was first detailed by Olsen *et al.* (1986).

The 16S rDNA is obtained from individual bacteria in the microbial community by genomic DNA extraction and PCR amplification with conserved primers for the 16S rDNA. Each 16S rDNA is separated from the others by cloning into *Escherichia coli*. Analysis involves determination of either full or partial 16S rDNA sequences from the activated sludge bacteria now stored in the *E. coli* cells, and comparative analysis of the sequence data. From the data, specific gene probes can be designed so that the cells in the microbial community can be quantified and visualised. A diagram outlining the method is found in Figure 6.

Probing for Microbial Community Structure Determination

The first description of gene probes from 16S rDNA data for *in situ* hybridisation experiments was by Giovannoni *et al.* (1988). The detection was by autoradiography from radioactively labelled probes but this was quickly superseded by a method employing fluorescently-labelled oligonucleotides and called "phylogenetic stains" (DeLong *et al.*, 1989). Oligonucleotides complimentary to the rRNA at varying levels of specificity from species to domain are labelled with fluorochromes and can enter bacterial cells. They bind to the ribosomes *via* base pairing and the probe-target interaction can be observed by viewing the sample by epifluorescence microscopy. Whole cells which have bound probes can be visualised and their morphology determined. Instruments such as the confocal laser scanning microscope makes the interpretation and documentation of results more convincing.

The fluorescence *in situ* hybridisation (FISH) method has clearly demonstrated the role of *Nitrospira* as a major member of the nitrite oxidising microbial community in flocs (Burrell *et al.*, 1998; Juretschko *et al.*, 1998; Schramm *et al.*, 1998). Previously it was thought that *Nitrobacter* was the major nitrite oxidiser and culture dependent methods were unable to clarify the situation. Juretschko *et al.* (1998) demonstrated that the nitrifiers appeared clustered and that the ammonia oxidiser clusters were closely juxtaposed to the nitrite oxidiser clusters.

In the field of biological phosphorus removal, *Acinetobacter* has been attributed with the P accumulating phenotype because it is easily isolated from EBPR reactors and the morphology of polyphosphate accumulating cells by staining resembled those of *Acinetobacter*. Additionally, some features of EBPR are found in strains of *Acinetobacter*. However, in most mixed cultures exhibiting EBPR, *Acinetobacter* cells are present in too small amounts to account for all the phosphorus removed from the wastewater, whereas bacteria from the beta Proteobacteria or the Actinobacteria have been linked with EBPR by their dominance as determined by FISH (Wagner *et al.*, 1994c; Kämpfer *et al.*, 1996; Bond *et al.*, 1999; Bond and Rees, 1999).

Another area in activated sludge microbiology where FISH has been used is in the identification of filamentous bacteria which are responsible for the common problems of bulking (Wagner *et al.*, 1994a; Erhart *et al.*, 1997) and foaming (de los Reyes *et al.*, 1997; Erhart *et al.*, 1997). Typically these filaments are identified by their *in situ* morphology and their reactions to staining techniques such as the Gram stain, Neisser stain and Sudan Black B stain. Morphology is a poor character to use in identification (Woese, 1987) and FISH probes give a more definite identification.

In situ Microbial Function of Activated Sludge Flocs

Although a lot more knowledge has been obtained about the identity of important microorganisms present in activated sludge systems over the past few years, very little is known about their physiology and activity. As most of the bacteria are not isolated, *in situ* methods must be applied for such studies. Methods such as *in situ* microautoradiography (MAR) are able to give insights into the types of substrates taken up by specific microorganisms. MAR was initially implemented with filamentous bacteria which could be identified by their morphology (Andreasen and Nielsen,

1997). The substrate feeding patterns of "*Microthrix parvicella*", a notoriously devastating filament in the activated sludge, were determined and its specific use of long chain fatty acids (LCFA) such as oleic acid *in situ* was discovered (Andreasen and Nielsen, 1998). In pure culture, this bacterium can use a range of carbon substrates but *in situ* it only competes for LCFA. However, by a combination of MAR and FISH the substrate feeding pattern of morphologically indistinct organisms whose identification can be facilitated by FISH can be determined (Lee *et al.*, 1999; Ouverney and Fuhrman, 1999).

Important information about microbial activity may also be obtained by microsensors such as microelectrodes. Additionally, microelectrodes can also be linked in MAR and FISH studies. Currently they have only been used with biofilm and fluidised bed samples (Schramm *et al.*, 1998; 1999), and results with their implementation to flocs are awaited.

ACKNOWLEDGEMENTS

We would like to thank Dr Phil Hugenholtz for the idea of Figure 6.

REFERENCES

Andreadakis A.D. (1993). Physical and chemical properties of activated sludge floc. *Water Res*, 27, 1707–1714.

Andreasen K., Nielsen P.H. (1997). Application of microautoradiography to study substrate uptake by filamentous microorganisms in activated sludge. *Appl Environ Microbiol*, 63, 3662–3668.

Andreasen K., Nielsen P.H. (1998). *In situ* characterization of substrate uptake by *Microthrix parvicella* using microautoradiography. *Water Sci Technol*, 37, 19–26.

Ardern E., Lockett W.T. (1914). Experiments on the oxidation of sewage without the aid of filters. *J Soc Chem Indust*, 33, 523–539.

Becari M., Mappelli P., Tandoi V. (1980). Relationships between bulking and physicochemical properties of activated sludges. *Biotechnol Bioeng*, 22, 969–979.

Blackall L.L. (1999). Bulking. In: Seviour R.J., Blackall L.L. (eds) *The Microbiology of Activated Sludge*. Kluwer Academic Publishers, London, pp. 147–160.

Blackall L.L., Marshall K.C. (1989). The mechanism of stabilization of actinomycete foams and the prevention of foaming under laboratory conditions. *J Ind Microbiol*, 4, 181–188.

Blackall L.L., Stratton H., Bradford D., Sjörup C., Del Dot T., Seviour E.M., Seviour R.J. (1996). "*Candidatus* Microthrix parvicella" – a filamentous bacterium from activated sludge sewage treatment plants. *Int J Syst Bacteriol*, 4, 344–346.

Bond P.L., Rees G.N. (1999). Microbiological aspects of phosphorus removal in activated sludge systems. In: Seviour R.J., Blackall L.L. (eds) *The Microbiology of Activated Sludge*. Kluwer Academic Publishers, London, pp. 227–256.

Bond P.L., Hugenholtz P., Keller J., Blackall L.L. (1995). Bacterial community structures of phosphate-removing and non-phosphate-removing activated sludges from sequencing batch reactors. *Appl Environ Microbiol*, 61, 1910–1916.

Bond P.L., Erhart R., Wagner M., Keller J., Blackall L.L. (1999). The identification of some of the major groups of bacteria in efficient and non-efficient biological phosphorus removal activated sludge systems. *Appl Environ Microbiol*, 65, 4077–4084.

Brown M.J., Lester J.N. (1980). Comparison of bacterial extracellular polymer extraction methods. *Appl Environ Microbiol*, **40**, 107–185.

Burrell P.C., Keller J., Blackall L.L. (1998). Microbiology of a nitrite-oxidizing bioreactor. *Appl Environ Microbiol*, **64**, 1878–1883.

Caccavo F., Frølund B., Kloeke F.v.O., Nielsen P.H. (1996). Deflocculation of activated sludge by the dissimilatory Fe(III)-reducing bacterium *Shewanella alga* strain Bry. *Appl Environ Microbiol*, **62**, 1487–1490.

Chiemchaisri C., Yamamoto K. (1993). Biological nitrogen removal under low temperature in a membrane separation bioreactor. *Water Sci Technol*, **28**, 325–333.

Daigger G.T., Norton L.E., Watson R.S., Crawford D., Sieger R.B. (1993). Process and kinetic analysis of nitrification in coupled trickling filter activated sludge processes. *Water Environ Res*, **65**, 750–758.

de los Reyes F.L., Ritter W., Raskin L. (1997). Group-specific small-subunit rRNA hybridization probes to characterize filamentous foaming activated sludge systems. *Appl Environ Microbiol*, **63**, 1107–1117.

Delanghe B., Nakamura F., Myoga H., Magara Y. (1994). Biological denitrification with ethanol in a membrane bioreactor. *Environ Technol*, **15**, 61–70.

DeLong E.F., Wickham G.S., Pace N.R. (1989). Phylogenetic stains: ribosomal RNA-based probes for the identification of single cells. *Science*, **243**, 1360–1363.

Eikelboom D.H. (1975). Filamentous organisms observed in activated sludge. *Water Res*, **9**, 365–388.

Eikelboom D.H. (1977). Identification of filamentous organisms in bulking activated sludge. *Prog Water Technol*, **8**, 153–162.

Eikelboom D.H., van Buijsen H.J.J. (1983). *Microscopic Sludge Investigation Manual*. TNO, Delft, The Netherlands.

Erhart R., Bradford D., Seviour R.J., Amann R.I., Blackall L.L. (1997). Development and use of fluorescent *in situ* hybridization probes for the detection and identification of "*Microthrix parvicella*" in activated sludge. *Syst Appl Microbiol*, **20**, 310–318.

Eriksson L., Steen I., Tendaj M. (1992). Evaluation of sludge properties at an activated sludge plant. *Water Sci Technol*, **25**, 251–265.

Farquhar G.J., Boyle W.C. (1971a). Identification of filamentous microorganisms in activated sludge. *J Water Poll Cont Fed*, **43**, 604–622.

Farquhar G.J., Boyle W.C. (1971b). Occurrence of filamentous microorganisms in activated sludge. *J Water Poll Cont Fed*, **43**, 779–798.

Frølund B., Griebe T., Nielsen P.H. (1995). Enzymatic activity in the activated-sludge floc matrix. *Appl Microbiol Biotechnol*, **43**, 755–761.

Frølund B., Palmgren R., Keiding K., Nielsen P.H. (1996). Extraction of extracellular polymers from activated sludge using a cation exchange resin. *Water Res*, **30**, 1749–1758.

Gehr R., Henry J.G. (1983). Removal of extracellular material, techniques and pitfalls. *Water Res*, **17**, 1743–1748.

Giovannoni S.J., DeLong E.F., Olsen G.J., Pace N.R. (1988). Phylogenetic group-specific oligodeoxynucleotide probes for identification of single microbial cells. *J Bacteriol*, **170**, 720–726.

Goodwin J.A.S., Forster C.F. (1985). A further examination into the composition of activated sludge surfaces in relation to their settlement characteristics. *Water Res*, **19**, 527–533.

Guan J., Waite T.D., Amal R., Bustamante H., Wukasch R. (1998). Rapid determination of fractal structure of bacterial assemblages in wastewater treatment: implications to process optimisation. *Water Sci Technol*, **38**, 9–15.

Halling-Sørensen B., Jørgensen S.E. (1993) *The Removal of Nitrogen Compounds from Wastewater*. Elsevier, Amsterdam, 456pp.

Hem L.J., Rusten B., Ødegaard H. (1994). Nitrification in a moving bed biofilm reactor. *Water Res*, **28**, 1425–1433.

Henze M., Harremoës P., Jansen J.C., Arvin E. (1995) *Wastewater Treatment*. Springer-Verlag, Heidelberg, Germany, 383pp.

Higgins M.J., Novak J.T. (1997a). Characterization of exocellular protein and its role in bioflocculation. *J Environ Engin*, **123**, 479–485.

Higgins M.J., Novak J.T. (1997b). The effect of cations on the settling and dewatering of activated sludges: laboratory results. *Water Environ Res*, **69**, 215–224.

Horan N., Eccles C.R. (1986). Purification and characterization of extracellular polysaccharides from activated sludge. *Water Res*, **20**, 1427–1432.

Jenkins D., Tandoi V. (1991). The applied microbiology of enhanced biological phosphate removal-accomplishments and needs. *Water Res*, **25**, 1471–1478.

Jenkins D., Richard M.G., Daigger G.T. (1993) *Manual on the Causes and Control of Activated Sludge Bulking and Foaming*. Lewis Publishers, New York, pp. 193.

Jorand F., Guicherd P., Urbain V., Manem J., Block J.C. (1995a). Hydrophobicity of activated sludge flocs and laboratory-grown bacteria. *Water Sci Technol*, **30**, 211–218.

Jorand F., Zartarian F., Thomas F., Block J.C., Bottero J.Y., Villemin G., Urbain V., Manem J. (1995b). Chemical and structural (2D) linkage between bacteria within activated sludge flocs. *Water Res*, **29**, 1639–1647.

Juretschko S., Timmermann G., Schmid M., Schleifer K.-H., Pommerening-Röser A., Koops H.-P., Wagner M. (1998). Combined molecular and conventional analyses of nitrifying bacterium diversity in activated sludge: *Nitrococcus mobilis* and *Nitrospira*-like bacteria as dominant populations. *Appl Environ Microbiol*, **64**, 3042–3051.

Kämpfer P., Erhart R., Beimfohr C., Bohringer J., Wagner M., Amann R. (1996). Characterization of bacterial communities from activated sludge – culture-dependent numerical identification versus in situ identification using group- and genus-specific rRNA-targeted oligonucleotide probes. *Microb Ecol*, **32**, 101–121.

Karapanagiotis N.K., Rudd T., Sterrit R.M., Lester J.N. (1989). Extraction and characterization of extracellular polymers in digested sewage sludge. *J Chem Technol Biotechnol*, **44**, 107–120.

Keiding K., Nielsen P.H. (1997). Desorption of organic macromolecules from activated sludge flocs: effect of ionic composition. *Water Res*, **31**, 1665–1672.

Klemetson S.L., Lang M.E. (1984). Treatment of saline wastewaters using a rotating biological contactor. *J Water Poll Cont Fed*, **56**, 1254–1259.

Kocianova E., Foot R.J., Forster C.F. (1992). Physicochemical aspects of activated sludge in relation to stable foam formation. *J Inst Water Environ Manag*, **6**, 342–350.

Lee N., Nielsen P.H., Andreasen K.H., Juretschko S., Nielsen J.L., Schleifer K.H., Wagner M. (1999). Combination of fluorescent in situ hybridisation and microautoradiography – a new tool for structure-function analyses in microbial ecology. *Appl Environ Microbiol*, **65**, 1289–1297.

Legrand V., Hourdet D., Audebert R., Snidaro D. (1998). Deswelling and flocculation of gel network: application to sludge dewatering. *Water Res*, **32**, 3662–3672.

Li D., Ganczarczyk J. (1993). Factors affecting dispersion of activated sludge flocs. *Water Environ Res*, **65**, 258–263.

Liss S.N., Droppo I.G., Flannigan D.T., Leppard G.G. (1996). Floc architecture in wastewater and natural riverine systems. *Environ Sci Technol*, **30**, 680–686.

Mikkelsen L.H., Keiding K. (1999). Equilibrium aspects of the effects of shear and solids content on aggregate deflocculation. *Adv Coll Interface Sci*, **80**, 151–182.

Mikkelsen L.H., Gotfredsen A.K., Agerbaek M.L., Nielsen P.H., Keiding K. (1996). Effects of colloidal stability on clarification and dewatering of activated sludge. *Water Sci Technol*, **34**, 449–457.

Mobarry B.K., Wagner M., Urbain V., Rittmann B.E., Stahl D.A. (1996). Phylogenetic probes for analyzing abundance and spatial organization of nitrifying bacteria. *Appl Environ Microbiol*, **62**, 2156–2162.

Morgan J.W., Forster C.F., Evison L. (1990). A comparative study of the nature of biopolymers extracted from anaerobic and activated sludge. *Water Res*, **24**, 743–750.

Nielsen A.T., Liu W.-T., Filipe C., Grady L., Molin S., Stahl D.A. (1999). Identification of a novel group of bacteria in sludge from a deteriorated biological phosphorus removal reactor. *Appl Environ Microbiol*, **65**, 1251–1258.

Nielsen P.H., Frølund B., Keiding K. (1996). Changes in the composition of extracellular polymeric substances in activated sludge during anaerobic storage. *Appl Microbiol Biotechnol*, **44**, 823–830.

Nielsen P.H., Jahn A. (1999). Extraction of EPS. In: Wingender J., Neu T.R., Flemming H.-C. (eds) *Microbial Extracellular Polymeric Substances*. Springer-Verlag, London, pp. 49–72.

Nielsen P.H., Keiding K. (1998). Disintegration of activated sludge flocs in presence of sulfide. *Water Res*, **32**, 313–320.

Novak L., Larrea L., Wanner J., Garcia-Heras J.L. (1993). Non filamentous activated sludge bulking in a laboratory scale system. *Water Res*, **27**, 1339–1346.

Olofsson A.C., Zita A., Hermansson M. (1998). Floc stability and adhesion of green-fluorescent-protein-marked bacteria to flocs in activated sludge. *Microbiology*, **2**, 519–528.

Olsen G.J., Lane D.J., Giovannoni S.J., Pace N.R., Stahl D.A. (1986). Microbial ecology and evolution: a ribosomal RNA approach. *Annu Rev Microbiol*, **40**, 337–365.

Ouverney C.C., Fuhrman J.A. (1999). Combined microautoradiography 16S rRNA probe technique for determination of radioisotope uptake by specific microbial cells in situ. *Appl Environ Microbiol*, **65**, 1746–1752.

Raunkjær K., Hvitved-Jacobsen T., Nielsen P.H. (1994). Measurement of pools of protein, carbohydrate and lipid in domestic wastewater. *Water Res*, **28**, 251–262.

Riffaldi R., Sartori F., Levi-Minzi R. (1982). Humic substances in sewage sludges. *Environ Poll*, **3**, 139–146.

Rudd T., Sterrit R.M., Lester J.W. (1983). Extraction of extracellular polymers from activated sludge. *Biotechnol Lett*, **5**, 327–332.

Santegoeds C.M., Muyzer G., de Beer D. (1998). Biofilm dynamics studied with microsensors and molecular techniques. *Water Sci Technol*, **37**, 125–129.

Schramm A., de Beer D., Wagner M., Amann R. (1998). Identification and activities *in situ* of *Nitrosospira* and *Nitrospira* spp. as dominant populations in an nitrifying fluidized bed reactor. *Appl Environ Microbiol*, **64**, 3480–3485.

Schramm A., de Beer D., van den Heuvel J.C., Ottengraf S., Amann R. (1999). Microscale distribution of populations and activities of *Nitrosospira* and *Nitrospira* spp. along a macroscale gradient in a nitrifying bioreactor: quantification by *in situ* hybridisation and the use of microsensors. *Appl Environ Microbiol*, **65**, 3690–3696.

Schramm A., Larsen L.H., Revsbech N.P., Ramsing N.B., Amann R., Schleifer K.H. (1996). Structure and function of a nitrifying biofilm as determined by *in situ* hybridization and the use of microelectrodes. *Appl Environ Microbiol*, **62**, 4641–4647.

Schuppler M., Mertens F., Schön G., Göbel U.B. (1995). Molecular characterization of nocardioform actinomycetes in activated sludge by 16S rRNA analysis. *Microbiology*, **141**, 513–521.

Seviour R.J. (1999). The normal microbial communities of activated sludge plants. In: Seviour R.J., Blackall L.L. (eds) *The Microbiology of Activated Sludge*. Kluwer Academic Publishers, London, pp. 76–98.

Seviour R.J., Lindrea K.C., Griffiths P.C., Blackall L.L. (1999). The activated sludge process. In: Seviour R.J., Blackall L.L. (eds) *The Microbiology of Activated Sludge*. Kluwer Academic Publishers, London, pp. 44–75.

Snidaro D., Zartarian F., Jorand F., Bottero J.-Y., Block J.-C., Manem J. (1997). Characterization of activated sludge flocs structure. *Water Sci Technol*, **36**, 313–320.

Soddell J. (1999). Foaming. In: Seviour R.J., Blackall L.L. (eds) *The Microbiology of Activated Sludge*. Kluwer Academic Publishers, London, pp. 161–202.

Urbain V., Block J.C., Manem J. (1993a). Bioflocculation in activated sludge. *Water Res*, **27**, 829–838.

Urbain V., Pys E., Block J.C., Manem J. (1993b). Composition and activity of activated sludge under starvation conditions. *Environ Technol*, **14**, 731–740.

van Loosdrecht M.C.M., Lyklema J., Norde W., Schraa G., Zender A.J.B. (1987). The role of bacterial cell wall hydrophobicity in adhesion. *Appl Environ Microbiol*, **53**, 1893–1897.

van Veen W.L. (1973). Bacteriology of activated sludge, in particular the filamentous bacteria. *Antonie Leeuwenhoek*, **39**, 189–205.

Wagner M., Amann R., Lemmer H., Schleifer K.-H. (1993). Probing activated sludge with oligonucleotides specific for proteobacteria: inadequacy of culture-dependent methods for describing microbial community structure. *Appl Environ Microbiol*, **59**, 1520–1525.

Wagner M., Aßmus B., Hartmann A., Hutzler P., Amann R. (1994b). *In situ* analysis of microbial consortia in activated sludge using fluorescently labelled, rRNA-targeted oligonucleotide probes and confocal scanning laser microscopy. *J Microsc (Oxf)*, **176**, 181–187.

Wagner M., Rath G., Amann R., Koops H.P., Schleifer K.H. (1995). *In situ* identification of ammonia-oxidizing bacteria. *Syst Appl Microbiol*, **18**, 251–264.

Wagner M., Erhart R., Manz W., Amann R., Lemmer H., Wedi D., Schleifer K.-H. (1994c). Development of an rRNA-targeted oligonucleotide probe specific for the genus *Acinetobacter* and its application for *in situ* monitoring in activated sludge. *Appl Environ Microbiol*, **60**, 792–800.

Wagner M., Amann R., Kämpfer P., Assmus B., Hartmann A., Hutzler P., Springer N., Schleifer K.-H. (1994a). Identification and in situ detection of Gram-negative filamentous bacteria in activated sludge. *Syst Appl Microbiol*, **17**, 405–417.

Wanner J. (1994a) *Activated Sludge Bulking and Foaming Control*. Technomic Publishing, Pennsylvania, 327pp.

Wanner J. (1994b). Activated sludge population dynamics. *Water Sci Technol*, **30**, 159–169.

Woese C.R. (1987). Bacterial evolution. *Microbiol Rev*, **51**, 221–271.

Zartarian F., Mustin C., Villemin G., AitEttager T., Thill A., Bottero J.Y., Mallet J.L., Snidaro D. (1997). Three-dimensional modeling of an activated sludge floc. *Langmuir*, **13**, 35–40.

Zita A., Hermansson M. (1994). Effects of ionic strength on bacterial adhesion and stability of flocs in a wastewater activated sludge system. *Appl Environ Microbiol*, **60**, 3041–3048.

Zita A., Hermansson M. (1997). Effects of bacterial cell surface structures and hydrophobicity on attachment to activated sludge flocs. *Appl Environ Microbiol*, **63**, 1168–1170.

Index